High Temperature Polymer Dielectrics

High Temperature Polymer Dielectrics

Fundamentals and Applications in Power Equipment

Edited by Jun-Wei Zha and Zhi-Min Dang

WILEY-VCH

The Editors

Prof. Jun-Wei Zha
University of Science and
Technology Beijing
No. 30 Xueyuan Road
Haidian
Beijing
China
100083

Prof. Zhi-Min Dang
Tsinghua University
Zhongguancun North Street
Beijing
China
100084

Library of Congress Card No.: applied for

British Library Cataloguing-in-Publication Data
A catalogue record for this book is available from the British Library.

Bibliographic information published by the Deutsche Nationalbibliothek
The Deutsche Nationalbibliothek lists this publication in the Deutsche Nationalbibliografie; detailed bibliographic data are available on the Internet at <http://dnb.d-nb.de>.

Print ISBN: 978-3-527-35182-4
ePDF ISBN: 978-3-527-84103-5
ePub ISBN: 978-3-527-84104-2
oBook ISBN: 978-3-527-84105-9

Typesetting: Straive, Chennai, India
Printing and Binding: CPI Group (UK) Ltd, Croydon, CR0 4YY

C9783527351824_101123

Contents

Preface

In recent decades, the research on high-temperature polymer dielectrics (HTPDs) has made great progress, which has laid the foundation for their application in the field of electrical insulation. HTPDs have excellent high temperature stability and environmental stability. With the continuous development of electrical insulation technology in the direction of high temperature, the demand for HTPDs is becoming increasingly urgent. The HTPDs are of great significance for achieving miniaturization, lightweight, and high capacity of electrical equipment. Currently, HTPD and its composites have played an important role in the field of national defense and civil high technology. All countries have carried out a lot of basic and applied research on HTPDs and classified them as strategic materials.

HTPD can generally be divided into intrinsic HTPD and composite HTPD. Common intrinsic HTPDs, such as polyimides, polysulfones, and polyphenylimidazoles, have typical thermally stable structures. Composite HTPDs typically include organic-inorganic composites and polymer-metal composites. There are many varieties of organic-inorganic composites with high temperature resistance and dielectric properties. High-temperature polymers compounded with silicon, boron, etc., have more remarkable comprehensive properties. This type of polymer as a transition species of organic materials to inorganic materials greatly enriched the polymer molecular structure composition. As a result, HTPD has been widely noticed and studied by numerous researchers and engineering experts in the field of modern insulation.

However, there are no existing specific books on HTPDs for electrical equipment. Based on the theoretical research value of HTPD in electrical equipment. First, this book outlines the preparation process and properties of representative HTPDs. Subsequently, focusing on the relationship between the structure and properties of HTPDs, this book highlights the applications of HTPDs in the field of electrical insulation, such as high-temperature energy storage capacitors, electric motors, encapsulation, printed circuit boards, new energy power equipment, aerospace electrical equipment, and other electrical insulation fields. Overall, the monomer synthesis, material preparation, characterization methods, properties, and applications of various HTPDs are comprehensively presented.

This book covers a wide range of topics rather than merely a single introduction to HTPDs, and it will provide an excellent guide to improving the performance of existing HTPDs, as well as the design and development of novel HTPDs. Furthermore, this book is intended for graduate students and researchers at universities and research institutes, along with engineers at related companies. It will work across academia, industry, education, and public service to its matched audience.

1

Overview of High-Temperature Polymers

Xue-Jie Liu, Mengyu Xiao, Wenjie Huang, Xing Yang, and Jun-Wei Zha

University of Science and Technology Beijing, Beijing Advanced Innovation Center for Materials Genome Engineering, School of Chemistry and Biological Engineering, 30 Xueyuan Road, Haidian District, Beijing 100083, P. R. China

1.1 Introduction

In recent years, the wide utilization of new energy sources has made it necessary for power equipment to operate in harsh environments such as high voltage, high power, and high temperature [1, 2]. Therefore, as an essential component of power equipment, dielectric materials have attracted more attention. Solid dielectrics used in power equipment can be divided into polymer dielectrics and inorganic dielectrics. Inorganic dielectric has high temperature stability but also has the disadvantages of low breakdown strength (E_b) and poor flexibility, which cause difficulties in large-scale preparation that cannot be ignored. Different from inorganic dielectrics, polymer dielectrics have the advantages of being lightweight, having good flexibility, and being easy to process [3]. Meanwhile, excellent dielectric properties (high E_b, low dielectric loss [tan δ]) make it widely used in power equipment. With the continued miniaturization and increased power output of electronic and power systems, many fields require polymer dielectrics for reliable operation in harsh environments. For example, control and sensing electronics near the shells of rockets and space shuttles require high-temperature dielectric materials to operate above 250 °C. In underground oil and gas exploration, operating temperatures exceed 200 °C [4]. Unfortunately, the poor thermal stability of traditional polymer dielectrics seriously threatens the reliable operation of power equipment and significantly shortens its life cycle. Therefore, secondary cooling equipment has been used to reduce the working temperature in high-temperature applications. However, secondary cooling is difficult to achieve in view of the extreme temperatures experienced by large facilities such as underground exploration and space stations. Thus, a more attractive strategy is to develop high-temperature-resistant polymer dielectrics that carry out long-term work at high temperatures. This strategy can improve system reliability, reduce costs, and eliminate the need for large cooling systems and the wiring and interconnections required for remote placement of electronics [5, 6].

High Temperature Polymer Dielectrics: Fundamentals and Applications in Power Equipment, First Edition.
Edited by Jun-Wei Zha and Zhi-Min Dang.

Currently, polymers such as polyethylene (PE) and polypropylene (PP) are commonly used, but their operating temperatures are less than 105 °C [7]. This fact can be attributed to the tendency of molecular chains to move or even to break under heating conditions. Polymers with high temperature resistance have been developed such as polyimide (PI), polyetherimide (PEI), and polyetheretherketone (PEEK) [8–14]. The favorable high temperature resistance is primarily caused by the ultra-strong molecular structures derived from the conjugated bonds (multiple bonds or aromatic rings) on the polymer backbone. Furthermore, the hydrogen bonds and cross-linked structures lead to strong interactions within and between the polymer chains, which is directly reflected in the elevated high temperature resistance [15]. In addition, the introduction of an inorganic filler can further optimize the high-temperature polymer [16]. Based on the above strategies, excellent polymer dielectrics are prepared through molecular structure modification and composite structure design to meet the urgent requirement of high temperatures for rapidly developing power equipment.

This chapter briefly introduces the history of high-temperature polymers. Then, the influence of parameters such as glass transition temperature (T_g), melting point (T_m), and thermal conductivity (λ) on the high temperature resistance of polymers is analyzed, and the method of determination of the above parameters is displayed. Subsequently, the research progress of high-temperature polymers and their composite materials is discussed. Finally, the future challenges of high-temperature polymer materials are presented. This chapter will instruct on the development of high-temperature polymers in power equipment.

1.2 Development of High-Temperature Polymers

Since the early twentieth century, the rapid development of commercially available synthetic polymers has led to their widespread usage in military and civilian applications. Research on polymers with high temperature resistance attained its peak in the 1950s. During the development of novel products, functional groups and segments were introduced into polymers for the improvement of high temperature resistance. It leads to poor processability and high costs, which hinder large-scale production. A few high-temperature polymers have been developed and used nowadays, such as PI, PEI, PEEK, and fluorene polyester (FPE). It has been found that they are characterized by aromatic and heterocyclic structures. Table 1.1 summarizes the thermal and electrical properties of high-temperature polymers, suggesting that the common characteristics of high-temperature polymers are high T_g and relatively low dielectric permittivity (ε_r) and tan δ [4, 17]. With the booming development of smart grids, aerospace, electronic packaging, and other technical fields, polymer dielectrics are expected to have excellent thermal stability, mechanical properties, dielectric properties, and long service lives. It is also developing toward being lightweight and low-cost to meet various special applications. For example, Fan et al. [18] introduced alumina (Al_2O_3) nanoparticles into the PEI matrix, and the $T_{d5\%}$ of the prepared composite films was increased from 479 to

Table 1.1 Thermal and electrical properties of high-temperature polymers [4, 17].

Polymer	T_g (°C)	Volume resistivity (Ω cm^{-1})	E_b (MV m^{-1})	ε_r	tan δ (1 kHz)
Kapton (PI)	360–410	2.3×10^{17}	154–303 (7.6–127 μm)	2.7–3.5	0.13–0.26
Upilex (PI)	285–500	10^{16}–10^{17}	147–320 (12.5–125 μm)	3.2–3.5	0.13–0.7
Ultem (PEI)	217–247	10^{17}	200 (25 μm)	3.15	0.12
Ketaspire (PEEK)	150	2.6×10^{16}	150 (50 μm)	3.1	0.3
Kepstan (PEKK)	162	10^{16}	84 (100 μm)	2.6	0.7
PTFE	327 (T_m)	$>10^{18}$	23.6 (2 mm)	2.1	0.03
PET	245–265 (T_m)	10^{15}	~500	3.3	0.3
Isaryl (FPE)	330	10^{15}–10^{17}	220–320 (100–120 μm)	3.2–3.5	0.31–0.7
Cyclotene (BCB)	>350	10^{18}	300 (10 μm)	2.75	0.12

525 °C. Additionally, the E_b also exhibited an increase from 422 to 503 MV m^{-1} at 150 °C. The wide bandgap of Al_2O_3 nanoparticles is the primary factor for the significant increase in E_b. Li et al. [19] incorporated polysulfone (PSF) with a high ε_r into PI matrix. While maintaining high-temperature stability (T_g = 285 °C), a dramatic improvement in the ε_r was obtained. When the mass fraction of PSF reached 40%, the composite film exhibited the highest ε_r of 6.4 at 1 kHz, versus 3.25 for pure PI.

1.3 Parameters of Polymers with High Temperature Resistance

The high temperature resistance of polymers determines their practical application. T_g is a vital factor affecting the high temperature resistance of a polymer. T_g is the critical temperature at which the chain segments of the noncrystalline phase of polymers transition from freezing to motion and depends on the difficulty of polymer chain rotation. Chain rotation relies on the types of chemical bonds (such as covalent bonds and hydrogen bonds) and polar groups, as well as the shape of the molecular chains [20]. Firstly, the strength of the covalent bonds plays a major role in the thermal stability of polymers. Table 1.2 lists various chemical bonds in polymers and their fracture energies, which indicates that the fracture energy of double bonds is much higher than that of single bonds [21]. High-temperature polymers are usually composed of aromatic ring structures because they provide

Table 1.2 Cleavage energy of representative chemical bonds.

Chemical bond	$C_{aryl}-C_{aliph}$	$C_{aryl}-C_{aryl}$	C=C	C–N	C=N	C=O
Cleavage energy (kcal mol^{-1})	93	100	145	70	147	84
Chemical bond	C—H	C—F	C—S	Si—O	Ti—O	B—O
Cleavage energy (kcal mol^{-1})	99	123	62	88	180	185

Source: Cottrell [21]/Elsevier.

more stable covalent bonds. Then, the C—F bond has bond energy of up to 126 kcal mol^{-1}, significantly higher than the C—H bond. Therefore, fluorinated polymers are more thermally stable than their corresponding hydrogen-substituted polymers. Furthermore, the presence of non-covalent bonds such as hydrogen bonds and van der Waals forces makes the polymer chains more attractive. This has a greater impact on the parameters such as T_g and modulus, so the polymer is more resistant to high temperature. In addition, flexible polymer chains are known to lower the T_g, therefore, polymers containing rigid groups (such as sulfone and cyano groups) have higher T_g than those containing ether bonds. And in the design of polymer molecular structure, the presence of Si, P, B, and F atoms in the main chains of polymers is beneficial to improving the high temperature resistance of polymers.

For crystalline polymers, the melting point (T_m) is more practical than T_g. T_m is the critical temperature at which crystalline polymers change from the crystalline state to the molten state [1, 4]. The molecular chains of crystalline polymers are tightly aligned during the crystallization process. It still has good mechanical properties when the temperature is much higher than T_g. The actual operating temperature is significantly elevated above T_g. Therefore, the crystallinity and regularity of the molecular structure are also one of the reasons for high temperature resistance of the polymer. For example, the oriented PI films show an aligned structure that facilitates their mechanical and thermal properties. Cross-linked polymers typically have superior resistance to high temperatures than linear polymers. The disruption of the molecular chains of cross-linked polymers is only completed when more than two chemical bonds are broken, which effectively suppresses the degradation process.

High T_g and T_m are not enough to protect the polymers from a series of failures caused by heat for a long time. The enhanced λ allows more rapid conduction of heat away from the polymer, reducing the risk of high temperatures. For most polymers, the main mechanism of heat conduction is generated by lattice vibration through phonons. The λ of the polymer can be obtained from the Debye equation, as shown below [22].

$$\lambda = \frac{C_p v l}{3} \tag{1.1}$$

where C_p is the specific heat capacity per unit volume, v is the phonon velocity, and l represents the phonon mean free path. Due to scattering with other phonons, defects, and grain boundaries, the l value of most polymers is quite low, essentially

Table 1.3 Thermal conductivity of high-temperature polymers and their composites [23–28].

Polymers	λ (W m^{-1} K^{-1})	Composite materials	λ (W m^{-1} K^{-1})
Polycarbonate (PC)	0.23	3 wt% Ag + 12 wt% rGO/PI	2.12
PEI	0.21	4.0 wt%MWCNTs @ PPD/PI	0.43
PI	0.27	30 wt%mBN/PI	0.696
PEEK	0.25	rGO-PI/BNNS-PI (total filler loading constant at 3.8 wt%)	1.49

less than 0.5 W m^{-1} K^{-1}. Due to the shortcomings of the complex synthesis process and low preparation efficiency of intrinsic thermally conductive polymers, researchers have gradually focused on the improvement of λ of polymer-based composite materials in recent years [23, 24]. Guo et al. [23] prepared PI composites containing highly aligned hybrid fillers with high thermal conductivity (1.49 W m^{-1} K^{-1}), which is a multifunctional dielectric material with excellent thermal management capabilities. Table 1.3 lists the λ values of polymers and their composites [23–28], where the λ of the polymers is between 0.2 and 0.5 W m^{-1} K^{-1}, as previously described. The λ of polymer-based composite materials has been significantly improved.

1.4 Thermal Analysis Technology

Due to the diversity of polymer characteristics, there are different methods and conditions for measuring thermal properties. A variety of thermal property analysis techniques have been used to evaluate the damage to polymers at high temperatures. Several common analysis techniques for polymers are described below.

1.4.1 Differential Scanning Calorimetry (DSC)

Differential scanning calorimetry (DSC) is a commonly used technique for measuring the T_g and T_m of materials. Huang et al. [29] characterized the T_g values of PEEKs containing different side chains by DSC tests, as shown in Figure 1.1a. Figure 1.1b depicts that carboxylated polymer (PEEK-COOH) achieved a maximum T_g of 213 °C. Due to the longer $-SO_2$ side chains and –CN side chains, the T_g of sulfonylated polymer (PEEK-SO_2) and cyanylated polymer (PEEK-CN) decreased to 156 and 155 °C, respectively.

1.4.2 Dynamic Thermomechanical Analysis (DMA)

Dynamic thermomechanical analysis (DMA) characterizes the mechanical properties of viscoelastic materials as a function of time, temperature, or frequency. On

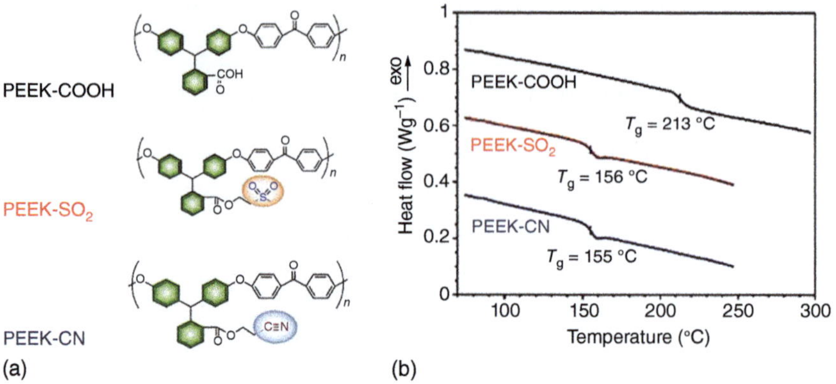

Figure 1.1 (a) Structure formula and (b) DSC of PEEK polymers [29]. Source: Huang et al. [29]. Reproduced with permission of Wiley-VCH.

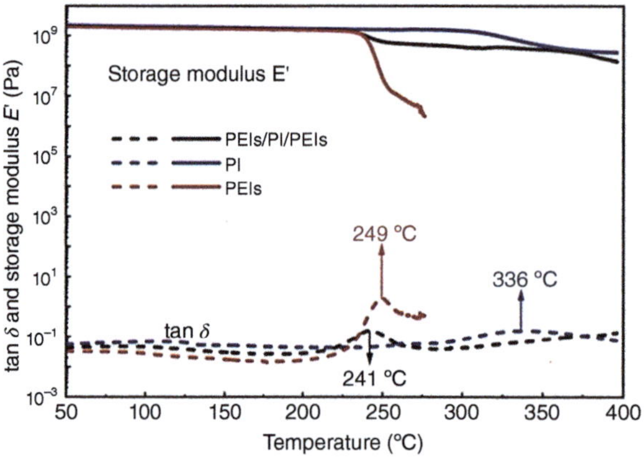

Figure 1.2 E' and tan δ as a function of temperature for PI, PEIs, and three-layer PEIs/PI/PEIs films [30]. Source: Niu et al. [30]. Reproduced with permission of Elsevier.

the one hand, DMA can be used to evaluate the mechanical behavior of materials by giving mechanical properties over a wide range of temperatures and frequencies. On the other hand, the T_g and secondary relaxation processes of polymers can be detected, which are closely related to the chain structure and condensed structure of polymers. Niu et al. [30] characterized the energy storage modulus (E′) and tan δ of PI, PEIs, and sandwich PEIs/PI/PEIs films by DMA tests. As shown in Figure 1.2, the T_g values of PI and PEI and sandwich PEIs/PI/PEIs reached 336, 249, and 241 °C, respectively, indicating excellent high temperature resistance.

1.4.3 Thermogravimetric Analysis (TGA)

TGA is used to characterize the variation of the sample mass in relation to the programming temperature; thus, it is used to analyze the thermal degradation temperature (T_d) and composition of the materials. Liu et al. [31] calculated the imidization degree (ID) of PI from TGA data. As shown in Figure 1.3a,b, the ID was obtained by

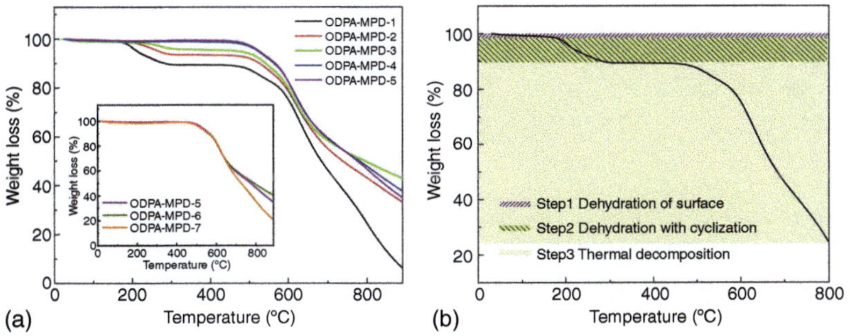

Figure 1.3 (a) TGA curves of the PI films with different IDs. (b) Three steps of the thermal weight loss process for the PI film were obtained from the TGA curve [31]. Source: Liu et al. [31]. Reproduced with permission of the Royal Society of Chemistry.

Table 1.4 Imidization degree and thermal property of the PI films [31].

ODPA-MPD	1	2	3	4	5	6	7
ID (%)	9	52	81	93	≈100	≈100	≈100
$T_{d5\%}$ (°C)	213	268	473	522	520	517	516
$T_{d10\%}$ (°C)	283	527	542	564	565	565	566

Source: Liu et al. [31]. Reproduced with permission of the Royal Society of Chemistry.

Eq. (1.2) based on the weight loss generated in the second step (cyclized dehydration of the PAA chain), and the results of T_d and ID are shown in Table 1.4.

$$\text{Weight loss\%} = \frac{M\left(2H_2O\right) \times (100\% - \text{ID})}{M\,(\text{PAA}) - M\left(2H_2O\right) \times \text{ID}} \times 100\% \tag{1.2}$$

1.4.4 Static Thermomechanical Analysis (TMA)

Thermomechanical analysis (TMA) refers to a technique for measuring the functional relationship between the deformation of the materials and the temperature and time under the action of a programmed temperature and non-vibration load, which can be used to measure the coefficient of expansion and phase transition temperature of materials. Figure 1.4c shows the coefficient of thermal expansion (CTE) of the colorless polyimide (CPI) and polyamideimide (PAI) films measured by TMA [32]. The CTE value of the CPI film is relatively high due to the bulky trifluoromethyl side groups. PAI films are obtained by the polymerization of diacids with diamines Figure 1.4a. Figure 1.4b demonstrates that intermolecular interactions between the amide units restrict the motion of the PAI chain segments. As the temperature approaches T_g, the films start to swell due to the increased mobility of the molecular chains. The PAI-1 (CTE = 15 ppm °C^{-1}) and PAI-2 (CTE = 21 ppm °C^{-1}) films have higher dimensional stability compared with CPI

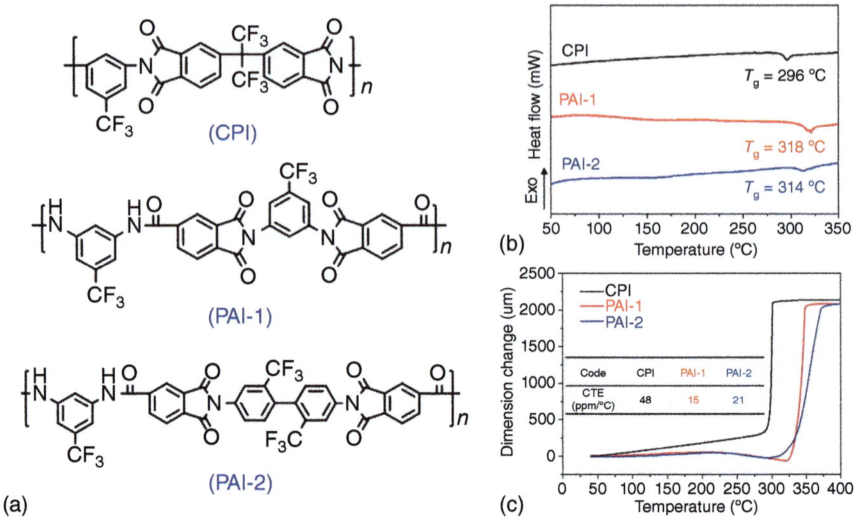

Figure 1.4 (a) Structure formula, (b) DSC, and (c) TMA of CPI and PAI films [32]. Source: Zuo et al. [32]. Reproduced with permission of the Royal Society of Chemistry.

films due to the rigid-rod amide units that increase the linearity and rigidity of the polymer chains.

1.4.5 Thermal Conductivity

Thermal conductivity reflects the ability of materials to conduct heat. Polymer materials with high thermal conductivity are conducive to emitting waste heat generated during operation, so they are beneficial to the optimization of electrical properties, life cycle, and reliability of power equipment. Zhou et al. [33] introduced a sandwich-structured PAI composite film as shown in Figure 1.5. The design of the layered structure and highly oriented functionalized BN bring about high

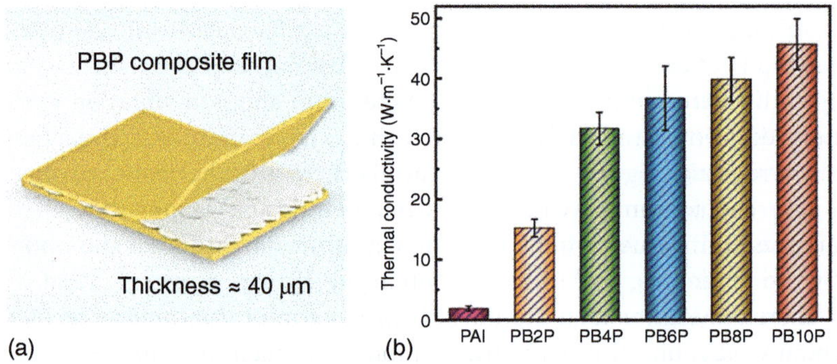

Figure 1.5 (a) Schematic diagram of the structure of PBP composite film. (b) $\lambda//$ of pure PAI and PBP composite films with different numbers of filler layers [33]. Source: Zhou et al. [33]. Reproduced with permission of Elsevier.

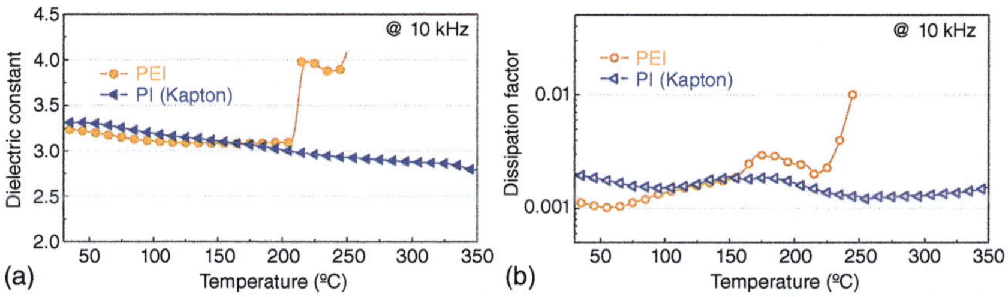

Figure 1.6 Temperature-dependent (a) ε_r and (b) tan δ of PI and PEI measured at 10 kHz [34]. Source: Li et al. [34]. Reproduced with permission of Nature Publishing Group.

in-plane thermal conductivity ($\lambda//$). The composite film achieved a high $\lambda//$ of $45.7\,W\,m^{-1}\,K^{-1}$ with 23 wt% BN loading.

1.4.6 Dynamic Dielectric Analysis (DEA)

Dynamic dielectric analysis (DEA) is a technique for analyzing the ε_r and tan δ of dielectric materials as a function of temperature and frequency under electric fields. As shown in Figure 1.6, the ε_r of PI steadily decreases from 3.3 to 2.8 and remains low tan δ (<0.3%) as the temperature increases from 25 to 350 °C [34]. Although the ε_r of PEI remains stable between 25 and 200 °C, its relatively low T_g causes the ε_r and tan δ of PEI to change dramatically with increasing temperatures above 200 °C. Furthermore, a significant relaxation peak in tan δ is observed at about 175 °C.

1.5 High-Temperature Polymer Materials

Compared with metal and ceramic materials, high-temperature polymers have the advantages of being lightweight, easy to process, and having tunable properties. This section describes high-temperature polymers and their composite materials that are widely used in electronic power systems.

1.5.1 Commercial High-Temperature Polymer

Figure 1.7 shows the structural formulas for several high-temperature polymers. PI is a thermosetting polymer containing imide rings, which is one of the most high-temperature-resistant polymer materials available in practice. PI is usually prepared through a traditional two-step process. Firstly, the dianhydride and diamine monomers form PAA through polycondensation. Then the PAA is converted to PI by the imide reaction. PI with different properties can be synthesized by selecting different types of diamine and dianhydride monomers. One of the most commercially valuable PI films is the Kapton film, made by DuPont. Kapton is synthesized from pyromellitic dianhydride (PMDA) and 4,4'-oxydianiline (ODA) monomers. Marketed since the mid-1960s, it has been widely used for high-temperature wire and cable insulation on aircraft with continuous operating

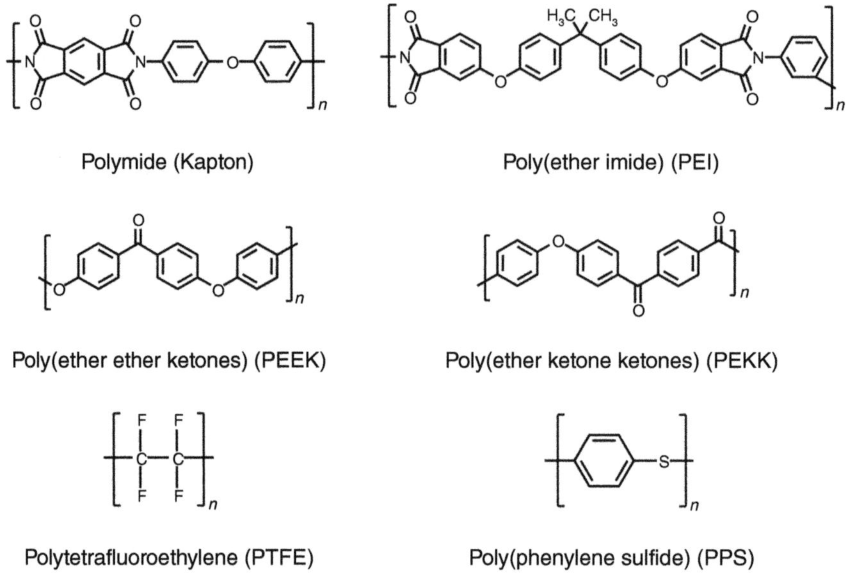

Figure 1.7 The structural formula for several common high-temperature polymers.

temperatures of 300–350 °C [35, 36]. Aromatic polyimides have strong rigidity, which brings difficulties for processing and manufacturing. In order to improve the processing properties of PI, researchers have introduced flexible groups (such as ether bonds and alkyl groups) into the PI molecular chains. For example, PEI is obtained by introducing flexible groups (such as ether bonds) onto the PI skeleton. The classic PEI is ULTEM, produced by SABIC using the disodium salt of bisphenol A and 1,3-bis(4-nitrophthalimido) benzene [37]. The existence of the flexible bond makes PEI transform into a thermoplastic polymer and obtain a lower $T_g = 217\,°C$. Thermoplastic PEI has excellent machining properties, including extrudable, injectable, and compressible molding, and is soluble in a variety of organic solvents, which provides conditions for application in different equipment [38].

PEEK is a semi-crystalline polymer material with ether bonds and ketone carbonyl groups in molecular chains, which is formed by the polymerization of 4,4′-difluorophenone and dibenzoate. PEEK not only has good high temperature resistance ($T_g \approx 150\,°C$), but also excellent mechanical strength, flame retardant, radiation resistance, insulation, and other advantages [5]. The semi-crystalline characteristics of PEEK make it have hot melting characteristics, which can be directly wounded with copper wire at high temperatures, and then the corresponding insulation products are prepared. The thermal properties of polyketone materials can be adjusted by changing the ratio of ether bonds and ketone carbonyl groups. For example, when the ratio of ether bond and ketone carbonyl group is changed from 2 : 1 in PEEK to 1 : 2, poly(ether ketone ketone) (PEKK) with a similar structure to PEEK is obtained. PEKK is generally polymerized from diphenyl ethers and benzene dicarboxylic acid halides [39]. The commercial PEKK named Kepstan, manufactured by Arkema, has a high T_g of 162 °C [40].

PTFE (polytetrafluoroethylene) can be prepared from tetrafluoroethylene monomers by polymerization. PTFE was accidentally discovered by Plunkett in 1938 [41]. As a semi-crystalline polymer, PTFE also has a high T_m of 327 °C, which gives it excellent thermal stability at 260 °C [42]. In addition, PTFE is widely used in the aerospace field due to its corrosion resistance, electrical insulation, and good anti-aging durability. Polyphenylene sulfide (PPS) is a thermoplastic crystalline resin composed of aromatic rings connected with sulfur. Although the T_g of PPS is only 120 °C, it can work in the range of 200–240 °C for a long time, and the maximum temperature can reach 260 °C for a short time. Moreover, the T_d of PPS in the air is higher than 500 °C. In addition to excellent thermal stability, PPS also has good corrosion resistance (second only to PTFE), electrical performance, and flame retardancy [43]. However, its high crystallinity (75%) and rigid molecular structure lead to its high brittleness and low toughness.

Although some widely used high-temperature polymers have been discussed above, they still have different shortcomings in practical application. Thus, it is necessary to develop novel high-temperature polymers. The recent research progress in high-temperature polymers and composite materials from the perspectives of molecular structure design and composite structure optimization is presented as shown below.

1.5.2 Molecular Structure Modification of High-Temperature Polymer

Homogeneous high-temperature polymer films can be obtained based on molecular structure modification strategies. There are extremely few defects in the film-forming process, which is conducive to large-scale preparation. Suppressing the motion of molecular chains is an effective strategy to make polymers with excellent high temperature resistance. This can be achieved by enhancing intermolecular forces, such as by building rigid and cross-linked structures. Researchers have made a lot of valuable progress in this field.

Most high-temperature-resistant polymers have rigid structures, and their molecular chains should contain aromatic rings or heterocyclic rings. In addition, the conjugate structure can be further designed on the main chains to improve the rigidity of polymers. Based on the above design concept, Pan et al. [44] incorporated rigid and asymmetric phenyl phthalazinone ether ketone molecules into the polymer skeleton to obtain poly(phthalazinone ether ketone) (PPEK). The T_g of PPEK was 250 °C, and $T_{d5\%}$ reached 490 °C. Although the presence of conjugated structures improves the high temperature resistance of polymers, the stacking of π–π structures leads to a reduction in bandgap, which deteriorates electrical properties. Figure 1.8a depicts that Wu et al. [45] synthesized polyoxafluoronorbornene (POFNB) through ring-opening metathesis polymerization reaction using Grubbs generation 2 catalyst. The aromatic structure was replaced by a saturated, fused bicyclic structure in the polymer. This nonplanar and non-conjugate rigid structure allows the polymer to achieve high temperature resistance while avoiding bandgap reduction due to π–π stacking between benzene rings. As shown in Figure 1.8b, POFNB achieved a high T_g of 186 °C and a wide bandgap of 4.9 eV. The wide

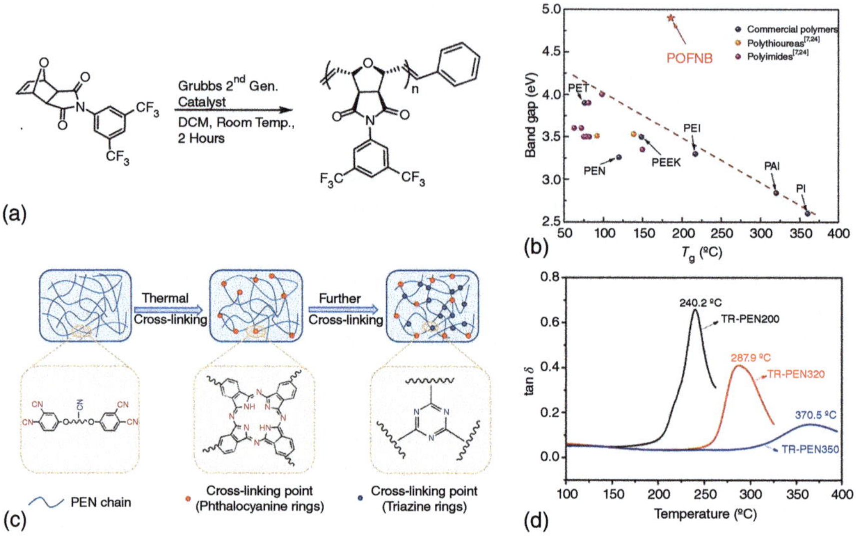

Figure 1.8 (a) Schematic of the synthesis process for POFNB. (b) The bandgap versus T_g for POFNB and high-temperature polymers [45]. Source: Wu et al. [45]. Reproduced with permission of Wiley-VCH. (c) Fabrication of the cross-linked PEN. (d) DMA of the PEN treated at different temperatures [46]. Source: You et al. [46]. Reproduced with permission of the American Chemical Society.

bandgap enables the conduction current of POFNB to be effectively restrained at high temperatures and in high fields, thus obtaining excellent insulation performance at 150 °C.

Cross-linking is a process in which polymers are linked by covalent bonds to form networks. There are substantial intermolecular forces in cross-linked polymers that significantly restrict the movement of molecular chains. For example, You et al. [46] synthesized self-cross-linking poly(arylene ether nitriles) (PEN) at different temperatures, and the high temperature resistance of PEN was significantly improved. As shown in Figure 1.8c, the phthalocyanine rings were formed by the cross-linking of phthalonitriles at the end of PEN at 320 °C, while the –CN groups on the side chains of PEN participate in the reaction to form triazine rings at 350 °C. Figure 1.8d illustrates that the 350 °C cross-linked PEN had a high T_g of 370 °C, and its operating temperature reached 300 °C.

Strong intermolecular interactions resulting from the introduction of polar groups in polymer chains have also been used to suppress the motion of molecular chains and improve high temperature resistance. Zhu et al. [47] synthesized a new type of diamine monomer with two ortho-position nitrile groups in the benzene ring through chemical reactions and obtained a series of PI after reacting with different dianhydrides. It was found that T_g was improved by the introduction of nitrile groups. Moreover, the strong interaction between nitrile groups enhances the mechanical properties of PI. Tang et al. [48] achieved the improvement of T_g and Young's modulus of polystyrene (PS) by introducing nitro groups. It is worth noting that the thermal decomposition of nitro groups results in a slight decrease in T_d of the polymer.

1.5.3 High-Temperature Polymer-Based Composite Materials

Polymers have great designability due to their diverse chemical structures. Moreover, it is also an effective way to enhance the high-temperature properties of traditional polymers by introducing fillers to prepare polymer-based composites.

Inorganic fillers have the advantages of excellent thermal stability and electrical properties, which can be combined with the advantages of machinability and high E_b of the polymers through the two-phase composite method. The thermal stability of inorganic/polymer composite materials is generally higher than that of pure polymers, which is more obvious at high temperatures. For example, Zahra et al. [49] added 5 wt% sepiolite nanowires into PEI matrix to prepare sepiolite/PEI composite materials. Compared with pure PEI, the T_g of composite material increased from 215 to 223 °C, and the $T_{d5\%}$ increased from 478 to 487 °C. At high temperatures, the polymers will be softened. It was found that the addition of carbon nanotubes (CNTs) enhanced the mechanical properties of polymers at high temperatures. Mamedov et al. [50] fabricated CNTs/PEI composites using a layer-by-layer deposition method. Compared with pure PEI, the modulus of the CNT/PEI composites increases from 0.3 to ~11 GPa, and strength increases from ~9 to 325 MPa. As shown in Figure 1.9a,b, Li et al. [34] thermally cross-linked divinyltetramethyldisiloxane-bis (benzocyclobutene) (BCB) in the presence of boron nitride nanosheets (BNNSs) to obtain c-BCB/BNNS nanocomposites ($T_g > 350$ °C). Due to the high thermal conductivity and wide bandgap of BNNS, the excellent thermal conductivity of 1.8 W m^{-1} K^{-1} and high-temperature insulation performance (403 MV m^{-1} at 250 °C) of c-BCB/BNNS were achieved.

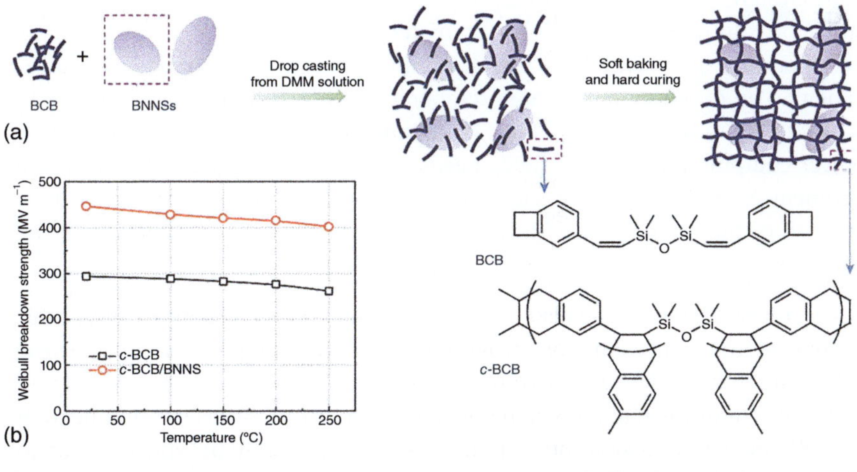

Figure 1.9 (a) Schematic of the synthesis process for c-BCB/BNNS. (b) Weibull breakdown strength of c-BCB and c-BCB/BNNS as a function of temperature [34]. Source: Li et al. [34]. Reproduced with permission of Nature Publishing Group.

The incorporation of inorganic fillers has improved the high temperature and electrical properties of the polymers. However, the introduction of a large number of inorganic fillers leads to poor interfacial compatibility between the two phases, which destroys the flexibility of nanocomposite materials. Compared to polymer-based nanocomposite materials, all-organic composite materials ensure great high-temperature properties while maintaining the inherent flexibility of polymers, which enhances the possibilities of high-temperature polymer applications. Polyacrylonitrile is further dehydrogenated to form a trapezoidal conjugated structure (PcLS) at 400–600 °C. Liao et al. [51] prepared PcLS/PI all-organic composite films by in-situ polymerization, which had an excellent $T_{5\%}$ of 600 °C. Zhang et al. [52] chose PEI and PI to obtain composite materials with excellent high-temperature insulation performance by co-blending method. The strong electrostatic interaction between positive and negative delocalized electrons was used to improve the packing density of molecular chains, thereby reducing voids and lowering the free volume. The composite material of 50 wt% PEI and 50 wt% PI not only had the T_g of 246 °C but also obtained excellent insulation performance (E_b of 550 MV m^{-1}) at 200 °C. Zhang et al. [53] designed a novel 1,4,5,8-naphthalenetetracarboxylic dianhydride (NTCDA)/PEI composite film. As a molecular semiconductor, NTCDA has high electron affinity, which significantly inhibits carrier migration, thus improving the high-temperature insulation performance of PEI. Compared with PEI film (E_b = 440 MV m^{-1}), the E_b of NTCDA/PEI composite film was 510 MV m^{-1} at 200 °C.

1.6 Summary and Outlook

In summary, researchers have conducted a lot of exploratory experiments on the structure design, preparation process, performance optimization, and application scenarios of high-temperature polymer materials. It is remarkable that many high-temperature polymers have been commercialized and applied to power equipment. The development of high-temperature polymers makes it possible to break through the bottleneck of their poor temperature resistance. However, as the power equipment in various fields needs to cope with a more severe working environment, higher requirements are put forward for high-temperature polymers. The growth of high-temperature polymer materials needs to pay attention to the following issues:

1. Based on the design of molecular structure, it is crucial to optimize the morphology of molecular chains through the preparation process for intrinsic high-temperature polymers. In order to ensure the uniform dispersion and reasonable distribution of the fillers in the polymer matrix, the surface modification of the fillers and the design of the multiple structures should be paid attention to in polymer-based composite materials.
2. It is crucial to explore and understand the inherent mechanisms of high-temperature polymers with the participation of various simulation models. On

the one hand, the molecular structure of the initial screen can be obtained by molecular simulation of the intrinsic polymers, which reduces the cost of the traditional "trial and error" method. On the other hand, simulation models are often used to understand the influence of two-phase interfacial compatibility on the properties of high-temperature polymer-based composite materials. In addition, the screening conditions for high-temperature polymers should not be limited to their high thermal stability and electrical properties. Film forming and mechanical properties should be considered together to greatly improve the possibility of future applications.

3. In the early development stage of high-temperature polymers, excessive attention is paid to their thermal stability while their processability is ignored. For example, the introduction of a large number of aromatic rings increases the thermal stability but makes the polymer insoluble and difficult to reprocess. The cooperative regulation mechanisms of processability and high temperature resistance should be established. In the long term, successful production of high-temperature polymers requires innovative structural design and preparation processes to balance the relationship between the above two parameters.
4. On the basis of the third issue, excellent electrical properties such as high insulation, corona resistance, and low loss should be maintained at high temperatures to meet the different requirements of each electrical field. It is necessary to strengthen the coordination between high-temperature polymers and application scenarios to provide targeted performance optimization.

References

1 Liu, X.J., Zheng, M.S., Chen, G. et al. (2022). High-temperature polyimide dielectric materials for energy storage: theory, design, preparation and properties. *Energy & Environmental Science* 15: 56–81.

2 Watson, J. and Castro, G. (2015). A review of high-temperature electronics technology and applications. *Journal of Materials Science - Materials in Electronics* 26: 9226–9235.

3 Tan, D.Q. (2019). Review of polymer-based nanodielectric exploration and film scale-up for advanced capacitors. *Advanced Functional Materials* 30: 1808567.

4 Li, Q., Yao, F.Z., Liu, Y. et al. (2018). High-temperature dielectric materials for electrical energy storage. *Annual Review of Materials Research* 48: 219–243.

5 Feng, Q.K., Zhong, S.L., Pei, J.Y. et al. (2022). Recent progress and future prospects on all-organic polymer dielectrics for energy storage capacitors. *Chemical Reviews* 122: 3820–3878.

6 McCluskey, F.P., Podlesak, T., and Grzybowski, R. (1997). *High Temperature Electronics*, 1–340. CRC Press.

7 Ghorbani, H., Christen, T., Carlen, M. et al. (2017). Long-term conductivity decrease of polyethylene and polypropylene insulation materials. *IEEE Transactions on Dielectrics and Electrical Insulation* 24: 1485–1493.

8 Wan, B.Q., Dong, X.D., Yang, X. et al. (2022). High strength, stable and self-healing copolyimide for defects induced by mechanical and electrical damages. *Journal of Materials Chemistry C* 10: 11307–11315.

9 Feng, Q.K., Dong, Q., Zhang, D.L. et al. (2022). Enhancement of high-temperature dielectric energy storage performances of polyimide nanocomposites utilizing surface functionalized MAX nanosheets. *Composites Science and Technology* 218: 109193.

10 Dong, X., Zheng, M., Wan, B. et al. (2021). Low-permittivity copolymerized polyimides with fluorene rigid conjugated structure. *Materials* 14: 6266.

11 Zha, J.W., Sun, F., Wang, S.J. et al. (2014). Improved mechanical and electrical properties in electrospun polyimide/multiwalled carbon nanotubes nanofibrous composites. *Journal of Applied Physics* 116: 134104.

12 Zha, J.W., Liu, Q., Dang, Z.M., and Chen, G. (2016). Tailored wide-frequency dielectric behavior of polyimide composite films with Ba_xSr_1-xTiO_3 Perovskites ceramic particles. *IEEE Transactions on Dielectrics and Electrical Insulation* 23: 113–120.

13 Dang, Z.M., Zhou, T., Yao, S.H. et al. (2009). Advanced calcium copper titanate/polyimide functional hybrid films with high dielectric permittivity. *Advanced Materials* 21: 2077–2082.

14 Liu, X.J., Zheng, M.S., Chi, Q.G. et al. (2022). High-temperature energy storage performances of "isomer-like" polyimide and its thermoplastic polyurethane blending system. *Journal of Materials Chemistry C* 10: 17326–17335.

15 Zhang, X.M., Liu, J.G., and Yang, S.Y. (2016). A review on recent progress of R&D for high-temperature resistant polymer dielectrics and their applications in electrical and electronic insulation. *Reviews on Advanced Materials Science* 46: 22–38.

16 Feng, Q.K., Pei, J.Y., Dong, Q. et al. (2021). Structural, electrical, and thermal features of polyimide composites filled with semiconductive MXene sheets. *Applied Physics Letters* 118: 262901.

17 Ho, J.S. and Greenbaum, S.G. (2018). Polymer capacitor dielectrics for high temperature applications. *ACS Applied Materials & Interfaces* 10: 29189–29218.

18 Fan, M.Z., Hu, P.H., Dan, Z.K. et al. (2020). Significantly increased energy density and discharge efficiency at high temperature in polyetherimide nanocomposites by a small amount of Al_2O_3 nanoparticles. *Journal of Materials Chemistry A* 8: 24536–24542.

19 Li, P., Yu, J.J., Jiang, S.H. et al. (2020). Dielectric, mechanical and thermal properties of all-organic PI/PSF composite films by in situ polymerization. *E-Polymers* 20: 226–232.

20 Hergenrother, P.M. (2003). The use, design, synthesis, and properties of high performance/high temperature polymers: an overview. *High Performance Polymers* 15: 3–45.

21 Cottrell, T.L. (1958). *The Strength of Chemical Bonds*, 2e, 1–3. London: Butterworths.

22 Chen, H., Ginzburg, V.V., Yang, J. et al. (2016). Thermal conductivity of polymer-based composites: Fundamentals and applications. *Progress in Polymer Science* 59: 41–85.

23 Guo, F., Shen, X., Zhou, J. et al. (2020). Highly thermally conductive dielectric nanocomposites with synergistic alignments of graphene and boron nitride nanosheets. *Advanced Functional Materials* 30: 1910826.

24 Dong, J., Cao, L., Li, Y. et al. (2020). Largely improved thermal conductivity of PI/BNNS nanocomposites obtained by constructing a 3D BNNS network and filling it with AgNW as the thermally conductive bridges. *Composites Science and Technology* 196: 108242.

25 Guo, Y., Ruan, K., Shi, X. et al. (2020). Factors affecting thermal conductivities of the polymers and polymer composites: a review. *Composites Science and Technology* 193: 108134.

26 Guo, Y., Yang, X., Ruan, K. et al. (2019). Reduced graphene oxide heterostructured silver nanoparticles significantly enhanced thermal conductivities in hot-pressed electrospun polyimide nanocomposites. *ACS Applied Materials & Interfaces* 11: 25465–25473.

27 Wang, X., Zhang, C., Du, Z. et al. (2017). Synthesis of non-destructive amido group functionalized multi-walled carbon nanotubes and their application in antistatic and thermal conductive polyetherimide matrix nanocomposites. *Polymers for Advanced Technologies* 28: 791–796.

28 Gu, J., Lv, Z., Wu, Y. et al. (2017). Dielectric thermally conductive boron nitride/polyimide composites with outstanding thermal stabilities via in-situ polymerization-electrospinning-hot press method. *Composites - Part A: Applied Science and Manufacturing* 94: 209–216.

29 Huang, W., Ju, T.X., Li, R.P. et al. (2022). High-k and High-temperature dipolar glass polymers based on sulfonylated and cyanolated poly(arylene ether)s for capacitive energy storage. *Advanced Electronic Materials* 1: 2200414.

30 Niu, Y.J., Dong, J.F., He, Y.F. et al. (2022). Significantly enhancing the discharge efficiency of sandwich-structured polymer dielectrics at elevated temperature by building carrier blocking interface. *Nano Energy* 97: 107215.

31 Liu, X.J., Zheng, M.S., Wang, G. et al. (2022). High energy density of polyimide films employing an imidization reaction kinetics strategy at elevated temperature. *Journal of Materials Chemistry A* 10: 10950–10959.

32 Zuo, H.T., Qian, G.T., Li, H.B. et al. (2022). Reduced coefficient of linear thermal expansion for colorless and transparent polyimide by introducing rigid-rod amide units: synthesis and properties. *Polymer Chemistry* 13: 2999–3008.

33 Zhou, S.S., Xu, T.L., Jin, L.Y. et al. (2022). Ultraflexible polyamide-imide films with simultaneously improved thermal conductive and mechanical properties: design of assembled well-oriented boron nitride nanosheets. *Composites Science and Technology* 219: 109259.

34 Li, Q., Chen, L., Gadinski, M.R. et al. (2015). Flexible high-temperature dielectric materials from polymer nanocomposites. *Nature* 523: 576–579.

35 Ghosh, M.K. and Mittal, K.L. (1996). *Polyimides: Fundamentals and Applications*, 1–6. New York: Marcel Dekker, Inc.

36 Odian, G. (1991). *Principles of Polymerization*, 313. New York: John Wiley & Sons, Inc.

37 Mark, H.F., Bikales, N.M., Overberger, C.G., and Menges, G. (1986). *Encyclopedia of Polymer Science and Engineering*, vol. 6, 94–131. New York: Wiley-Interscience.

38 Li, H., Zhou, Y., Liu, Y. et al. (2021). Dielectric polymers for high-temperature capacitive energy storage. *Chemical Society Reviews* 50: 6369–6400.

39 Cheng, S.Z.D., Ho, R.M., Hsiao, B.S., and Gardner, K.H. (1996). Polymorphism and crystal structure identification in poly(aryl ether ketone ketone)s. *Macromolecular Chemistry and Physics* 197: 185–213.

40 Pan, J., Li, K., Li, J. et al. (2009). Dielectric characteristics of poly(ether ketone ketone) for high temperature capacitive energy storage. *Applied Physics Letters* 95: 022902.

41 Seymour, R.B. and Kirshenbaum, G.S. (1986). *High Performance Polymers: Their Origin and Development*, 261–266. New York: Elsevier Science Publishing Co, Inc.

42 Wunderlich, B. (1980). *Macromolecular Physics, Vol. 3: Crystal Melting*, vol. 3, 48. New York: Academic Press, Inc.

43 Yan, P., Yang, F., Xiang, M. et al. (2020). New insights into the memory effect on the crystallization behavior of poly(phenylene sulfide). *Polymer* 195: 122439.

44 Pan, J., Li, K., Chuayprakong, S. et al. (2010). High-temperature poly(phthalazinone ether ketone) thin films for dielectric energy storage. *ACS Applied Materials & Interfaces* 2: 1286–1289.

45 Wu, C., Deshmukh, A.A., Li, Z. et al. (2020). Flexible temperature-invariant polymer dielectrics with large bandgap. *Advanced Materials* 32: 2000499.

46 You, Y., Liu, S., Tu, L. et al. (2019). Controllable fabrication of poly(arylene ether nitrile) dielectrics for thermal-resistant film capacitors. *Macromolecules* 52: 5850–5859.

47 Zhu, T., Yu, Q., Zheng, W. et al. (2021). Intrinsic high-k–low-loss dielectric polyimides containing ortho-position aromatic nitrile moieties: reconsideration on Clausius–Mossotti equation. *Polymer Chemistry* 12: 2481–2489.

48 Tang, X., Din, C., Yu, S. et al. (2022). Synthesis of dielectric polystyrene via one-step nitration reaction for large-scale energy storage. *Chemical Engineering Journal* 446: 137281.

49 Tabatabaei-Yazdi, Z. and Mehdipour-Ataei, S. (2015). Poly(ether-imide) and related sepiolite nanocomposites: investigation of physical, thermal, and mechanical properties. *Polymers for Advanced Technologies* 26: 308–314.

50 Mamedov, A.A., Kotov, N.A., Prato, M. et al. (2002). Molecular design of strong single-wall carbon nanotube/polyelectrolyte multilayer composites. *Nature Materials* 1: 190–194.

51 Liao, X., Ding, Y., Chen, L. et al. (2015). Polyacrylonitrile-derived polyconjugated ladder structures for high performance all-organic dielectric materials. *Chemical Communications* 51: 10127–10130.

52 Zhang, Q., Chen, X., Zhang, B. et al. (2021). High-temperature polymers with record-high breakdown strength enabled by rationally designed chain-packing behavior in blends. *Matter.* 4: 2448–2459.
53 Zhang, B., Chen, X.M., Pan, Z. et al. (2022). Superior high-temperature energy density in molecular semiconductor/polymer all-organic composites. *Advanced Functional Materials* 1: 2210050.

2

Basic Principles of Dielectrics

Anastasios Chr. Patsidis and Georgios Chr. Psarras

University of Patras, School of Natural Sciences, University Campus of Rion, Patras 26504, Greece

2.1 Introduction

Following the classification of materials according to their electrical properties, and especially according to the ease of charge migration through them, materials are classified as conductors, semiconductors, and insulators. The ease of charge migration within a material is expressed via conductivity, which is a material property, and its values range over twenty-six orders of magnitude from conductors to insulators. The electrical properties of materials are exploited in numerous applications and without exaggeration can be considered as the basis of our everyday life and our technological civilization.

2.2 Definition of Dielectrics

Dielectric materials or dielectrics are non-conductive materials that are polarized under the influence of an external field. Exerting an external field on a dielectric material forces positive charges in the field's direction and negative charges in the opposite direction. Displacement of charges is short being at the order of magnitude of the atomic diameters. No macroscopic charge migration takes place, and the positive and negative charge separation is present in any elementary volume of the material. The whole material takes the form of a dipole with equal, in absolute values, positive and negative charges at the two-edge surfaces. The net charge of the material remains zero. The occurring effect is completely recovered after removing the applied field. The electrostatic field existing within dielectrics is the resultant of the applied field and any locally developed fields. It should be noted that dielectrics represent a wider class of materials than insulators, and typically refer to materials with an energy band gap higher than 3 eV.

Polymers are, in their majority, typical dielectric materials, because of their low conductivity and ability to be polarized. The superior insulating properties of most polymers, their low weight, the easy fabrication procedure, their corrosion and

High Temperature Polymer Dielectrics: Fundamentals and Applications in Power Equipment, First Edition.
Edited by Jun-Wei Zha and Zhi-Min Dang.

environmental resistance, their thermomechanical stability, and their relative low cost are responsible for a vast number of applications. Polarization is related to the distortion and alignment of polar groups, molecules, parts of the macromolecular chain, and the build-up of charges at the interface of blends, semicrystalline, and heterogeneous systems. All these effects give rise to the dielectric relaxation phenomena. By studying these processes, valuable knowledge for molecular mobility, phase transitions, conductivity mechanisms, interfacial effects, occurring interactions, and functional performance can be derived for polymers and polymer-based composites.

2.3 Dipole Moment and Types of Dielectric Materials

An electric dipole is composed of a positive and a negative charge or distribution of charges, having equal absolute values, with their centers being at a distance. A dipole represents the simplest polarization unit. Electric dipole moment is defined via Eq. (2.1) and its measure is the product of the absolute value of the positive or negative charge (q) with the distance between the two centers (r). An electric dipole moment is a vector beginning from the center of the negative charge and pointing at the center of the positive charge as shown in Figure 2.1.

$$\mu = qr \tag{2.1}$$

In a dielectric material, different kinds of dipoles could exist. These dipoles should differ in dipoles' moment, because of different charges and/or distances between the charges. In this case, the total dipole moment is obtained by employing Eq. (2.2):

$$M = \sum_i q_i r_i \tag{2.2}$$

where subscript i denotes the number of different dipoles.

Dielectric materials are classified into two categories – (i) polar dielectrics, and (ii) non-polar dielectrics. In polar dielectrics the molecules exhibit permanent dipole moment since the centers of the distribution of positive and negative charges are apart from each other. The non-symmetrical molecules belong to this category.

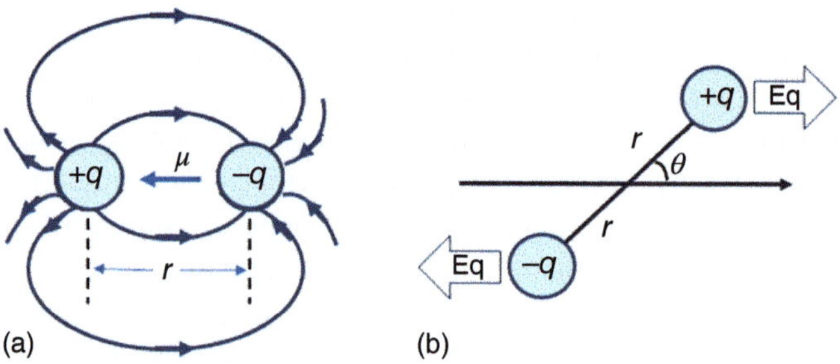

Figure 2.1 Schematic configuration of an electric dipole. (a) In the absence and (b) in the presence of an external field.

The dipole moment depends on the difference in electronegativity between the atoms of the molecule. Typical examples of polar dielectrics are H_2O and HCl. In non-polar dielectrics, molecules exhibit symmetry in the distribution of the positive and negative charges. Because of the symmetrical arrangement in space, of the positive and negative charges, the centers of their distributions coincide, leading to no dipole moment. Typical examples of non-polar dielectrics are CH_4 and CO_2.

2.4 Polarization and Dielectric Permittivity

Polarization (***P***) is a vector expressing the total dipole moment of a substance upon unit volume. Its measure is given by Eq. (2.3):

$$P = \frac{M}{V} \tag{2.3}$$

where V is the sample's volume. The units of polarization in SI (C m^{-2}) are the same as the surface charge density (σ).

The intensity of the electric field ***E***, or simple electric field, is determined as the ratio of the force exerted on a charge upon the charge's quantity, when the charge is at an interior point of the area of the field, via Eq. (2.4):

$$\vec{E} = \frac{\vec{F}}{q} \tag{2.4}$$

A third vector quantity related to the electric field and polarization, given by Eq. (2.5), is the electric displacement (***D***).

$$\vec{D} = \varepsilon_0 \vec{E} + \vec{P} \tag{2.5}$$

where ε_0 is the permittivity of free space ($\varepsilon_0 = 8.854 \times 10^{-12}$ F m^{-1}). In a parallel plate capacitor configuration, with dielectric material, and under the influence of a constant voltage between the electrodes of the capacitor, electric displacement occurs only from the free charges at the electrodes. Polarization occurs from the induced charges at the edge surfaces of the sample, while the electric field is related to both free and induced charges.

Atoms do not possess permanent electric dipole moment, because of the symmetrical distribution of the electron cloud around the positive nucleus. The situation changes with the application of an external field, which causes a distortion in the arrangement of the electron cloud. The resulting non-symmetrical distribution of the negative charge around the nucleus forms a dipole moment in the direction of the field, and thus induced polarization is acquired by the atom.

In the case of a polar dielectric substance, and in the absence of an external field, permanent dipoles are randomly oriented leading to a net zero total dipole moment and polarization. The application of an external field forces all kinds of permanent dipoles to be aligned with the field. Interactions between dipoles and between dipoles and their environment, as well as the increase in temperature, disrupts the perfect orientation. When an external field is applied permanent and induced dipoles tend to be oriented parallel to the field, contributing both to the occurring polarization.

The static value of dielectric permittivity, also known as relative dielectric constant, is a characteristic property for each dielectric material, which is related to the field vectors as shown in Eq. (2.6):

$$\varepsilon = \frac{D}{\varepsilon_0 E} = 1 + \frac{1}{\varepsilon_0}\frac{P}{E} = 1 + \chi \tag{2.6}$$

The quantity (χ) is the electric susceptibility and is considered the measure of the developed polarization by a specific field.

2.5 Polarization Under Static Electric Field

2.5.1 Dielectric Permittivity and Polarizability

Consider a polar dielectric with n permanent dipoles per unit volume inside an electric field, which causes an equal number of induced dipoles. If μ and m_{ind} are the dipole moments of a permanent and an induced dipole, respectively, and assuming negligible interactions between dipoles, the achieved polarization is parallel to the field and can be expressed with Eq. (2.7):

$$P = \left\langle \sum_{i=1}^{n} (\mu_i + m_{\text{ind},i}) \right\rangle \tag{2.7}$$

Polarizability or coefficient of polarization (α) is a microscopic electrical quantity, which expresses the ability of atoms, molecules, or even groups of a dielectric material to be polarized, responding to an applied field. In the case of linear dielectrics (dielectrics under the influence of low fields), polarizability is defined via Eq. (2.8):

$$\alpha = \frac{m_i}{E_i} \tag{2.8}$$

where m_i is the dipole moment of the i type of dipoles, E_i is the local field which acts on the dipole and not the externally exerted one.

Polarization (and thus polarizability) in a dielectric material might occur because of various physical origins. Polarizability and polarization are, mainly, classified in – distortional (or deformational) and orientational.

(a) Distortional (or deformational) polarizability is further subdivided into electronic and ionic or atomic. This type of polarization results from the deformation (relative displacement) of the distribution of charges between electron orbits and nuclei (electronic polarization) or ions (ionic or atomic polarization), due to the application of an external electric field (Figure 2.2). Distortional polarization is characterized as a "fast" process (related to low relaxation time) because of the low inertia of the involved dipoles and is not influenced by temperature [1–3].
(b) Orientational polarizability occurs in dielectric substances which contain permanent dipoles. Dipoles, initially randomly distributed, tend to be aligned

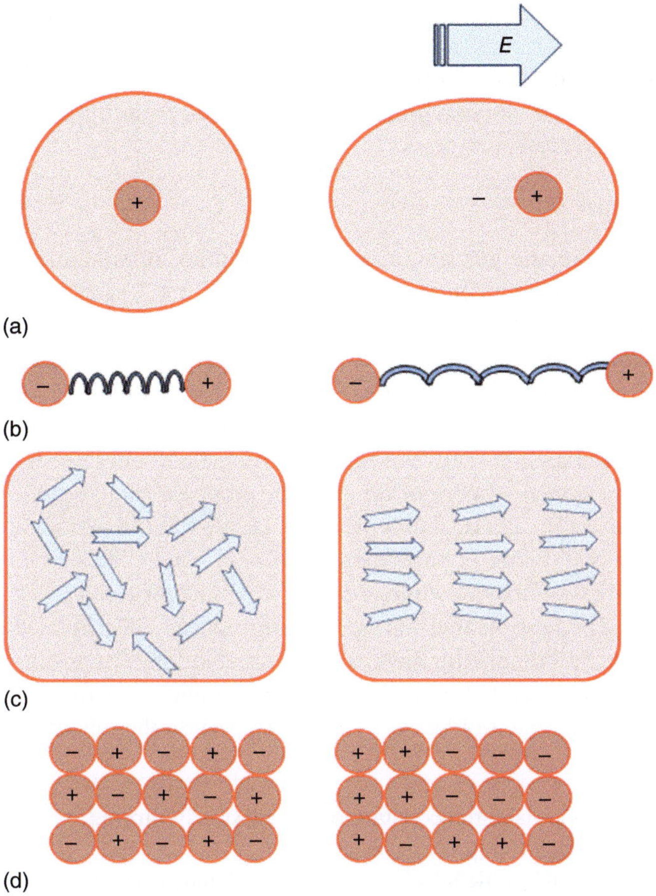

Figure 2.2 Schematic representation of the types of polarization. (a) Electronic, (b) atomic/ionic, (c) orientational, (d) space charge (interfacial) polarization.

with the applied field. This process is relatively slower than distortional polarization (characterized by higher relaxation time or times) because of the enhanced inertia of the permanent dipoles, which vary with their structure, mass, and molecular environment. Orientational polarization depends on the temperature, since temperature affects the random motion of polar groups and the occurring interactions between dipoles [1–3].

A common characteristic in distortional and orientational polarizability is the fact that both are assigned to charges bound to atoms, molecules, or groups of the dielectric material. In addition, unbounded charges, which might be present in the material, are influenced by the applied field. These charge carriers are able to migrate, usually, short distances within the dielectric, interact between themselves and with dipoles, and finally, are trapped in the system's interface or grain boundaries, where they build-up dipoles. This effect, known as space charge (or interfacial) polarization, is significantly present in heterogenous or composite systems [4, 5]. The resulting dipoles exhibit increased inertia in being oriented to the applied field. However, if sufficient time is given to them to align with the field, they contribute considerably to the system's polarization and capacitance.

In a dielectric material, all types of polarization could be present, and thus the total polarizability (α) can be written as follows:

$$\alpha = \alpha_e + \alpha_a + \alpha_{or} + \alpha_{sp} \tag{2.9}$$

where α_e, α_a, α_{or}, and α_{sp} are the electronic, atomic (or ionic), orientational, and space charge (interfacial) polarizabilities, respectively. Figure 2.2 provides a schematic representation of all four types of polarization.

Macroscopic polarization and the static value of dielectric permittivity (usually referred to as dielectric constant) are dependent on the polarizabilities of all occurring mechanisms and the permanent dipole moments of the polar groups. The first step for deriving relationships between the macroscopic and microscopic quantities was based on the statistic theory of orientation, proposed by Langevin for the orientation of permanent magnetic moments [1, 6]. Debye was the first who applied this method to the orientation of permanent dipoles in a polar dielectric, by considering the occurring torque of a couple of forces which aligns the dipoles parallel to the applied field. The possible thermal agitation introduces a disturbance to the orientation of dipoles by influencing their random distribution. The resulting polarization is a counterbalance between dipoles' rotation and their mutual interactions [1, 2, 6]. The development of Debye's theory is based on the following physical assumptions:

(i) The applied electric field is considered constant or is very slowly varying. Thus, equilibrium of all types of polarization can be achieved.
(ii) The dipole moment of polar groups is considered independent from the temperature and the applied field. This condition is strong in substances with low density such as polar gases.
(iii) Prior to the application of the external field, permanent dipoles could attain any possible orientation.
(iv) Permanent dipole moments are distributed, in relation to the direction of the applied field, according to the Boltzmann statistics.
(v) The employed dielectric material is considered isotropic.

By adopting the above considerations, it is possible to avoid the complications arising from the interactions between dipoles and their molecular environment, which are present in condensed matter and to derive an expression for the average moment of the permanent dipoles [1, 2, 6]. Thus, for low-strength fields, the average dipole moment is given by Eq. (2.10):

$$\langle \mu \rangle_{E_r} = \frac{\mu^2 E_r}{3 k_B T} \tag{2.10}$$

where, μ is the dipole moment, E_r the actual field applied to the dipole, T is temperature and k_B, the Boltzmann's constant. If n dipoles per unit volume, each with moment μ, are present in the material the resulting polarization can be expressed as:

$$P_{\text{or}} = \frac{n\mu^2 E_r}{3k_B T} \tag{2.11}$$

Equation (2.11) accounts only for the contribution of permanent dipoles. However, the applied field induces additional components of polarization, because of electronic and atomic/ionic polarizability. Equation (2.12) includes the influence of distortional polarizability, and thus the total average dipole moment can be written as:

$$\langle m \rangle_{E_r} = \left(\alpha_e + \alpha_a + \frac{\mu^2}{3k_B T} \right) E_r \tag{2.12}$$

and the resulting total polarization is:

$$P = n \left(\alpha_e + \alpha_a + \frac{\mu^2}{3k_B T} \right) E_r \tag{2.13}$$

Substituting Eq. (2.13) by Eq. (2.6) gives a relationship between the macroscopic parameter of static dielectric permittivity and atomic/molecular microscopic quantities:

$$\varepsilon_s - 1 = \frac{1}{\varepsilon_0} n \left(\alpha_e + \alpha_a + \frac{\mu^2}{3k_B T} \right) \tag{2.14}$$

2.5.2 Dipole's Local Electric Field

In low pressure polar gases, polar entities can be reasonably considered as non-interacting. In this case, the externally applied field can be reckoned as equal to the one actually exerted on polar molecules and groups. The situation becomes different, in the case of dense polar gases, polar fluids, and even more in solid state, where the field acting on dipoles is substantially different from the externally applied one. This deviation is related to the dipole's molecular environment and the distribution of charges in its adjacent surroundings. For the determination of local field, the spherical model of Clausius and Mosotti is used [1, 6]. In this model, a molecular dipole is considered to be at the center of a spherical cavity, at the position where the internal field has to be evaluated (Figure 2.3a). The spherical cavity is part of a dielectric medium which is placed between the opposite charged electrodes of a plate capacitor. The radius of the spherical cavity is sufficiently smaller than the dielectric material's dimensions and, at the same time, substantially larger than the dimension of the molecule. By these means, the material surrounding the spherical cavity is regarded as a continuous and homogeneous medium with constant value of static dielectric permittivity equal to (ε_s). In the interior of the spherical cavity, individual polar molecules exist. Therefore, the field in the interior of the cavity is determined as the vector sum of – (i) the field related to the charges at the electrodes of the capacitor and also to the induced charges at the material's contact faces with the electrodes (E_1), (ii) the field (E_2) generated by the induced

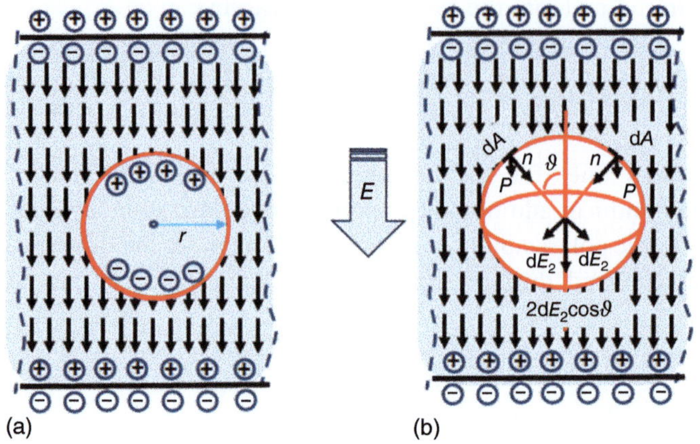

Figure 2.3 (a) Dielectric with spherical cavity according to the Clausius and Mosotti/Lorentz model. (b) The geometrical arrangement used, in the same model, for the determination of the local (internal) field.

charges at the spherical cavity's surface, and (iii) the field (E_3) related to the dipoles inside the cavity [1, 6].

By employing the Gauss law, the field E_1 can be easily proved equal to the external applied one E_0, Eq. (2.15):

$$E_1 = E_0 = \frac{q}{\varepsilon_0 \varepsilon_s S} \tag{2.15}$$

where q is the sum of the free and induced charges at the electrode/dielectric material interface (that is the total charge inside the Gaussian surface) and S the electrode's area. The field E_2 is generated by the charges induced at the surface of the cavity. By considering the normal component of the developed polarization and the charge carried by an infinitesimal area dA of the cavity (Figure 2.3b), the field E_2 can be determined by the integration of Eq. (2.16):

$$\mathrm{d}E_2 = \frac{P\cos\theta}{4\pi\varepsilon_0 r^2}\mathrm{d}A \tag{2.16}$$

where, P is the polarization, r the radius of the spherical cavity, dA the surface element, θ the angle between polarization and the normal unit vector, and ε_0 the permittivity of free space. By expressing dA in terms of r and θ the value of E_2 can be obtained [1, 6]:

$$E_2 = \int_0^{\pi} \frac{P\cos^2\theta}{\varepsilon_0 \pi r^2} 2\pi r^2 \sin\theta \cdot \mathrm{d}\theta = \frac{1}{3}\frac{P}{\varepsilon_0} \tag{2.17}$$

Finally, the field E_3 inside the cavity is considered as equal to zero, since the resultant polarization at the center of the sphere is zero, because of the mutually canceled contributions of the randomly distributed polar molecules [1].

By employing the results obtained via the spherical model, the field inside (E_i) the dielectric medium can be determined according to Eq. (2.18):

$$E_i = E_1 + E_2 + E_3 = E_1 + \frac{P}{3\varepsilon_0} = E_0 + \frac{P}{3\varepsilon_0} \tag{2.18}$$

Recalling Eq. (2.6) and substituting to Eq. (2.18) the following relationship is obtained:

$$E_i = \frac{E_0}{3}(\varepsilon_s + 2) \tag{2.19}$$

The latter relates the unknown field inside the dielectric medium with the applied field and the macroscopically determined quantity of ε_s. However, the field E_i is not in general equal to the field E_r which is actually exerted on a dipole. Dipoles modify their surrounding space by developing non-homogeneous fields, which are applied to adjacent atoms, molecules, and groups resulting in non-homogeneous polarization. In turn, the generated polarization exerts a reaction field to the initial dipoles [1].

2.5.3 Models for Static Dielectric Permittivity: Debye, Onsager, Kirkwood, and Fröhlich

The polarization of a dielectric material is the sum of distortional and orientational polarization. The contribution of distortional polarization can be expressed as:

$$P_\alpha = \sum_j n_j \alpha_{\alpha j}(E_i)_j \tag{2.20}$$

where n_j is the number of atoms or molecules per unit volume, $\alpha_{\alpha j}$ is the sum of electronic and atomic/ionic polarizabilities, and E_i the local field, the index j refers to the possible different types of atoms or molecules. The contribution of orientational polarization is given by Eq. (2.21):

$$P_{\text{or}} = \sum_j n_j \langle \mu_j \rangle \tag{2.21}$$

where $\langle \mu_j \rangle$is the average value of the dipole moment of the jst dipole and n_j the number of dipoles per unit volume. The influence of temperature upon orientational polarization is shown by employing Eq. (2.10) in Eq. (2.21), and by these means it is obtained:

$$P_{\text{or}} = \sum_j n_j \frac{\mu_j^2}{3k_B T}(E_r)_j \tag{2.22}$$

Considering Eqs. (2.20) and (2.22), the total polarization can be expressed via Eq. (2.23):

$$P = \sum_j n_j \left[\alpha_{\alpha j}(E_i)_j + \frac{\mu_j^2}{3k_B T}(E_r)_j \right] \tag{2.23}$$

Substituting polarization with its equal term $(\varepsilon_s - 1)\varepsilon_0 E_0$ the Debye's equation for static fields is obtained:

$$(\varepsilon_s - 1)\varepsilon_0 E_0 = \sum_j n_j \left[\alpha_{\alpha j}(E_i)_j + \frac{\mu_j^2}{3k_B T}(E_r)_j \right] \tag{2.24}$$

In this relation, E_0 represents the external field and E_i, E_r the internal fields, which in general differ, because E_r accounts also for the mutual interactions

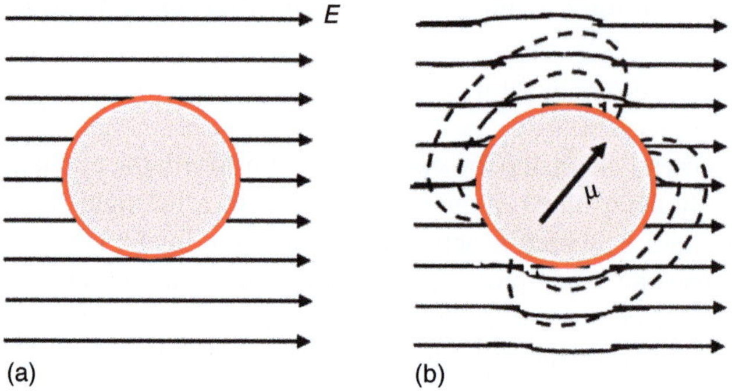

Figure 2.4 Spherical cavity and field's lines in a dielectric medium according to the: (a) Clausius and Mosotti/Lorentz model, and (b) Onsager model.

between dipoles. In the case that these interactions can be neglected $E_i \cong E_r$ and by recalling Eq. (2.19) the following expression is obtained:

$$\frac{\varepsilon_s - 1}{\varepsilon_s + 2} = \frac{1}{3\varepsilon_0}\sum_j n_j\left[\alpha_{\alpha j} + \frac{\mu_j^2}{3k_B T}\right] \quad (2.25)$$

Since Eq. (2.25) has been derived by neglecting the occurring interactions between the dipoles themselves, and the dipoles and their surroundings, its validity appears to be limited in condensed matter. However, it remains useful because it correlates the static value of dielectric permittivity ε_s with the molecular and atomic quantities of n_j, $\alpha_{\alpha j}$, and μ_j.

Onsager was the first person who tried to extend the validity of Eq. (2.25) in condensed phase polar dielectrics [7]. His model was based on a better approximation of the local field acting on a molecule. He modified the spherical model of Clausius and Mosotti by reducing the size of the cavity to the molecular size and by putting a single dipole in its center (Figure 2.4). The cavity in the Mosotti or Lorentz approach is a mathematical concept and thus the field's lines appear not to be deflected (Figure 2.4a). In contrast in the Onsager approximation, the cavity is considered real and therefore, the field's lines are deflected (Figure 2.4b). Onsager assumed that the polarizability of the dipole occupying the sphere is isotropic and its interaction with the surrounding medium can be described via the polarization of the medium caused by the dipole's field. In addition, the polarization of the medium "reacts back" on the molecular dipole. By these means, Onsager proposed that the local field acting on the dipole, located at the center of the cavity, is the sum of two fields, the cavity field, and the reaction field. The cavity field was determined as the field present at the center of an empty cavity when an external field is applied to the dielectric medium. The reaction field was determined as the additional field occurring at the center of the cavity because of the polarization of the surrounding medium due to the presence of a molecular dipole at the center [1, 2, 8, 9]. Under these conditions, the Onsager relation takes the form of Eq. (2.26):

$$\frac{(\varepsilon_s - \varepsilon_\infty)(2\varepsilon_s + \varepsilon_\infty)}{\varepsilon_s(\varepsilon_\infty + 2)^2} = \frac{N\mu_0^2}{9\varepsilon_0 k_B T} \quad (2.26)$$

where ε_{∞} is the permittivity measured at high frequencies (optical) where any contribution from permanent dipoles' orientation can be neglected, N the number of molecular dipoles per unit volume, and μ_0 the permanent dipole moment. Onsager theory is valid in non-associating rigid dipoles. The model assumes that any molecule is surrounded by a continuum medium, considering thus only long-range interactions and neglecting any short-range interactions. For these reasons, it accounts only for weak molecular interactions, failing to describe strong interactions and interactions due to induced dipoles, occurring in organic compounds and polymers [2, 3, 8, 9].

Kirkwood developed a theory, via statistical-mechanical consideration, taking into account the interactions between neighboring dipoles [10]. The short-range interactions between molecules were described by introducing a correlation function (g). However, in his initial approach, the deformational polarization of the molecules was omitted and later was empirically described by a term added to his relation. Kirkwood's model for static polarization is expressed via Eq. (2.27):

$$\frac{(\varepsilon_s - 1)(2\varepsilon_s + 1)}{9\varepsilon_s} = \frac{N}{9\varepsilon_0}\left(a_e + a_a + \frac{g\mu_0^2}{3k_B T}\right) \tag{2.27}$$

where g is the correlation factor related to molecular interactions. When no interactions occur, g takes the value 1. A value of $g > 1$ indicates that a specific dipole tends to orient the nearest dipoles in a parallel direction, while a value $g < 1$ indicates orientation of the neighboring dipoles in the antiparallel direction. Assuming that the reference dipole is surrounded by n_z neighboring molecules, $\overline{\cos\phi}$ is the average of $\cos\phi$ of the angle ϕ made by the reference dipole and one of its nearest neighbors', thus the correlation factor can be expressed as:

$$g = 1 + n_z\overline{\cos\phi} \tag{2.28}$$

Fröhlich generalized the Onsager and Kirkwood models by including both the orientation correlation between molecules (short-range interactions) and the deformational polarization. In his approach, Fröhlich considered a homogeneous continuum medium with static dielectric permittivity ε_s in which a spherical macroscopic region of volume V, with N dipoles per unit volume is included. The latter is regarded as a continuum with dielectric permittivity ε_{∞} and the included dipoles follow statistical-mechanical laws [11]. The general form of Fröhlich's relation is:

$$\frac{(\varepsilon_s - \varepsilon_{\infty})(2\varepsilon_s + \varepsilon_{\infty})}{\varepsilon_s(\varepsilon_{\infty} + 2)^2} = \frac{Ng\mu_0^2}{9\varepsilon_0 k_B T} \tag{2.29}$$

where g is the correlation function, which can be expressed as:

$$g = 1 + \sum_{j,j\neq 1}^{N_V} \overline{\cos\phi_{ij}} \tag{2.30}$$

where N_V is the number of dipoles within the spherical region and $\overline{\cos\phi_{ij}}$ is the average of $\cos\phi_{ij}$ of the angle ϕ made by the reference dipole i and the jth nearest neighbor. Fröhlich's equation can be employed in describing the static dielectric response of liquid and solid polymers, by modifying correlation function g. In polymers, repeat units can be considered as the basic dipole unit, which contribute equally to the average polarization of the spherical region. Moreover, in solid polymers, entanglements of the polymer chains are present. By considering x repeat units in a polymer chain and a polymer with high molecular weight, in order to neglect the effect of end groups, the correlation function takes the form [2, 8, 9]:

$$g' = 1 + \sum_{j,j\neq 1}^{x} \overline{\cos \phi'_{ij}} + \sum_{k}^{x} \overline{\cos \phi''_{ik}} \tag{2.31}$$

where $\overline{\cos \phi'_{ij}}$ is the average of $\cos\phi'_{ij}$ of the angle ϕ' made by the reference unit i and the jth unit in the same polymer chain and $\overline{\cos \phi''_{ik}}$ is the average of $\cos\varphi''_{ik}$ of the angle ϕ'' made by the reference unit i and the kth unit in the same polymer chains which do not contain the unit i.

2.6 Polarization Under Time Varying Electric Field

Up to this point, the conducted approach for the dielectrics' response to an applied field refers to static or non-time varying fields. However, the application of a time varying or frequency dependent electric field on dielectric materials affects their response in such ways which could be beneficial to both scientific research and applications. Broadband dielectric spectroscopy (BDS) is the main experimental technique employed for studying the performance of dielectric materials under ac conditions. The frequency of the applied field could vary in a wide range from mHz to GHz. Exerting an alternating field on a dielectric medium causes the orientation of permanent and induced dipoles parallel to the field. However, this alignment is not instantaneous and is not conducted simultaneously for the various types of dipoles. Each kind of dipole exhibits a characteristic delay in its orientation, which is related to its nature and the occurring interactions. Thus, the resulting polarization is frequency dependent.

At very high frequencies, also referred to as optical frequencies, the wavelength of the applied electric field is comparable to the dielectric sample's dimensions. At this extreme edge of frequencies, polarization is determined by recording the variation of the electromagnetic waves' transmission velocity or the material's refractive index. According to the electromagnetic theory, the phase velocity (v) in a dielectric medium is expressed via Eq. (2.32):

$$v = \frac{c}{n'} = \frac{c}{\sqrt{\mu'\varepsilon'}} \tag{2.32}$$

where c is phase velocity in vacuum, n', ε', and μ' are the real parts of refractive index, dielectric permittivity, and magnetic susceptibility, respectively, at the measured frequency. For dielectric materials with $\mu' = 1$, Eq. (2.32) gives $n' = \sqrt{\varepsilon'}$. By these means, the well-known Clausious–Mosotti equation takes the form:

$$\frac{(n')^2 - 1}{(n')^2 + 2} = \frac{1}{3\varepsilon_0}\sum_{j} n_j[\alpha_{aj} + \alpha_{ej}] \tag{2.33}$$

where α_{aj} and α_{ej} are the atomic/ionic and the electronic polarizabilities respectively. Equation (2.33) includes the influence of the rotations and vibrations of atoms, ions, or electrons, and does not account for the contribution of permanent dipoles, because the alternation of the field is too high. Permanent dipoles, due to their inertia, are not able to follow the rapid alternation of the field.

2.6.1 Frequency Dependent Dielectric Permittivity – Debye's Equations

The response of a dielectric material to an externally applied ac field depends on the amplitude and frequency of the field as well as on the material's temperature. In addition, the recorded response reflects the molecular structure of the material, while providing insight information for molecular phenomena, interfacial effects, polarization and conduction mechanisms, phase changes and other physical effects [4, 5]. The orientation of permanent dipoles (i.e. molecular dipoles, or polar groups) is considered as a slow process, in comparison to the distortional polarization, and thus is characterized by enhanced relaxation time. Relaxation time is a characteristic time interval implying the delay of a system in responding to an external stimulus. If the alternation of the field is sufficiently low, permanent dipoles are able to be aligned with field, overcoming their inertia, reaching an equilibrium state correlated to the maximum value of polarization. This state corresponds to the maximum value of dielectric permittivity (ε_s-known as static value of permittivity or dielectric constant). On the contrary, at a very fast alternation of the field, dipoles do not attain sufficient time for their orientation and the measured polarization includes the contribution only from atomic/ionic and electronic polarizability (distortional effects). This state is expressed by the minimum value of dielectric permittivity (ε_∞). Debye, with his original work in 1929 [6], expressed the dispersion of dielectric permittivity between these two extreme states.

Considering the application of an ac field having the form of Eq. (2.34) on a dielectric material:

$$E = E_0 e^{i\omega t} \tag{2.34}$$

where E_0 is the amplitude and ω the angular frequency of the applied field. The material's response alternates its direction, and an angular phase difference (δ) occurs, because of the dipoles' delay in following the field. The electric displacement can now be expressed as:

$$D = D_0 e^{i(\omega t - \delta)} \tag{2.35}$$

where D_0 is the amplitude of electric displacement. Recalling Eq. (2.6) and substituting Eqs. (2.34) and (2.35) it derives

$$D_0 e^{i(\omega t - \delta)} = \varepsilon^* \varepsilon_0 E_0 e^{i\omega t} \tag{2.36}$$

where ε^* is the complex dielectric permittivity. The latter is defined according to Eq. (2.37) as:

$$\varepsilon^* = \varepsilon' - i\varepsilon'' \tag{2.37}$$

where ε' is the real and ε'' the imaginary part of complex dielectric permittivity, ε'' is also referred as loss factor, or loss index, or simply dielectric loss of the material. Phase difference is related to the dissipation factor, or loss tangent δ of a dielectric material which is defined as:

$$\tan\delta = \frac{\varepsilon''}{\varepsilon'} \tag{2.38}$$

Its physical meaning expresses the ratio of the dissipated energy upon the stored one at every charging cycle. The development of Debye's theory was based on the assumptions– (i) the exponential approach of equilibrium state, and (ii) the application of the superposition principle.

The relation between the time dependent electric field (E) and electric displacement (D) in a dielectric material with known values of dielectric permittivity at very low (ε_s) and very high (ε_∞) frequencies is given by Eq. (2.39) [3, 5, 12]:

$$\tau\frac{dD(t)}{dt} + D(t) = \tau\varepsilon_0\varepsilon_\infty\frac{dE(t)}{dt} + \varepsilon_0\varepsilon_s E(t) \tag{2.39}$$

where τ is the relaxation time of the process. Employing Eqs. (2.34) and (2.35) into the above relation results in the following expression:

$$\frac{D(t)}{\varepsilon_0 E(t)} = \frac{\varepsilon_s + \varepsilon_\infty(i\omega t)}{1 + i\omega t} \tag{2.40}$$

Generalizing Eq. (2.6) to express time varying quantities and employing Eq. (2.40) the dispersion equation of Debye is derived:

$$\varepsilon^*(\omega) = \varepsilon_\infty + \frac{\varepsilon_s - \varepsilon_\infty}{1 + i\omega\tau} \tag{2.41}$$

Using the definition of complex permittivity, the Debye's dispersion equation, Eq. (2.41), can be resolved into the real and imaginary parts, as follow:

$$\varepsilon'(\omega) = \varepsilon_\infty + \frac{\varepsilon_s - \varepsilon_\infty}{1 + \omega^2\tau^2} \tag{2.42}$$

and

$$\varepsilon''(\omega) = \frac{(\varepsilon_s - \varepsilon_\infty)\omega\tau}{1 + \omega^2\tau^2} \tag{2.43}$$

Eliminating the term $\omega\tau$ between Eqs. (2.42) and (2.43) a semi-circle function is derived:

$$\left(\varepsilon' - \frac{\varepsilon_s + \varepsilon_\infty}{2}\right)^2 + \varepsilon'' = \left(\frac{\varepsilon_s - \varepsilon_\infty}{2}\right)^2 \tag{2.44}$$

Plotting the imaginary part of dielectric permittivity as a function of the real one, at a specific frequency for a Debye process, gives a perfect semi-circle with its center lying on x-axis. Graphs of this type, $\varepsilon'' = f(\varepsilon')$, are referred as Cole–Cole plots and are used for identifying if a process follows a pure Debye behavior.

2.6.2 Models of Dielectric Relaxations

Debye's relaxation theory, as expressed via Eqs. (2.41)–(2.43), describes only one process which is characterized by a single relaxation time, neglecting at the same time the interactions between dipoles or between dipoles and their surroundings. By these means, they can describe very well the dielectric response of polar gases and liquids [3]. However, significant deviations from the experimentally determined dielectric behavior of solid dielectrics and the Debye's theoretical curves are observed [1–3, 5, 8, 9]. Solid dielectrics exhibit more than one relaxation process because of the various types of permanent and induced dipoles which they possess. Moreover, in solid state or condensed matter, the occurring interactions between charges and dipoles, and the variation of the local field inside the material causes a distribution of relaxation times. The observed discrepancies from the ideal Debye behavior were addressed by the introduction of several semi-empirical dielectric relaxation models. In all cases, the Debye equations were modified in order to include a type of distribution of relaxation times. In the first attempt, Cole and Cole proposed a symmetric distribution of relaxation times [13], which is expressed by the introduction of an exponent in the denominator of the Debye equations. The modified expression is:

$$\varepsilon^*(\omega) = \varepsilon_\infty + \frac{\varepsilon_s - \varepsilon_\infty}{1 + (i\omega\tau)^{1-\alpha}} \tag{2.45}$$

where the exponent $(1 - \alpha)$ represents a measure of the symmetric distribution of relaxation times and at the limiting value $\alpha = 0$, Eq. (2.45) reduces to Debye equation. Resolving the real and imaginary parts leads to the following relations:

$$\varepsilon' = \varepsilon_\infty + \frac{(\varepsilon_s - \varepsilon_\infty)\left[1 + (\omega\tau)^{1-\alpha} \sin\frac{\alpha\pi}{2}\right]}{1 + (\omega\tau)^{2(1-\alpha)} + 2(\omega\tau)^{1-\alpha} \sin\frac{\alpha\pi}{2}} \tag{2.46}$$

and

$$\varepsilon'' = \frac{(\varepsilon_s - \varepsilon_\infty)(\omega\tau)^{1-\alpha} \cos\frac{\alpha\pi}{2}}{1 + (\omega\tau)^{2(1-\alpha)} + 2(\omega\tau)^{1-\alpha} \sin\frac{\alpha\pi}{2}} \tag{2.47}$$

Cole–Cole model can be better fitted to experimental data and elimination of the term $\omega\tau$ results in a semi-circle with its center below the x-axis. The Cole–Cole model is equivalent to a set of discrete Debye type relaxations with relaxation times symmetrically distributed about the central value (τ) of Eq. (2.45). The skewing factor $(1 - \alpha)$ reflects the broadness of the relaxation curves and is a measure of the interactions between dipoles and their environment.

Aiming to achieve a better description of experimental data, Davidson and Cole proposed their model, by introducing an exponent in the whole denominator of the Debye equations [14]. The Davidson–Cole model is expressed by the following equation:

$$\varepsilon^*(\omega) = \varepsilon_\infty + \frac{\varepsilon_s - \varepsilon_\infty}{(1 + i\omega\tau)^\gamma} \tag{2.48}$$

where the parameter γ is a measure of the non-symmetrical distribution of relaxation times and its values range between 0 and 1. The model reduces to a Debye process for $\gamma = 1$. In the Davidson–Cole model, the loss peak does not occur at $\omega_{max}\,\tau = 1$, as predicted by the Debye model, but at:

$$\omega_{max}\,\tau = \tan\left[\frac{1}{(\gamma + 1)}\frac{\pi}{2}\right] \tag{2.49}$$

Separating the real and imaginary parts of dielectric permittivity from Eq. (2.48) gives the expressions:

$$\varepsilon' = \varepsilon_\infty + (\varepsilon_s - \varepsilon_\infty)(\cos\varphi)^\gamma \cos\gamma\varphi \tag{2.50}$$

and

$$\varepsilon'' = (\varepsilon_s - \varepsilon_\infty)(\cos\varphi)^\gamma \sin\gamma\varphi \tag{2.51}$$

where

$$\tan\varphi = \omega\tau \tag{2.52}$$

It should be noted that the plot of the $\varepsilon'' = f(\varepsilon')$ function is now a skewed ellipsoid.

Havriliak and Negami generalized the aforementioned models by employing both exponents in the denominator of the Debye equations [15]. By these means, they introduced a superposition of symmetric and non-symmetric distribution of relaxation times, approaching even better the behavior of a solid dielectric. The Havriliak–Negami model is expressed as:

$$\varepsilon^*(\omega) = \varepsilon_\infty + \frac{\varepsilon_s - \varepsilon_\infty}{[1 + (i\omega\tau)^{1-\alpha}]^\gamma} \tag{2.53}$$

or

$$\varepsilon' = \varepsilon_\infty + \frac{(\varepsilon_s - \varepsilon_\infty)\cos\gamma\varphi}{\left[1 + (\omega\tau)^{2(1-\alpha)} + 2(\omega\tau)^{1-\alpha}\sin\frac{\alpha\pi}{2}\right]^{\gamma/2}} \tag{2.54}$$

and

$$\varepsilon'' = \frac{(\varepsilon_s - \varepsilon_\infty)\sin\gamma\varphi}{\left[1 + (\omega\tau)^{2(1-\alpha)} + 2(\omega\tau)^{1-\alpha}\sin\frac{\alpha\pi}{2}\right]^{\gamma/2}} \tag{2.55}$$

In the Havriliak–Negami model, the maximum dielectric loss occurs at:

$$\omega_{max} = \frac{1}{\tau}\left[\sin\frac{(1-\alpha)\pi}{2(1+\gamma)}\right]^{1/\gamma}\left[\sin\frac{(1-\alpha)\gamma\pi}{2[1+(1-\alpha)]}\right]^{-1/(1-\alpha)} \tag{2.56}$$

The majority of the contemporary softwares for fitting dielectric data is based on the Havriliak–Negami model.

2.6.3 Effect of Temperature

When a dielectric material is subjected to an ac electric field, thermal agitation affects the alignment of dipoles (permanent and induced) with the direction of the field, as well as the motion of unbounded charges, which might exist in the material. The occurring dielectric losses are related to the delay of the dipoles in following the field and to the attempt of migrating, between equilibrium sites, of the charges. Energetically, dipoles and charges can be considered in continuous thermal fluctuations within potential wells. The rising of temperature enhances the fluctuations and, in many cases, influences the misalignment with the field's direction. Potential wells define energy barriers for dipoles and charges, which can be surmounted if the temperature is sufficiently high and thermal agitation provides the necessary activation energy. In this case, dipoles and charges can migrate to another well or equilibrium site. These new positions are locally adjacent or are at a close energy level. The rate of the jumps, over the potential barriers, which are executed by the activated dipoles and charges (or in other words the average number of jumps per unit time), at temperature T, can be written as [2]:

$$K_0 = \frac{k_B T}{h} \exp\left(-\frac{\Delta G}{RT}\right) \tag{2.57}$$

where k_B is the Boltzmann's constant, h the Planck's constant, R the gas constant, and ΔG the free energy of relaxation process. Free energy is related to temperature via Eq. (2.58):

$$\Delta G = \Delta H - T \cdot \Delta S \tag{2.58}$$

where ΔH is the activation energy of the relaxation and ΔS the entropy of activation. Relaxation time (τ) can be defined as the reverse quantity of the jumps rate (K_0):

$$\tau = \frac{h}{k_B T} e^{\frac{\Delta H}{k_B T}} e^{-\frac{\Delta S}{R}} \tag{2.59}$$

or in logarithmic form:

$$\ln \tau = \frac{\Delta H}{k_B T} - \ln T + \left(\ln \frac{h}{k_B T} - \frac{\Delta S}{R}\right) \tag{2.60}$$

Equation (2.60) implies that the main influence of temperature is exerted by the term $(1/T)$, since the quantity in the parenthesis is considered as temperature independent and the effect of the factor $\ln\tau$ is limited. Following these conditions, it is expected that the graph of the function $\ln\tau = f(1/T)$ should provide a linear relationship, at least at a good approximation. By these means, the dependence of relaxation time upon temperature can be expressed as:

$$\tau = \tau_0 e^{\frac{E_A}{k_B T}} \tag{2.61}$$

where τ_0 is a pre-exponential factor indicating the relaxation time at very high temperature, which is also considered as practically independent from temperature, and E_A the activation energy of the process. The relation of the frequency where the loss peaks occur, as a function of temperature, is given by Eq. (2.62) [4, 5]:

$$f_{\max} = f_0 e^{-\left(\frac{E_A}{k_B T}\right)} \tag{2.62}$$

Equation (2.62) is derived from Eq. (2.61) if the condition $\omega_{\max}\tau = 1$ is considered as valid, which is a feature of processes characterized by a single relaxation time and an activation energy independent from temperature. This Arrhenius type behavior is not a common characteristic of all dielectric relaxation processes. Several processes exhibit a variable activation energy and a non-Arrhenius performance [16–19]. The observed discrepancy can be resolved by employing the Vogel–Fulcher–Tammann equation [20] which takes the form:

$$\tau = \tau_0 e^{\frac{AT_0}{T - T_0}} \tag{2.63}$$

or

$$f_{\max} = f_0 e^{-\left(\frac{AT_0}{T-T_0}\right)} \tag{2.64}$$

where A is a constant being a measure of activation energy and T_0 is the Vogel temperature or ideal glass transition temperature.

2.7 Conduction Phenomena in Dielectrics

Dielectric materials are non-conductive. They are characterized by a large energy bandgap and typically are ceramics, glasses, polymers, and polymer or ceramic matrix composites. Dielectrics could be crystalline or disordered. As a consequence, conductivity values remain low for dielectrics. Conduction processes are related to the presence of charge carriers and their ability to migrate within a material. The concentration of mobile charge carriers in dielectrics is low, and according to their origin could be classified in– (i) intrinsic charges (intrinsic conductivity), when carriers are provided by the material's chemical structure, (ii) extrinsic (extrinsic conductivity), in this case charges are introduced by impurities or doping elements, and (iii) injection charges (injection conductivity) when the carriers are injected by the electrodes contacts, at the material-electrode interface. Typical charge carriers are electrons, holes, positive, and negative ions. In some cases, polarons and bipolarons could also interfere in the conduction processes. In a basic approach, the total conductivity can be expressed via Eq. (2.65):

$$\sigma_T = q(\mu_n n + \mu_p p) + q(\mu_- n_- + \mu_+ n_+) = \sigma_{\text{el}} + \sigma_{\text{ion}} \tag{2.65}$$

where n and p are the concentrations of electrons and holes, μ_n and μ_p the electrons and holes mobilities, n_-, n_+, μ_-, and μ_+ the concentrations and mobilities of the negative and positive ions, respectively; σ_{el} and σ_{ion} correspond to electronic and ionic conductivity, respectively. Charge carriers transport is governed by the potential barriers of the wells or sites where charges are located and by the intense of the applied field and temperature. Impurities, crystal structure defects, like vacancies and Frenkel or Schottky defects, interface between crystalline and amorphous regions, interface in composites and heterogeneous materials, and conductive reinforcing phases facilitate charge migration. The occurring processes involve a "trapping-detrapping" procedure, where carriers' motion is restricted to specific sites by potential barriers which can be overcome by sufficient thermal agitation, or via lowering the barriers by the applied field. Under the influence of low or moderate fields conductivity enhances with temperature according to Eq. (2.66):

$$\sigma(T) = \sigma_0 \exp\left(-\frac{E_{\text{A}}}{k_{\text{B}} T}\right) \tag{2.66}$$

where σ_0 is a pro-exponential factor and E_{A} the activation energy. Activation energy of the process might not be constant and in this case Eq. (2.66) takes the form:

$$\sigma(T) = \sigma_0 \exp\left(-\frac{BT_0}{T - T_0}\right) \tag{2.67}$$

where B is a parameter considered as the measure of activation energy and T_0 is the, so called, Vogel temperature or ideal glass transition temperature.

The occurring conductivity (σ_{ac}) under the influence of an ac field expresses the set of all energy dissipative processes in the dielectric. In these are included, the actual ohmic conductivity caused my migrating charge carriers on isolated or adjacent conductive sites or clusters and the frequency dielectric dispersions [1]. The frequency dependent conductivity can be determined via the dielectric loss as follows:

$$\sigma^*(\omega) = i\varepsilon_0\omega\varepsilon^*(\omega) = i\varepsilon_0\omega(\varepsilon' - i\varepsilon'') = \varepsilon_0\omega\varepsilon'' + i\varepsilon_0\omega\varepsilon' = \sigma' + i\sigma'' \tag{2.68}$$

σ_{ac} is calculated from the real part of $\sigma^*(\omega)$:

$$\sigma_{ac}(\omega) = \varepsilon_0\omega\varepsilon'' \tag{2.69}$$

where, $\varepsilon_0 = 8.85 \times 10^{-12}$ F · m^{-1} is the permittivity of the free space and $\omega = 2\pi f$ the angular frequency.

In the low frequency range, the alternation of the field is slow and charge carriers are forced to migrate over relatively large distances from site to site. However, their transport is restricted by the insulating material and only the sufficiently agitated carriers are able to take place in the conduction process. Thus, a limited number of charges participates in the conduction process and conductivity attains low values. In this region, the effect of temperature is significant. In contrast, in the high frequency region, the alternation of the field is fast, and an increased number of carriers participates in the conduction process, although the migration distances are very short. Carriers transport forward and backwards between adjacent sites and conductivity increases because of the enhanced number of participating charges. In this region, the effect of temperature is negligible. In many types of dielectric materials, and especially in those which can be considered as disordered solids, the variation of ac conductivity with frequency, at constant temperature, is in accordance with the well-known "ac universality law" [21, 22]:

$$\sigma_{ac}(\omega) = \sigma_{dc} + A(\omega)^s \tag{2.70}$$

where σ_{dc} is the limiting value of ac conductivity when frequency tends to be zero, and A, s parameters related to the applied temperature and the specific material [23, 24]. Figure 2.5 provides an example for the frequency dispersion of σ_{ac}.

In many disordered dielectrics the predominant conduction mechanism is hopping. Hopping refers to the sudden displacement of a charge carrier from one site to another neighboring. In general, hopping mechanism includes both jumps over a potential barrier and quantum mechanical tunnelling [21, 25]. Various models have been developed aiming to describe the hopping conduction mechanism in disordered dielectrics. The Variable Range Hopping (VRH) model originally proposed by Mott [26, 27] for amorphous semiconductors, sufficiently describes the variation of conductivity with temperature, under dc conditions or at constant frequency [28–31] via the equation:

$$\sigma(T) = \sigma^0 \exp\left[-\left(\frac{T^0}{T}\right)^{\gamma}\right] \tag{2.71}$$

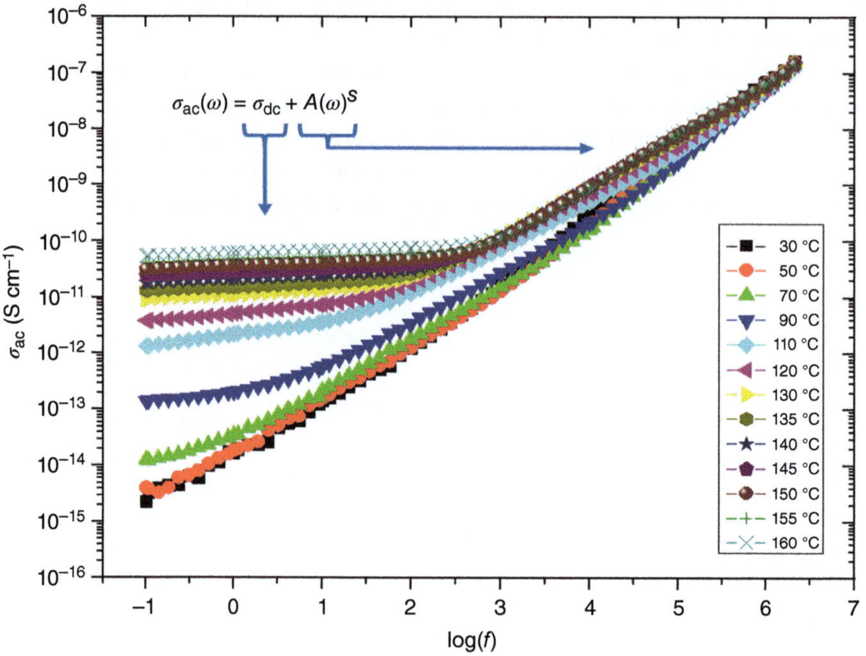

Figure 2.5 Typical dispersion of ac conductivity with frequency, at various temperatures. Data are from an epoxy resin/barium titanate (5 wt%) nanocomposite.

where the parameter σ^0 is a pro-exponential factor, T^0 is the characteristic temperature which determines the thermally activated hopping between localized states at different energies and is considered as a measure of disorder [32, 33], and the exponent (γ) indicates the dimensionality (d) of the transport process via the equation $\gamma = 1/(1 + d)$, where $d = 1, 2, 3$.

The random free-energy barrier model (also referred to as the symmetric hopping model) proposed by Dyre [23, 34–36], provides mathematical relationships for the frequency dependent conductivity, at constant temperature [4, 28]. According to this model, σ_{ac} is given by:

$$\sigma'(\omega) = \sigma_{ac} = \frac{\sigma_{dc}\omega\tau \arctan(\omega\tau)}{\frac{1}{4}\ln^2(1 + \omega^2\tau^2) + (\arctan \omega\tau)^2} \tag{2.72}$$

where, σ_{dc}, ω and τ are the direct current conductivity, the angular frequency, and the relaxation time respectively.

2.8 Active Dielectrics

As is very well-known, dielectric materials are polarized when an external electric field is applied. Active dielectrics, which sometime are also called functional dielectrics, are a peculiar category of dielectrics, since they are able to be polarized in the absence of an external electric field. They are subdivided into three groups namely, piezoelectrics, ferroelectrics, and pyroelectrics. Crystalline solids can be categorized in 32 classes of crystal symmetry point groups. Of the 32 classes, 11 are centrosymmetric and 21 are non-centrosymmetric [9]. The centrosymmetric systems are non-polar, because of the symmetric distribution of charges, and thus do not exhibit any dipole moment. One class of the remaining 21 non-centrosymmetric systems is non-polar due to other symmetric elements, and thus the 20 crystal systems which are characterized by lack of symmetry in their unit cell and structure constitute the group of active dielectrics.

In piezoelectric materials there is a coupling between mechanical and electrical energy or parameters. Piezoelectrics, in the absence of an external electric field, and under the influence of a mechanical stress or strain, acquire induced dipole moments in their unit cells because of the displacement of charges, and thus an overall polarization occurs. The final result is the development of a potential difference across the material. Piezoelectrics are characterized by the reverse effect, when an external electric field is applied, mechanical stress and strain is generated resulting in the deformation of the material. The direct piezoelectric effect can be described via Eq. (2.73):

$$T_{i=1,2,\ldots,6} = \sum_{j=1,2,\ldots,6} c_{ij} S_j \tag{2.73}$$

where (T) is the mechanical stress, (S) the mechanical strain, and subscripts i, j accounts for normal and shear stresses, and c_{ij} are the 36 stiffness coefficients which form a matrix. Of the 36 coefficients only 21 are independent because $c_{ij} = cj_i$. The applied stress generates polarization according to Eq. (2.74):

$$P_{i=1,2,3} = \sum_{j=1,2,\ldots,6} d_{ij} T_j \tag{2.74}$$

where P_i indicates the components of polarization vector to the x, y, z axes, and d_{ij} is the matrix of electromechanical coupling coefficients, also referred as piezoelectric coefficients. This matrix is also symmetrical ($d_{ij} = d_{ji}$) and for this reason, the 18 coefficients are reduced to 15.

The reverse effect is described by the Eqs. (2.75) and (2.76):

$$P_{i=1,2,3} = \sum_{j=1,2,3} (\varepsilon_{ij} - 1)\varepsilon_0 E_j \tag{2.75}$$

where the subscripts i, j denote the anisotropic nature of the material, E_j is the applied electric field and ε_{ij} are the components of dielectric permittivity. The generated strain in the reverse effect is:

$$S_{j=1,2,\ldots,6} = \sum_{i=1,2,3} d_{ij} E_i \tag{2.76}$$

The piezoelectric coefficients are parameters which associate mechanical strain with electric field at constant applied mechanical stress, and polarization with mechanical stress at constant applied electric field.

$$d_{ij} = \left(\frac{\partial P_i}{\partial T_j}\right)_E = \left(\frac{\partial S_j}{\partial E_i}\right)_T \tag{2.77}$$

Ferroelectrics is a class of dielectric materials which can be polarized when no electric field or mechanical load is applied. Although the name of these materials refers to iron (*ferrum* in Latin) they include no iron atoms. They obtained their name from the analogy in the form of the curve of polarization versus electric field with the curve of magnetization versus magnetic field in the ferromagnetic materials. Ferroelectrics, because of the low symmetry in their crystal structure, exhibit a polar axis. Thus, they possess domains which exhibit spontaneous polarization. The effect of ferroelectricity is the spontaneous orientation of dipoles because of mutual interactions [1]. Ferroelectric crystals exhibit two or more polarization states, the direction of these states can vary from domain to domain. However, in the absence of an external field, the vector sum of all spontaneous polarization components (summing the contributions of all domains) is zero. Under the application of an external field, material's polarization follows a linear dependence on the field only for weak fields. As the intensity of the field increases, polarization rises, also following a non-linear form, since the spontaneous components of the domain are aligned parallel to the field. Ferroelectric material reaches its maximum polarization, called saturation polarization, when all spontaneous polarization vectors are oriented according to the electric field. Upon the removal of the field, polarization does not become zero and a remanent polarization is recorded. Varying the applied field between its value at the saturation point and to its opposite value forms a $P = f(E)$ curve exhibiting a hysteresis loop. The required field for polarization to become zero is called coercive field and is temperature dependent [5]. The resulting hysteresis loop indicates the required energy for reversing the direction of spontaneous vectors of the domains, which is dissipated as heat.

Ferroelectric materials undergo a structural/phase transition at a characteristic temperature called Curie temperature, (T_C). Below this temperature ferroelectrics are in a low symmetry, polar phase, known as ferroelectric phase. Above the critical temperature the material transposes to a high symmetry, non-polar phase named paraelectric phase. Ferroelectric to paraelectric transition is a first order transition, and for this reason is characterized by a discontinuous change of the polarization at the transition temperature [37, 38]. In the paraelectric phase, spontaneous polarization does not exist. Applying a strong field, when the material is in the paraelectric phase, it induces polarization to the whole crystal. The subsequent cooling of the ferroelectric material at a temperature below T_C, under the influence of the field, results in maintaining the polarization. Switching of the field at this temperature leaves the polarization unaffected. This process is known as the poling procedure. If the material is re-heated at a temperature above the critical one, without the application of electric field, the material turns to its paraelectric state [5].

Dielectric permittivity as a function of temperature for a ferroelectric material shows a sharp peak at the critical temperature. At a temperature above the critical one, permittivity varies according to the Curie–Weiss law:

$$\varepsilon'(T) = \frac{\mathrm{C}}{T - T^{\mathrm{C-W}}} \tag{2.78}$$

where C is the Curie constant and $T^{\mathrm{C-W}}$ the Curie–Weiss temperature which is equal to the critical temperature only in the case of a continuous transition.

The third group of active dielectrics is pyroelectrics. In pyroelectrics, polarization is caused by the uniform heating or cooling of the whole material. Crystal structures with polar axes are both pyroelectric and piezoelectric. When an electric field E is applied on a pyroelectric material, the resulting electric displacement can be expressed as:

$$D = \varepsilon_0 E + P_{\text{total}} = \varepsilon_0 E + (P_{\text{s}} + P_{\text{ind}}) = \varepsilon_0 \varepsilon E + P_{\text{s}} \tag{2.79}$$

where P_{total} is the overall polarization, which can be separated in spontaneous (P_{s}) and induced (P_{ind}) parts, and ε is the dielectric constant [39].

Differentiating Eq. (2.79), under constant electric field, with respect to temperature gives the generalized pyroelectric coefficient (P_{gen}):

$$P_{\text{gen}} = \frac{\partial D}{\partial T} = \frac{\partial P_{\text{s}}}{\partial T} + \varepsilon_0 E \frac{\partial \varepsilon}{\partial T} \tag{2.80}$$

The latter can be written as:

$$P_{\text{gen}} = p + \varepsilon_0 E \frac{\partial \varepsilon}{\partial T} \tag{2.81}$$

where $p = \frac{\partial P_{\text{s}}}{\partial T}$ is the true pyroelecrtic coefficient [39].

Pyroelectrics are used in thermal imaging and sensing devices, such as I-R or human body thermal radiation.

2.9 Polymers as Dielectric Materials

2.9.1 Dielectric Relaxations in Polymers

During the last century, polymers were revealed as an important class of engineering material. Although the initial scientific interest was focused on their synthesis research and thermomechanical characterization, the application of polymers as insulating materials, in cables, rotating loading members, gear and timing belts and others, raised the necessity for studying their electrical properties. Typically, polymers are insulating materials because of the low concentration of free charges. Thus, under the influence of an electric field, polymers can be polarized and dielectric phenomenon occur. Macromolecular chains might include polar segments, polar branches, or side groups, which upon exerting an external field tend to be aligned parallel to the field's direction. In some cases, polarization of the whole macromolecular chain might be present. The application of a field could cause polymer chain bending, twisting of segments or groups around carbon-carbon bonds, and rotation of branches and side groups. During these re-arrangements in space, the polar entities are interacting with their molecular environment and typically dielectric relaxation effects are observed. All types of polarization (i.e. electronic, atomic/ionic, orientational, and interfacial) could be present in the dielectric spectra of polymers, depending on the frequency of the applied field and the molecular structure of the examined polymer. The ease of the orientation of polymer's polar parts reflects the interactions and restrictions exerted by the local molecular environment. For example, polar parts connected to flexible macromolecules follow the direction of the field more easily than dipoles connect to stiff chains. Moreover, in semicrystalline polymers, polar parts and groups are restricted in their orientation by the crystalline phase. The occurring relaxation processes are related to the type and the characteristics of the specific dipoles (permanent or induced, size, charge, mass, molecular environment) and the recorded dielectric loss peaks are labeled with low case Greek alphabet letters. In plots of loss peaks (either of permittivity loss index, or tanδ, or imaginary part of electric modulus) as a function of temperature, at constant frequency, the process recorded at the higher temperature is labeled as α, the next one as β, and so on.

As α-relaxation process is denoted, in most cases, the process related to the transition from the stiff glassy state, of the amorphous polymers or the amorphous regions of semicrystalline polymers, to the soft rubber-like state upon heating. This transition takes place at a characteristic temperature (although it is a characteristic temperature range) known as glass to rubber transition temperature (T_g). At temperatures lower than T_g polymer chains are frozen and stiff, exhibiting enhanced mechanical properties and brittleness. As temperature approaches T_g large segments of the macromolecular chains are thermally activated, their mobility is enhanced, and they relax in synergetic manner. In the same temperature range, polymer becomes softer, its mechanical performance degrades, and the mobility of polar parts is facilitated resulting in increased polarization and dielectric permittivity. Consequently, in dielectric loss spectra, a relaxation peak is recorded, which is typically labeled as α-relaxation process. It should be noted that in the literature there are cases where the term α-relaxation does not correspond to the glass to rubber transition [40–43].

Relaxations labeled as β- and γ-process are considered as secondary relaxations, since they are local, weaker, and characterized by lower relaxation times and activation energy. The re-arrangement of polar side groups, such as, C–O, –OH, and –CN of the main polymer chain corresponds to β-relaxation. In addition, γ-process is related to motions of $(CH_2)_n$ units, which might be present in the soft part of the macromolecular chain and involve the C–O polar groups [44, 45]. The motion of the $(CH_2)_n$ units is also regarded as "crankshaft" motion.

Although not so often, two additional relaxation processes have been reported for polymeric materials. Normal-mode relaxation is the first one, corresponding to a translational motion of the whole chain. This process exhibits extended length scale, longer relaxation time, and results by adding the molecular dipole vectors which are parallel to the polymer's repeat unit. The outcome is the end-to-end vector of the resultant dipole moment of the whole polymer chain, which is unaffected by the segmental motion of the macromolecules and strongly dependent on polymer's molecular weight [3, 46]. The second process, rarely observed, is δ-relaxation. This is the faster mechanism (characterized by the lowest relaxation time), is recorded at very low temperatures, and its physical origin has been ascribed to defect dipoles in the crystalline phase [43].

2.9.2 Heterogenous Systems – Polymer Matrix Composites

Dielectric materials and especially polymers could be regarded as heterogeneous systems because of the presence of additives, plasticizers, and reinforcing phase or phases [44, 47, 48]. In these cases, the dielectric material can be treated as a heterogeneous mixture and its electrical properties is a combination or mixing of the respective properties of the constituents' phases, their concentration, their distribution, and their geometrical characteristics. In polymer matrix composite materials and in semicrystalline polymers, an interface occurs between the existing phases. The extent of the interfacial area increases dramatically with the diminishing of the dimensions of the employed inclusions. For this reason, nanocomposites are characterized by huge interfacial areas. The application of an external electric field forces unbounded charges to be accumulated at the interface of the system, where they form induced dipoles. These charges are present in the composite systems from the fabrication procedure or have been injected during service life or tests. The size and the distribution of the formed dipoles is related to the shape and dimensions of the inclusions. Induced dipoles attempt to follow the orientation of the applied field, providing an additional contribution to the system's polarization, known as Interfacial Polarization (IP) or Maxwell–Wagner–Sillars (MWS) effect. The effect of IP is strengthened when the electrical properties (conductivity and permittivity) of the composite's phases vary considerably and in the case of nanocomposites because of the enormous extent of interfacial area [49–53]. Since the dimensions of the inclusions are larger than the atomic ones, dipoles at the interface exhibit inertia in being aligned with the field and the phenomenon becomes predominant at low frequencies (sufficient time is given to be oriented) and at high temperatures (thermal agitation provides energy for the orientation) [4, 49].

IP or MWS effect is a double process because it includes charge migration at the interface and dipoles orientation, which of the two components will be the dominating one depends on the specific material and conditions of testing or servicing. IP can be mathematically described via the following equations [1, 12, 44]:

$$\varepsilon'(\omega) = \varepsilon_\infty + \frac{(\varepsilon_s - \varepsilon_\infty)\omega\tau}{1 + \omega^2\tau^2} \tag{2.82}$$

$$\varepsilon''(\omega) = \frac{\sigma}{\varepsilon_0\omega} + \frac{(\varepsilon_s - \varepsilon_\infty)\omega\tau}{1 + \omega^2\tau^2} \tag{2.83}$$

where ε_s, ε_∞, are the values of the real part of dielectric permittivity at low and high frequencies, σ the conductivity of the composite, τ the relaxation time, ω the angular frequency of the field, and ε_0 the dielectric permittivity of free space. The occurring parameters ε_s, ε_∞ and σ in Eqs. (2.82) and (2.83) are dependent upon the permittivity (ε_i), conductivity (σ_i), and volume fraction (v_i) of the constituents, respectively:

$$\varepsilon_s = \frac{\sum_i v_i \varepsilon'_i / \sigma_i^2}{\left(\sum_i v_i / \sigma_i\right)^2} \tag{2.84}$$

$$\varepsilon_\infty = \frac{1}{\sum_i v_i / \varepsilon'_i} \tag{2.85}$$

$$\sigma = \frac{1}{\sum_i v_i/\sigma_i} \tag{2.86}$$

Interfacial area has an important influence upon the electrical response of composites and especially nanocomposites. Initial considerations are treating interface as an area of negligible thickness, which acts as a boundary condition between the properties of the reinforcing phase and the matrix. In this approach interface has no properties of its own. Nowadays, it is accepted that interface has a narrow thickness of few nm and properties different from both the matrix and the reinforcing phase, and thus is considered as an additional phase to the composite (interphase) [54]. Targeting the determination of dielectric permittivity of the interface/interphase, a model has been developed based on energy density [54]. The energy density stored in a composite system should be the sum of the energy density stored at each constituent of the system. The theoretical value of energy density for a nanocomposite can be determined by employing the simple rule of mixtures:

$$U_{\mathrm{comp}}^{\mathrm{theor}} = V_{\mathrm{f}} U_{\mathrm{f}} + (1 - V_{\mathrm{f}}) U_{\mathrm{m}} \tag{2.87}$$

where U_{f}, U_{m}, are the energy density stored in the reinforcing phase and the matrix, and V_{f} is the volume fraction of the reinforcing phase. U_{f} and U_{m} can easily be found if the dielectric permittivity of the matrix and filler is known via Eqs. (2.88) and (2.89):

$$U_{\mathrm{f}} = \frac{1}{2}\varepsilon_0 \varepsilon_{\mathrm{f}}' E^2 \tag{2.88}$$

$$U_{\mathrm{m}} = \frac{1}{2}\varepsilon_0 \varepsilon_{\mathrm{m}}' E^2 \tag{2.89}$$

where $\varepsilon_{\mathrm{f}}'$, $\varepsilon_{\mathrm{m}}'$ are the dielectric permittivities of the filler and the matrix, and E is the applied electric field. The employed permittivity values should be determined at the same frequency and temperature, while the field's intensity should remain constant. The energy density of the composite can be evaluated by substituting the experimentally found values of permittivity for the nanocomposite $(\varepsilon_{\mathrm{com}}')$ in Eq. (2.90):

$$U_{\mathrm{com}}^{\mathrm{exp}} = \frac{1}{2}\varepsilon_0 \varepsilon_{\mathrm{com}}' E^2 \tag{2.90}$$

The energy density stored at the interface/interphase is the difference between $U_{\mathrm{com}}^{\mathrm{exp}}$ and $U_{\mathrm{comp}}^{\mathrm{theor}}$:

$$U_{\mathrm{inter}} = U_{\mathrm{comp}}^{\mathrm{exper}} - U_{\mathrm{comp}}^{\mathrm{theor}} \tag{2.91}$$

Having calculated the energy density stored at the interface/interphase its permittivity can be found by employing Eq. (2.92):

$$U_{\mathrm{inter}} = \frac{1}{2}\varepsilon_0 \varepsilon_{\mathrm{inter}}' E^2 \Rightarrow \varepsilon_{\mathrm{inter}}' = \frac{2U_{\mathrm{inter}}}{\varepsilon_0 E^2} \tag{2.92}$$

Equation (2.92) provides a good approximation for the dielectric permittivity of the interface/interphase, which will be more accurate if the volume fraction of the interface/interphase is taken into account.

2.10 Thermal Properties of Dielectrics

The term thermal properties or thermal effects of materials refers to the behavior exhibited by a material under varying temperature conditions or under constant temperature as a function of time. When a material absorbs or emits energy in the form of heat, several physical phenomena could take place. These include phase changes (i.e. melting or solidifying), variation of temperature, change in its dimensions, heat conduction, and development of internal mechanical stresses (which could lead to creep or fatigue failure). The main physical quantities used for describing the thermal response or thermal properties of materials are heat capacity and thermal conduction.

2.10.1 Heat Capacity

Heat capacity is considered one of the most important characteristics of materials. Heat capacity implies the amount of heat required in order to raise the temperature of a given amount of material by 1 K. The molar heat capacity at constant volume (C_V) and at constant pressure (C_P) are defined according to Eqs. (2.93) and (2.94):

$$C_V = \left(\frac{\partial Q}{\partial T}\right)_V = \left(\frac{\partial U}{\partial T}\right)_V \tag{2.93}$$

$$C_P = \left(\frac{\partial Q}{\partial T}\right)_P \tag{2.94}$$

where U is the internal energy of one mole of the material, T the temperature, Q the amount of heat and V, P volume and pressure, respectively. Heat capacity per unit mass is called specific heat capacity (c_V, c_P).

In a classical approach, each atom of a solid material is regarded as an oscillator, which vibrates around its equilibrium position. Atoms experience forces, the measure of which is proportional to their displacement with respect to the equilibrium position. The total energy of an atom is the sum of its kinetic and potential energy. The vibrational energy (u) in all three directions per atom is expressed via Eq. (2.95):

$$u = 3\left(\frac{k_B T}{2} + \frac{k_B T}{2}\right) = 3k_B T \tag{2.95}$$

where k_B is the Boltzmann constant. Recalling that $k_B = R/N_A$, where R is the universal gas constant and N_A the Avogadro's number, a single mole of a solid material can be considered as a system with simple one-dimensional harmonic oscillators. By these means, the internal energy of the system is given by Eq. (2.96):

$$U = 3N_A k_B T = 3RT \tag{2.96}$$

Combining Eqs. (2.93) and (2.96) the so-called law of Dulong-Petit is derived, Eq. (2.97):

$$C_V = 3R \tag{2.97}$$

The Dulong–Petit law appears to be in agreement with experimental results from many solid materials in a wide range of temperature, and especially at high temperatures. However, it deviates significantly from the experimental values of C_V at low temperatures. In the low temperature range, the extension of which depends on the specific material, values of C_V decrease being also temperature dependent [55–57]. When temperature tends to 0 K the heat capacity at constant volume for non-metallic materials, such as dielectrics, becomes proportional to T^3 [55, 56].

Einstein attempted to address the occurred discrepancy between experimental data and the Dulong–Petit law at low temperatures. In his model, Einstein suggested that the crystal lattice of a solid containing N vibrating atoms can be considered as a system of $3N$ independent linear harmonic oscillators. Each atom is regarded as corresponding to three oscillators, vibrating to the three perpendicular directions in space (x, y, and z) with the same frequency, independently from all its atomic neighbors. Oscillators in this model differ from the classical ones since they can acquire only discrete energy values. This energy can be expressed as shown in Eq. (2.98):

$$E = nh\nu = n\hbar\omega \tag{2.98}$$

where n is an integer, h the Planck's constant, ν the vibration frequency, $\omega = 2\pi\nu$ the angular frequency, and $\hbar = h/2\pi$. Thus, the quantized average energy of a harmonic oscillator is given by Eq. (2.99):

$$\overline{E} = \frac{\sum_{n=0}^{\infty} n\hbar\omega e^{-\frac{n\hbar\omega}{k_B T}}}{\sum_{n=0}^{\infty} e^{-\frac{n\hbar\omega}{k_B T}}} \tag{2.99}$$

For low values of the exponential term $e^{-\frac{n\hbar\omega}{k_B T}}$ and by expanding in series both numerator and denominator of Eq. (2.99), the latter takes the form:

$$\overline{E} = \frac{\hbar\omega}{e^{\frac{\hbar\omega}{k_B T}} - 1} \tag{2.100}$$

In the case of high temperatures, that is $k_B T \gg \hbar\omega$, Eq. (2.100) leads to the Dulong–Petit law.

The heat capacity under constant volume for a system with N_A (where N_A the Avogadro's number) identical quantum harmonic oscillators becomes:

$$C_V = \left(\frac{\partial U}{\partial T}\right)_V = \left(\frac{\partial (N_A\overline{E})}{\partial T}\right)_V = 3N_A k_B \left(\frac{\hbar\omega}{k_B T}\right)^2 \frac{e^{\frac{\hbar\omega}{k_B T}}}{\left(e^{\frac{\hbar\omega}{k_B T}} - 1\right)^2} \tag{2.101}$$

The form of Eq. (2.101) can be simplified by introducing the so-called Einstein temperature $\theta_E = \hbar\omega/k_B$. For many solids θ_E ranges between 100 and 300 K. Thus Eq. (2.101) is rewritten as:

$$C_V = 3N_A k_B \left(\frac{\theta_E}{T}\right)^2 \frac{e^{\frac{\theta_E}{T}}}{\left(e^{\frac{\theta_E}{T}} - 1\right)^2} \tag{2.102}$$

Equation (2.102) is in good agreement with the classical result and experimental data at high temperatures and in fairly good agreement, definitely better than the classic model, with materials' data in general. At the other extreme edge of temperature, at very low temperatures, Eq. (2.102) fails to describe the temperature dependence of C_V deviating from the experimental results.

In Einstein's model, all atoms are regarded as independent harmonic oscillators vibrating with the same frequency and not interacting with their environment. In addition, he assumed that there is no upper limit for the oscillators (or phonon) energies. Debye realized that these considerations are at the base of the discrepancies with experimental data and proposed modifications into Einstein's model. Debye suggested that a solid can be treated as an elastic continuum, since in solid state atoms are firmly bounded with each other and probably vibrate as an entity. In his second suggestion, Debye introduced a maximum frequency in the range of the vibrating frequencies. By these means, Debye proposed the following expression for internal energy:

$$U = \frac{3k_B{}^4T^4}{2\pi^2\hbar^3\bar{c}^3}\int_0^{x_m}\frac{x^3\mathrm{d}x}{\mathrm{e}^x - 1} \tag{2.103}$$

where $x_m = \hbar\omega_m/k_BT$, ω_m is the maximum frequency of the lattice vibrations, and $\bar{c}$ is the average velocity of elastic waves in the solid. By introducing Debye temperature $\theta_D = \hbar\omega_m/k_B$, in an analogous interpretation as in the Einstein model, and differentiating Eq. (2.103) heal capacity at constant volume is obtained:

$$C_V = 9N_Ak_B\left(\frac{T}{\theta_D}\right)^3\int_0^{x_m}\frac{e^x x^4\mathrm{d}x}{(\mathrm{e}^x - 1)^2} \tag{2.104}$$

The above equation for heat capacity can be written in the following form by introducing the Debye function $D\left(\frac{\theta_D}{T}\right)$:

$$C_V = 3N_Ak_BD\left(\frac{\theta_D}{T}\right) \tag{2.105}$$

where the Debye function is equal to:

$$D\left(\frac{\theta_D}{T}\right) = 3\left(\frac{T}{\theta_D}\right)^3\int_0^{\frac{\theta_D}{T}}\frac{\left(\frac{\theta_D}{T}\right)^4 e^{\frac{\theta_D}{T}}}{\left(e^{\frac{\theta_D}{T}} - 1\right)^2}d\left(\frac{\theta_D}{T}\right) \tag{2.106}$$

Debye function cannot be calculated accurately and therefore various approximations are made. Approximations are significantly simplified in the extreme case of very high ($T \gg \theta_D$) and very low ($T \ll \theta_D$) temperatures. At high temperatures ($x_m \ll 1$) from the expression (2.104) the Dulong–Petit law can be approximated, while at low temperatures ($x_m \gg 1$) the following expression for heat capacity can be obtained:

$$C_V = \frac{12}{5}\pi^4N_Ak_B\left(\frac{T}{\theta_D}\right)^3 \tag{2.107}$$

Equation (2.107) is the Debye cube law. Its agreement with experimental data is limited at very low temperatures, sometimes at $T \ll \theta_D/50$ [55], where the crystal lattice experiences oscillations of sufficient long wavelength, significantly larger than the lattice parameters. Discrepancies between Debye's theory and experimental data are attributed in the assumptions that all elastic waves in the lattice travel with the same speed, disregarding any distribution of speed, and that the limiting maximum ω_m for all waves in the lattice remains the same. An additional reason for these deviations in polymeric materials is the neglect of the anisotropy of the interactions in macromolecular chains.

2.10.2 Thermal Conduction of Dielectrics

Thermal or heat conduction is an energy transfer process. The presence of a temperature gradient in a solid will result in the generation of heat flux, which will exist until the gradient become equal to zero. Energy will be "diffused" from the sites with higher temperatures to the sites with lower temperature until temperatures become equal. In the case of maintaining artificially a constant temperature gradient, a steady-state heat flux, constant in time, will be developed in the solid. Everyday experience indicates that metals are characterized by higher thermal conduction than insulators, although diamond is a striking exception.

Consider a solid rod of cross section A, with one edge at constant temperature T and the other one be heated constantly at $\frac{dQ}{dt}$. After achieving dynamic equilibrium, at two points away from the rod's edges, separated by dx, a temperature difference dT is measured. Thermal conductivity is defined as:

$$\kappa = -\frac{1}{A}\frac{dQ}{dt}\frac{dx}{dT} \Longrightarrow \kappa = -\frac{dq}{dt}\frac{dx}{dT} \tag{2.108}$$

where, dq/dt is the thermal flux (amount of heat per unit time and per unit area). Relative to the mechanism of heat transfer, solids can be classified in three categories: (i) nonmetals, where heat transfer takes place as a result of lattice vibrations, (ii) metals, where electrons play the primary role for heat conduction, and (iii) solids with low conduction, such as alloys, where thermal conductivity is a combination of the contributions from free electrons and lattice vibrations (phonons). In general, thermal conductivity includes contributions from both mechanisms, free electrons (κ_e), and lattice vibrations (κ_{lat}):

$$\kappa = \kappa_e + \kappa_{lat} \tag{2.109}$$

Pure metals have a lot of free electrons, and their contribution is the predominant mechanism for thermal conduction. In dielectric materials, ionic crystals, and solids with covalent bonds, the phonon contribution is responsible for the total thermal conduction, because of the absence of free electrons.

Suppose that the two points in the rod, mentioned previously, are in a solid crystalline dielectric material, thermal flux can be written as:

$$\frac{dq}{dt} = -c_V \rho v_x \Delta T \tag{2.110}$$

where c_V is the heat capacity per unit mass of the crystal, ρ its density, v_x phonon velocity in the x direction, and ΔT the temperature difference. Suggesting that the space separation between the two points is equal to the length of the mean free path l of the phonons and $l = \upsilon_x \tau$, Eq. (2.110) can be rewritten as:

$$\frac{dq}{dt} = -c_V \rho \upsilon_x \frac{dT}{dx} l \Longrightarrow \frac{dq}{dt} = -c_V \rho \upsilon_x^2 \frac{dT}{dx} \tau \qquad (2.111)$$

where τ is the mean time between two successive phonon collisions. The term υ_x^2 in the previous equation can be replaced by $\overline{\upsilon^2}/3$, because of symmetry and the equivalency of velocity in all three axes. $\overline{\upsilon^2}$ is the average value of the sum of the squared phonon velocities. Thus, Eq. (2.111) becomes:

$$\frac{dq}{dt} = -c_V \rho \frac{\overline{\upsilon^2}}{3} \frac{dT}{dx} \tau \qquad (2.112)$$

Assuming that, approximately all phonons have equal velocities, the root mean square velocity becomes equal to the common phonon velocity and Eq. (2.112) takes the following expression:

$$\frac{dq}{dt} = -c_V \rho \frac{\upsilon l}{3} \frac{dT}{dx} \qquad (2.113)$$

Employing Eqs. (2.108) and (2.113) leads to the thermal conductivity of a crystalline solid dielectric material:

$$\kappa = \frac{c_V \rho \upsilon l}{3} \qquad (2.114)$$

The determination of thermal conductivity of amorphous dielectric solids presents considerable difficulties because of the absence of symmetry and long-range order. The disordered arrangement of atoms in an amorphous solid is responsible for the variation of the material's elastic properties from point to point. Elastic waves propagated in such a solid might undergo scattering in a way similar with the propagation in a medium with random inhomogeneities. Phonons scattering and the variation of their velocity from one point to another raise difficulties in determining the mean free path l. In amorphous polymers, thermal conductivity increases with temperature up to the glass transition temperature and diminishes after T_g. Several theoretical and semi-empirical models proposed for the thermal conductivity of amorphous and semi-crystalline polymers are beyond the scope of the present chapter [55].

2.11 Concluding Remarks

The name, as well as the majority of the applications of dielectric materials, is related to their electrical properties. This chapter attempts to summarize the basic concepts and theories regarding the electrical performance of dielectrics under dc and ac electric field. The peculiar behavior of the active dielectrics is briefly examined. In addition, basic concepts referring to thermal properties (heat capacity and thermal conduction) of dielectrics are also discussed.

Acknowledgements

G. C. Psarras dedicates the present work to the memory of Georgios N. Tsamis a man of culture, politeness, visions, and open society, but mostly a valuable friend.

References

1 von Hippel, A.R. (1995). *Dielectrics and Waves*. Boston: Artech.
2 Bartnicas, R. (1983). Dielectric loss in solids. In: *Engineering Dielectrics, vol IIA, Electrical Properties of Solids Insulating Materials: Molecular Structure and Electrical Behaviour* (ed. R. Bartnicas and E.E. Eichhorn), 3–67. ASTM.
3 Blythe, A.R. (1980). *Electrical Properties of Polymers*. Cambridge: Cambridge University Press.
4 Psarras, G.C. (2010). Conductivity and dielectric characterization of polymer nanocomposites. In: *Polymer Nanocomposites: Physical Properties and Applications* (ed. S.C. Tjong and Y.-M. Mai), 31–69. Cambridge: Woodhead Publishing Limited.
5 Psarras, G.C. (2018). Fundamentals of dielectric theories. In: *Dielectric Polymer Materials for High-Density Energy Storage* (ed. Z.-M. Dang), 11–57. Elsevier.
6 Debye, P. (1945). *Polar Molecules*. Dover Publications.
7 Onsager, L. (1936). Electric moments of molecules in liquids. *Journal of the American Chemical Society* 58: 1486–1493.
8 McCrum, N.G., Read, B.E., and Williams, G. (1967). *Anelastic and Dielectric Effects in Polymeric Solids*. New York: John Wiley and Sons.
9 Kao, K.C. (2004). *Dielectric Phenomena in Solids*. Elsevier.
10 Kirkwood, J.G. (1939). The dielectric polarization of polar liquids. *The Journal of Chemical Physics* 7: 911–919.
11 Frohlich, H. (1958). *Theory of Dielectrics*. Oxford: Clarendon.
12 Tsangaris, G.M., Psarras, G.C., and Kouloumbi, N. (1998). Electric modulus and interfacial polarization in composite polymeric systems. *Journal of Materials Science* 33: 2027–2037.
13 Cole, K.S. and Cole, R.H. (1941). Dispersion and absorption in dielectrics I. Alternating current characteristics. *The Journal of Chemical Physics* 9: 341–351.
14 Davidson, D.W. and Cole, R.H. (1950). Dielectric relaxation in glycerine. *The Journal of Chemical Physics* 18: 1417.
15 Havriliak, S. and Negami, S. (1967). A complex plane representation of dielectric and mechanical relaxation processes in some polymers. *Polymer* 8: 161–210.
16 Psarras, G.C., Soto, A., Voyiatzis, G.A. et al. (2006). Dielectric and conductivity processes in poly(ethylene terephthalate) and poly(ethylene naphthalate) homopolymers and copolymers. *Journal of Polymer Science: Part B: Polymer Physics* 44: 3078–3092.
17 Kontos, G.A., Soulintzis, A.L., Karahaliou, P.K. et al. (2007). Electrical relaxation dynamics in TiO_2-polymer matrix composites. *Express Polymer Letters* 1 (12): 781–789.
18 Psarras, G.C., Gatos, K.G., Karahaliou, P.K. et al. (2007). Relaxation phenomena in rubber/layered silicate nanocomposites. *Express Polymer Letters* 1 (12): 837–845.

19 Patsidis, A. and Psarras, G.C. (2008). Dielectric behaviour and functionality of polymer matrix-ceramic $BaTiO_3$ composites. *Express Polymer Letters* 2 (10): 718–726.

20 Kremer, F. and Schonhals, A. (ed.) (2003). *Broadband Dielectric Spectroscopy*. Berlin: Springer.

21 Böttger, H. and Bryskin, U.V. (1985). Hopping conduction in solids. *Verlag Akademie* 169–213.

22 Jonscher, A.K. (1992). *Universal Relaxation Law*. London: Chelsea Dielectrics Press Chapter 5.

23 Dyre, J.C. (1998). The random free-energy barrier model for ac conduction in disordered solids. *Journal of Applied Physics* 64 (5): 2456–2468.

24 Tsangaris, G.M., Psarras, G.C., and Manolakaki, E. (1999). DC and AC conductivity in polymeric particulate composites. *Advanced Composites Letters* 8 (1): 25–29.

25 Long, A.R. (1991). Hopping conductivity in the intermediate frequency regime. In: *Hopping Transport in Solids* (ed. M. Pollak and B. Shklovskii), 207–231. Amsterdam: Elsevier.

26 Mott, N.F. (1987). *Conduction in Non-Crystalline Materials*. Oxford: Clarendon Press.

27 Mott, N.F. (1990). *Metal-Insulator Transitions*. London: Taylor & Francis.

28 Psarras, G.C. (2006). Hopping conductivity in polymer matrix –metal particles composites. *Composites Part A-Applied Science* 37: 1545–1553.

29 Psarras, G.C. (2009). Conduction processes in percolative epoxy resin/silver particles composites. *Science of Advanced Materials* 1: 101–106.

30 Pontikopoulos, P.L. and Psarras, G.C. (2013). Dynamic percolation and dielectric response in multiwall carbon nanotubes/poly(ethylene oxide) composites. *Science of Advanced Materials* 5: 1–7.

31 Manika, G.C. and Psarras, G.C. (2020). $SrTiO_3$/epoxy nanodielectrics as bulk energy storage and harvesting systems: the role of conductivity. *ACS Applied Energy Materials* 3: 831–842.

32 Mandal, P., Neumann, A., Jansen, A.G.M., and Wyder, P. (1997). Temperature and magnetic-field dependence of the resistivity of carbon-black polymer composites. *Physical Review B* 55: 452–456.

33 Joo, J., Long, S.M., Pouget, J.P. et al. (1998). Charge transport of the mesoscopic metallic state in partially crystalline polyanilines. *Physical Review B* 57: 9567–9580.

34 Dyre, J.C. and Shrøder, T.B. (2000). Universality of ac conduction in disordered solids. *Reviews of Modern Physics* 72 (3): 873–892.

35 Dyre, J.C. and Schrøder, T.B. (2002). Hopping models and ac universality. *Physica Status Solidi B* 230 (1–5): 5–13.

36 Schrøder, T.B. and Dyre, J.C. (2002). Computer simulations of the random barrier model. *Physical Chemistry Chemical Physics* 4: 3173–3178.

37 Kittel, C. (1996). Dielectrics and ferroelectrics. In: *Introduction to Solid State Physics*, 455–485. Hoboken, NJ, USA: John Wiley & Sons Inc.

38 Manika, G.C., Andrikopoulos, K.S., and Psarras, G.C. (2020). On the ferroelectric to paraelectric structural transition of $BaTiO_3$ micro-/nanoparticles and their epoxy nanocomposites. *Molecules* 25: 2686.

39 Singh, J. (2005). *Smart Electronic Materials*. Cambridge: Cambridge University Press.

40 Starkweather, H.W. Jr., (1981). Simple and complex relaxations. *Macromolecules* 14 (5): 1277–1281.

41 Boyd, R.H. (1985). Relaxation processes in crystalline polymers: experimental behavior-a review. *Polymer* 26 (3): 323–347.

42 Sauer, B.B., Avakian, P., Flexman, E.A. et al. (1997). A.C. Dielectric and TCS studies of constrained amorphous motions in flexible polymers including poly(oxymethylene) and miscible blends. *Journal of Polymer Science B: Polymer Physics* 35 (13): 2121–2132.

43 Karahaliou, P.K., Kerasidou, A.P., Georga, S.N. et al. (2014). Dielectric relaxations in polyoxymethylene and in related nanocomposites: identification and molecular dynamics. *Polymer* 55: 6819–6826.

44 Hedvig, P. (1977). *Dielectric Spectroscopy of Polymers*. Bristol: Adam Hilger Ltd.

45 Korzhenko, A., Tabellout, M., and Emery, J.R. (1999). Influence of a metal-polymer interfacial interaction on dielectric relaxation properties of polyurethane. *Polymer* 40: 7187–7195.

46 Hernández, M., Carretero-González, J., Verdejo, R. et al. (2010). Molecular dynamics of natural rubber/layered silicate nanocomposites as studied by dielectric relaxation spectroscopy. *Macromolecules* 43 (2): 643–651.

47 Boyd, R.H. and Liu, F. (1997). Dielectric spectroscopy of semicrystalline polymers. In: *Dielectric Spectroscopy of Polymeric Materials* (ed. J.P. Runt and J.J. Fitzgerald), 107–136. Washington, DC: American Chemical Society.

48 Steeman, P.A.M. and van Turnhout, J. (2003). Dielectric properties of inhomogeneous media. In: *Broadband Dielectric Spectroscopy* (ed. F. Kremer and A. Schönhals), 495–522. Berlin: Springer.

49 Mathioudakis, G.N., Patsidis, A.C., and Psarras, G.C. (2014). Dynamic electrical thermal analysis on zinc oxide/epoxy resin nanodielectrics. *Journal of Thermal Analysis and Calorimetry* 116: 27–33.

50 Patsidis, A.C., Kalaitzidou, K., and Psarras, G.C. (2014). Graphite nanoplatelets/polymer nanocomposites: thermomechanical, dielectric, and functional behaviour. *Journal of Thermal Analysis and Calorimetry* 116: 41–49.

51 Vryonis, O., Anastassopoulos, D.L., Vradis, A.A., and Psarras, G.C. (2016). Dielectric response and molecular dynamics in epoxy-$BaSrTiO_3$ nanocomposites: effect of nanofiller loading. *Polymer* 95: 82–90.

52 Stavropoulos, S.G., Sanida, A., and Psarras, G.C. (2020). A comparative study on the electrical properties of different forms of carbon allotropes – epoxy nanocomposites. *Express Polymer Letters* 14 (5): 477–490.

53 Konstantinou, A.C., Patsidis, A.C., and Psarras, G.C. (2021). Boron nitride/epoxy resin nanocomposites: development, characterization and functionality. *Journal of Thermal Analysis and Calorimetry* 45: 2925–2933.
54 Sanida, S.A., Stavropoulos, S.G., Speliotis, T., and Psarras, G.C. (2017). Development, characterization, energy storage and interface dielectric properties in $SrFe_{12}O_{19}$/epoxy nanocomposites. *Polymer* 120: 73–81.
55 Perepechko, I. (1980). *Low-Temperature Properties of Polymers*. Moscow: Mir Publishers.
56 Fredriksson, H. and Åkerlind, U. (2008). *Physics of Functional Materials*. Wiley.
57 Hofmann, P. (2009). *Solid State Physics*. Wiley-VCH.

3

High-Temperature Energy Storage Polymer Dielectrics for Capacitors

Zongliang Xie[1,2], He Li[1], Zongren Peng[2], and Yi Liu[1]

[1]*Lawrence Berkeley National Laboratory, The Molecular Foundry, One Cyclotron Road, Berkeley, California 94720, USA*
[2]*Xi'an Jiaotong University, State Key Laboratory of Electrical Insulation and Power Equipment, 28 Xianning West Road, Xi'an, Shaanxi 710049, China*

3.1 Introduction

Achieving carbon neutrality is an imminent global mission, which requires the rational design and optimization of renewable energy internet at multiple levels and scales [1, 2]. The grid integration of renewable energy generation, such as wind power, solar energy, and geothermal energy, brings about urgent demands for high-performing energy storage devices [3, 4]. In particular, as one of the most fundamental passive components, the dielectric energy storage capacitor is present in almost every electronic circuit, which is essential for power quality regulation [5, 6]. Although dielectric capacitors have relatively low stored energy density (defined as maximum energy stored per unit mass) compared to other energy storage devices, the inherent advantage of the highest power density makes the dielectric capacitor stand out among the existing energy storage devices [7–9]. Specifically, power density is defined as output power per unit mass, which governs the response speed of energy conversion. Traditional batteries and electrochemical capacitors are capable of storing high energy density (>100 Wh kg^{-1}) and are extensively used for power supplies in the power grid. However, they exhibit low power density (<10^4 W kg^{-1}), caused by inevitable hysteresis during energy conversion between electrical energy and chemical energy [10, 11]. On the contrary, dielectric capacitors can discharge energy in microseconds owing to their purely physical charging and discharging processes, resulting in a much higher power density (>10^5 W kg^{-1}). Energy storage devices such as dielectric capacitors with high power density are indispensable in high-power applications, including pulse power devices, electric vehicles, and spacecraft [12].

Moreover, dielectric capacitors have presently occupied ~50% of the global capacitor market [13]. Compared to dielectric ceramics, polymer-based dielectrics possess characteristics of excellent electrical insulating performance and mechanical flexibility, as well as lightweight, low-cost, and facile processing, which are

High Temperature Polymer Dielectrics: Fundamentals and Applications in Power Equipment, First Edition. Edited by Jun-Wei Zha and Zhi-Min Dang.

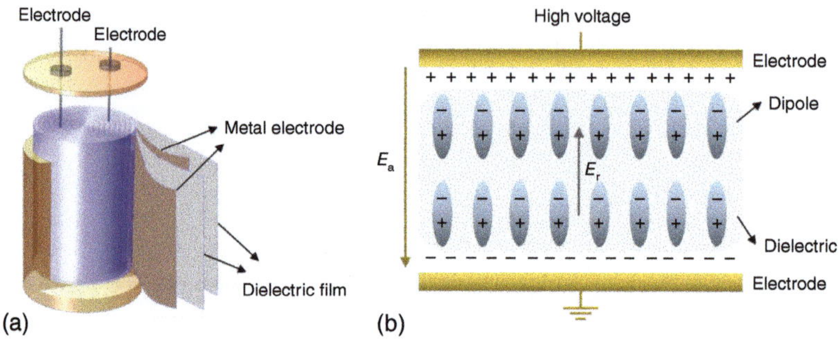

Figure 3.1 Schemes of (a) dielectric capacitor structure and (b) internal energy storage process.

promising materials for flexible dielectric capacitors, especially for those operating under high electric fields [13–18]. The current in-service dielectric capacitors are mostly fabricated by winding thin polymer films along with metal electrodes into a capacitor core. Diagrams of the roll-to-roll processed capacitor structure and energy storage process in a dielectric capacitor are shown in Figure 3.1 [19]. As demonstrated in Figure 3.1b, with an applied electric field (E_a) on the electrodes, an equivalent internal reverse field (E_r) with a value equal to minus E_a (i.e. $E_r = -E_a$) is instantly formed after the reorientation of intrinsic dipoles. As a result, the energy can be electrostatically stored at the electrode/polymer interfaces. When the external power supply is removed, the accumulated charges at the electrode/dielectric interfaces are accelerated under E_r; in this way, the stored energy can be rapidly released.

The stored energy density decides the energy storage capacity of a dielectric capacitor. In other words, the larger stored energy density signifies the higher capability of a capacitor to store and convert renewable energy resources. In principle, the stored energy density of a dielectric material is proportional to the electric displacement (D) and the square of the applied electric field. More specifically, D is positively correlated with the dielectric constant (K); the maximal applied electric field is determined by the material's dielectric breakdown strength (E_b). A high E_b not only provides margins for polymer dielectrics to operate under extreme electrical conditions but is also beneficial for the miniaturization of equipment. Therefore, simultaneously high K and E_b are always desired for a dielectric material to achieve high energy density. In addition, due to the presence of dielectric and conduction losses in the capacitors practical use, the discharging energy is usually smaller than the charging energy. The discharged energy density (U_e) is defined as the difference between stored energy density (U_s) and energy loss (U_l). Accordingly, the charge–discharge efficiency (η) is defined as $\eta = (U_e/U_s) \times 100\%$ (see Figure 3.2a).

Most dielectric polymers possess K values smaller than 4. For example, ~2.25 for biaxially oriented polypropylene (BOPP), ~3 for polyphenylenesulfide (PPS), and polyethylene (PC) [13, 20]. The relatively low K value is one of the fundamental challenges limiting the energy density of polymeric dielectric capacitors. Although ferroelectric polymers, represented by polyvinylidene fluoride (PVDF), exhibit higher K values (e.g. ≥10) than conventional linear dielectric polymers, the large

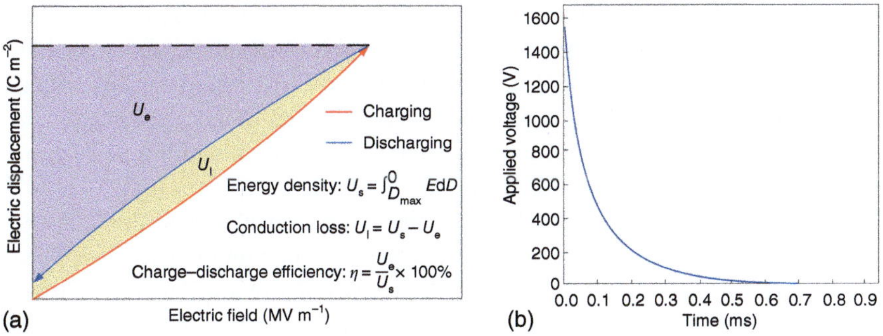

Figure 3.2 Schematic of (a) electric displacement–electric field (*D–E*) loop and (b) discharging curve of a dielectric material.

ferroelectric loss significantly reduces energy efficiency in dielectric capacitor applications [21–23]. More importantly, the aforementioned dielectric polymers fail to meet the increasing demands of heat-resistant dielectric capacitors working under high-temperature electrification conditions, including electrified aircraft (180–300 °C in converters near turbine engines), aerospace power systems (up to 250 °C), and integrated microelectronics (~150 °C in inverters) [13, 24, 25]. The present benchmark dielectric polymer for electrostatic film capacitors, BOPP, has a high E_b over 700 MV m^{-1} and can display a reasonably high U_e of about 5 J cm^{-3} at room temperature despite its notoriously low K [13]. However, the poor temperature resistance (long-term working temperature ≤85 °C) essentially excludes its direct use under those harsh electrification conditions (e.g. temperatures ≥150 °C). For instance, the η of commercial BOPP film decreases steeply from 96.2% to 87.4% and 68.5% with increasing temperatures from 25 to 80 and 120 °C, respectively, at an electric field of 400 MV m^{-1} [26]. As exemplified in hybrid electric vehicles, an extra cooling system has been employed to reduce temperatures to ~70 °C to accommodate BOPP-based film capacitors in the power inverter. This inevitably adds additional mass and volume, as well as reduces the energy efficiency of the power system of vehicles. Thus, the exploration and development of heat-resistant polymer dielectrics with well-retained performance over a wide temperature range are of great significance to high-temperature dielectric capacitors.

Polymers with high glass transition temperatures (T_g), such as polyetherimide (PEI) (T_g ~217 °C), fluorene polyester (FPE, T_g ~330 °C), and polyimide (PI) (T_g ~360 °C), are considered as promising dielectric polymers for high-temperature energy storage capacitors and have attracted much research interest [13, 25, 27, 28]. Although a high T_g can ensure the mechanical stability of the polymer backbone at high temperatures, the thermally assisted charge injection and migration increase exponentially with the increase in temperature and applied electric field, which lowers the voltage endurance capability (e.g. E_b) of dielectric polymers. Meanwhile, the steep rise in electrical conductivity originated from severer charge injection and transport at elevated temperatures, bringing about larger energy loss and leading to decreases in U_e and η. To this end, extensive nanoengineering efforts have been made to improve the energy storage performance of high-T_g polymer-based dielectrics. For example, nanoparticles with large K values, such as $BaTiO_3$, have been incorporated into the polymer matrices to yield high-K nanocomposites [29]; nanocoatings with wide bandgaps, such as Al_2O_3 and SiO_2, have been introduced onto the surface of polymer films to repulse charge injection [30]. Furthermore, synthetic efforts have also led to encouraging progress in developing high-performing polymer dielectrics operating under electrical and thermal extremes [31].

In this chapter, we will first introduce some important electrical and thermal parameters of dielectric materials for high-temperature polymer capacitor applications. The noteworthy commercial high-T_g dielectric polymers will be summarized subsequently. The following parts will focus on the recent progress in the design and development of high-temperature dielectric polymers and polymer nanocomposites, as classified as follows: (i) The preparation of randomly dispersed polymer/inorganic nanofiller composites; (ii) The design of core@shell-structured nanofillers for polymer composites; (iii) The development of polymer composites with multilayer structures; and (iv) The synthesis of novel polymers. Furthermore, the advantages and limitations of the current approaches to high-temperature polymer dielectrics will be elaborated. Challenges and future research opportunities will be highlighted at the end of the chapter.

3.2 Basic Parameters of High-Temperature Capacitor Materials

3.2.1 Electrical Characteristics of High-Temperature Capacitor Materials

3.2.1.1 Energy Storage Parameters

Energy storage density, also known as stored energy density (U_s), refers to stored energy per unit volume inside dielectrics in J cm^{-3}. Theoretically, U_s is the integral of electric displacement (D) and applied electric field (E_a), which can be expressed as:

$$U_s = \int E_a \mathrm{d}D \tag{3.1}$$

Under actual working conditions, the discharged energy density (U_e) matters and represents the actual energy storage ability of capacitors, which is typically smaller than U_s since there exists energy loss (U_l). The U_e and charge–discharge efficiency (η) can be calculated as:

$$\begin{aligned} U_e &= U_s - U_l \\ \eta &= \frac{U_e}{U_s} \times 100\% \end{aligned} \tag{3.2}$$

As shown in Figure 3.2a, the red line represents the charging process, and the blue line is the discharging procedure. The enclosed purple area is the overall discharged energy density U_e. The energy loss U_l is the surrounding area, i.e. the yellow one between the charging and discharging curves.

During the discharging process of capacitors with load accessed, the load resistance (R_L) varies under different working conditions and is usually larger than the equivalent series resistance (ESR) of dielectric materials. The discharging power density (P_e) (usually called power density) can be expressed as:

$$P_e = \frac{U(t)^2}{R_L} \tag{3.3}$$

where U(t) is the time-dependent voltage (in V) applied to R_L.

It is worth mentioning that the discharging time depends on the time constant $\tau = R_L C$. With capacitance (C) in order of pF and R_L in order of kΩ, the discharging time of dielectric polymers is within μs, satisfying demands for dielectric capacitors used in pulse power equipment. The schematic of discharging curve is shown in Figure 3.2b.

3.2.1.2 Dielectric Constant and Dielectric Loss

Dielectrics refer to materials in which charges are bonded by intramolecular or intermolecular forces, while bonded charges redirect/relocate and establish induced electric dipoles (in other words, dielectrics are polarized) under the action of the applied electric field E_a. The electric polarization P_{polar} is defined as total dipole moments in dielectrics per unit volume, which can be expressed as follows in isotropic dielectrics:

$$P_{polar} = N\alpha E_i = (K - 1)\varepsilon_0 E_a \tag{3.4}$$

where N is the number of molecules per unit volume in m^{-3}, α is molecular polarizability in F m^2, E_i is the effective electric field applied on a molecule in V m^{-1}, and ε_0 is the vacuum permittivity, which equals to 8.85×10^{-12} F m^{-1}.

The dielectric constant K can thus be expressed as follows (Clausius equation):

$$K = \frac{N\alpha E_i}{\varepsilon_0 E_a} + 1 \tag{3.5}$$

There are mainly five different modes of polarization in dielectrics: electronic displacement polarization (P_{elec}), ionic displacement polarization (P_{ion}), orientational polarization (P_{orien}), thermionic polarization (P_{therm}), and interfacial polarization (P_{inter}). The polarization processes are related to molecular chain displacement, either chain movement or rotation, which is closely related to frequency and temperature.

The electronic displacement polarization P_{elec} refers to elastic displacements of outer electronic clouds of molecules, atoms, or ions under the applied electric field, which only last 10^{-15} to 10^{-16} s with vibration frequency in the optical frequency range. The electronic polarizability α_{elec} (in F m^2) of atoms with an atomic radius of r_e (in m) can be calculated as:

$$\alpha_{elec} = 4\pi\varepsilon_0 {r_e}^3 \tag{3.6}$$

Similarly, ionic displacement polarization P_{ion} exists in ionic crystals, referring to elastic displacements between heteroions within 10^{-12} to 10^{-13} s, and ionic polarizability α_{ion} (in F m^2) can be expressed as:

$$\alpha_{ion} = \frac{4\pi\varepsilon_0 {r_i}^3}{n - 1} \tag{3.7}$$

where r_i is the distance between negative and positive ion centers in m, and n is the lattice constant.

Orientation polarization P_{orien} exists in polar molecules with intrinsic dipole moments, in which dipoles tend to orientate toward the electric field direction. Subject to molecular thermal motions, the polarization establishment time can be as long as 10^{-6} to 10^{-2} s, and orientation polarizability α_{orein} can be approximately calculated as:

$$\alpha_{orien} = \frac{\mu_0^2}{3K_B T} \tag{3.8}$$

where μ_0 is inherent dipole moment in C m, which is formed by the asymmetric distribution of positive and negative charges in polar molecules, K_B is the Boltzmann constant equal to 1.38×10^{-23} J K^{-1}, and T is the temperature in K.

Another temperature-dependent relaxation polarization is thermionic polarization P_{therm}, which is caused by ion migration after overcoming the potential barrier of adjacent molecules under thermal vibration, even without an applied electric field. The relaxation time $\tau_{therm} = \left(e^{u/K_B T}\right)/2\nu$ is estimated to be around 10^{-6} to 10^{-2} s with an activation energy u of 10^{-19} J and an intrinsic vibration frequency ν of 10^{12} Hz.

Interfacial polarization P_{inter}, also known as Maxwell–Wagner–Sillars polarization, occurs at interfaces between different materials and is one of the determining processes for the energy storage behavior of dielectrics. The space charges migrate and accumulate at interfaces, thus forming macroscopic dipole moments due to uneven charge distribution. The establishment of such a process usually takes a long time (>10^{-6} s) according to the time constant defined by $\tau = R_L C_a$, as mentioned above. For a double-layer dielectric with dielectric constants of K_1 and K_2, electrical conductivities of γ_1 and γ_2, and thicknesses of d_1 and d_2, the accumulated charge σ_{inter} between layers can be expressed as:

$$\sigma_{inter} = \frac{\varepsilon_0(\gamma_1 K_2 - \gamma_2 K_1)}{\gamma_1 d_2 + \gamma_2 d_1} \tag{3.9}$$

The combined action of all five polarization processes determines the K of polymer dielectrics in energy storage capacitors. Under high temperatures, relaxation polarization can be affected by the increase in molecular thermal motions, resulting in slight variations of K for dielectrics at different temperatures. Dielectric loss, on the other hand, increases a lot with temperature rise. For dielectrics, mainly two types of dielectric losses exist, i.e. ohmic conduction loss and relaxation polarization loss. The first one comes from direct current conductance in nonideal insulators. The second loss always occurs under alternating current when current lags behind voltage during a slow relaxation polarization process. The overall dielectric loss P_{loss} can be calculated as the sum of ohmic conduction loss P_{ohmic} and relaxation polarization loss tanδ:

$$P_{loss} = P_{ohmic} + \tan\delta = \frac{\gamma}{\omega\varepsilon_0 K} + \frac{(\varepsilon_s - \varepsilon_\infty)\omega\tau}{K(1+\omega^2\tau^2)} \tag{3.10}$$

where γ is the electrical conductivity in S m^{-1}, ω is the angular frequency in s^{-1}, ε_s is the static dielectric constant under a constant electric field, and ε_∞ is the dielectric constant at optical frequency.

In all, the polarization processes directly determine K and, in turn, affect U_e of polymer dielectrics, while the relaxation processes impact energy loss U_l. Hence, high K and low P_{loss} are desired to achieve high U_e and high η in polymer dielectrics. In addition, polymers usually display large variations in K and tanδ around their T_g due to the enormous changes in molecular structures and states. Thus, high T_g is typically considered as one of the essential parameters to secure high thermal stability for high-temperature dielectrics. Yet, even though high-T_g polymers can retain stable K and tanδ with increasing temperature, challenges still exist since thermally assisted charge injection and migration can cause steep rises in electrical conductivity, which would subsequently lower the voltage endurance and energy storage performance of dielectric materials.

3.2.1.3 Dielectric Breakdown Strength

As mentioned above, besides dielectric constant K, another decisive factor is dielectric breakdown strength E_b. For polymers, higher E_b provides a greater voltage rise margin and is valuable for lightweight and miniaturization of dielectric capacitors. The dielectric breakdown process is complicated and depends on intrinsic material properties (e.g. electrical conductivity and thermal conductivity) and external factors (e.g. sample thickness, temperature, electrode conditions, and electrical stress conditions). The three main dielectric breakdown mechanisms for dielectric polymers under high temperatures are electronic breakdown, thermal breakdown, and electromechanical breakdown, which indicate the coupled effect of electro, thermal, and mechanical forces on the dielectric breakdown strength of polymers, respectively.

The electronic breakdown usually occurs under high electric fields, where electrons are excited from the valence band to the conduction band and gain kinetic energy under electric force. The collision ionization occurs, forming a steep rise in free electrons and destroying dielectric structures. For soft dielectric materials with a relatively small Young's modulus, the film thickness can be reduced under Maxwell electrostatic stress, which would further lead to the increase of the endured electric field. The critical condition of electromechanical breakdown E_{em} is when electrostatic stress equals to elastic stress, depicted as follows:

$$E_{em} = 0.606\left(\frac{Y}{\varepsilon_0 K}\right)^{1/2} \tag{3.11}$$

where Y is the Young's modulus in Pa.

Since Young's modulus decreases at high temperatures (especially at T_g), electromechanical breakdown exists in most polymer dielectrics in high-temperature regions. Due to the low thermal conductivity and inferior heat dissipation ability of dielectric polymers, the gradually accumulated heat energy would eventually lead to thermal breakdown when the heat generation rate exceeded heat loss. A diagram of the dielectric breakdown mechanism is shown in Figure 3.3. The

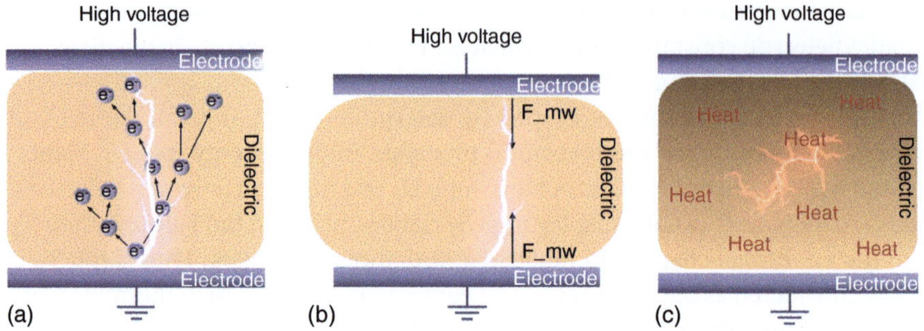

Figure 3.3 Schematic of (a) electronic breakdown, (b) electromechanical breakdown, in which F_mw refers to Maxwell electric stress under E_a, and (c) thermal breakdown.

rapid decline of dielectric breakdown strength under high temperatures seriously reduces application feasibility and equipment reliability. Until now, the dominant mechanisms of dielectric breakdown remain unclear, which are governed by complicated conditions such as temperature, film thickness, voltage ramp, and electrical conduction. The two-parameter Weibull distribution, expressed as $P(E_a) = 1 - \exp(-[E_a/E_{wb}])^{\beta}$, has been widely used for the evaluation of dispersed breakdown data. E_{wb} represents the Weibull breakdown strength, which corresponds to a 63.2% probability of breakdown. β is the shape parameter that evaluates the dispersity of data. A larger β represents more stable testing results.

3.2.1.4 Electrical Conduction and Charge Injection

As a matter of fact, both dielectric breakdown strength E_b and dielectric loss P_{loss} is closely related to electrical conduction and charge injection processes in polymer dielectrics. For ideal insulating dielectrics, electrons are localized in the valence band, and there is no free charge inside the material. Under high electric fields and high temperatures, however, charges are able to overcome barriers and gain enough energy to be injected from electrodes into polymer dielectrics, known as Schottky emission, as follows:

$$J_{sch} = AT^2 \exp\left(\frac{-\omega_{sch}}{K_B T}\right) \exp\left(\frac{e}{K_B T}\sqrt{\frac{e|E_a|}{4\pi\varepsilon_0 K}}\right) \tag{3.12}$$

where J_{sch} is the injection current density at the electrode in A m^{-2}, A is the Richardson constant, which equals to 1.2×10^6 A m^{-2} K^{-2}, e is the elementary charge equal to 1.6×10^{-19} C, and ω_{sch} is the Schottky injection barrier at the electrode/polymer interface in eV.

When the potential barrier is sufficiently thin, the electron wave function can penetrate through the barrier due to Fowler–Nordheim tunneling. To suppress the injection current from electrodes that become larger under high electric fields and high temperatures, embedding a charge-blocking layer to repulse charge injection from the electrode to the polymer dielectric proves to be a feasible method (to be discussed in Chapter 3.5).

With certain concentrations of injected charges, the electrical conduction current is formed due to charge migration under the action of E_a, following Ohmic law:

$$J_{ohm} = \gamma_a E_a \tag{3.13}$$

where J_{ohm} is the conduction current density inside the dielectric in A m^{-2}, and γ_a is the electrical conductivity of polymer dielectrics in S m^{-1}.

When the injection current exceeds the conduction current under a higher electric field, the space charge limited current (SCLC) J_{SCLC} is formed and consists of drift current and diffusion current. Like Schottky emission at the electrode/dielectric interfaces, a similar thermal–electrical-assisted charge migration process exists inside polymers, named Poole–Frenkel emission conduction with migration barrier of ω_{PF}. In most semi-crystalline polymer dielectrics composed of crystalline and amorphous regions, charge carriers tend to hop between conduction bands in different crystalline regions, especially under high electric fields and high temperatures. A hopping electrical conduction current J_{hop} is thus formed and is essentially the combined action of thermal excitation and tunneling processes where trapped carriers inside polymers "hop" from one trap site to another. The hopping conduction current considering a single trapping level of depth ω_{hop} is as follows:

$$J_{hop} = 2Ne\lambda v \exp\left(-\frac{\omega_{hop}}{K_B T}\right) \sinh\left(\frac{\lambda e E_a}{2K_B T}\right) \tag{3.14}$$

where λ is the hopping distance in m and v is the attempt-to-escape frequency in s^{-1}.

The schematics of the abovementioned charge injection and electrical conduction processes are shown in Figure 3.4. Due to the existence of these conduction processes, part of the stored charges is dissipated in the form of electrical conduction current across the polymer dielectrics, leading to an increase in U_l, drop in η, and potential risks of thermal instability. Moreover, the steep rise in electrical conduction current degrades the insulation properties of polymer dielectrics, resulting in decreased E_b. Based on research findings [32–34], the electrical hopping conduction and Poole–Frenkel emission conduction processes are closely related to trap energy levels in polymer dielectrics. Deeper charge traps can impede charge injection and restrict charge transport, thereby reducing electrical conduction current. Thus, introducing deep traps by incorporating microscopic/macroscopic interfaces into polymer matrices is feasible to suppress electrical conduction and retain high E_b and moderate U_l at high temperatures.

3.2.2 Thermal Characteristics of high-Temperature Capacitor Materials

Thermal characteristics of polymer dielectrics directly determine their energy storage performance under elevated temperatures. The glass transition temperature (T_g) is usually thought of as the primary factor for evaluating the potential of dielectric polymers to be used for high-temperature capacitor materials. T_g refers to the temperature when the transition occurs from a glassy state to a highly elastic state. At temperatures near T_g, the displacement and rotation of molecular chains

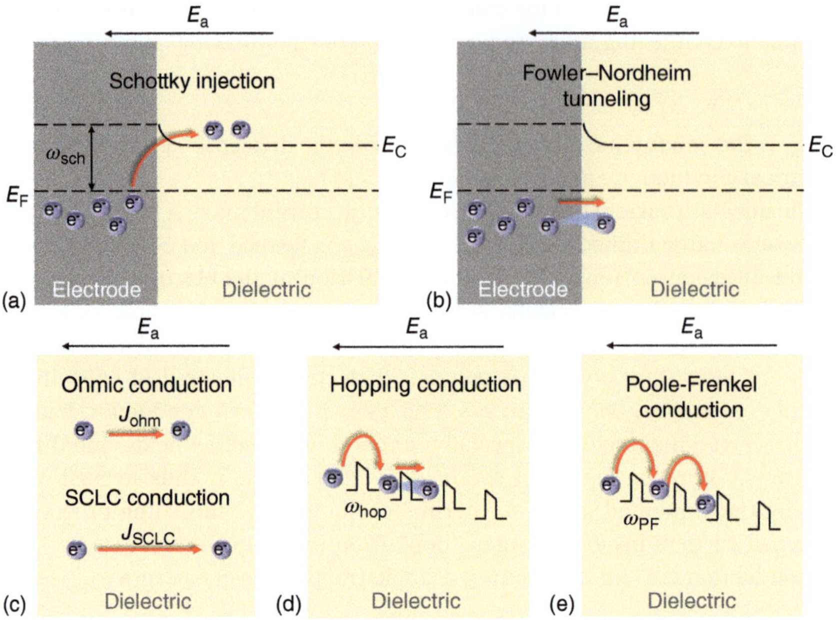

Figure 3.4 Schematic of charge injection processes under E_a including (a) Schottky injection and (b) Fowler–Nordheim tunneling, in which E_F is Fermi level and E_c is conduction band of dielectric. Schematic of electrical conduction processes under E_a including (c) Ohmic conduction, SCLC conduction, (d) hopping conduction, and (e) Poole–Frankel conduction.

of polymer dielectrics increase the free volume and lower mechanical strength, which in turn accelerates the occurrence of electron avalanches.

BOPP is the most commonly used commercial dielectric polymer in high-voltage and high-frequency applications [13, 20, 35] due to its excellent electrical resistance (E_b of ~700 MV m^{-1}, electrical conductivity of about 1×10^{-16} S m^{-1} at room temperature) with a stable dielectric constant (K ~2.25) and tanδ ~0.02% at 1 kHz below 100 °C). The maximum U_e of about 5 J cm^{-3} can be achieved in BOPP at room temperature [34]. Yet, the maximum long-term working temperature of BOPP is limited to 85 °C, restricting its direct use for high-temperature energy storage film capacitors. A breakdown failure of benchmark BOPP film occurs after 10 200 charging/discharging cycles at 120 °C and 200 MV m^{-1} [34], which incurs potential hazards in practical high-temperature electrification applications.

Since the low heat resistance of BOPP fails to meet the trending demands of high-temperature capacitor materials, high-T_g polymer dielectrics such as PEN (T_g ~125 °C), poly(ether ether ketone) PEKK (T_g ~155 °C), polycarbonate (PC, T_g ~150–160 °C), PEI (T_g ~217 °C), poly(phthalazinone ether ketone) (PPEK, T_g up to 250 °C), polytetrafluoroethylene (PTFE) (T_g ~260 °C), FPE(T_g ~330 °C), and Kapton PI (T_g ~360 °C) [13, 36] have been examined and chosen as the polymer substitutes to BOPP. A thin PTFE film below 5 μm was fabricated by researchers [37] through a casting process of the suspension containing PTFE nanoparticles, and the resulting polymer exhibited great high-temperature energy storage performances with U_e of 1.08 J cm^{-3} and η of 94% at 200 °C. PEN is a copolymer of ethylene glycol and 2,6-naphthalene dicarboxylic acid (NDA), which possesses U_e and η of 1.41 J cm^{-3} and 69%, respectively, at 150 °C and 350 MV m^{-1}, but exhibits a temperature-dependent K and tanδ [34]. PC is one kind of bisphenol A-based aromatic polymer synthesized from carbonic acid and dihydric alcohols. PC possesses stable K ~3 and tanδ ~0.1% at 1 kHz at temperatures from 25 to 125 °C and exhibits U_e of 1.29 J cm^{-3}, η of 47% at 150 °C, 400 MV m^{-1} [34, 38]. PEI, Kapton PI, and FPE have higher T_g (>200 °C) than PEN and PC and possess relatively high K and high E_b under high temperatures (150 °C) [38].

Another important thermal factor is thermal conductivity, which represents heat-conducting ability between molecular chains or crystalline/amorphous phases in dielectric polymers. For most electrically insulating polymers with low electrical conductivities (<1 × 10^{-14} S m^{-1}), the thermal conductivities are usually low (~0.2–0.5 W m^{-1} K^{-1}). Under high temperatures, heat accumulation could be faster than the heat dissipation process, resulting in further temperature rise inside the polymer-based dielectric capacitors. To facilitate heat dissipation across the polymer dielectrics, thermally conductive nanoparticles such as Al_2O_3, AlN, and BN have been loaded into the polymer matrices to yield polymer nanocomposites with enhanced thermal conductivity [39, 40]. The inherent high thermal conductivity in these inorganic materials is primarily owing to lattice vibrations that are beneficial for phonon heat transfer. In addition, it has been found that constructing interconnected thermally conductive frameworks in polymer matrices is a very effective strategy for obtaining high thermal conductivity polymer dielectrics [41–45].

A summary of key dielectric and energy storage parameters of the abovementioned polymers is illustrated in Table 3.1.

3.2.3 Self-Clearing Abilities of High-Temperature Capacitor Materials

For practical capacitors, e.g. BOPP-based film capacitors, both sides of the polymer dielectric film are metalized with thin layers of metals (mostly zinc or aluminum) as electrodes. Under severe conditions of high temperatures and high electric stresses, dielectric breakdown may occur at weak points. In such a breakdown event, the surge current flowing near breakdown sites generates Joule heating and results in the melting and vaporization of metals, which are later oxidized and blown away by the electric arc. The partial loss of metal electrodes leads to the exposure of circular insulating regions around the breakdown points on the dielectric surface, leaving behind a working capacitor that is operational under the same conditions (as illustrated in Figure 3.5). The spontaneous "clearing" of the breakdown sites following a breakdown event is known as the so-called self-clearing of film capacitors. The loss usually increases after each self-clearing process and eventually reaches a critical failure point, which largely determines the service reliability of dielectric polymer film capacitors.

However, unsuccessful self-clearing is not uncommon in film capacitors. When that happens, the breakdown channels are covered by conducting or

Table 3.1 Summary of key dielectric and energy storage parameters of commercial dielectric polymers.

Ref.	—	[13]	[13]	[13]	[13]	[13]	[13]	[13]	[13]	[13]
Polymer		BOPP	Kapton® PI	Ultem® PEI	PC	PEKK	PPEK	PTFE	PEN	FPE
η (%)	**Condition**	120 °C, 10 Hz	150 °C, 10 Hz	150 °C, 10 Hz	150 °C, 10 Hz	140 °C, 100 Hz	190 °C, 100 Hz	200 °C, 100 Hz	150 °C, 10 Hz	150 °C, 10 Hz
	Value	52.1	55.7	33.7	47	~35	~44	94	68.8	40.8
Maximum U_e ($J\ cm^{-3}$)	**Condition**	120 °C, 10 Hz	150 °C, 10 Hz	150 °C, 10 Hz	150 °C, 10 Hz	140 °C, 100 Hz	190 °C, 100 Hz	200 °C, 100 Hz	150 °C, 10 Hz	150 °C, 10 Hz
	Value	1.28	0.82	1.6	1.29	2.37	2.1	1.08	1.41	1.2
T_g (°C)	**Value**	< 0	~360	~217	~150–160	155	~250	—	~125	~330
E_b ($MV\ m^{-1}$)	**Condition**	120 °C, 5.8 μm	150 °C, 9–12 μm	150 °C, 8–14 μm	—	140 °C, ~25 μm	190 °C, 20–40 μm	200 °C, 4–5 μm	—	—
	Value	496	314	439	—	420	441	350	—	—
tanδ (%) at 1 kHz, RT	**Value**	~0.01	~0.1	~0.1	0.1	~0.3	0.63	~0.15	~0.1–0.2	~0.1
***K* at 1 kHz, RT**	**Value**	~2.25	~3.3–3.4	~3.2	~3	3.6	3.5	~2.17	~2.9–3	~3.5

Note that all η were achieved under the same E_a of the maximum U_e. RT refers to room temperature, which is 20 °C for BOPP and 25 °C for all other polymers.

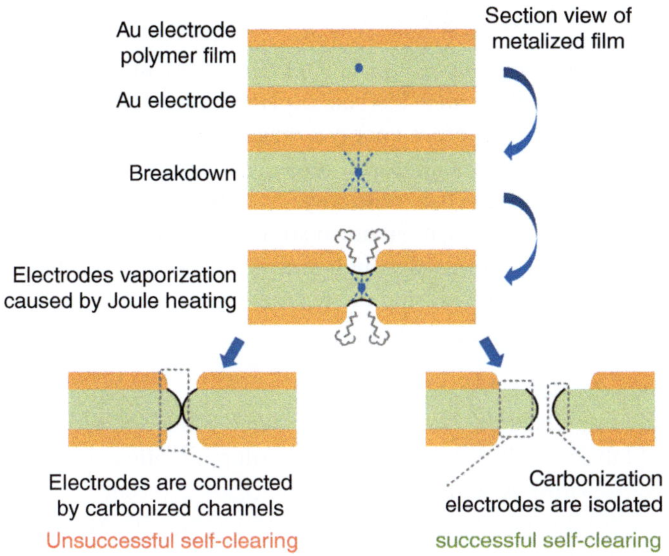

Figure 3.5 Schematic of successful and unsuccessful self-clearing processes. Source: Reproduced from Li et al. [13] with permission from the Royal Society of Chemistry.

semi-conducting materials such as carbon. Under such conditions, electrodes are connected by carbonized channels, and the whole film capacitor will fail (illustrated in Figure 3.5). Empirically, polymers with lower carbon contents tend to exhibit better self-clearing performance due to their lower tendency to form carbonaceous conducting channels, though there lacks a fundamental understanding of the self-clearing behavior of dielectric polymers, which calls for in-depth investigations in the future.

3.3 Randomly Dispersed Polymer/Inorganic Nanofiller Composites

Since the first publication in 1994 by Lewis [46], nanodielectrics have shown great performance in enhancing breakdown strength, suppressing space charge, and reducing dielectric loss [47]. Based on this, the method of incorporating nanofillers into polymer matrices has been widely explored for achieving better energy storage performances. For high-temperature capacitors, the key is to realize high U_e and η under harsh temperature conditions meanwhile concurrently maintaining low energy loss and Joule heating to ensure the stability of capacitors during long-term operation. In this section, the filler categories and their functions on high-temperature energy storage performances are summarized to offer a global perspective on the doping strategy for designing advanced dielectric polymers.

3.3.1 Composites Filled with Insulating Fillers

To date, doping inorganic insulating fillers into polymer matrices like PI, PEI, or cross-linked divinyltetramethyldisiloxane bis(benzocyclobutene) (*c*-BCB) has been proven to be an effective way to improve energy storage performance at high temperatures. Inorganic insulating fillers possess a wide bandgap and excellent electrical properties (high E_b, high electric resistivity, and charge suppression ability), which are beneficial for increasing breakdown strength, reducing conduction loss, and inhibiting charge injection under high temperatures. The common doping strategies include solution processing and melt extrusion, depending on the polymer matrices' solubility and melting temperature.

3.3.1.1 Composites Filled with Metal Oxides

The commonly used insulating fillers can be divided into metal oxides and nitrides. Shen and coworkers [48] fabricated PEI/Al_2O_3 composites (doping ratios of 0.5, 1, 2, and 3 vol%, particle size ~20 nm) by solution casting. The U_e of 3.70 J cm^{-3} and η of 90.1% were achieved in PEI/1 vol%-Al_2O_3 composites at 150 °C and 500 MV m^{-1}. The dielectric constants and dielectric losses remained stable at 1 kHz over a temperature range from 25 to 200 °C. The enhanced energy storage performance was attributed to the wide bandgap and high E_b of Al_2O_3 nanoparticles, which were effective for reducing conduction loss, inhibiting charge injection from electrodes, improving Young's modulus, and enhancing the Weibull breakdown strength of polymer composites. In another work of Shens group [49], dipolar groups of $-CF_3$ and $-SO_2-$ were introduced into PEI chains to increase electrical polarization, thus improving the energy storage performance of PEI/Al_2O_3 composites. It was found that PEI nanocomposites with $-CF_3$ groups or $-SO_2-$ groups display optimal U_e (3.32 J cm^{-3}, ~3.2 J cm^{-3}, respectively) and high η (92.5%, ~86%, respectively) at 150 °C. The incorporation of functional groups with high dipole moments effectively increases orientational polarization. It greatly enhances *K*, leading to improved energy storage densities of such modified PEI-based composites under high temperatures and electric fields. Li and coworkers [50] investigated how different crystalline forms of Al_2O_3 impacted the electric energy storage performance of PEI/Al_2O_3 composites. The team fabricated PEI/α-Al_2O_3 and PEI/γ-Al_2O_3 composites (doping ratios of 0.05, 0.1, 0.3, 0.5, and 1 vol%, with the particle size of ~20 nm). It was found that γ-Al_2O_3 was more effective in improving U_e and η of PEI-based nanocomposite. A U_e of ~4.2 J cm^{-3} along with a η of >70% were achieved in the PEI/0.3 vol%-γ-Al_2O_3 composites at 150 °C, 550 MV m^{-1}). Compared to α-Al_2O_3, γ-Al_2O_3 can form more/deeper traps due to its special defective spinel structure with oxygen polyhedral vacancies that localize charges and thus enhance the electric field endurance of polymer nanocomposites. Another recently published work [51] reported the cross-linked PEI/Al_2O_3 composites exhibiting superior energy storage capacities with an U_e of ~4.5 J cm^{-3} and a η of over 90% (PEI/0.3 vol%-Al_2O_3 composites) at 150 °C, 500 MV m^{-1}. Nanoscale infrared spectroscopy based on AFM-IR revealed that the addition of Al_2O_3 nanofillers was beneficial for the vibration of the carbonyl dipole and the movement of polymer chains, thus improving the *K* of polymer nanocomposites.

For a better understanding of the effect of nanofiller morphology on high-temperature capacitive performance, Wang and coworkers [32] developed a class of *c*-BCB/γ-Al_2O_3 composites, in which three different nanostructures of nanofillers were chosen: 0-D nanoparticles (particle size of ~20 nm), 1-D nanowires (mean length of 80 nm and mean diameter of 5 nm), and 2-D nanoplates (average thickness of 30 nm and width of 800–1000 nm in hexagonal shape). Nanoplates were found to be the most effective in providing high resistance to dielectric breakdown as well as dispersing local electric fields throughout the polymer matrix. The resultant *c*-BCB/γ-Al_2O_3 nanoplate composites delivered the greatest energy storage performance with respect to those *c*-BCB composites loaded with 0-D and 1-D nanofillers. At 150 °C and 500 MV m^{-1}, the *c*-BCB/7.5 vol%-γ-Al_2O_3 nanoplate composite displayed a U_e of 4.07 J cm^{-3} and a η of ~85%. The advantage of 2-D nanofiller in enhancing dielectric energy storage performance has been reaffirmed in a similar study, in which MgO nanoplate-filled PI composites exhibited better performance than nanoparticle- and nanowire-filled PI composites with the same filler loadings [52]. As a result, a high U_e of 4.78 J cm^{-3} was achieved in PI/0.5 vol%-MgO nanoplates at 433 MV m^{-1}, 150 °C.

In addition to Al_2O_3, some other commonly used metal oxides with wide bandgaps also show great performance in enhancing the high-temperature energy storage performance of polymer dielectrics. Wang and coworkers [53] fabricated PEI/HfO_2 composites (doping ratios of 1, 3, 5, 7, 9, particle size of 61–80 nm) and discovered enhanced dielectric properties of polymer composites. The U_e of 2.82 J cm^{-3} was achieved in PEI/3 vol%-HfO_2 composites at 150 °C, 500 MV m^{-1}. The difference between bandgaps of the PEI matrix and HfO_2 was calculated to be 2.6 eV, which was considered as the deep trap capable of capturing/scattering mobile charges and reducing conduction current density in polymer dielectrics. Therefore, enhanced U_e and η compared to neat PEI were realized in the PEI/HfO_2 nanocomposites under high temperatures. The Shen group [54] developed PEI/SiO_2 composites (doping ratios of 1, 3, 5, and 7 vol%, particle size of ~25 nm) and directly measured the charge behavior of polymer composites with nanoscale resolution under various temperatures (25–150 °C) by using scanning kelvin probe microscopy. It was found that at relatively high temperatures, charges had gained enough energy to overcome electrical potentials and migrate in particle-free PEI. Comparatively, charge injection and transport were greatly inhibited in PEI/5 vol%-SiO_2 composites, which enhance E_b and suppress conduction current. Consequently, a maximum U_e of 6.30 J cm^{-3} coupled with a high η of 90.5% was obtained in PEI/5 vol%-SiO_2 composites at 150 °C and 600 MV m^{-1}.

3.3.1.2 Composites Filled with Nitrides and Fluorides

Nitrides and fluorides with large bandgaps are potent candidates as insulating fillers into polymer matrices for high-temperature energy storage capacitors. Wang and coworkers [38] designed flexible, high-temperature *c*-BCB-based polymer nanocomposites using boron nitride nanosheet (BNNS). Outstanding and stable energy storage capacities were achieved at harsh temperature conditions up to 250 °C. For example, at 150 °C and 400 MV m^{-1}, the U_e exceeded 2.2 J cm^{-3} with a η

of >90%. The incorporation of BNNS greatly suppressed electrical conduction loss, which is of great significance for energy storage capacitors where thermally and electrically stimulated charges are effectively impeded under high temperatures. Moreover, Young's modulus and Weibull breakdown strength were concurrently improved upon the incorporation of BNNS. More encouragingly, the presence of 10 vol% BNNS improved the thermal conductivity of the *c*-BCB matrix by about 8 times. According to phase-field simulations, the BNNS-doped nanocomposites exhibited much better heat dissipation capability than those unfilled commercial high-T_g polymers under high electric fields. The corresponding capacitors displayed improved operational reliability under electrical and thermal extremes. Pan and coworkers [55] chose BNNS as doping fillers and fabricated PEI/BNNS composites (doping ratios of 2, 4, 6, and 8 vol%, layer thickness ~2.7 nm) through the solution casting method. A much higher E_b (701.2 MV m^{-1}) was achieved for the nanocomposite with an optimal doping ratio, which compared favorably to that of pure PEI (550.4 MV m^{-1}). This improvement was due to the incorporated BNNS with high electric resistivity and wide bandgap, which efficiently released electrode-oriented electric fields, suppressed charge injection from electrodes, and decreased current density. The resulting PEI/4 vol%- BNNS exhibited the optimized discharge energy density (U_e of 3.43 J cm^{-3}) and charge–discharge efficiency (η of ~90.1%) under 150 °C, 500 MV m^{-1}. Li and coworkers [56] synthesized PI-PAA copolymer through thermal imidization, followed by doping with BNNS (doping ratios of 0.05, 0.1, 0.2, and 0.3 vol%). The tiny-quantity doping showed a great effect on the energy storage performance of polymer nanocomposites. An optimized U_e of 7.4 J cm^{-3} at 150 °C, 527 MV m^{-1} was reached in the PI-PAA/0.1 vol%-BNNS (ratio of PI:PAA is 0.87 : 0.13). In addition, the cycling stability of thus-synthesized composite film was evaluated. After 20 000 charging–discharging cycles, there was almost no degradation in energy storage properties.

Besides BNNS, wide-bandgap fluorides have also been demonstrated as effective doping fillers for improving the energy storage properties of high-T_g dielectric polymers. The Wang group [57] synthesized calcium fluoride (CaF_2) nanoparticles with an ultrawide bandgap (~12.1 eV) and a moderate K (~10). PI/CaF_2 nanocomposites were fabricated with filler contents of 1, 3, 5, and 7 vol%. The incorporation of the CaF_2 nanofillers simultaneously increased K and E_b. As a result, at 150 °C, PI composite film with 3 vol% CaF_2 exhibited an improved U_e of 2.68 J cm^{-3} at 455.4 MV m^{-1}. Moreover, the addition of other fluorides (e.g. LaF_3 and MgF_2) into the PI matrix also gave rise to enhanced U_e and η at elevated temperatures.

Wang and coworkers [30] compared the energy storage characteristics of PI-based composites doped with different inorganic insulating fillers (Al_2O_3, HfO_2, TiO_2, BNNS) (Figure 3.6a). As shown in Figure 3.6b, the addition of nanofillers increased the K of PI while retaining their tanδ at relatively low values (≤0.005). The ratio of the increase in K of the nanocomposites followed the trend of fillers' K values, that is, $TiO_2 > HfO_2 > Al_2O_3 >$ BNNS. Meanwhile, collective enhancement in E_b and electrical resistivity was achieved in these PI-based nanocomposites (Figure 3.6c,d). The nanocomposites displayed enhanced U_e and η than those of neat PI. The combination of a large bandgap and a moderate K associated with Al_2O_3 and

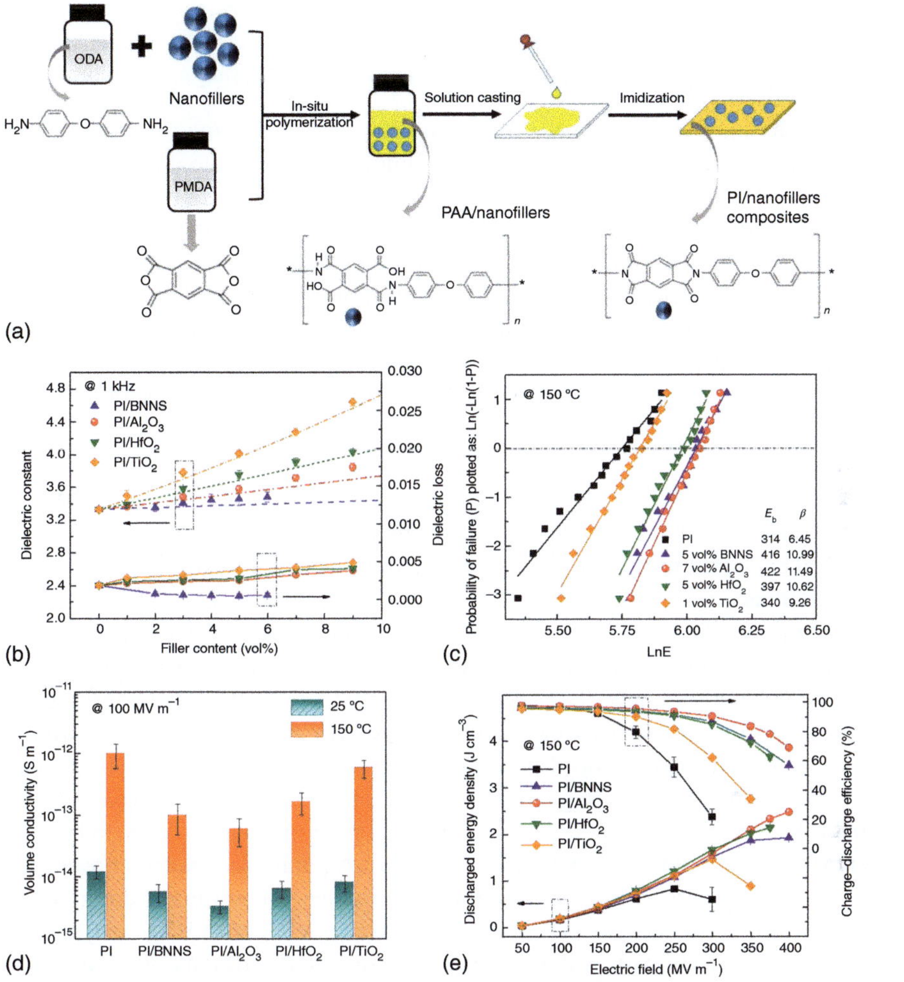

Figure 3.6 (a) Schematic of in situ preparation of the PI-based polymer nanocomposites, in which pyromellitic dianhydride (PMDA) and 4, 4-diaminodiphenyl ether (4,4′-ODA) were condensed in the presence of inorganic nanofillers including Al_2O_3, HfO_2, TiO_2, and BNNS. (b) Dielectric constant and loss of the PI nanocomposites as a function of filler content at 25 °C and 1 kHz. The dashed lines represent the calculated effective dielectric constant K. (c) Weibull breakdown strength of PI and the PI nanocomposites with the optimized filler contents measured at 150 °C, in which β is the shape parameter evaluating the scatter of the breakdown data. (d) Volume conductivity under 100 MV m^{-1} of PI and the PI nanocomposites at 25 and 150 °C. (e) Discharged energy density and charge–discharge efficiency of PI and the PI nanocomposites with 5 vol% BNNS, 7 vol% Al_2O_3, 5 vol% HfO_2, and 1 vol% TiO_2 at 150 °C [30]. Source: Reproduced from Ai et al. [30] with permission from John Wiley and Sons.

HfO_2 fillers contributed synergistically to the superior capacitive performance. The highest U_e (~2.4 J cm^{-3} at 400 MV m^{-1}) was obtained in PI/Al_2O_3 composites with an optimized filler content of 7 vol% under 150 °C (Figure 3.6e).

3.3.2 Composites Filled with Conductive or Semi-Conductive Fillers

Besides designing polymer-based nanocomposites with insulating fillers, another strategy is to incorporate conductive or semi-conductive fillers into polymer dielectric matrices. For one thing, the large K of such fillers can significantly increase the K of polymer composites, thus showing great potential in improving energy storage capacity. For another, the E_b of polymer composites can be enhanced under certain filler ratios. This is primarily due to the traps generated at organic/inorganic interfaces, provided that the filler ratio is much lower than the percolation threshold of the polymer composites. Percolation theory describes the effective electric conductivity of the composites (σ_{eff}) near the insulator–conductor transition as follows:

$$\begin{aligned} f_c > f_{nano} &: \sigma_{eff} \propto (f_c - f_{nano})^{-s'} \\ f_c < f_{nano} &: \sigma_{eff} \propto (f_{nano} - f_c)^{t'} \end{aligned} \tag{3.15}$$

where f_c is the percolation threshold, f_{nano} is the volume fraction of nanofiller, t' is the critical exponent in the conductive phase, and s' is the critical exponent in the insulation phase.

Nanoscale metals such as Cu, Al, Ag, and Ni nanoparticles have been introduced into different polymer matrices to form nanocomposites with enhanced energy storage performances. Salmeron and coworkers [58] fabricated polystyrene (PS)/Au nanocomposites with oleylamine-capped Au nanoparticles by solution processing. PS composites doped with tiny contents of Au nanoparticles showed large enhancement in E_b (e.g. 576.6 MV m^{-1} of PS/0.02 vol% Au composites versus 395.2 MV m^{-1} of neat PS) and obtained an improved U_e of 4.76 J cm^{-3} at 600 MV m^{-1} at room temperature. Similarly, Ag nanodots were chosen as conductive nanofillers in another study [59], in which the optimally configured poly(vinylidene fluoride-*co*-hexafluoropropylene) [P(VDF-HFP)]/Ag nanocomposite (with 0.05 vol% Ag) exhibited greatly increased E_b (674 MV m^{-1}) and U_e (~26.6 J cm^{-3} at 735 MV m^{-1}) at room temperature. Low-content metal nanodot doping strategies also greatly improved high-temperature energy storage capabilities in polymer nanocomposites. Wang and coworkers [60] reported hierarchically structured PC-based nanocomposites with sputtered Au nanoparticles. It was found that Au nanodots enhanced K from 2.85 of pure PC to 3.49 of PC/Au nanocomposite that was fabricated with a sputtering time of 100 s, while the composites retained a low tanδ of less than 2% across a wide temperature range from 20 to 200 °C. The E_b was enhanced due to charge-capturing capability of introduced deep traps and the coulomb blockade effect of Au nanodots on charge transportation, which were desired for achieving high U_e (6.25 J cm^{-3} at 550 MV m^{-1}) under a high temperature of 150 °C.

MXene is a 2-D inorganic compound synthesized from MAX (M for transition metal; A for main group elements such as Al, Ga, and Si; X for C or N), which was first reported and fabricated in 2011 [61]. MXene has an electrical conductivity similar to transition metal carbides and possesses surface-OH groups, which ensures its good compatibility with most polymer matrices. Dang and coworkers [62] used surface-modified MAX fillers to improve energy storage characteristics of polymer dielectrics and fabricated PI/MAX nanocomposites (doping ratios of 1, 3, 5, and 7 wt%). The addition of surface-modified MAX nanosheets improved K (3.02 for pure PI versus 3.7 for PI/5 wt%-MAX) and E_b (~299 MV m^{-1} for pure PI versus ~320 MV m^{-1} for PI/1 wt%-MAX) of PI-based nanocomposites at 150 °C. A maximum U_e of 1.19 J cm^{-3} accompanied by 61.2% of η was achieved in the PI-based nanocomposite filled with 1 wt% MAX nanosheets. Also, the high thermal conductivity of 2D MAX nanosheets was beneficial for thermal conduction and heat dissipation of the polymer/MAX composites operating under high electric fields. Chen and coworkers [63] also adopted the strategy of doping carbonitride and fabricated a library of PI/MXene composites with doping ratios of 0.5, 1, 2, and 3 wt%. The maximum U_e of ~2 J cm^{-3} was achieved in PI/0.5 wt% MXene composites at 400 MV m^{-1}, 150 °C. The large surface area and the heterojunction structure of oxidized MXene contributed to the enhancement of the E_b of the PI matrix. The difference in electron affinity between TiO_2 (decorated on MXene surface) and PI acted as electron traps to capture injected charges, which could form an inverse electric field for inhibiting charge transport and reducing charge losses of the nanocomposites.

Compared to conducting fillers, semi-conductive fillers typically exhibit lower K and losses. In addition, polymer composites filled with semi-conductive fillers usually have higher threshold values of electric conductivity than those filled with conducting fillers. As the most commonly used high-K semi-conductive ceramic filler, $BaTiO_3$ has a great effect on increasing the K of dielectric polymers, and it has received extensive attention. Shen and coworkers [64] chose 0-D $BaTiO_3$ nanoparticles as doping fillers and prepared PI-based nanocomposites with improved K. The K can be enhanced from ~2.8 of pure PI to ~7.2 of PI/9 vol%-$BaTiO_3$ under elevated temperatures from 25 to 150 °C, while $\tan\delta$ was reduced (0.045 for pure PI versus ~0.01 for PI/9 vol%-$BaTiO_3$) at 150 °C because of restrained chain mobility by $BaTiO_3$ fillers. However, the addition of $BaTiO_3$ decreased E_b and U_e compared to neat PI. In the following work from the same group [29], 1-D $BaTiO_3$ nanofibers were prepared via electrospinning. PI/$BaTiO_3$ nanofiber composites were fabricated by a solution casting and thermal imidization process. Aside from improving K and reducing dielectric loss, 1-D $BaTiO_3$ nanofibers greatly improved E_b and high-temperature U_e compared to $BaTiO_3$ nanoparticles. As a result, the maximum U_e of 2.98 J cm^{-3} was achieved in PI/1 vol%-$BaTiO_3$ at 150 °C and 400 MV m^{-1}. The interfaces between nanofillers and polymer matrix were capable of trapping mobile charges to reduce charge migration, while the in-plane distributed nanofibers induced charge transport in the in-plane direction to decrease the out-of-plane electric conduction current of the PI nanocomposites. $SrTiO_3$ is another semi-conductive ceramic filler that has raised research interests. Pan and coworkers [65] fabricated PEI/$SrTiO_3$ nanocomposites with filler contents of 1, 3, 5, 7, and 10 vol%. The K increased with the addition of $SrTiO_3$ nanoparticles. The highest U_e of 3.18 J cm^{-3} along with a η of 85% were achieved in PI/5 vol%-$SrTiO_3$ composites at 400 MV m^{-1} under 150 °C. Besides, improved thermal conductivity and faster thermal diffusion were observed in such PI-based nanocomposites due to the higher thermal conductivity of $SrTiO_3$ nanoparticles (~10 W m^{-1} K^{-1}) than that of PEI matrix (~0.22 W m^{-1} K^{-1}).

3.3.3 Composites Co-filled with Both Insulating and Semi-Conductive Fillers

Different strategies of doping insulating or semi-conductive fillers into polymer matrix have shown positive effects on improving the energy storage capabilities of polymer-based nanocomposites. However, K and E_b generally show a negative correlation in conventional polymer nanocomposites. High E_b can be obtained in polymer/insulating-filler composites, while K is limited due to the relatively small K of insulators. On the contrary, incorporating semi-conductive fillers into polymer matrices has a better effect in increasing K but is more likely to decrease E_b, especially when the filler loading is high.

To solve this dilemma and overcome restrictions on enhancing the U_e of polymer nanocomposites, Wang and coworkers [66] designed a class of PEI-based nanocomposite consisting of two different fillers with complementary functionalities: insulating BNNSs with high dielectric strength and semi-conducting $BaTiO_3$ nanoparticles with large electric polarization. The concurrent enhancements in both K and E_b were achieved in the resulting ternary PEI/$BaTiO_3$/BNNS nanocomposites with optimized filler feeding ratios, along with a low tanδ (<0.01) across a wide range of temperatures. An enhanced E_b of 547 MV m^{-1} under 150 °C and an increased K of 3.74 at 10^3 Hz were obtained in PEI/1.27 vol%-$BaTiO_3$/6.05 vol%-BNNS composites. As a result, an U_e of 2.92 J cm^{-3} can be achieved at 450 MV m^{-1}, 150 °C.

3.4 Core@Shell-Structured Nanofillers for Polymer Composites

To date, nanofillers have exhibited great effectiveness in enhancing the electric properties and energy storage characteristics of polymer-based nanocomposites. Yet, several existing problems limit the further development and mass production of polymer nanocomposites. For instance, the agglomeration of nanoparticles weakens local dielectric properties and causes poor compatibility between the organic and inorganic interfaces. In addition, electric field distortion at filler/matrix interfaces occurs as a result of the divergence of K (under alternating current conditions) or the difference in electrical conductivity (under direct current conditions) in different phases. A promising strategy to balance the advantages and drawbacks of nanofiller doping is to design core@shell-structured nanofillers with specific functions for polymer composites. Both organic and inorganic materials have attracted research interest in their use as coating shells onto inorganic nanoparticle cores.

3.4.1 Organic Shells for Inorganic Nanoparticles

To improve compatibility between inorganic nanoparticles and organic polymer matrix, a feasible method is surface grafting of nanoparticles to construct core–shell structures. Li and coworkers [67] designed interface-modulated nanoparticles by tethering a 10 nm PP-g-mah layer onto surface-modified MgO nanoparticles (~150 nm), which were subsequently used to fabricate PP/MgO/PP-g-mah nanocomposites. The fabrication of nanoparticles with different interface designs and the corresponding polymer nanocomposites are shown in Figure 3.7a (the TEM figures are shown in Figure 3.7b–d. The polar anhydride groups of PP-g-mah contributed to the enhancement in K (2.47 in PP/MgO/PP-g-mah nanocomposites with filler content of 3 wt% versus 2.22 in pristine PP) at a high temperature of 120 °C, while tanδ was reduced from 0.086% to 0.06%. The E_b was enhanced, and a maximum U_e of 1.66 J cm^{-3} could be achieved in PP/MgO/PP-g-mah nanocomposites at 400 MV m^{-1}, 120 °C with the η of >90%. Deep traps were directly measured through Kelvin probe force microscopy, indicating that the interface modification brings about more local traps in PP/MgO/PP-g-mah nanocomposites, which is beneficial for restricting charge transport and enhancing electric resistivity under high temperatures.

Inspired by spider silk, Zhang and coworkers [68] utilized BCB as a "bridging agent" at the interface between poly(aryl ether sulfone) (DPAES) polymer matrix and BNNS nanofillers. Both the high T_g (~180 °C) and polar -SO_2- groups of DPAES contributed to the improved high-temperature dielectric properties (K ~3.7–4.0 and tanδ of 0.006 at 150 °C) of DPAES/BNNS@BCB nanocomposites. An U_e of 4.2 J cm^{-3} and a η of ~86% were obtained in the DPAES nanocomposites with core@shell-structured BNNS@BCB nanofillers (10 vol%) at 500 MV m^{-1}, 150 °C. Such outstanding energy storage capacity could be attributed to the suppression effect of locally cross-linked bonds on polymer chain motions and inhibition of charge hopping processes under high temperatures and high electric fields. In addition, the covalent bonding between the BNNS and the DPAES served as deep traps that could reduce electric conduction. In another work from the same group [69], the BNNS core was replaced by $BaTiO_3$ (BT) with a larger K value. The resulting DPAES/BT@BCB nanocomposites have smaller tanδ values compared to those of DPAES/BT nanocomposites due to improved nanoparticle dispersion in the polymer matrix. An enhanced E_b of 442 MV m^{-1} was obtained in optimized core–shell structured composites with locally cross-linked organic–inorganic interfaces. At the same time, a maximum U_e of 3.1 J cm^{-3} was achieved under 400 MV m^{-1}, 150 °C. It was concluded that improved dispersion of nanoparticles, limited polymer chain motion, and a strengthened interface contributed synergistically to reduce electrical conduction loss and enhance energy storage capabilities of DPAES/BT@BCB composites under high electric fields and high temperatures. In a follow-up study [70], DPAES/BT@BCB nanocomposites with different BT particle sizes were fabricated. It was found that the use of 100 nm-sized $BaTiO_3$ particles gave the highest U_e under 150 °C. Weaker electric field distortion in DPAES/100 nm-BT@BCB nanocomposites, as derived from finite element simulation, was attributed to being responsible for the enhanced E_b.

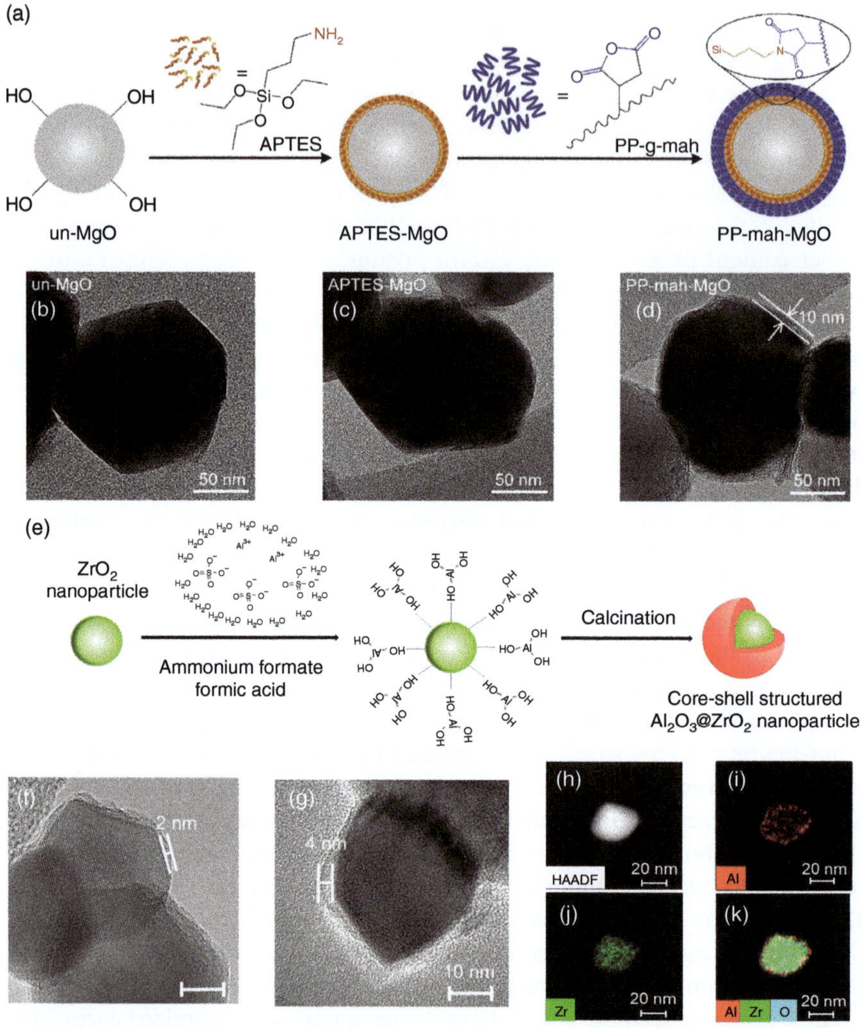

Figure 3.7 (a) Schematic of the preparation of APTES-MgO and PP-mah-MgO nanoparticles. The MgO nanoparticles (un-MgO, ~150 nm in diameter) were firstly modified with 3-aminopropyltriethoxysilane (APTES), yielding the amino-functionalized MgO nanoparticles (termed as APTES-MgO). TEM images of (b) un-MgO, (c) APTES-MgO, and (d) PP-mah-MgO nanoparticles [67] (e) Schematic of the synthesis of the core@shell-structured $Al_2O_3@ZrO_2$ nanoparticles. TEM of (f) $Al_2O_3@ZrO_2-1$ and (g) $Al_2O_3@ZrO_2-2$ nanoparticle. (h) High-angle annular dark field (HAADF) image of $Al_2O_3@ZrO_2$ nanoparticle. (i–k) EDS of Al, Zr, and superposition of Al, Zr, and O elements in the nanoparticle [33]. Source: Zhou et al. [67]/reproduced with permission from Elsevier. Ren et al. [33]/reproduced from John Wiley & Sons.

The same group [71] also employed poly (aryl ether ketone) (DPAEK) as polymer matrix and fabricated cross-linked *c*-DPAEK/BN@BCB nanocomposites. *c*-DPAEK/BN@BCB nanocomposite exhibited an U_e of 3.83 J cm^{-3} and a η of 90.6% at 500 MV m^{-1}, 150 °C. According to thermally stimulated depolarization current (TSDC) results, the nanocomposites containing core@shell-structured BN@BCB nanofillers showed much deeper traps, implying that cross-linking at the interface between organic polymer matrix and inorganic fillers is effective in inhibiting conductive loss, thus increasing both U_e and η of polymer dielectrics.

3.4.2 Inorganic Shells for Inorganic Nanoparticles

Aside from issues of nanofiller aggregation and compatibility between phases, the large contrast in K between inorganic fillers and organic polymer matrix may cause severe local electric field distortion and bring about potential risks of partial discharge or even dielectric breakdown. The design of polymer nanocomposites with core@shell nanoparticles with inorganic shells has been proposed and demonstrated. Materials with moderate K were chosen as surface buffer shells for mitigating the contrast of K and homogenizing field distribution between inorganic cores and polymer matrices.

Wang and coworkers [33] proposed all-inorganic core–shell structured nanofillers of ZrO_2@Al_2O_3. The nanoparticles with a core diameter of ~25 ± 5 nm and a shell thickness of ~2 nm were prepared via a sol–gel method followed by a post-annealing process (Figure 3.7e). The formation of core–shell nanoparticles was supported by TEM (Figure 3.7f,g, in high-angle annular dark field (HAADF) mode) and EDS studies (Figure 3.7h–k). An improved E_b of 585 MV m^{-1} was obtained in PEI/11 vol%-ZrO_2@Al_2O_3 composites at a high temperature of 150 °C, which showed ~33% and ~12% improvement compared to pristine PEI and PEI/11 vol%-ZrO_2 composites, respectively. The PEI/11 vol%-ZrO_2@Al_2O_3 composites delivered the highest U_e of 5.19 J cm^{-3} with a η of >80% at 525 MV m^{-1}, 150 °C. Upon the incorporation of inorganic shells, the maximum electric field near the ZrO_2@Al_2O_3 nanoparticle (i.e. 213 MV m^{-1}) was ~15% lower than that around the ZrO_2 nanoparticle (i.e. 245 MV m^{-1}), which in turn inhibited electrical tree initiation from the most vulnerable region. Also, the presence of the Al_2O_3 shell layer in the core@shell nanoparticles was found to be capable of inducing more trap sites at ZrO_2–Al_2O_3 and Al_2O_3–PEI interfaces, which effectively lowered hopping conduction current and inhibited space charge accumulation. In another work by Wang and coworkers [72], core@shell-structured $BaTiO_3$@SiO_2 nanofibers (with SiO_2 layer thickness of ~10 nm) were synthesized through electrospinning and a modified sol–gel method. The PI/$BaTiO_3$@SiO_2 composites were fabricated accordingly. It was proposed that the wide bandgap SiO_2 shell introduced deep traps, and such deep traps could restrain the Schottky injection process, further forming a charge-accumulated region near the electrodes and establishing a built-in electric field in the opposite direction to the applied field. A recent study [73] reported an improved energy storage capacity in a class of PEI nanocomposites filled with rationally designed core@shell-structured PEI/Ag@Al_2O_3 nanofibers, which exhibited the highest U_e of 6.1 J cm^{-3} and a high η of 89.2% at 413.4 MV m^{-1} under 150 °C. The K was enhanced with the addition of Ag@Al_2O_3 nanofibers owing to additional interfacial polarizations. In addition, the E_b was improved from 344.2 MV m^{-1} of pure PEI to 414.8 MV m^{-1} of the nanocomposite with the optimal filler loading.

The above examples demonstrate the proof-of-principle for designing high-temperature energy storage polymer dielectrics by doping polymers with core@shell-structured nanoparticles. Due to their improved physical compatibility and the introduction of deep traps at interfaces, polymer nanocomposites incorporating novel inorganic nanostructures are promising for high-performance energy storage dielectrics, especially under high temperatures and high electric fields. In the next section, another prevailing method to improve the energy storage capacities of polymer dielectrics, i.e. building layered polymer composite films, will be thoroughly discussed.

3.5 Layered Polymer Composites

For improving energy storage capabilities under high temperatures, researchers have explored different approaches to increase the dielectric constant, enhance electrical insulation strength (e.g. increasing E_b and electric resistivity), and reduce energy loss (including dielectric loss and conduction loss). Achieving simultaneous optimization of the aforementioned parameters is quite challenging; thus, most of the existing research has focused on improving one or two specific dielectric parameters to develop high-performing polymer dielectrics. Taking electric resistivity as an example, restriction of migrating charges and inhibition of conduction current have been widely accepted as important measures for enhancing E_b and η. The approach of nanofiller doping into polymer matrices has been demonstrated to be effective in restricting charge transport, as summarized and discussed in Sections 3.3 and 3.4. Another promising way is to introduce functional layers onto or/and between dielectric polymer layers to inhibit charge injection from electrode/polymer interfaces or repulse "internal" transport at dielectric/dielectric interfaces. This section will summarize different strategies for constructing layered polymer composites and their effectiveness in enhancing the energy storage performance of polymer dielectrics at elevated temperatures.

3.5.1 Polymer Films with Inorganic Layers

In order to improve U_e under high temperatures, coating inorganic layers with a wide bandgap and high thermal conductivity has raised great research interest recently. Wang and coworkers [74] modified the surface of PEI films using chemical vapor deposited (CVD) hexagonal BN nanosheets. The thickness of the BN layer was dependent on the growth time of CVD, e.g. ~5 nm after 20 minutes of growth on a Cu substrate. The energy loss derived from *D–E* loops greatly decreased from 74.2% in pristine PEI to 10.8% in coated BN/PEI/BN films under 150 °C. Meanwhile, electric resistivity in coated PEI film was more than one order of magnitude higher than that in pure PEI. The introduced coating layers exhibited improved capability to block injected charge from electrodes (including Schottky emission and tunneling) owing to the formed potential barrier at metal/BN and BN/PEI interfaces. BN has also been used as surface coating materials in PC-based films [75] by Chi and coworkers. The BN film was first prepared by electrospinning, and BN-coated PC films were then fabricated through hot pressing. The high U_e of 3.39 J cm^{-3} was achieved in the symmetrical 1 μm-BN/PC/1 μm-BN film under 500 MV m^{-1}, 150 °C. The surface BN layer showed a positive effect in obstructing leakage current and improving heat dissipation, along with an increase in dielectric constant due to higher K of BN (4–5) compared to the PC matrix (~3). Yet, the effectiveness of surface coating was thickness-dependent. With a further increase in surface layer thickness, the leakage current increased rapidly, and the energy storage performance became worse than that in pure PC. Such performance degradation was attributed to the increase in defects in the coating layer, which rendered the polymer composite vulnerable to high temperatures and high electric fields. The Cao group [76] adopted montmorillonite (MMT) nanosheets as the inorganic layers for PAI films. With a doping ratio of 40 wt% per layer, the 4-layer-coated MMT/PAI/MMT nanocomposites exhibited the best U_e of 2.9 J cm^{-3} along with increased efficiency under elevated electric fields at 150 °C. The optimized MMT content led to impressive enhancement of the electrical insulating strength of the coated PAI film, thanks to MMT's large bandgap and high resistivity, which were beneficial for inhibiting charge injection from the electrodes. In another study from the same group [77], a self-assembled nanocoating approach was applied to fabricate polymer films with surface-coated MMT nanosheets. Three commercially available polymer films (BOPP, PI, and low-density polyethylene (LDPE) were coated with MMT-containing layers. A remarkable improvement in E_b for all three types of polymer films was achieved. The barrier height of the metal–nanocoating–polymer junction was calculated to be larger than that of the original metal–polymer interface, which was effective in suppressing charge injection and limiting tunneling across energy barriers. Studies of nanoscale charge behavior through KPFM suggested that MMT's relatively high surface conductivity could facilitate charge dissipation along the in-plane direction, reinforcing the insulation strength of dielectric polymers.

Apart from inorganic surface coatings, sandwich-structured polymeric films with a functional interlayer are also effective in improving the dielectric and energy storage performance of polymer dielectrics. For example, Chi and coworkers [78] designed layered PEI film composites containing an inorganic MgO layer (with different MgO doping ratios of 1, 3, 5, and 7 wt%). The intercalated MgO layer was capable of repulsing leakage current across the composite film. The PEI/MgO/PEI composite with 5 wt% MgO exhibited a 13% enhancement in E_b (573.1 MV m^{-1}) when compared to that of pure PEI (506.7 MV m^{-1}) at 150 °C. Also, the K value increased from ~3.15 of pristine PEI to ~3.5 of the PEI/MgO/PEI upon the introduction of the MgO layer with a K of ~9.8. Consequently, an U_e of 4.13 J cm^{-3}, together with a high η of over 90%, was obtained at 500 MV m^{-1}, 150 °C. The formed interfacial potential barrier was associated with the difference in Fermi level between PEI and MgO, which was beneficial for blocking charge transport. Meanwhile, a local electric field was built due to the different charged states of the PEI matrix and MgO layer, which was inverse to the applied external electric field and efficient in blocking electrons. In another work, Wang and coworkers [79] introduced a middle layer of graphene fluoride (GF) nanosheet between two cross-linked aramid nanofiber (PANF) polymer layers to form PANF/GF/PANF composite films. The rationally designed structure combined the advantages of outer PANF layers (e.g. high K of ~6.7 at 10^5 Hz, 200 °C) and middle GF layers (high E_b of 1000 MV m^{-1} and outstanding Young's modulus of 0.3 TPa), which endowed PANF/GF/PANF composite with an extraordinary U_e of 11.7 J cm^{-3} under harsh conditions of 400 MV m^{-1}, 200 °C. Additionally, η reaches 87.2% under the same testing condition, owing to the reduced conduction loss enabled by the charge-blocking GF layer. Such a coating strategy guides the fabrication of high-performance polymer-based energy storage capacitors operating under harsh conditions.

In addition to three-layered polymer composites, multi-layered high-temperature capacitor dielectric films have also raised great interest. Wang and coworkers [80] fabricated metal-oxide/PI/metal-oxide/PI/metal-oxide laminated films through a layer-by-layer solution casting method, in which Al_2O_3, ZrO_2, and MgO were chosen as three wide-bandgap inorganic materials for reinforcing coatings. The 5-layer PI-based composites consisting of one middle layer and two surface layers of Al_2O_3 exhibited an order of magnitude decrease in electric conduction current and 70% enhancement in E_b compared to pristine PI at 200 °C (i.e. 438.6 MV m^{-1} versus 257.3 MV m^{-1}). The reduced conduction loss and enhanced E_b contributed to an improved U_e of ~2.2 J cm^{-3} and a η of ~80% at 350 MV m^{-1}, 200 °C, which far exceeded the performance of three-layer Al_2O_3/PI/Al_2O_3 structured composite film. This novel multilayer design successfully realized the simultaneous suppression of charge injection from electrodes and mobile charges inside the composite. Deng and coworkers [81] adopted a similar design concept and prepared multilayer structured composite films with alternating PI and BNNS layers. All films with different coating layers showed similar energy storage performance at room temperature. The 5-layer PI/BNNS/PI/BNNS/PI dielectric film exhibited an U_e of 3.98 J cm^{-3} and a η of ~85% at 500 MV m^{-1}, 150 °C. The multiple embedded BNNS layers could significantly restrict the migration of electrons and lower leakage current density under high temperatures and high electric fields. It was also revealed that multilayered PI-BNNS composites showed a better inhibition effect on leakage current than PI composites with randomly dispersed BNNS. Chi and coworkers [82] designed PC-based multilayer composites with two different coating materials, SiO_2 and BN. The fabricated SiO_2/BN/PC/BN/SiO_2 composite films showed a high U_e of ~5 J cm^{-3} under 600 MV m^{-1}, 150 °C, which could operate 50 000 continuous charge and discharge cycles without failure at 100 °C, 200 MV m^{-1}. Specifically, this design selects SiO_2 as the outside layer directly contacting metal electrodes, which has a wider bandgap than BN (8.3 eV vs. 5.56 eV). The higher injection barrier at the metal–SiO_2 interface was thus in place to restrict charge injection. Simultaneously, BN was chosen as the second layer to form a built-in inverse field at the SiO_2–BN interface, further inhibiting charge transfer between inorganic layers. Through coating inorganic barrier layers with different energy band structures on the surface of the polymer film, carrier migration was effectively regulated to achieve higher energy storage performance.

To verify the generality of surface coating on native polymers [34], the plasma-enhanced chemical vapor deposition (PECVD) and roll-to-roll processing methods were utilized for the facile production of SiO_2-coated polymer films based on different polymer platforms (BOPP, PEI, PEN, PI, PC, and FPE). The SiO_2 coating layer (~180 nm) enhanced high-temperature energy storage capacities of all polymer films. Further analysis on SiO_2/BOPP/SiO_2 films revealed that the higher LUMO level and lower HOMO level of SiO_2 relative to BOPP enabled larger energy gaps for electrons and holes (4.3 and 4.7 eV in SiO_2 versus 3. and 1.3 eV in BOPP). The increase in energy barrier height proved to be effective in inhibiting space charge injection. As a result, the charge density decreased from 4.11 to 1.08 C m^{-3} at 120 °C after being coated with SiO_2 layers. Another work reported by Tan and coworkers [83] compared the effect of Al_2O_3 coating on BOPP and PEI films through the atomic layer deposition (ALD) method. The 80 nm-Al_2O_3 layers effectively improved the energy storage capacities of polymer-based composites. The resultant PP/ Al_2O_3 film showed a maximum U_e of 4.1 J cm^{-3} (600 MV m^{-1}, 140 °C), and the PEI/ Al_2O_3 film showed an optimized U_e of ~2.5 J cm^{-3} (600 MV m^{-1}, 150 °C). The improvement in dielectric strength was also associated with the increased Young's modulus of dielectric polymer films after adding alumina coatings. Li and coworkers [84] investigated the effect of various oxide coating layers, including SiO_2, Al_2O_3, HfO_2, and TiO_2, on high-temperature energy storage performance of PEI-based composites. According to the energy band structures at interfaces of metal/coating layers, SiO_2 with the highest bandgap (8.08 eV) formed the largest electron injection barrier height of 3.28 eV. In comparison, the smallest barrier height at Au/TiO_2 interface was calculated to be 0.3 eV. Among all the oxide layer-coated PEI films, the Al_2O_3/PEI/Al_2O_3 sample exhibited the best high-temperature energy storage capability (e.g. U_e of ~4 J cm^{-3}, η of ~80% at 550 MV m^{-1} and 150 °C). The study suggested that parameters such as bandgap, K, and electrical conductivity of the deposition layers should be collectively considered when designing high-performance capacitor films with inorganic coatings.

3.5.2 Sandwich-Structured Nanocomposites

In addition to coating and intercalating inorganic layers for dielectric polymer films, multilayered polymer composite films with nanofillers spatially distributed within constituent layers also showed improved performance relative to native polymers. Wang and coworkers [85] designed sandwich-structured BCB nanocomposites using *c*-BCB/BNNS as outer layers and *c*-BCB/$BaTiO_3$ as a middle layer (SEM image shown in Figure 3.8a). The incorporation of BNNSs in the outer layers efficiently impeded charge injection from electrodes, blocked the current path, and reduced electrical conduction. Meanwhile, the introduced $BaTiO_3$ nanoparticles in the middle layer incurred a much-improved K. The coupled effect of different layers in

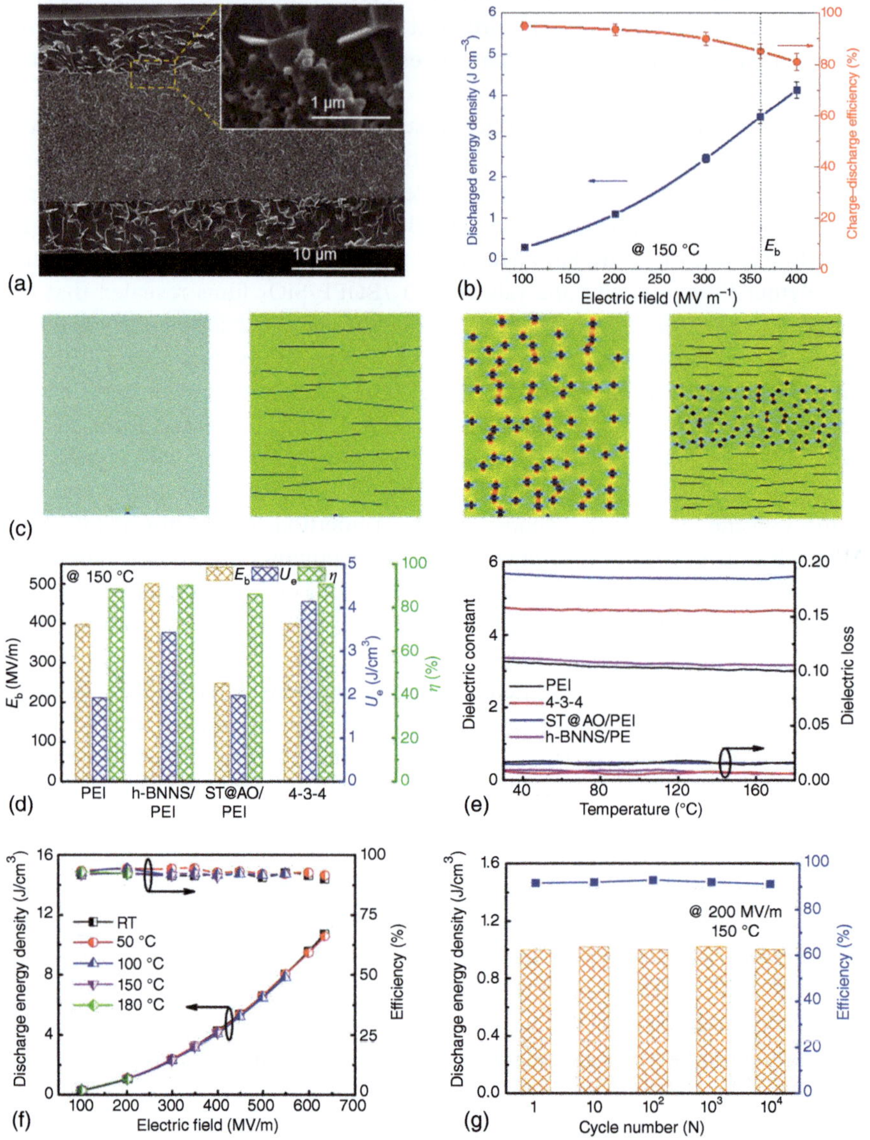

Figure 3.8 (a) Cross-sectional SEM images of SSN-25. (b) U_e and η of SSN-25 at 150 °C as a function of electric field. [85] (c) The electric field distribution (at the initial point of electrical tree growth) of PEI, 4 vol% BNNS/PEI, 3 vol% ST@AO/PEI, and the sandwich-structured 4–3–4 nanocomposite. (d) Summary of maximum E_b, discharged energy density, and efficiency of single-layer PEI, BNNS/PEI, ST@AO/PEI, and sandwich-structured 4–3–4 nanocomposites at 150 °C. (e) The temperature dependence of the dielectric constant and dielectric loss of single-layer PEI, BNNS/PEI, ST@AO/PEI, and sandwich-structured 4–3–4 nanocomposites. (f) U_e and η of the 4–3–4 nanocomposites at various temperatures. (g) U_e and η of 4–3–4 (1–10 000 cycles at 150 °C and 200 MV m^{-1}) as a function of cycle number [86]. Source: Li et al. [85]/reproduced from National Academy of Science. Reproduced from Chen et al. [86] with permission from the Royal Society of Chemistry.

sandwich-structured composites accounted for the great dielectric and capacitive performance at high temperatures. As a result, the highest U_e of ~4 J cm^{-3} at 150 °C was achieved in the triple-layered composite with the optimized doping ratio (10 vol% BNNS for outer layers and 25 vol% $BaTiO_3$ for middle layer) (Figure 3.8b).

In a recently published work [86], similar strategies were adopted to construct (PEI/BNNS)/(PEI/$SrTiO_3$)/(PEI/BNNS) composites (structure and electric field distribution as shown in Figure 3.8c) through solution casting. The sandwich-structured polymer behaved well in balancing the two opposing parameters of K and E_b under high temperatures (Figure 3.8d,e), with BNNSs in the outer composite layers inhibiting charge injection and $SrTiO_3$ nanoparticles in the inner layer contributing to an improved K. The tri-layer composite film with the optimized doping ratios of 4 vol% BNNS in the outer layers and 3 vol% $SrTiO_3$ in the middle layer yielded an improved U_e of 4.14 J cm^{-3} at 400 MV m^{-1}, 150 °C (Figure 3.8f). The low energy loss and high mechanical properties (e.g. modulus of 0.94 GPa) of the sandwich-structured nanocomposites enabled remarkable cyclability. The U_e of the sandwich composite film remained virtually unchanged (~1 J cm^{-3}) after 10 000 cycles with η of >90% under an applied electric field of 200 MV m^{-1} at a high temperature of 150 °C (Figure 3.8g). Zhang and coworkers [87] designed (DPAES/BN)/(DPAES/BT)/(DPAES/BN) composites, in which BCB groups were adopted as organic shells grafted on BN and BT fillers. The sandwich-structured composites yielded a great U_e of 4.64 J cm^{-3} and a η of 84% at 150 °C, 500 MV m^{-1}. It was postulated that the introduced dielectric transition interlayer played the role of "dielectric gradient", mitigating the local electric field distribution and blocking electric tree growth. Also, the mechanical strength of overall composites was improved due to the interlayer cross-linked networks, which were beneficial for enhancing electric resistance against electromechanical failure.

3.6 Novel Polymers and All-Organic Polymer Composites

The above-mentioned strategies, including doping functional nanofillers and constructing layered composites based on commercial high-T_g polymer matrices such as PI, PEI, and PC, have been proven to be effective in improving high-temperature dielectric properties and energy storage capabilities. Yet, challenges still exist for industrial-scale mass production considering materials costs and laborious synthesis steps, especially for those with complex designs with multiple doping fillers and coating layers (even with promising lab data of U_e and η). A thriving research focus has been directed toward all-organic polymer composites, spanning novel polymers, all-organic polymer composites with organic fillers, or sandwiched composite films with different organic layers. In this chapter, progress on novel polymers as well as all-organic polymer composites will be summarized and analyzed to provide a broader perspective in designing extraordinarily high-temperature dielectric capacitors. In addition, all-organic layered polymer composites will be discussed, along with underlying mechanisms and design strategies for improving energy storage capabilities.

3.6.1 Novel Polymers

3.6.1.1 Structural Modification of Commercial High-T_g Polymers

A direct way of designing novel polymers for high-temperature energy storage capacitors is to modify the structures of commercially available high-T_g dielectrics such as PEI (T_g of ~217 °C) and PI (T_g of ~360 °C). Zha and coworkers [31] modified the imidization process of polyamic acid to generate PI with different imidization degrees (Figure 3.9a). Partial imidization led to simultaneous enhancement in K and

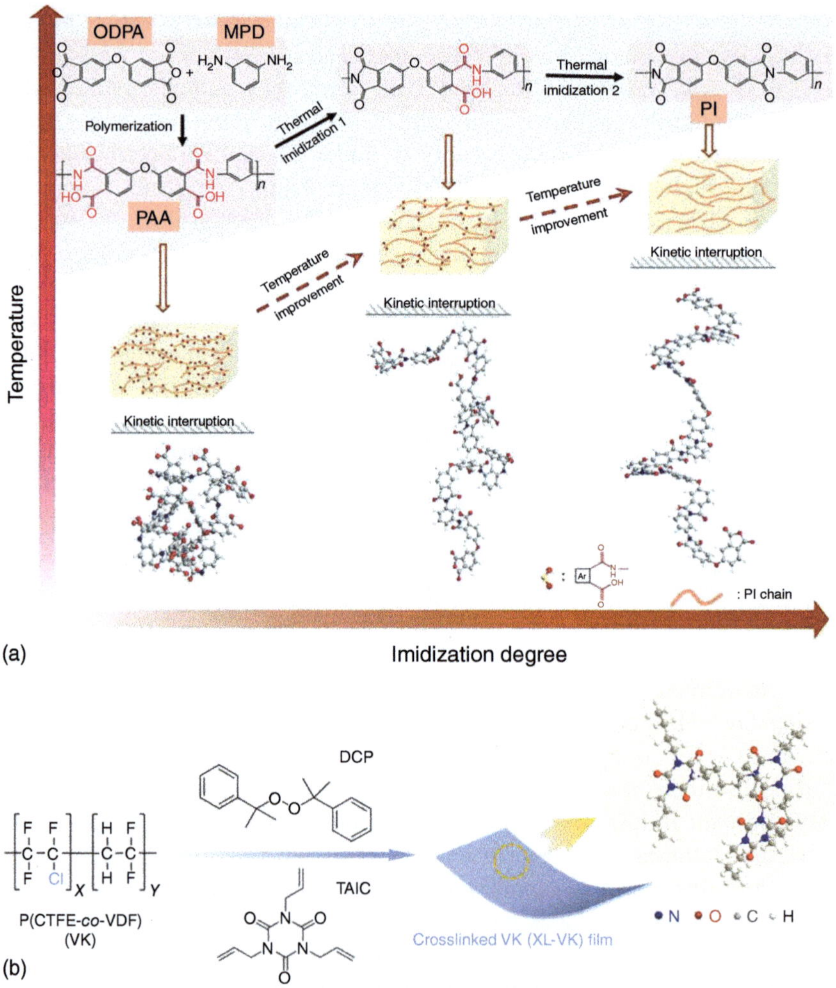

Figure 3.9 (a) Schematic of the synthesis process for the ODPA-MPD films with different degrees of imidization. [31] (b) Schematic of the preparation of XL-VK films [88]. Source: Reproduced from Liu et al. [31] with permission from The Royal Society of Chemistry. Adapted from Li et al. [88].

E_b, due to the presence of more polar amide and carboxylic acid groups that also function as deep traps. Moreover, increasing imidization degree resulted in higher T_g (from 169 °C to 290 °C with imidization degree increasing from 9% to ~100%) and thermal decomposition temperature, owing to the rising rigidity of the molecular chains and decrease in chain mobility. With an optimized imidization degree of 81%, the modified PI polymer yielded a great U_e of 3.9 J cm^{-3} at 550 MV m^{-1}, 150 °C, exceeding most commodity polymers such as PEI, PEN, and FPE. Sotzing and coworkers [89] focused on the structural modification of PEI by introducing *para*-phenylenediamine as a defect to lower the rotation energy barrier around the N (imide) – C (aromatic) bond while maintaining an almost unchanged T_g. The modified PEI with 5% of *para*-phenylenediamine exhibited a high η of over 90% at 400 MV m^{-1}, 200 °C. In addition, the synthesized polymer exhibited stable energy storage performance (η of over 90%) after 100 000 cycles of charging and discharging processes. It was suggested that the structural modification strengthened the polymer chain interactions and modified the localized states, leading to a lowered hopping distance, impeded charge carrier transport, and suppressed electric conduction.

3.6.1.2 Cross-linked Polymers

For commercial dielectric polymers, the large K and great U_e are typically accompanied by high energy loss and low η, especially under high temperature and high electric field conditions. One feasible remedy is to introduce various degrees of cross-linking in polymer films to restrict molecular chain movement and reduce dielectric loss. An earlier study by the Wang group [90] first reported the positive effect of cross-linking networks on high-temperature energy storage performance, in which P(VDF-CTFE) (VK) exhibited enhanced U_e with high η after being cross-linked with triallyl isocyanurate (TAIC). In a recent work from the same group [88], a novel polymer dielectric material was realized by melting the mixture of P(CTFE-VDF), initiator dicumyl peroxide (DCP), and cross-linker TAIC (Figure 3.9b). The cross-linked fluoropolymer (XL-VK) with optimized initiator and cross-linker concentrations exhibited an enhanced E_b of 494 MV m^{-1} at 150 °C, compared to that of pure P(CTFE-VDF) (251 MV m^{-1}). Consequently, an U_e of 2.67 J cm^{-3} with η of >90% was achieved at 400 MV m^{-1}, 150 °C, and the η remained as high as >97.7% with an applied electric field of 200 MV m^{-1} at elevated temperatures from 25 °C to 150 °C. It was postulated that the molecular trapping centers created by the cross-linked structures can effectively inhibit charge migration and reduce conduction loss at high temperatures and high electric fields. The cross-linking strategy has also been adopted for PEI-based polymers. Taking advantage of the thermally activated reactivity of phenylethynyl groups, Zhang and coworkers [91] prepared cross-linked PEIs from the 4-phenylethynylphthalic anhydride (PEPA)-terminated ether imide oligomers. The dielectric loss was significantly reduced in cross-linked PEI with the β-relaxation peak shifting at higher temperatures, resulting in increased dielectric performance compared to pure PEI. A high U_e of 3.6 J cm^{-3} and a high η of 96.5% were achieved at 500 MV m^{-1}, 150 °C when the degree of cross-linking reached 10%. The decrease in interchain spacing after cross-linking may have restricted the movement of the molecular chains and subsequently limited the β-relaxation process.

The Cheng group [92] selected epoxy resin as the base polymer and introduced the polar electronegative halogens as well as asymmetric halogen phenyl pendant groups to increase the permanent dipole moment of cross-linked polymer. The cured product of bisphenol-A epoxy resin, 3,4,5-trifluoroaniline (curing agent), and 4,4′-methylenedianiline (crosslinking agent) displayed a high U_e of 3.1 J cm^{-3} and a high η of over 93% at 100 °C. It was found that the introduction of fluorine groups effectively enhanced the K and increased the density of trapped electrons, resulting in reduced energy loss and boosted η. In another work from this group [93], a new strategy was adopted by introducing asymmetric molecular chains in cross-linked epoxy networks. The optimized structure was obtained by controlling the mole ratio of alicyclic amine (ACA) and polyether amine (PEA) in the ACA-PEA curing agent, where rigid ACA moiety accounted for high thermal stability and PEA moiety with rotatable C–O bonds provided the flexibility needed for roll-to-roll film processing. The fabricated epoxy resin film with an ACA/PEA ratio of two displayed the best U_e of ~1 J cm^{-3} and a η of over 90% at 120 °C. Besides, the great self-healing ability of this novel cross-linked epoxy resin, with a defect removal time of <10 μs after the breakdown, helped enhance the reliability and cycle life of the epoxy film capacitor.

3.6.1.3 Wide Bandgap Polymers

As mentioned before, the steep rise in leakage current and charge injection/transport mainly accounts for the deterioration of electric strength, increase in energy loss, and decrease in E_b, further leading to a drop in U_e and η. Even for materials with great thermal properties, such as PEI (with T_g of 217 °C) and PI (with T_g of ~360 °C), U_e and η exhibit obvious decay under 150 °C, which is due to the inevitable thermal-assisted Schottky injection from electrodes and accelerated electron thermal motions, as well as the increase in electric conduction loss and dielectric loss.

To date, wide bandgap polymers have shown excellent performance in achieving enhanced high-temperature dielectric strength and well-performed energy storage capabilities. The higher energy gap greatly inhibits electron excitation from the valence band and elevates charge injection barrier height. Studies on synthesizing wide-bandgap polymers have thus been extensively conducted. However, there exists an empirical inverse correlation between bandgap and T_g in most commercial high-T_g polymers because the aromatic repeat units employed in the backbone of high-T_g polymers incur lower bandgaps. Exceptions have been demonstrated by Cao and coworkers [94], who developed a series of nonaromatic polyoxafluoronorbornene (POFNB) polymers with fused bicyclic repeat units in the backbone. POFNB showed a large bandgap (4.9 eV) and high T_g (186 °C), attributable to its rigid bicyclic structure and the nonconjugated polymer backbone. Such structural features could incur great thermal stability and restrict charge migration between and along polymer chains. As a result, a high U_e of ~5.7 J cm^{-3} was achieved at 150 °C, and the conduction loss remained low even at 150 °C. To enhance the K of POFNB (K ~2.5 in the original work) and further improve its high-temperature energy storage performance, the same group [95] synthesized an isomer of POFNB, named as *m*-POFNB, by moving the trifluoromethyl substituent group from the *para*-position of the pendant phenyl groups to the *meta*-position. The decreased symmetry in *m*-POFNB was deemed responsible for stronger polarizability that led to enhanced dipolar relaxation and a higher dielectric constant ($K \sim 3$) while having little effect on the bandgap (4.84 eV) and T_g (178 °C). The nonpolar structure was beneficial for decoupling the orientation of dipoles from the segmental motion and lowering polarization loss. The highest U_e of 7.44 J cm^{-3} was obtained at 700 MV m^{-1}, 150 °C. In a recently published work from the same group [26], the substitution effect was comprehensively studied by systematically varying the location of -CF_3 on the pendant phenyl group, giving rise to *o*-POFNB (*ortho*-substituted), *m*-POFNB (*meta*-substituted), and *p*-POFNB (*para*-substituted), respectively. Similar to the earlier findings about *m*-POFNB and *p*-POFNB, the rigid and nonconjugated bicyclic backbones in *o*-POFNB resulted in a high T_g and a large bandgap. All three isomers of POFNB achieved a good balance between bandgap and T_g, among which *o*-POFNB showed the largest bandgap (~5 eV) and highest T_g (244 °C) and exhibited a high U_e of 8.3 J cm^{-3} at 800 MV m^{-1}, 150 °C. Moreover, at a condition of 200 °C and 200 MV m^{-1}, *o*-POFNB exhibited a stable U_e of 0.6 J cm^{-3} coupled with a high η over 91% after 100 000 cycles of charging and discharging.

3.6.1.4 Polymers with Polar Groups

Besides synthesizing high-T_g polymers with wide bandgaps to enhance electrical strength, another feasible way of improving U_e is to incorporate polar groups into polymers to enhance orientational polarization and increase the K. Typically, there are four types of polarization in polymer dielectrics as mentioned earlier: electronic polarization P_{elec}, ionic polarization P_{ion}, orientation polarization P_{orien} and interfacial polarization P_{inter} (also known as space charge polarization). The common approaches to enhancing P_{elec}, such as bond conjugation, inevitably lower the bandgap and decrease E_b. The increase in space charge polarization should also be avoided for high-frequency applications due to the consequent high dielectric losses, which lowers η and causes potential risks of insulation failure in the long run.

On the other hand, rationally introduced polar groups in polymers are effective in achieving both high K and restricted dielectric loss. Zhu and coworkers reported [96] the synthesis of sulfonylated poly(2,6-dimethyl-1,4-phenylene oxide) (SO_2-PPO), which showed a higher K than S-PPO (5.9 versus 3.5) due to the rotation polarization of its methylsulfonyl side groups that were introduced by oxidation of S-PPO. Stable performance of U_e was achieved at 300 MV m^{-1} at different temperatures from 20 °C (U_e ~2.5 J cm^{-3}) to 150 °C (U_e ~2.3 J cm^{-3}), while the low tanδ of 0.003 was maintained due to the high rigidity of the PPO backbone. In another example from the same group [97], the dipolar sulfonyl side groups were combined with the rigid contorted polymer backbone in polymers of intrinsic microporosity (PIMs) to achieve both high K and relatively low dielectric losses. Novel sulfonylated PIMs (SO_2-PIMs) were synthesized and cast into films from chloroform solution, displaying a promising K of 5.3–6.0 and a low $\tan\delta$ of 0.005. A high U_e of ~2.3 J cm^{-3} was achieved with a η of over 90% at 150 °C. It was proposed that large pores in PIMs were beneficial for sulfonyl side group rotation, while the rigid polymer backbone and the high-T_g ensured a relatively low dielectric loss under high temperatures. A similar strategy was adopted in another work by Zhu and coworkers [98], in which two dipolar groups ($-SO_2CH_3$ and $-CH_2CH_2CN$) were introduced as side groups of various poly(aryl ether)s. The dipolar groups played a big role in achieving desirable dielectric properties (K of ~4.4–5.6, $\tan\delta$ of ~0.003–0.004) and U_e [e.g. maximum of ~4 J cm^{-3} in cyanolated poly(ether ketone) (PEEK)-CN)] under 150 °C.

Zhang and coworkers [99] exercised a different method to prepare high-performance dielectric polymers with polar groups embedded in the backbone instead of side groups. A single-component polymer, 1,4-poly(ether fluoromethyl naphthalene amide) (PNFA) was prepared by condensing 1,4-naphthalenedioic acid and 2,2-bis[4-(4-amino phenoxy)phenyl] hexafluoropropane using the Yamazaki phosphorylation reaction. The subtle improvement lies in the curly packed chain structure, in which the larger chain spacing and shorter axial distance endow higher K and E_b, and the presence of naphthalene well suppresses the dielectric loss at low frequencies. As a result, a high K of ~4 and an enhanced E_b of 577 MV m^{-1} were achieved at 150 °C, along with an outstanding U_e of 4.5 J cm^{-3} at 150 °C, 550 MV m^{-1}.

Overall, the design of high-temperature dielectric polymers typically concerns two key aspects: enhancing E_b and increasing K, both through molecular structural tunings. Based on these, some more work has been done with the design of all-organic polymer composites by leveraging the complementary advantages of different organic components.

3.6.2 All-Organic Polymer Composites

3.6.2.1 Polymer Blends

For ideal dielectric polymers, large bandgaps typically endow great insulating properties since they require more energy to directly "excite" charge carriers from the valence band to generate current. The intrinsic E_b of polymers is defined as insulation failure under the coupling action of electrical and mechanical forces, which depends on K and Young's modulus (described by Eq. (3.3–3.11)). Yet, the presence of weak points, such as free volume and structural disorder, increases mean free path and charge carrier mobility, thereby lowering the dielectric strength of polymers. In an effort to achieve higher E_b and better energy storage performance of polymer dielectrics under high temperatures, the Zhang group [100] adopted the strategy of polymer blending and designed chain-packing-structured PI/PEI polymer blends. The electron-deficient phenyls in PI matched well with the relatively electron-rich phenyls in PEI (Figure 3.10a), which ensured great miscibility of the two polymers to form polymer alloys with more extended chain-packing morphology and increased chain-packing density (Figure 3.10b,c]). The 1 : 1 PI/PEI blend exhibited an enhanced E_b of ~520 MV m^{-1} under 150 °C compared to that

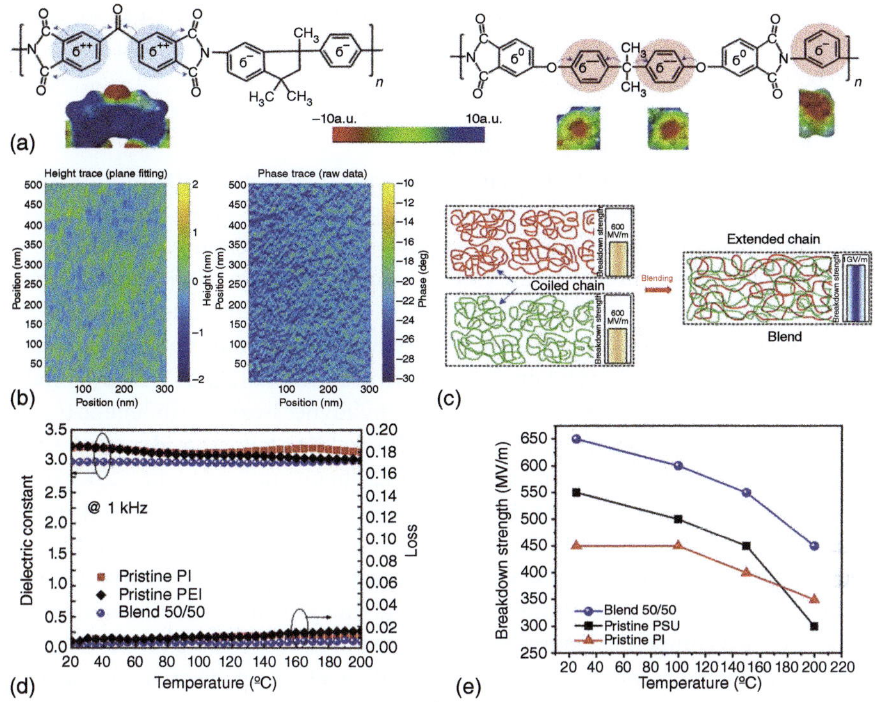

Figure 3.10 (a) Schematic of chemical structures of polyimide and polyetherimide, which illustrates the electron-deficient phenyl groups in PI and electron-rich phenyl groups in PEI, contributing to the strong electrostatic interaction between the two polymers. The color mappings indicate the electrostatic potential of the highlighted groups on the van der Waals surface. (b) Taping-mode AFM data for the PI/PEI (50/50) blend, topography (left, the color bar ranges from −3 to 3 nm), and phase (right) images of a scanning area of 300 × 500 nm. (c) Schematic illustration of the PI/PEI blend strategy in enhancing the electrostatic interaction and engineering the extended chain packing, resulting in a compact structure with fewer weak points, thus leading to a high E_b over a broad temperature range. (d) Temperature-dependent dielectric constant and tanδ at 1 kHz of PI, PEI, and the 50–50% blend. (e) Temperature dependence of breakdown strengths of the PI, PSU, and their 50–50% blend [100]. Source: Zhang et al. [100]/reproduced with permission from Elsevier.

of pristine PI (~370 MV m^{-1}) and PEI (~320 MV m^{-1}), along with an improved U_e of 1.7 J cm^{-3} and reduced tanδ of <0.01 (Figure 3.10d) at 150 °C, 500 MV m^{-1}. Another polymer blend film, poly(1,4-phenylene ether-sulfone) (PSU)/PI, showed improvement in E_b (~560 MV m^{-1} under 150 °C) as well (Figure 3.10e). In another study, Dang and coworkers [101] prepared all-organic PI/ P(VDF-HFP) blends through in situ polymerization, using PI as the polymer matrix and P(VDF-HFP) as the disperse phase. The incorporation of P(VDF-HFP) significantly improved the K of the polymer blend with only a slight increase in dielectric loss. Meanwhile, the introduction of deep traps altered the original trap distributions and greatly improved E_b. Among all, the 5 : 1 proportioned blend exhibited the best energy storage performance at 150 °C, reaching a high U_e of 2.25 J cm^{-3} at 382 MV m^{-1}, along with a notable η of 76.9%. Moreover, the polymer blend showed great cycle stability and operational reliability under harsh environments (150 °C, 200 MV m^{-1}) after 30 000 charge–discharge cycles.

3.6.2.2 Polymer Doped with Organic Fillers

As stated in Section 3.4, nanofiller doping has shown great potential in impeding charge migration and enhancing electric strength, thereby being beneficial for achieving high energy storage performance under high temperatures. Unlike most studies that focused on inorganic nanofillers, Li and coworkers [102] demonstrated the design of all-organic polymer composites with low concentrations (0.25–0.75 vol%) of organic semiconductors (three commercial molecular semiconductors: 2,2′-[[6,6,12,12-Tetrakis(4-hexylphenyl)-6,12-dihydrodithieno[2,3-d : 2′,3′-d′]-s-indaceno[1,2-b : 5,6-b′]dithiophene-2,8-diyl]bis[methylidyne(3-oxo-1H-indene-2,1(3H)-diylidene)]]bis[propanedinitrile] (ITIC), [6,6]-Phenyl C61 butyric acid methyl ester (PCBM) and 2,2′,9,9′-Tetrakis(1-pentylhexyl)-[5,5′-bianthra[2,1,9-def:6,5,10-d′e′f′]diisoquinoline]-1,1′,3,3′,8,8′,10,10′(2H,2′H,9H,9′H)-octone (DPDI) (Figure 3.11f–h). The PEI/PCBM composite exhibited concurrently high U_e of ~3 J cm^{-3} and high η of ~90% at 200 °C. Through nanoscale characterization of potential distribution at electrode/polymer interface (Figure 3.11a–e) and analysis of electrical conduction mechanisms, two competing factors were proposed to be responsible for the electric resistance of polymer composites: (i) the positive effect of deep traps formed by introducing molecular semiconductors with higher electron affinities than polymer matrix, (ii) the negative effect of charge excitation from semiconductors with relatively lower bandgaps compared to polymer matrix. At ultra-low volume concentrations, the former factor dominated, yielding improved energy storage performance for the all-organic polymer composite at an optimized composition.

3.6.2.3 All-Organic Layered Polymer Composites

The construction of all-organic layered polymer composites was conducted by Zhu and coworkers in 2011 [103]. The sandwich-structured PC/PVDF/PC composites exhibited a great U_e (~11 J cm^{-3}) at room temperature, which indicated that the layered structure and PC blocking layers prevented the charge migration from one layer to the next. Multiple all-organic layered polymer composites, such as polyethylene terephthalate (PET)/polymethyl methacrylate (PMMA)/P(VDF-HFP) [104], PC/PVDF [105, 106], and PC/PVDF [107, 108], have shown great performance in enhancing dielectric properties or/and energy storage capabilities at relatively low temperatures (<100 °C). The simultaneous improvement in E_b and electric resistivity verifies the effectiveness of designing all-organic layered polymer composites, which also demonstrates their potential in high-temperature energy storage capacitors. A recent work by Zhu group [109] reported the high-temperature electric properties of all-organic alternately layered PC/PVDF polymers. The breakdown and lifetime Weibull analyses suggested that charge injection was significantly reduced in sandwich-structured polymer composites. More importantly, HTPC/PVDF multilayer films exhibited a high η of 95% at 140 °C, 200 MV m^{-1}. Yu and coworkers [110] chose PTFE as the polymer matrix and fabricated EP/PTFE

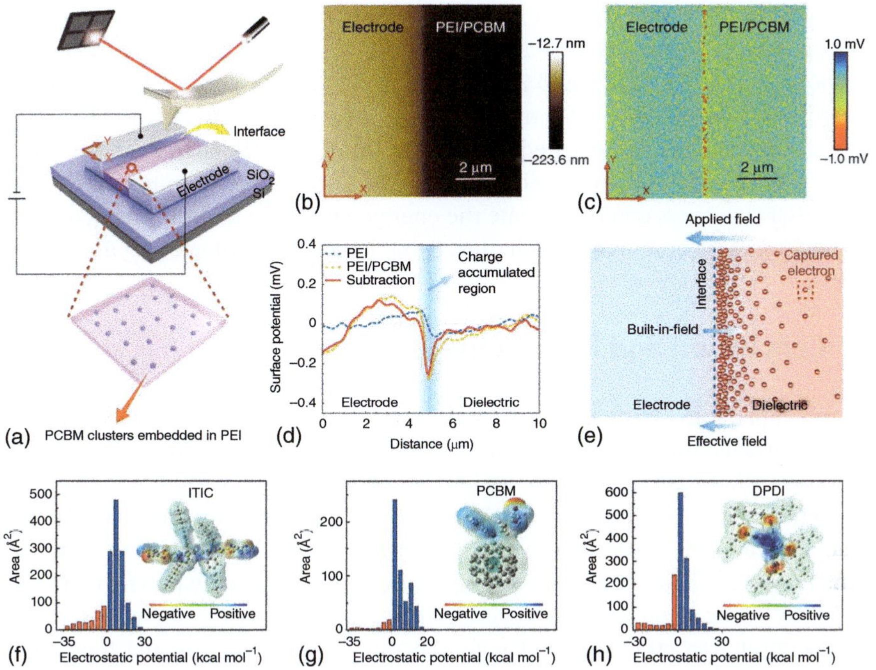

Figure 3.11 (a) Schematic diagram of the KPFM testing. The circuit is used for charge injection prior to the measurement and is disconnected during the measurement. (b) Topography of the PEI/PCBM composite. (c) Surface potential mapping of the PEI/PCBM composite. (d) Surface potential profiles along the horizontal axis. The amplitude of surface potential is averaged along the y direction in the surface potential mappings. The subtraction (red curve) of surface potential between the pristine PEI and the PEI/PCBM composite excludes the influence of the different energy levels between the electrode and the polymer. (e) Schematic illustration of the built-in electric field formed by the molecular semiconductors captured electrons. (f–h) Electrostatic potential distribution and area percentage in each electrostatic potential range of ITIC, PCBM, and DPDI [102]. Source: Yuan et al. [102]/Springer Nature/CC BY 4.0. CC BY 4.0.

layered composites, in which epoxy was able to fill in the pores at the PTFE surface. The bi-layer composites exhibited an U_e of ~1.8 J cm^{-3} with a high η of over 98% at 400 MV m^{-1}, 150 °C, proving the positive effect of surface engineering in smoothing surface electric field strength and inhibiting charge injection.

The Wang group [111] selected high-T_g polymers to fabricate PEI/PI/PEI-structured all-organic dielectric films. The AFM-IR spectra showed the unique molecular structure at PEI/PI interfaces, which was believed to be responsible for the formation of deep traps located in the interfacial region, leading to local charge accumulation and building an interfacial barrier to block charge migration. As a result, the tri-layered composite yielded an enhanced E_b of 374 MV m^{-1} and improved U_e of ~3 J cm^{-3} at 150 °C, meanwhile displaying stable performance after 100 000 cycles at 200 °C, 200 MV m^{-1}. Recently, Chi and coworkers [112] reported a novel all-organic dielectric polymer composite exhibiting high energy storage performance under high temperatures. The molecular semiconductor ITIC was chosen as the doping filler for the multilayered (PEI/ITIC)/PEI composite. The optimized PEI/(PEI/0.5 vol% ITIC)/PEI/(PEI/0.5 vol% ITIC)/PEI composites, which occupied 28, 15, 14, 15, and 28% of volume for each respective layer, showed the best-performing U_e of ~6.3 J cm^{-3} at 150 °C. It was proposed that suitable trap density can capture injected/excited electrons in the path accelerated by the electric field, contributing to the great performance of all-organic composites. Also, the spaced PEI/ITIC layers could dissipate the energy within the dielectric breakdown path and impede electric tree growth, thus being favorable for increasing E_b.

3.7 Conclusion and Perspective

Dielectric capacitors are extensively used in grid-connected energy systems and modern microelectronics. The majority of existing polymer dielectrics for capacitors, however, have yet to meet the stringent requirements for high-temperature, high-electric field electrification applications. The steep rise in thermal-assisted leakage current inevitably leads to weakened electric-withstanding ability and lower E_b, which severely limits the energy storage capability of dielectric polymers under high temperatures. Even worse, surged electric conduction loss leads to a decrease in charge–discharge efficiency and potential reliability hazards in the long run. Inspiringly, great strides have been made in the field of high-temperature polymer dielectrics over the past decade. In particular, polymer nanocomposites (usually with high-T_g polymer matrices) are widely considered by researchers as promising materials. The incorporated nanofillers are capable of scattering/capturing free charges and impeding electric conduction current within polymer dielectrics. Meanwhile, nanofillers play the role of "functional tools" for the balance or reinforcement of several key factors related to U_e (e.g. dipole moment, bandgap) and η (e.g. electric conductivity, trap distribution), resulting in enhancement in E_b or/and K of polymer nanocomposites with low dielectric losses at high temperatures. Another research hotpot is to rationally construct layered polymer composites with multiple functional layers, which has been effective in hindering current paths and modulating electrical properties to deliver enhanced high-temperature energy storage performance. Notable effort has also been devoted to the development of novel high-T_g polymers with promising intrinsic E_b and K through molecular-level engineering.

Yet, there is room for improvement as several key issues remain to be addressed. There are gaps in fundamental understanding and accurate measurement of the electrical properties of nanoscale heterogeneous interfaces in polymer nanocomposites and layered polymer composites. Moreover, the correlation between microscopic composite structures and macroscopic high-temperature electrical properties remains unclear, which obstructs the further design and exploration of high-performing polymer dielectrics. In addition, low-cost fabrication of large-scale thin films of dielectric materials for capacitors remains to be realized.

Future development of high-temperature polymer dielectrics could leverage both advanced simulations and advanced characterization techniques. Multiscale simulations, from molecular-level modeling to mesoscopic phase-field simulation and macroscopic finite element simulation, are powerful tools for understanding fundamental processes associated with dielectric energy storage, such as charge transport, heat dissipation, self-clearing, and electrical breakdown. Theoretical insights have to be validated by experimental feedback, which requires advanced characterization techniques that are applicable to various kinds of interfaces, sometimes buried, with the desirable temporal and spatial resolution. The combined computational and characterization efforts will inform the rational design of next-generation dielectric polymers and polymer nanocomposite for high-temperature, high-electric field electrostatic film capacitor applications.

References

1 Li, C., Yang, Y., Xu, G. et al. (2022). Insulating materials for realising carbon neutrality: opportunities, remaining issues and challenges. *High Voltage* 7: 610.

2 Wang, F., Harindintwali, J.D., Yuan, Z. et al. (2021). Technologies and perspectives for achieving carbon neutrality. *The Innovation* 2: 100180.

3 Carrasco, J.M., Franquelo, L.G., Bialasiewicz, J.T. et al. (2006). Power-electronic systems for the grid integration of renewable energy sources: a survey. *IEEE Transactions on Industrial Electronics* 53: 1002.

4 Gur, T.M. (2018). Review of electrical energy storage technologies, materials and systems: challenges and prospects for large-scale grid storage. *Energy & Environmental Science* 11: 2696.

5 Sarjeant, W.J., Zirnheld, J., and MacDougall, F.W. (1998). Capacitors. *IEEE Transactions on Plasma Science* 26: 1368.

6 Tan, Q., Irwin, P., and Cao, Y. (2006). Advanced dielectrics for capacitors. *IEEJ Transactions on Fundamentals and Materials* 126: 1153.

7 Chu, B., Zhou, X., Ren, K. et al. (2006). A dielectric polymer with high electric energy density and fast discharge speed. *Science* 313: 334.

8 Yang, M., Li, Q., Zhang, X. et al. (2022). Surface engineering of 2D dielectric polymer films for scalable production of high-energy-density films. *Progress in Materials Science* 128: 100968.

9 Feng, Q.-K., Zhong, S.-L., Pei, J.-Y. et al. (2022). Recent progress and future prospects on all-organic polymer dielectrics for energy storage capacitors. *Chemical Reviews* 122: 3820.

10 Armand, M. and Tarascon, J.-M. (2008). Building better batteries. *Nature* 451: 652.

11 Simon, P. and Gogotsi, Y. (2008). Materials for electrochemical capacitors. *Nature Materials* 7: 845.

12 Barshaw, E.J., White, J., Chait, M.J. et al. (2007). High Energy Density (HED) Biaxially-Oriented Poly-Propylene (BOPP) capacitors for pulse power applications. *IEEE Transactions on Magnetics* 43: 223.

13 Li, H., Zhou, Y., Liu, Y. et al. (2021). Dielectric polymers for high-temperature capacitive energy storage. *Chemical Society Reviews* 50: 6369.

14 Zha, J.-W., Zheng, M.-S., Fan, B.-H., and Dang, Z.-M. (2021). Polymer-based dielectrics with high permittivity for electric energy storage: a review. *Nano Energy* 89: 106438.
15 Dang, Z.-M., Yuan, J.-K., Yao, S.-H., and Liao, R.-J. (2013). Flexible nanodielectric materials with high permittivity for power energy storage. *Advanced Materials* 25: 6334.
16 Palneedi, H., Peddigari, M., Hwang, G.-T. et al. (2018). High-performance dielectric ceramic films for energy storage capacitors: progress and outlook. *Advanced Functional Materials* 28: 1803665.
17 Burn, I. and Smyth, D.M. (1972). Energy storage in ceramic dielectrics. *Journal of Materials Science* 7: 339.
18 Wang, G., Lu, Z., Li, Y. et al. (2021). Electroceramics for high-energy density capacitors: current status and future perspectives. *Chemical Reviews* 121: 6124.
19 Ducharme, S. (2009). An inside-out approach to storing electrostatic energy. *ACS Nano* 3: 2447.
20 Ho, J.S., Greenbaum, S.G., and Appl, A.C.S. (2018). Polymer capacitor dielectrics for high temperature applications. *ACS Applied Materials & Interfaces* 10: 29189.
21 Li, H., Liu, F., Fan, B. et al. (2018). Nanostructured ferroelectric-polymer composites for capacitive energy storage. *Small Methods* 2: 1700399.
22 Guo, M., Jiang, J., Shen, Z. et al. (2019). High-energy-density ferroelectric polymer nanocomposites for capacitive energy storage: enhanced breakdown strength and improved discharge efficiency. *Materials Today* 29: 49.
23 Zhu, L. and Wang, Q. (2012). Novel ferroelectric polymers for high energy density and low loss dielectrics. *Macromolecules* 45: 2937.
24 Fan, B., Liu, F., Yang, G. et al. (2018). Dielectric materials for high-temperature capacitors. *IET Nanodielectrics* 1: 32.
25 Li, Q., Yao, F.-Z., Liu, Y. et al. (2018). High-temperature dielectric materials for electrical energy storage. *Annual Review of Materials Research* 48: 219.
26 Deshmukh, A.A., Wu, C., Yassin, O. et al. (2022). Flexible polyolefin dielectric by strategic design of organic modules for harsh condition electrification. *Energy & Environmental Science* 15: 1307.
27 Liu, X.-J., Zheng, M.-S., Chen, G. et al. (2022). High-temperature polyimide dielectric materials for energy storage: theory, design, preparation and properties. *Energy & Environmental Science* 15: 56.
28 Yang, M., Ren, W., Guo, M., and Shen, Y. (2022). High-energy-density and high efficiency polymer dielectrics for high temperature electrostatic energy storage: a review. *Small* 18: 2205247.
29 Hu, P., Sun, W., Fan, M. et al. (2018). Large energy density at high-temperature and excellent thermal stability in polyimide nanocomposite contained with small loading of $BaTiO_3$ nanofibers. *Applied Surface Science* 458: 743.
30 Ai, D., Li, H., Zhou, Y. et al. (2020). Tuning nanofillers in in situ prepared polyimide nanocomposites for high-temperature capacitive energy storage. *Advanced Energy Materials* 10: 1903881.

31 Liu, X., Zheng, M., Wang, G. et al. (2022). High energy density of polyimide films employing an imidization reaction kinetics strategy at elevated temperature. *Journal of Materials Chemistry A* 10: 10950.

32 Li, H., Ai, D., Ren, L. et al. (2019). Scalable polymer nanocomposites with record high-temperature capacitive performance enabled by rationally designed nanostructured inorganic fillers. *Advanced Materials* 31: 1900875.

33 Ren, L., Li, H., Xie, Z. et al. (2021). High-temperature high-energy-density dielectric polymer nanocomposites utilizing inorganic core–shell nanostructured nanofillers. *Advanced Energy Materials* 11: 2101297.

34 Zhou, Y., Li, Q., Dang, B. et al. (2018). A scalable, high-throughput, and environmentally benign approach to polymer dielectrics exhibiting significantly improved capacitive performance at high temperatures. *Advanced Materials* 30: 1805672.

35 Tan, D.Q. (2020). Review of polymer-based nanodielectric exploration and film scale-up for advanced capacitors. *Advanced Functional Materials* 30: 1808567.

36 Pan, J., Li, K., Chuayprakong, S. et al. (2010). High-temperature poly(phthalazinone etherketone) thin films for dielectric energy storage. *ACS Applied Materials & Interfaces* 2: 1286.

37 Luo, S., Yu, J., Ansari, T.Q. et al. (2020). Elaborately fabricated polytetrafluoroethylene film exhibiting superior high-temperature energy storage performance. *Applied Materials Today* 21: 100882.

38 Li, Q., Chen, L., Gadinski, M. et al. (2015). Flexible high-temperature dielectric materials from polymer nanocomposites. *Nature* 523: 576.

39 Li, H., Ren, L., Zhou, Y. et al. (2020). Recent progress in polymer dielectrics containing boron nitride nanosheets for high energy density capacitors. *High Voltage* 5: 365.

40 Huang, X., Jiang, P., and Tanaka, T. (2011). A review of dielectric polymer composites with high thermal conductivity. *IEEE Electrical Insulation Magazine* 27: 8.

41 Chen, J., Huang, X., Zhu, Y., and Jiang, P. (2017). Cellulose nanofiber supported 3D interconnected BN nanosheets for epoxy nanocomposites with ultrahigh thermal management capability. *Advanced Functional Materials* 27: 1604754.

42 Chen, Y., Zhang, H., Chen, J. et al. (2022). Thermally conductive but electrically insulating polybenzazole nanofiber/boron nitride nanosheets nanocomposite paper for heat dissipation of 5G base stations and transformers. *ACS Nano* 16: 14323.

43 Chen, X., Wu, K., Zhang, Y. et al. (2022). Tropocollagen-inspired hierarchical spiral structure of organic fibers in epoxy bulk for 3D high thermal conductivity. *Advanced Materials* 34: 2206088.

44 Yao, Y., Sun, J., Zeng, X. et al. (2018). Construction of 3D skeleton for polymer composites achieving a high thermal conductivity. *Small* 14: 1704044.

45 Pan, G., Yao, Y., Zeng, X. et al. (2017). Learning from natural nacre: constructing layered polymer composites with high thermal conductivity. *ACS Applied Materials & Interfaces* 9: 33001.

46 Lewis, T.J. (1994). Nanodielectrics. *IEEE Transactions on Dielectrics and Electrical Insulation* 1: 812.

47 Liu, P., Xie, Z., Pang, X. et al. (2022). Space charge behavior in epoxy-based dielectrics: progress and perspective. *Advanced Electronic Materials* 8: 2200259.

48 Fan, M., Hu, P., Dan, Z. et al. (2020). Significantly increased energy density and discharge efficiency at high temperature in polyetherimide nanocomposites by a small amount of Al_2O_3 nanoparticles. *Journal of Materials Chemistry A* 8: 24536.

49 Ren, W., Pan, J., Dan, Z. et al. (2020). High-temperature electrical energy storage performances of dipolar glass polymer nanocomposites filled with trace ultrafine nanoparticles. *Chemical Engineering Journal* 420: 127614.

50 Fu, J., Yang, M., Wang, R. et al. (2022). Improvement of high-temperature energy storage performance in polymer dielectrics by nanofillers with defect spinel structure. *Materials Today Energy* 29: 101101.

51 Zhu, L., Zhang, Y., Xu, W. et al. (2022). Crosslinked polyetherimide nanocomposites with superior energy storage achieved via trace Al_2O_3 nanoparticles. *Composites Science and Technology* 223: 109421.

52 Wang, P., Guo, Y., Zhou, D. et al. (2022). High-temperature flexible nanocomposites with ultra-high energy storage density by nanostructured MgO fillers. *Advanced Functional Materials* 32: 2204155.

53 Ren, L., Yang, L., Zhang, S. et al. (2021). Largely enhanced dielectric properties of polymer composites with HfO_2 nanoparticles for high-temperature film capacitors. *Composites Science and Technology* 201: 108528.

54 Sun, B., Hu, P., Ji, X. et al. (2022). Excellent stability in polyetherimide/SiO_2 nanocomposites with ultrahigh energy density and discharge efficiency at high temperature. *Small* 18: 2202421.

55 Chen, H., Pan, Z., Wang, W. et al. (2021). Ultrahigh discharge efficiency and improved energy density in polymer-based nanocomposite for high-temperature capacitors application. *Composites Part A: Applied Science and Manufacturing* 142: 106266.

56 Dai, Z., Bao, Z., Ding, S. et al. (2022). Scalable polyimide-poly(amic acid) copolymer based nanocomposites for high-temperature capacitive energy storage. *Advanced Materials* 34: 2101976.

57 Li, L., Dong, J., Hu, R. et al. (2022). Wide-bandgap fluorides/polyimide composites with enhanced energy storage properties at high temperatures. *Chemical Engineering Journal* 435: 135059.

58 Pieters, P.F., Lainé, A., Li, H. et al. (2022). Multiscale characterization of the influence of the organic–inorganic interface on the dielectric breakdown of nanocomposites. *ACS Nano* 16: 6744.

59 Huang, X., Zhang, X., Ren, G. et al. (2019). Non-intuitive concomitant enhancement of dielectric permittivity, breakdown strength and energy density in percolative polymer nanocomposites by trace Ag nanodots. *Journal of Materials Chemistry A* 7: 15198.

60 Li, S., Dong, J., Niu, Y. et al. (2022). Enhanced high-temperature energy storage properties of polymer composites by interlayered metal nanodots. *Journal of Materials Chemistry A* 10: 18773.

61 Naguib, M., Kurtoglu, M., Presser, V. et al. (2011). Two-dimensional nanocrystals produced by exfoliation of Ti_3AlC_2. *Advanced Materials* 23: 4248.

62 Feng, Q., Dong, Q., Zhang, D. et al. (2022). Enhancement of high-temperature dielectric energy storage performances of polyimide nanocomposites utilizing surface functionalized MAX nanosheets. *Composites Science and Technology* 218: 109193.

63 Yu, S., Ding, C., Liu, Y. et al. (2022). Enhanced breakdown strength and energy density over a broad temperature range in polyimide dielectrics using oxidized MXenes filler. *Journal of Power Sources* 535: 231415.

64 Sun, W., Lu, X., Jiang, J. et al. (2017). Dielectric and energy storage performances of polyimide/$BaTiO_3$ nanocomposites at elevated temperatures. *Journal of Applied Physics* 121: 244101.

65 Miao, W., Chen, H., Pan, Z. et al. (2021). Enhancement thermal stability of polyetherimide-based nanocomposites for applications in energy storage. *Composites Science and Technology* 201: 108501.

66 Li, H., Ren, L., Ai, D. et al. (2020). Ternary polymer nanocomposites with concurrently enhanced dielectric constant and breakdown strength for high-temperature electrostatic capacitors. *InfoMat* 2: 389.

67 Zhou, Y., Yuan, C., Wang, S. et al. (2020). Interface-modulated nanocomposites based on polypropylene for high-temperature energy storage. *Energy Storage Materials* 28: 255.

68 Xu, W., Liu, J., Chen, T. et al. (2019). Bioinspired polymer nanocomposites exhibit giant energy density and high efficiency at high temperature. *Small* 15: 1901582.

69 Liu, J., Shen, Z., Xu, W. et al. (2020). Interface-strengthened polymer nanocomposites with reduced dielectric relaxation exhibit high energy density at elevated temperatures utilizing a facile dual crosslinked network. *Small* 16: 2000714.

70 Liu, J., Zhang, Y., Wang, Z. et al. (2020). Optimizing electric field distribution via tuning cross-linked point size for improving the dielectric properties of polymer nanocomposites. *Nanoscale* 12: 12416.

71 Jin, L., Liu, J., Zhu, L. et al. (2021). Crosslinked poly (aryl ether ketone)/boron nitride nanocomposites containing a stable chemical bonding structure as high temperature dielectrics. *Composites Science and Technology* 213: 108949.

72 Dong, J., Hu, R., Niu, Y. et al. (2022). Enhancing high-temperature capacitor performance of polymer nanocomposites by adjusting the energy level structure in the micro-/ meso-scopic interface region. *Nano Energy* 99: 107314.

73 Lin, Q., Hou, Y., Zheng, Y. et al. (2022). Poly(ether imide)-based nanocomposites with low fraction of hierarchical Ag@AO nanofiber for high-temperature energy storage. *ACS Applied Energy Materials* 5: 2329.

74 Azizi, A., Gadinski, M.R., Li, Q. et al. (2017). High-performance polymers sandwiched with chemical vapor deposited hexagonal boron nitrides as scalable high-temperature dielectric materials. *Advanced Materials* 29: 1701864.

75 Liu, G., Zhang, T., Feng, Y. et al. (2020). Sandwich-structured polymers with electrospun boron nitrides layers as high-temperature energy storage dielectrics. *Chemical Engineering Journal* 389: 124443.

76 Wang, Y., Li, Z., Wu, C. et al. (2022). Polyamideimide dielectric with montmorillonite nanosheets coating for high-temperature energy storage. *Chemical Engineering Journal* 437: 135430.

77 Zhang, B., Liu, J., Ren, M. et al. (2021). Reviving the "Schottky" barrier for flexible polymer dielectrics with a superior 2D nanoassembly coating. *Advanced Materials* 33: 2101374.

78 Zhang, T., Yang, L., Zhang, C. et al. (2022). Polymer dielectric films exhibiting superior high-temperature capacitive performance by utilizing an inorganic insulation interlayer. *Materials Horizons* 9: 1273.

79 Vu, M.C., Kang, H., Park, P.J. et al. (2022). Scalable graphene fluoride sandwiched aramid nanofiber paper with superior high-temperature capacitive energy storage. *Chemical Engineering Journal* 444: 136504.

80 Dong, J., Hu, R., Xu, X. et al. (2021). A facile in situ surface-functionalization approach to scalable laminated high-temperature polymer dielectrics with ultrahigh capacitive performance. *Advanced Functional Materials* 31: 2102644.

81 Zhang, K., Ma, Z., Fu, Q., and Deng, H. (2022). Multi-layered boron nitride/polyimide high-temperature capacitor dielectric film. *Materials Today Energy* 29: 101093.

82 Liu, G., Lei, Q., Feng, Y. et al. (2022). High-temperature energy storage dielectric with inhibition of carrier injection/migration based on band structure regulation. *InfoMat* 5: e12368.

83 Wu, X., Liu, Y., Lin, X. et al. (2022). Atomic layer deposition coated polymer films with enhanced high-temperature dielectric strength suitable for film capacitors. *Surfaces and Interfaces* 28: 101686.

84 Cheng, S., Zhou, Y., Li, Y. et al. (2021). Polymer dielectrics sandwiched by medium-dielectric-constant nanoscale deposition layers for high-temperature capacitive energy storage. *Energy Storage Materials* 42: 445–453.

85 Li, Q., Liu, F., Yang, T. et al. (2016). Sandwich-structured polymer nanocomposites with high energy density and great charge–discharge efficiency at elevated temperatures. *Proceedings of the National Academy of Sciences of the United States of America* 113: 9995.

86 Chen, H., Pan, Z., Cheng, Y. et al. (2022). Ultrahigh charge–discharge efficiency and high energy density of a high-temperature stable sandwich-structured polymer. *Journal of Materials Chemistry A* 10: 1579.

87 Zhao, D., Cai, Q., Zhu, X. et al. (2022). Multilayer dielectric nanocomposites with cross-linked dielectric transition interlayers for high-temperature applications. *ACS Applied Materials & Interfaces* 14: 42531.

88 Li, H., Gadinski, M.R., Huang, Y. et al. (2020). Crosslinked fluoropolymers exhibiting superior high-temperature energy density and charge–discharge efficiency. *Energy & Environmental Science* 13: 1279.

89 Alamri, A., Wu, C., Mishra, A. et al. (2022). Improving the rotational freedom of polyetherimide: enhancement of the dielectric properties of a commodity high-temperature polymer using a structural defect. *Chemistry of Materials* 34: 6553.

90 Khanchaitit, P., Han, K., Gadinski, M.R. et al. (2013). Ferroelectric polymer networks with high energy density and improved discharged efficiency for dielectric energy storage. *Nature Communications* 4: 2845.

91 Zhang, Y., Liu, Z., Zhu, L. et al. (2020). Enhanced discharged efficiency and high energy density at elevated temperature in polymer dielectric via manipulating relaxation behavior. *CCS Chemistry* 2: 1169.

92 Mao, J., Wang, S., Cheng, Y. et al. (2022). High energy storage density and efficiency achieved in dielectric films functionalized with strong electronegative asymmetric halogen-phenyl groups. *Chemical Engineering Journal* 444: 136331.

93 Chen, S., Meng, G., Kong, B. et al. (2020). Asymmetric alicyclic amine-polyether amine molecular chain structure for improved energy storage density of high-temperature crosslinked polymer capacitor. *Chemical Engineering Journal* 387: 123662.

94 Wu, C., Deshmukh, A.A., Li, Z. et al. (2020). Flexible temperature-invariant polymer dielectrics with large bandgap. *Advanced Materials* 32: 2000499.

95 Wu, C., Deshmukh, A.A., Yassin, O. et al. (2021). Flexible cyclic-olefin with enhanced dipolar relaxation for harsh condition electrification. *Proceedings of the National Academy of Sciences of the United States of America* 118: e2115367118.

96 Zhang, Z., Wang, D.H., Litt, M.H. et al. (2018). High-temperature and high-energy-density dipolar glass polymers based on sulfonylated poly(2,6-dimethyl-1,4-phenylene oxide). *Angewandte Chemie, International Edition* 57: 1528.

97 Zhang, Z., Zheng, J., Premasiri, K. et al. (2020). High-j polymers of intrinsic microporosity: a new class of high temperature and low loss dielectrics for printed electronics. *Materials Horizons* 7: 592.

98 Huang, W., Ju, T., Li, R. et al. (2023). High-κ and high-temperature dipolar glass polymers based on sulfonylated and cyanolated poly(arylene ether)s for capacitive energy storage. *Advanced Electronic Materials* 9: 2200414.

99 Zhou, C., Xu, W., Zhang, B. et al. (2022). Curly-packed structure polymers for high-temperature capacitive energy storage. *Chemistry of Materials* 34: 2333.

100 Zhang, Q., Chen, X., Zhang, B. et al. (2021). High-temperature polymers with record-high breakdown strength enabled by rationally designed chain-packing behavior in blends. *Matter* 4: 2448.

101 Feng, Q., Liu, D., Zhang, Y. et al. (2022). Significantly improved high-temperature charge-discharge efficiency of all-organic polyimide composites by suppressing space charges. *Nano Energy* 99: 107410.

102 Yuan, C., Zhou, Y., Zhu, Y. et al. (2020). Polymer/molecular semiconductor all-organic composites for high-temperature dielectric energy storage. *Nature Communications* 11: 3919.

103 Mackey, M., Schuele, D.E., Zhu, L. et al. (1954). Reduction of dielectric hysteresis in multilayered films via nanoconfinement. *Macromolecules* 2012: 45.

104 Yin, K., Zhou, Z., Schuele, D.E. et al. (2016). Effects of interphase modification and biaxial orientation on dielectric properties of poly(ethylene terephthalate)/poly(vinylidenefluoride-co-hexafluoropropylene) multilayer films. *ACS Applied Materials & Interfaces* 8: 13555.

105 Chen, X., Tseng, J., Treufeld, I. et al. (2017). Enhanced dielectric properties due to space charge-induced interfacial polarization in multilayer polymer films. *Journal of Materials Chemistry C* 5: 10417.

106 Liu, G., Feng, Y., Zhang, T. et al. (2021). High-temperature all-organic energy storage dielectric with the performance of self-adjusting electric field distribution. *Journal of Materials Chemistry A* 9: 16384.

107 Yin, K., Zhang, J., Li, Z. et al. (2019). Polymer multilayer films for high temperature capacitor application. *Journal of Applied Polymer Science* 136: 47535.

108 Chen, X., Li, Q., Langhe, D. et al. (2020). Achieving flat-on primary crystals by nanoconfined crystallization in high-temperature polycarbonate/poly(vinylidene fluoride) multilayer films and its effect on dielectric insulation. *ACS Applied Materials & Interfaces* 12: 44892.

109 Ju, T., Chen, X., Langhe, D. et al. (2022). Enhancing breakdown strength and lifetime of multilayer dielectric films by using high temperature polycarbonate skin layers. *Energy Storage Materials* 45: 494.

110 Luo, S., Ansari, T.Q., Yu, J. et al. (2021). Enhancement of dielectric breakdown strength and energy storage of all-polymer films by surface flattening. *Chemical Engineering Journal* 412: 128476.

111 Niu, Y., Dong, J., He, Y. et al. (2022). Significantly enhancing the discharge efficiency of sandwich-structured polymer dielectrics at elevated temperature by building carrier blocking interface. *Nano Energy* 97: 107215.

112 Feng, M., Feng, Y., Zhang, C. et al. (2022). Ultrahigh energy storage performance of all-organic dielectrics at high-temperature by tuning the density and location of traps. *Materials Horizons* 9: 3002.

4

Review on High-Temperature Polymers for Cable Insulation: State-of-the-Art and Future Developments

Youcef Kemari[1,2], Guillaume Belijar[2], Zarel Valdez-Nava[1], Frédéric Forget[3], and Sombel Diaham[1]

[1] *Université de Toulouse, CNRS, INPT, UPS, LAPLACE Institute, 118 Route de Narbonne, Toulouse 31062, France*
[2] *Institut de Recherche et Technologie (IRT) Saint Exupéry, 3 Rue Tarfaya, Toulouse 31400, France*
[3] *Central Research and Technology – 1XRE, Airbus Operations S.A.S, 18 rue Marius Terce, Toulouse 31025, France*

4.1 Brief History of Cables Development and Insulating Materials

According to many, the first recorded use of an insulated cable or wire dates back to 1812 when Baron Schilling invented rubber-varnish-insulated wires to detonate mineral ores at St. Petersburg, Russia [1]. In the middle of the nineteenth century, trans-1,4-polyisoprene (a natural latex material also known as Gutta-percha) and rubber materials were extensively used for telegraph cables for more than three decades [2, 3]. Afterwards, vulcanized rubber was introduced in 1860 as insulation for cables. It had better insulating properties than those of other materials in that period [1, 3]. In 1880, Oliver Heaviside introduced and patented the coaxial cable design that opened the door for the development of telegraph and telephone cables [4].

Thereafter, the need to expand and improve transmission networks of telephone, telegraph as well as electric lighting systems was the driving force behind the development of insulated power cables in the second half of the nineteenth century [5]. In 1882, Thomas Alva Edison introduced his first electric lighting network in New York City. However, the first insulating materials mentioned above were not satisfactory for lighting feeders due to aging problems at relatively high temperatures. Edison decided to design underground DC cables operated at 110 V [6, 7]. These cables consisted of iron street pipes in which were placed jute-wrapped copper bars and bituminous compounds added to fill the interstices [6, 8].

A few years later, a new era began in 1890 in the United Kingdom (UK) with the appearance of a concentric design for paper-insulated cables by Dr. Ferranti, operated at 10 kV in AC [6, 9], and then, first screened cables appeared in 1917.

High Temperature Polymer Dielectrics: Fundamentals and Applications in Power Equipment, First Edition. Edited by Jun-Wei Zha and Zhi-Min Dang.

The most prominent materials for underground power cables at the beginning of the twentieth century were fluid-impregnated paper and polyvinylchloride (PVC) which were first used in 1903 in Germany [6]. In addition to underground and aerial bundled cables, PVC continued to be used for many years for large commercial and military aircrafts and the automotive industry until the popularity of PVC insulated high-voltage cables began to decrease in the late 1970s [10].

For economic, environmental, and safety concerns, a considerable progress has occurred in improving dielectric materials. However, the basics of cable design did not change significantly over the past century [11]. After being developed in 1937, polyethylene (PE) was used in cables for the first-time during World War II, before it succeeded in quickly replacing rubber and PVC materials in many applications [12]. The development of cross-linked polyethylene (XLPE) in 1963 marked a significant advancement in the production of modern cables designed for operating temperatures below 90 °C [13]. XLPE remained the most popular insulation for many applications, mainly for underground, submarine, and nuclear power plants (NPPs) cables besides Ethylene Propylene Rubber (EPR). XLPE and EPR could be used with flame-retardant, and the nominal working temperature required in cables does not exceed 90 °C [13, 14].

The 1930s also witnessed the accidental discovery of polytetrafluoroethylene (PTFE) by Dr. Roy J. Plunkett (DuPont industry). PTFE, also known with the trade name Teflon®, has been one of the largest selling high-temperature polymers [15, 16]. Since then, immense research efforts were geared toward discovering new fluorinated polymers. Besides, DuPont invented a typical polyimide (commercially named as Kapton®) in 1955 for high-temperature uses [17, 18]. Since the beginning of the 1960s, the quest for high-temperature materials (for above 150 °C use) grew rapidly due to the significant evolution in propulsion system technologies for aviation and aerospace sectors [18, 19].

The first patent of PTFE insulated cable was assigned to W.L. Gore & Associates in 1964 and could be used in the Apollo Space Program for the first moon landing [20]. The sixties and the early seventies also saw the discovery of a number of new fluoropolymers including fluorinated ethylene propylene (FEP), perfluoroalkoxy (PFA), ethylene tetrafluoroethylene (ETFE, also referred as Tefzel®) and polyvinyldifluoride (PVDF) [20–22]. These polymers quickly broke into the market of high-temperature materials used in signal transmission and power cables in the 1980s [15]. In NPPs, temperature exceeds 200 °C in the primary cooling loop, where water is pumped into the reactor core. Cables in this part must have a service life of 40–60 years and they can be equipped with a jacket made of reinforced mica tapes and PTFE coated with glass braids for high-temperature uses [23].

In the 1970s, Kapton polyimide (4,4′-oxydiphenylene-pyromellitimide), which exhibits exceptional thermal stability, was widely used as cable insulation in space shuttles as well as in civil and military aircraft. Although the arc-tracking vulnerability of Kapton insulated cables presented (and still do) a serious concern in the 1980s as the insulation damage was blamed by the U.S. military for some fire accidents and crashes. Extensive investigations by the U.S. Navy Safety Center in Virginia in the early 1980s revealed 791 cable and connector related issues [24–26].

Consequently, the U.S Navy banned Kapton use in cables for aircraft applications in 1988. After further research, the gradual phasing out of the polyimide in cables was adopted as well by the National Aeronautics and Space Administration (NASA), the U.S. Federal Aviation Administration (FAA), and the British Civil Aviation Authority (CAA) [27]. Therefore, aircraft industries started searching for alternatives to polyimide in cables, mainly for high-temperature applications. Unfortunately, such an insulation system, made of two polyimide tapes, is not as immune as has been thought, and Kapton degradation and failure could cause deadly accidents e.g. the New York TWA Flight 800 crash of Boeing 747 in 1996, the downing of Swissair 111 in 1998, and the F-22 crash in 2012 [28–30]. A typical alternative (referred as KT) was a tape-wrapped composite polyimide (Kapton coated with a thin layer of FEP) paired with sintered PTFE as top-layer in order to limit the exposure of Kapton to moisture and reduce arc-tracking risks [24]. In addition, the ratio between polyimide and PTFE thicknesses determines the resistance to arc-tracking. A new version of the insulation system, known as TKT (Teflon/Kapton/Teflon), was introduced in Boeing 737s and 757s aircraft in the early 1990s before it was adopted by Airbus in 2006 for cables that must withstand elevated temperatures up to 260 °C [18, 28]. Indeed, Boeing made a choice to use cross-linked ethylene tetrafluoroethylene (XL-ETFE) massively and KT/TKT for some applications. NASA tested ETFE and XL-ETFE as potential alternatives to Kapton, although, they were found to be limited for maximum use temperatures of 150 and 200 °C, respectively [31]. Additionally, both lacked the robustness of polyimide, and could generate a significant quantity of toxic smoke when they burn [18, 26]. Despite these hurdles, these technologies are still in use today in some commercial civil aircraft [32].

Indeed, the ongoing energy transition of aeronautic power systems is being carried out in two main stages. The first stage (for More Electric Aircraft) corresponds to the replacement of non-propulsive systems (such as turbine blades cooling, electric braking, deicing, and environmental control systems.) by electrical ones. The second stage corresponds to the electrification of propulsive functions (full electric for small aircraft and hybridization for regional aircrafts) which involves a strong increase in power densities. Therefore, MW-class electrical machines are required in addition to the components of the harnesses (cables, connectors, and busbars) with high power. This strong increase in power goes by increasing the electric current values.

As the required power might not be compatible with the standard voltage, higher voltages coupled with high frequencies will be needed to keep the current at compatible levels [33]. The combination of these two factors, high current with elevated voltage levels, has many implications, including potentially high temperatures.

The operating voltage in today's commercial aircraft is relatively low, compared to other industrial applications. For instance, Boeing 787 and Airbus 350 have electrical systems that operate under ±270 V DC (540 V DC) and/or 115/230 V AC [30, 34]. Although insulating materials can face harsh environmental conditions where the temperature could exceed 200 °C, the use of high-temperature cables on board hybrid aircraft and space shuttles remain limited to several zones as in the engine compartments. They can also be employed in order to convoy high electric current, which implies high temperatures. Consequently, the search for new high-temperature materials for power cables is considered today as a matter of key concern for future electrified aircraft. Moreover, increased voltage levels up to 3 kV DC are expected today by original equipment manufacturers (OEMs) or even higher up to ±4.5–12 kV DC for the propulsion systems [35, 36].

In oil and gas applications, electrical components can operate at elevated temperatures that may reach 150 °C [37]. In addition to fluoropolymers, polysulfone (PSU) produced in 1965 has been widely used in high-temperature power cables dedicated to gas and oil recovery operations besides PE and PVC for low temperatures. PSU can also be found in some wires used in automotive and aerospace. The second-most popular insulating material in the oil industry for high-temperature cables is polyetheretherketone (PEEK), a member of the polyketone family [38]. PEEK was first introduced in 1967 and then commercialized in 1982 by Imperial Chemical Industries [39].

The most dominant cable insulation are low-temperature materials such as XLPE, EPR/EPDM, and PVC [11]. In NPPs in the USA for example, they presented 72% of insulating materials in cables [40]. The maximum operating temperature of those materials can range from 70 °C up to about 150 °C depending on the material type and grade. It can be said that a material is considered as high-temperature when it retains usable properties at a temperature of 150 °C or higher. The use of high-temperature cables remains very limited in the previously mentioned sectors because of their high prices. They are necessary near hot areas such as in the containment of nuclear reactors. They might be inevitable for specific uses with high radiation doses or aggressive chemical contaminants. The timeline history of cables and some insulating polymers was briefed in this introduction. Figure 4.1 provides some of the important historical milestones in the development of various high-temperature materials and their applications in power cables.

Our main objective through the first part of this chapter was to show the rapidly growing interest for such materials across different industrial applications in order to achieve carbon neutrality and higher power density with a reduced cost. In Sections 4.2 and 4.3, we will highlight the design requirements of power cables and review the relevant properties of the most popular high-temperature polymers. The issues, forthcoming improvements, and trends will be addressed.

4.2 Technologies of Modern Power Cables

4.2.1 Designs and Manufacturing Processes

4.2.1.1 Wrapping Technology

There are two types of cable manufacturing processes that are in wide use nowadays, wrapping and extrusion. Wrapping was the first known technique in history, and it was the most common in the early twentieth century when oil-impregnated paper tape insulated cables were invented for power transmission and distribution networks [23]. In a simple wrapped cable, the polymer tape is wrapped helically

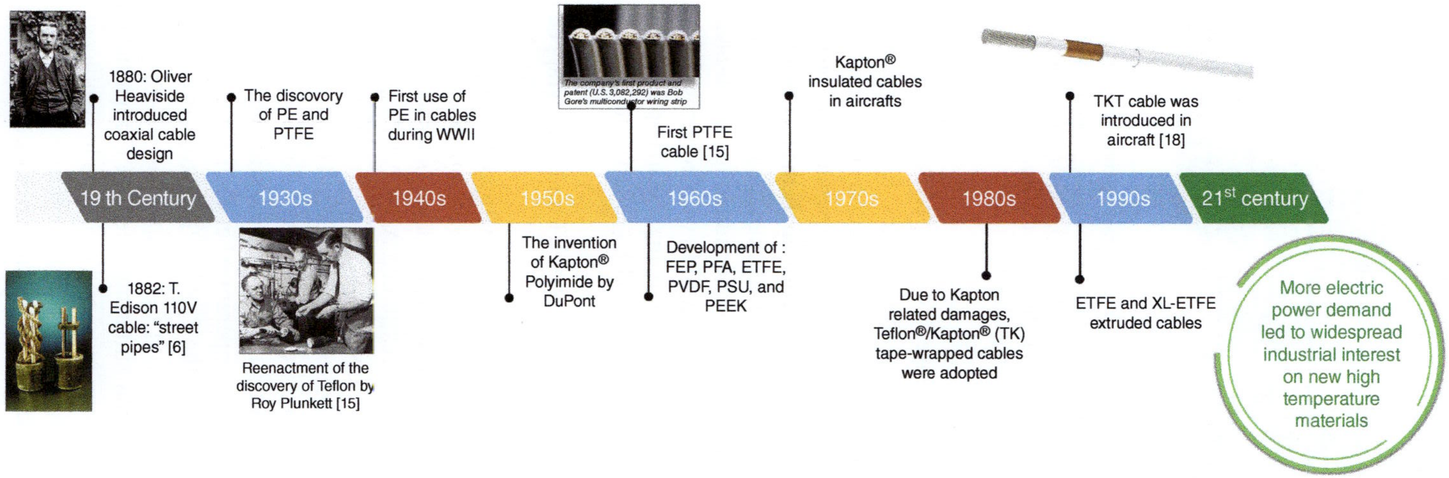

Figure 4.1 Historical milestones in the development of high-temperature insulating polymers for power cables. Images: Top left: Oliver Heaviside photo, Source: IET Archive/Wikimedia Commons/Public Domain. Top middle: image of first PTFE cable, Source: Reproduced from Ebnesajjad [15]/with permission from Elsevier. Top right: TKT cable, Source: Reproduced from Lopez [18]. LIDSEN Publishing Inc./CC BY 4.0. Bottom left: T. Edison 110 V cable, Source: Reproduced with permission from Orton [6]. © 2013/IEEE. Bottom right: Teflon discovery, Source: Reproduced from Ebnesajjad [15]/with permission from © 2017 Elsevier.

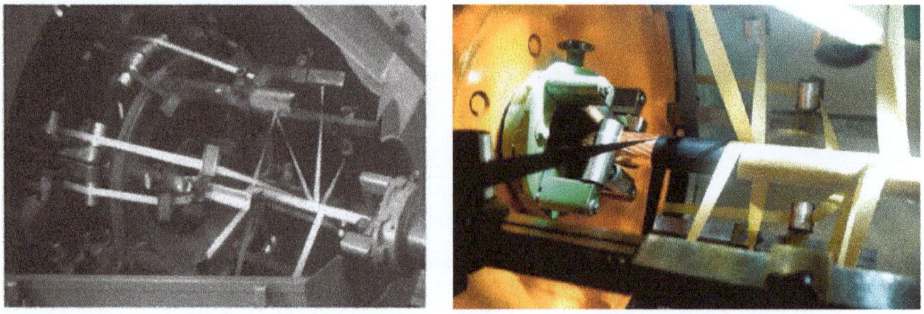

(a) (b)

Figure 4.2 Examples of wrapping machines: (a) paper taping machine for underground cables, used with permission from Taylor and Francis Group LCC – Books, from [42]; permission conveyed through Copyright Clearance Center, Inc. (b) wrapping of semi-conductor layer and insulation tape onto the conductor core. Source: Worzyk [43]/Reproduced with permission from Springer Nature.

around the conductor in one or more layers with a specified overlap [41]. Figure 4.2 shows two different examples of wrapping machines. In both machines, the tape passes around a pulley rotating in the opposite direction of the tape feeds. The rotation is regulated according to the tension of the tape, which must be well controlled by brake devices on every pulley in order to tightly apply the tape over the conductor. In the second machine (Figure 4.2b), the insulating tape is being wrapped around the conductor core after adding a layer of semiconducting carbon-black tape. This design can be found in very few cables used in NPPs with polyethylene terephthalate (PETP) for low temperature applications. Wrapping is also applied in various types of aeronautical power cables insulated with PTFE (Teflon) and Kapton for uses up to 260 °C [44].

In Table 4.1 are presented several examples of KT and TKT wrapped cables, the most common designs on board the aircraft.

It is worth noting that power cables are AWG 8 (American Wire Gauge) size or larger (with numbers lower than 8). Other designs, of course, exist. There are also Teflon, KT or TKT insulated cables with additional layers such as PTFE coated glass braid or special composites for high fire resistance. Also, PVDF tapes can be found in some cables such as in BMS13-58, for instance. A special treatment can be applied to ensure the mechanical stability of the insulation tapes and that the insulation tape is well-coated around the cable conductive core without wrinkling. In practice, PTFE tapes are generally sintered by heating at a temperature sufficiently higher than the melting one (>327 °C, generally at a temperature between 360 and 380 °C) [15, 45]. PTFE sintering allows the adhesion of the FEP (or PFA) layer covering the side(s) of Kapton tape with the conductor and with PTFE tape. This process is analogous to the forming of powdered ceramics by sintering. It leads PTFE particles to coalesce and fuse resulting in a drop of crystallinity, and therefore in significant changes of mechanical behavior of PTFE.

Wrapping technique has its advantages and disadvantages. It combines the pros of PTFE and Kapton and shows better arc-resistance performance than XL-ETFE cable.

Table 4.1 Examples of power wrapped cables used in aeronautic and aerospace for high temperatures.

Cable image		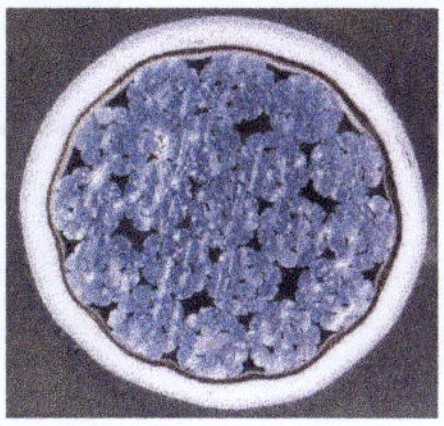	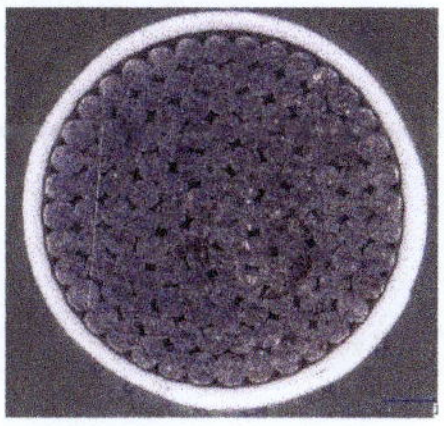
Cable type	AD	AZ	DR
Size	AWG 8	AWG 3	AWG 8
Conductor material	Nickel-plated Copper Clad Aluminum	Un-plated aluminum	Nickel-plated copper
Rated voltage	115 V AC in pressurized and non-pressurized areas and 230 V AC in pressurized area	230 V AC in non-pressurized area	115 V AC in pressurized and non-pressurized areas and 230 V AC in pressurized area
Insulation	(Kapton/Teflon) KT	(Teflon/Kapton/Teflon) TKT	(Kapton/Teflon) KT
Temperature ranges	−65 to 180 °C	−65 to 180 °C	−65 to 260 °C

Source: Reproduced from Diaw [41] with permission from the Author.

Tape-wrapping is suitable for thin insulation thicknesses, providing better weight benefit with good mechanical and electrical performance. Although wrapping process is relatively expensive compared to extrusion. For instance, according to [31], a PTFE and Kapton cable, with a small size of AWG 22, offers a 6.7% weight save compared to an AWG 22 XL-ETFE extruded cable. The weight save can reach 16.39% for larger size (AWG 0) power cables. It requires an AWG 0 KT cable to carry a current of 540–550 A at an operating temperature of 260 °C and an ambient temperature of 20 °C (temperature gradient of 240 °C). In contrast, a larger size of AWG 00 XL-ETFE extruded cable is needed for the same current carrying capacity where the cable is in operating temperature of 200 °C (temperature gradient of 180 °C). Although, wrapping can cause several issues within the insulation system design as can be noticed in the images illustrated in Figure 4.3.

First, variability of the insulation thickness is unavoidable, due to the overlap where some parts of the cable could present one more layer than others. Nevertheless, the wrapping process guarantees to have a minimum thickness of insulation that is well known. Moreover, air voids standing between different insulation layers and between the conductor and the insulation by tape wrapping can be the site of partial discharges (PDs) that may lead to severe degradation of the insulation system [47]. Furthermore, the polyimide used in wrapped cables is commonly coated with a

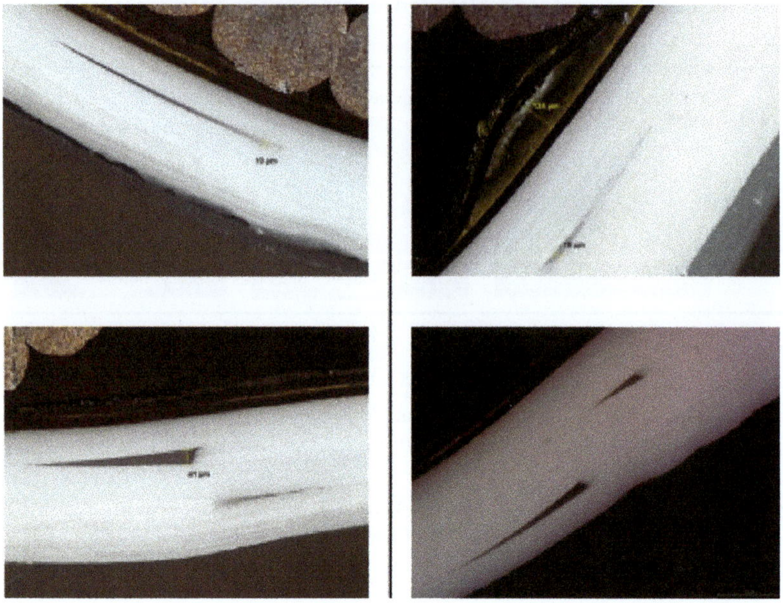

Figure 4.3 Air voids between insulating layers in different wrapped KT cables, Source: Reproduced from Lebey et al. [46]/with permission, from IEEE.

thin layer of FEP or ETFE acting as an adhesive to glue the insulating layers together. Unfortunately, air voids are created between insulating tapes after the wrapping process, thereby promoting the generation of PD and their mechanical strength might decline [48].

4.2.1.2 Extrusion Technology

First extruded power cables date back to the 1970s with the use for XLPE material for metropolitan networks in the USA [1]. The extrusion is a thermomechanical manufacturing process, which consists of sheathing the conductive core of the cable by injecting at least one insulation layer, also known as a single screw extrusion. Figure 4.4 shows a scheme of a typical single extrusion line for cable production. Prior to extrusion, the conductor must be preheated before passing into the die for moisture elimination and to better adhere to the insulating material. On the other

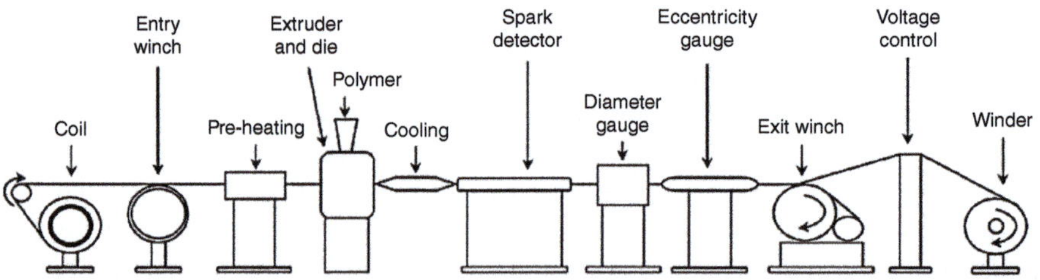

Figure 4.4 Scheme of cable extrusion line [49]. Source: Reproduced with permission from John Wiley and Sons.

Figure 4.5 Different images of a triple extrusion line production of an aerospace cable. Source: Courtesy of Nexans.

side, the extrusion screw has four main parts, feeding, melting, mixing, and metering zones. The polymer is melted and mixed inside the extruder. The molten polymer is pushed with a controlled pressure, and then the polymer once inside the die, coats the conductor. Thereafter, the insulating and sheathing materials have to be cooled inside cooling troughs, commonly by water aspersion, before reaching the winder. During this operation, different measurements are carried out to assess the quality of the process. The main measured parameters are– the cold insulation diameter, insulation concentricity, and insulation continuity testing by the electric spark tester [50].

Double and triple extrusions, termed as co-extrusion, can also be made where two or three layers are applied simultaneously on the conductor. The production line is similar to that of a simple extrusion, except for the die where additional extruders are present. In fact, multiple molten insulations are extruded from different extruders that feed into one pressure assembly-die (Figure 4.5) [51]. The extrusion technique has good flexibility and is suitable for long cable lengths with thick insulation. Semiconducting material can be extruded over the conductor and/or over the insulation layer to avoid air gaps between the conductor and the insulation and reduce local electric field enhancement [7, 52]. Although, the formation of voids in the insulation bulk is dangerous as PD inception would be very likely. The extrusion technique requires excellent material purity, a smooth surface, and a good concentricity, which are extremely complicated to achieve today with many high-temperature materials due to their high melting temperature (above 300 °C) and high melt viscosity. Figure 4.6 represents different examples of extruded cables.

There are a very limited number of insulating polymers that meet the requirements in terms of suitable thermos-mechanical properties for extrusion process. For example, the melt viscosity of PTFE is 1–100 GPa s^{-1}, which is 10^6 times high for melt processing by extrusion or injection molding [1, 15]. The process of applying PTFE insulation to cables can also be achieved through a method called "paste extrusion," inspired from ceramic processing. First, PTFE powder is mixed with a hydrocarbon lubricant, which acts as an aid to the extrusion process. The mixture is then placed

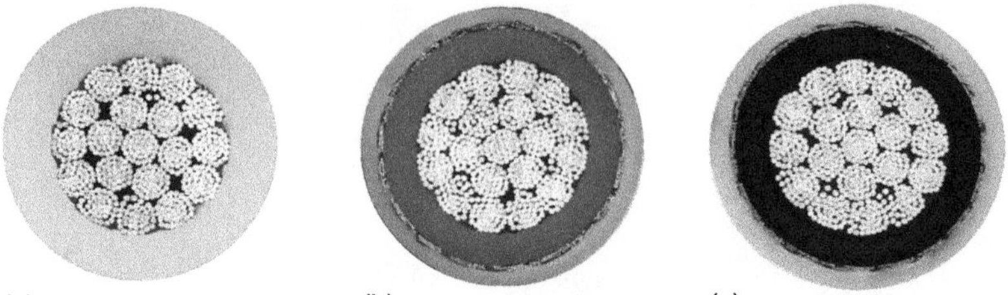

Figure 4.6 Cross-sections of different extruded cables: (a) One-layer PFA insulated cable, Source: Courtesy of Nexans. (b) Shielded cable with PFA insulation and thin outer semiconductor layer, Source: Courtesy of Nexans. (c) Insulated PFA cables with triple extrusion, Source: Courtesy of Nexans.

inside a ram extruder, where it is pushed through a die under consistent pressure (1–3 MPa). After passing through several ovens for drying, sintering, and cooling, the lubricant is removed through extraction in a hot solvent bath. It is worth noting that cables insulated with PTFE through paste extrusion have lower cut-through resistance compared to wrapped cables insulated with PTFE tapes [20].

The melt viscosity of FEP is almost 10^8 times lower than that of PTFE, which places it among the melt-processible thermoplastic polymers in extruded cable operating up to 250 °C [53]. Among other extrudable fluoropolymers, ETFE (Tefzel) and XL-ETFE are still being used in cables by many aeronautic and aerospace companies up to 200 °C. Some investigations carried out by Lectromec company show that extruded XL-PTFE cable (AS22759/34) has better performances against PD than a tape wrapped one with PTFE and Kapton (AS22759/87) [54]. For much higher temperatures, PFA polymer has been also considered as a typical candidate for high temperature cables and significant attention is being paid by many researchers and industries to study its properties and performances in the aeronautic field [46, 55]. On the other hand, PEEK is among the high-temperature extrudable materials, and it is used extensively in cables within containment vessel of NPPs. Polyamide imide (PAI), thermoplastic polyimide (TPI), and polyether-imide (PEI) are among the main extrudable varieties of polyimide for extrusion process [56]. Today, continued research interest has been conducted on improving these materials to satisfy the requirements of future nuclear cables as well as in automotive and oil and gas industries [14, 38].

4.2.1.3 Micro-Multilayer Multifunctional Electrical Insulation (MMEI) System

The micro-multilayer multifunctional electrical insulation (MMEI) is a new technology introduced in the framework of the NASA-targeted Convergent Aeronautics Solutions research strategy. The MMEI concept (Figure 4.7) is for the main incentive to achieve lightweight insulation systems with high performances (high breakdown voltage, PD resistance, moisture barriers, and thermal dissipater layers.). In fact, decreasing the layer thickness reduces the probability of defects or imperfection leading to higher electrical breakdown voltages.

It has been found that the best performance from MMEI systems is achieved with a thickness of at least 0.15 mm, formed by multiple thin layers of known

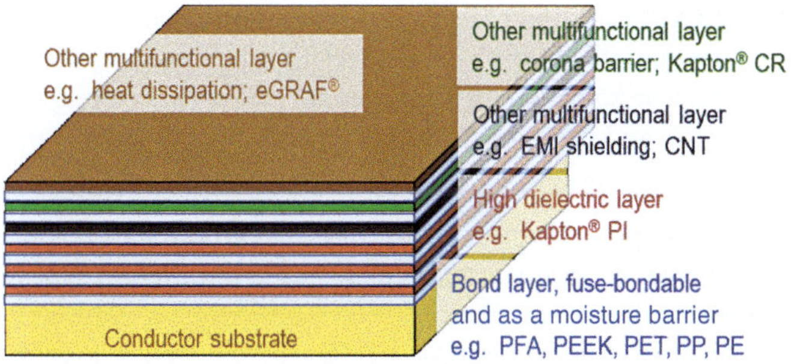

Figure 4.7 Illustration of the micro-multilayer multifunctional electrical insulation design, © 2021 IEEE. Source: Reprinted, with permission from Borghei and Ghassemi [30].

high-temperature materials. The MMEI structures were found to be highly effective at higher voltages above 20 kV, their application is also relevant for low voltage applications. Moreover, some factors must be considered, including processing defects, electrical stress, frequency, and temperature [57]. Initial investigations on different MMEI structures using known high-temperature materials have revealed a substantial improvement in voltage breakdown with the multi-layer configuration: $[1 \times HN/1 \times PFA]_{10}/1 \times HN$, where HN is a type of Kapton polyimide and PFA is used as a bonding layer.

This configuration allows up to 91% breakdown voltage increase compared to a single layer of Kapton with the same thickness. This has been translated to a reduction of 98% in insulation thickness for the same voltage breakdown. Furthermore, $[0.3 \times HN/(1 \times PFA/1 \times HN)_{10}/(1 \times PFA/1 \times CR)_2/2 \times PEEK/0.3 \times HN]$ configuration was determined as an MMEI candidate for future prototypes due to its exceptional dielectric performance (CR is a type of Kapton). Additionally, configurations with multi-functionalities such as EMI shielding, thermal dissipation, and mechanical strength are still under development.

Efforts have been undertaken to adopt MMEI structures for developing full-scale prototypes of high-performance power cables. In the light of this, GORE (W. L. Gore and Associates, Inc., Landenberg, PA) designed and fabricated a unique HV power pod cable under the auspices of NASA-GRC. As introduced in [58], the cable was designed by implementing a MMEI system of 17 layers and it is capable of carrying 0.25 MW at 15 kV (targeting 40 kV) and can withstand high service temperatures up to 260 °C. The flat cable pod consists of six identical conductor pods insulated by GORE's proprietary PTFE-PTFE composite and surrounded by a corona-resistant PTFE jacket. Experiments are being continued to evaluate other characteristics and to improve the actual designs.

4.2.2 Insulating Material Considerations for High-Temperature Cables

In practice, cables are generally designed to have a service life of at least 30 or 40 years. Besides, insulating polymers for power cables, as well as for other electrical systems, can be distinguished into several thermal classes. The thermal (or temperature) class is defined according to IEC 60216-1 standard as the temperature below which there is no modification of the material properties, and their lifetime surpasses 20 000 hours. The so-called high-temperature polymers belong to thermal classes over H (>180 °C) and they should fit a number of requirements in order to be termed so [38, 59]:

- The polymer must maintain its properties for more than 10 000 hours at 177 °C and it should be stable at this temperature after being exposed to different types of stresses (mechanical, chemical, and electrical.).
- A thermal decomposition temperature superior to 450 °C where a weight loss of 5% is passable (heating rate of 2.5 °C min^{-1} by dynamic thermos-gravimetric analysis).
- When the material is under a load of 1.52 MPa, the heat deflection temperature should be higher than 177 °C for a 10% deflection of the specimen.
- A high-temperature insulating polymer with aromatic content and relatively rigid segments should exhibit high glass temperature ($T_g \geq 150$ °C).

These criteria, in addition to the process-ability of the polymer in wrapping or extrusion processes, further narrows the list of insulating polymers that can be directed toward high-temperature power cables. A hierarchical presentation can be found in Figure 4.8 where the most employed polymers in cables are classified according to their thermal performances. In this chapter, we will shed more light on the properties of materials in the top two categories of the pyramid. It is also worth noting that the distinguishing features of polymers in those two categories are their high price and costly manufacturing process [60]. Also, some grades of the materials within the second tier of the pyramid, such as PEEK and PEKK, can be suitable for applications that require high-temperature resistance beyond 220 °C.

The development of cables for high-temperature uses must take into account other harsh environmental conditions and different constraints and forces, thus

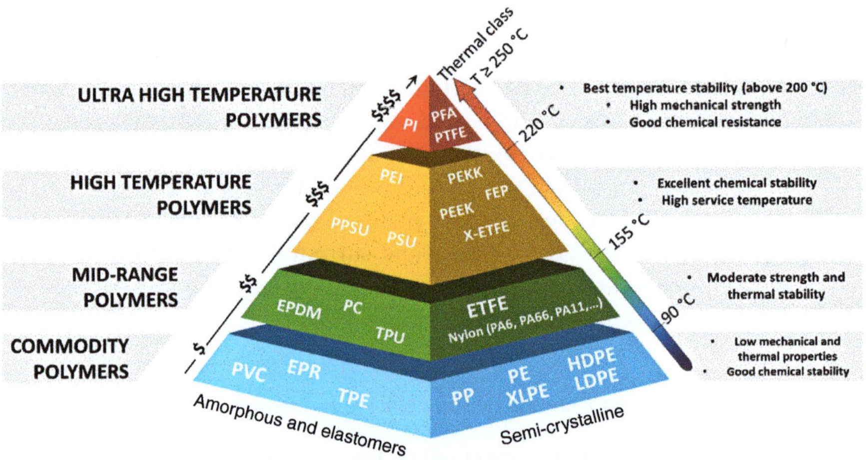

Figure 4.8 Pyramid of most-used insulating polymers in power cables according to operating temperatures.

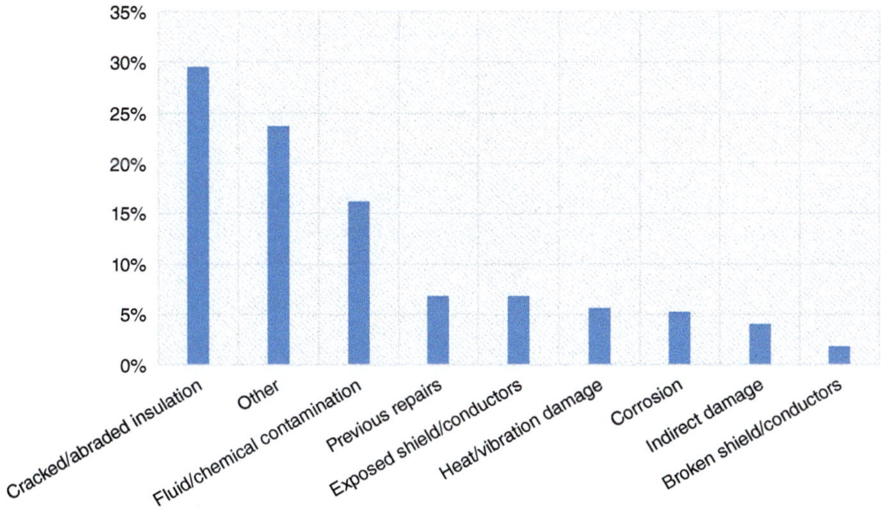

Figure 4.9 Stressors and conditions leading to the degradation of cable insulation in aircraft.

high-performance properties of insulation are required. In second-generation NPPs, for example, cables are divided essentially into two categories– safety and non-safety cables. The first category (referred to as 1E) includes three different classes– K1, K2, and K3.

The first two classes are mounted in the containment of nuclear reactors. K1 and K2 cables must be able to operate normally under high temperature and under the seismic load of safe shutdown, which requires excellent materials. K3 cables, however, are installed outside the containment where the maximum service temperature is below 90 °C [14]. For aviation applications, Figure 4.9 shows the most frequent stressors present in aircraft, responsible for the degradation of the cable insulation. These data are approximate, and they have been extracted from [24] according to an Intrusive Inspection Report. It can be noticed that cable failure is due mainly to cracked and abraded insulation, chemical contamination in addition to other problems resulting from previous stressors that may have been present.

According to U.S. Navy data investigations on cable failure modes for aircraft (1980–1999), primary modes of failure were related to mechanical properties, chafing damages (37%), and broken wires (11%), as illustrated in Figure 4.10 [27]. Short circuits including arcing incidents were the cause of 18% of failures in cable systems and interconnections.

An excellent high-temperature insulation must require a good resistance against the above constraints and stressors and maintain the maximum of their performances up to the maximum operating temperature. High-temperature cables are special ones, and they should meet special requirements depending on their application. The most important requirements will be highlighted in the following points.

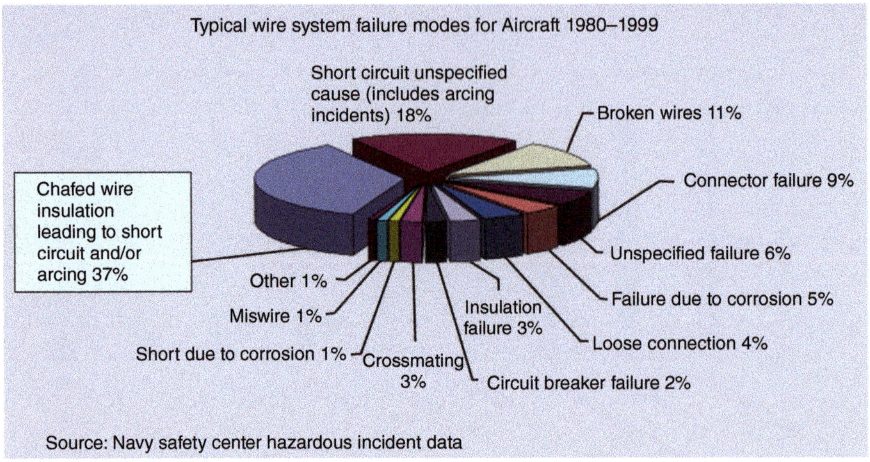

Figure 4.10 Failure modes of cable systems in aircraft according to Navy Safety Center Hazardous Incident Data 1980–1999. Source: Reproduced with permission from Brusso [27] IEEE.

4.2.2.1 Electrical Requirements

As long as electrical characteristics are concerned, high-temperature materials must have sufficiently high dielectric strength for medium and high-voltage use. Nowadays, the vast majority of high-temperature cables are designed for low-voltage stresses and their insulation thickness is adequate for the desired dielectric breakdown strength. Although elevated temperature remains one of the most critical influencing parameters on dielectric properties of polymers and the changes in such properties must be carefully considered for the cable design. The breakdown voltage for example can significantly decrease at elevated temperatures.

At low voltage, the properties of importance are dissipation factor, volume resistivity, and dielectric constant. This last presents the ability of the material to "store" charges while the dissipation factor is a measure of the energy lost by heat. An excellent high-temperature insulation is one that exhibits a low dielectric constant (generally between 2 and 3.5) and very low losses (typical values are 0.005–0.02 at 1 kHz) and this over the entire operating temperature range. For instance, PVDF is an excellent temperature-resistant fluoropolymer; however, their known piezoelectric properties and high dielectric constant (8.2–12 at 1 kHz) make it not really a good choice for use as an electrical insulation under high voltages. The most common standards used for dielectric measurements are ASTM D150, IEC 60250 for breakdown tests and ASTM D149, IEC 60243 to characterize the dielectric constant and dissipation factor of a material. The DC conductivity/resistivity test method can be found in ASTM D257 and IEC 60093 standards.

The most relevant electrical properties of high-temperature insulating materials will occupy a large, separated part at the end of this chapter to review the most recent researches with the latest findings.

4.2.2.2 Thermal Requirements

In order to verify the compatibility of insulating material with the recommended operating temperature range, a thermal endurance test must be performed as described in Section 13 of the ASTM-D-3032-21A standard, for example. Such a test is also known as the "Relative Thermal Life and Temperature Index" test. The standard provides a procedure for determining the temperature index and thermal end time versus temperature curves for flexible electrical insulating materials. The purpose of the test is to evaluate the cable lifetime by subjecting it to various temperatures and plotting the time curve as a function of temperature in accordance with the Arrhenius law. For instance, for the DR (KT) cable series that operates at 260 °C, the requirement is to demonstrate that the cable insulation can withstand a minimum of 10 000 hours at 260 °C and 100 000 hours at 200 °C.

Furthermore, the required insulation thickness in high-temperature cables can be based on thermal and mechanical considerations. Thermal conductivity is a key challenging factor that needs to be promoted, and high glass transition or melting temperatures might not be enough to prevent damage to the polymer caused by heating. High thermal conductivity, defined as high speed of heat transfer through a material, can improve heat dissipation and reduce potential energy losses. It can reduce the temperature gradient in the insulation, which may have an advantageous effect on limiting space charge accumulation. Poor thermal conductivity of the material can provide higher temperature rises, and hence, reduce insulation lifetime according to the Arrhenius equation (10 °C rise of operating temperature cuts insulation lifetime by half). Low thermal conductivity can also affect the quality of melt processing of insulating polymers [1, 9]. Moreover, materials with high thermal conductivity are less likely to have voids when being extruded, as the entire polymer cross-section is cooling fast and shrinking at the same time at the end of the extrusion process [61].

In practice, insulating polymers have a thermal conductivity below 0.5 W m^{-1} K^{-1}. Two approaches are widely adopted to enhance thermal conductivity. The first consists of changing the molecule structure of the polymer by increasing its crystal structure (also referred to as the higher-order structure) [62]. In the second approach, additives and nano-fillers (usually with high thermal conductivity) can be incorporated in order to provide the required thermal conductivity of the insulation. However, high concentrations of these additives may increase dielectric losses and electrical conductivity as well as degrade mechanical properties. In addition, several parameters must be taken into account in the filling process such as– the loading amount, the shape and size of added particles, their orientation and dispersion, and the interface status. Inorganic fillers such as aluminum oxide (Al_2O_3), magnesium oxide (MgO), silicon oxide (SiO_2), zinc oxide (ZnO), boron nitride (BN), aluminum nitride (AlN), silicon nitride (Si_3N_4), and silicon carbide (SiC), have been extensively studied for thermal conductivity enhancement in electrical insulating material [62, 63]. The most common fillers for electrical insulation can be found in [64] with their thermal conductivity values at room temperature.

Thermal conductivity and the permittivity of fillers may have also advantageous effects on mechanical and electrical properties at high temperatures. On the other

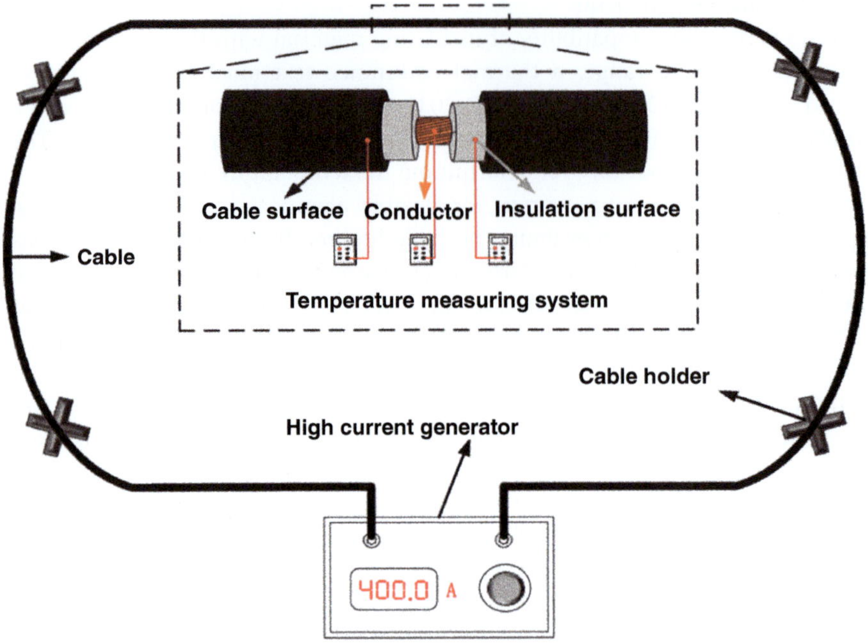

Figure 4.11 Scheme of an experimental setup used for temperature measurement of cables. Source: Reproduced with permission from Du et al. [66] IEEE.

hand, conductive particles such as metals, carbon (graphene and carbon nanotubes) and graphite are not satisfactory for electrical insulation [65]. For measuring the thermal conductivity of an insulating material, ASTM C177 is the standard test. On the other hand, the thermal characteristics of cables are generally studied using the experimental setup shown in Figure 4.11.

Both ends of the cable are connected to high current generator where the output current could be adjusted according to the needs of the experiment. The temperature distribution on each insulation surface can be measured with time by T type thermocouples inserted in different radial depths. The results can be then used to calculate the ampacity of the cable according to the IEC 60287 standard. The cable ampacity represents the current value in which the cable operates at its maximum temperature under a steady state under given ambient conditions.

4.2.2.3 Mechanical Requirements

Physical structure of the polymer (such as the crystalline or amorphous structure and the motion of the polymer chain and branches) dictates its mechanical properties. Important properties of high-temperature materials include tensile strength, elongation at break, flexural strength, and Young's moduli. Parameters affecting the strength of insulation are molecular weight, cross-linking, and crystallinity.

High-temperature cable design must consider the different stresses and vibrations, which may occur on the cable at elevated temperatures. Mechanical stress includes vibration, acceleration loads, and potential damage during installation and maintenance. Cable chafing can have a detrimental effect on cable reliability as concluded by the U.S. Navy (Figure 4.10). Cables rubbing against each other or their fixtures over time typically cause this phenomenon.

For extrudable materials, good mechanical characteristics impart sufficiently high tensile strength, coefficient of friction, flexural modulus, and elongation at break. These properties are less important for the wrapping process. Fluoropolymers, used in cables, have a low coefficient of friction, between 0.08 and 0.4, which is safe enough to limit the degradation of the cable insulation. Meanwhile, flexural modulus is around 500 MPa for FEP, PTFE, PFA, and about 1200 MPa for ETFE, depending on the grade and the formulation of the material [12]. These values are considered sufficiently low to provide a good level of flexibility to the cable, which results in easier installation. In contrast, a flexural modulus of PEEK is 3600 MPa at 20 °C and 2200 MPa at 150 °C [67], which is relatively high and limits the installation of the cable in specific aircraft zones, where cables need to be installed with tight bending radii.

Furthermore, many research works have shown that elongation at break is the most reliable feature reflecting the degradation state of a cable. In practice, when 50% decrease of the elongation at the break of the insulating material is obtained after thermal exposure, the material is considered reaching the end of its life [68]. Tensile tests are performed on specific machines and the most employed standards for those mechanical characterizations on insulating samples are ASTM D1708, D638, D790, and IEC 60216.

Moreover, thermal expansion of polymers due to a change in cable temperature may lead to a sudden sagging of the cable, hence, there is an increased risk of a fault current accident. Additionally, air voids can be formed in wrapped cables between the insulation and conductor due to thermal expansion or between insulting layers due to the shrinking of the inner one. It is critical that the insulation remains partially discharge free by avoiding gas-filled gaps and contamination at interfaces, mainly at low air pressure as discussed earlier (Section 2.1.1). Due to those reasons, shrinkage and thermal expansion coefficients should be reduced and kept stable with temperature increase. As for thermal properties, inorganic fillers can be added to insulating materials to tune their mechanical characteristics. In some cases, braids made of glass or textile fibers can be set on the outside of the cable for better mechanical resistance.

4.2.2.4 Environmental and Functional Considerations

Cable insulation properties should be selected based on their industrial application and the environment in which the component will be installed. High-temperature cables for aeronautic and aerospace can be exposed to UV radiation and aggressive chemicals. Ionizing radiation and water vapor can also be serious issues for cables used in NPPs. Whereas cables for petrochemical uses are installed in an environment that is heavily polluted with hydrocarbons that might be very aggressive and lead to cable failure. Designs involving two or more insulating layers are commonly adopted in low-voltage cables. In this case, the outer layer of the cable insulation, as the main insulation, has to meet a number of safety, functional, and reliability requirements.

4.2.2.5 Flame Resistance

It is defined as the ability of a material to prevent the propagation of a flame after combustion. The flammability is a measure of this ability that should meet standards. There are different parameters that can determine the flammability of a material, the major parameters are– ease of ignition (flammability), resistance to propagation (flame spread), heat of combustion (heat release), smoke emission, and toxic and corrosive gas evolution [9]. These parameters are critical for high-temperature cables in many use applications; power plants, vehicles (such as cars, trains, ships, and aircraft), and tunnels.

As far as aeronautic cables are concerned, the certification authorities (EASA and FAA) are required to carry out a flammability test, as described in ABD0031 (EN3475-407) standard for example, on all cables, particularly on high-temperature ones. The standardized flame is applied for 30 seconds to the cable, after removing the flame on the cable, it must go out in less than 30 seconds, and it must not spread over more than 76 mm. It is also necessary to distinguish the fire resistance test EN3475-408 which also applies to high-temperature cables, which have a fire resistance function. There are two notions that need to be well distinguished, the notion of fire resistant 5 minutes under 1100 °C and the notion of fireproof 15 minutes under 1100 °C. Meanwhile, the cable must retain its properties under thermal stress and vibration. Flame- resistant cables are made of special materials, like those that have been mentioned previously (PTFE and PI/PTFE) and Mica material can be particularity added to ensure better fire resistance. The tests defined in standards also measure the heat released during combustion in addition to smoke characteristics and fire growth rate. Cables can be then classified by a letter symbol or series of letter symbols according to the characteristics they meet. IEC 331 is one of the most widely used tests to investigate the fire resistance of cables. The test establishes whether a cable can maintain electrical circuit integrity at temperatures ranging from 750 °C up to 950 °C for up to three hours.

It should be noted that some high-temperature cables, as in NPPs for instance, are permitted to burn but the flame must not propagate. A material with excellent flame retardancy is rated UL94 V-0, where UL94 (Underwriters Laboratories test standard UL 94) is the name of the testing method. Materials in the top two categories of the pyramid in Figure 4.8 have this rate. In fact, most used high-temperature polymers such as polyimides and fluoropolymers are intrinsically resistant to flame. Flame-retardant additivities such as alumina thrihydrate (ATH) and magnesium hydroxide (MDH) can be added to improve flame resistance of the insulation. This can also be achieved by reinforcing the cable with glass fiber.

Various parameters are investigated by laboratory tests to provide more information on flammability–

- The LOI (limited oxygen index): Measures the minimum percentage of oxygen in nitrogen that will support candle-like burning of a specimen for at least three minutes. The cable insulation should require a level of oxygen higher than the one present in the atmosphere (21%). A material with an oxygen index of 26 or higher is defined as self-extinguishing according to ISO 4589 and ASTM D2863 standards.

- The Fire Performance Index (FPI): Defined as the ratio of time to ignition and the peak rate of heat release. The most used standards for FPI calculation and evaluation are ASTM E1354, NFPA 271, and ISO 5660 1-2.

In addition, various international standards provide different methods to measure flame propagation on cables for different configurations– IEC 60332-1-3, IEEE 383, IEEE 1202, UL 1581 HB, and UL 1685.

4.2.2.6 Smoke Evolution

Smoke due to fire is a critical factor for cables installed in enclosed spaces. Standards such as IEC 61034-1, IEC 61034-2, NES 711, and BS 7622-1 define smoke evolution tests. The Arapahoe smoke chamber is the most known laboratory test to evaluate the smoke evolution of a cable sample. The NBS smoke chamber test is commonly employed to measure the density of the generated smoke from a square sample according to ASTM E662-79. It is worth noting that halogen-containing insulating materials, such as fluoropolymers, exhibit a high flame resistance, however, they generate toxic fumes (perfluoroisobutylene, PFIB for example) during combustion. For this reason, the Grumman Corporation and NASA banned the use of XL-ETFE material in the 1980s. Nowadays, high-temperature cables with low smoke, halogen-free and flame-retardant compound are expected to grow fast in the next few years.

4.2.2.7 Toxicity

The toxicity of gases generated from burning insulating polymers depends on the type of material and additives used, together with the surrounding environment. Even if it is very complicated to measure, toxicity is an important factor to consider regarding human safety. The toxic hazard that can be determined in laboratories is generally indicated by LC_{50}. However, it remains challenging to simulate the conditions in real fire accidents of cables during laboratory tests, in order to analyze the toxicity of gases. The limits of the toxicity of gases from smoke are quantified in AD0031 standard: $\leq$100 ppm for hydrogen fluoride, sulfur dioxide, and nitrous gases, $\leq$150 ppm for hydrogen chloride and hydrogen cyanide, $\leq$1000 ppm for carbon monoxide.

4.2.2.8 Corrosivity

Corrosivity is an important topic for the safety of electronic devices as well as communication equipment mounted near power cables. Different standards, such as IEC 60754 and BS 6425, provide techniques to evaluate the evolution of produced halogen acid gas (except hydrofluoric acid). The acidity degree of gases can be determined as well where the measured pH must be higher than 4.3. The insulation must generate less than 0.5 wt% of halogen acid gas to be considered as zero halogen. Besides, (DEC-0611/C) measuring technique is used in many countries to assess the resistance changes of electronic circuits during combustion.

4.2.2.9 Heat Release

This can indicate the burning behavior of the cable insulation when being exposed to different radiant heat levels. This technique, referred to as the cone calorimeter, allows to measure the total amount of heat release, the peak rate of heat release, the time to peak rate of heat release, smoke release data, and weight loss due to combustion. The experimental setup of the full-scale fire test by IEC 60332-1-2 and EN 50399 standards is shown in Figure 4.12 where heat and smoke releases and flame ignitability of four vertical cables are measured. The results of heat release rate and flame spread tests can print out a classification of the cable. Seven classes of cables are defined– A_{ca} (No contribution to fire), $B1_{ca}$ (Low contribution), $B2_{ca}$ (Medium), C_{ca} (Significant), D_{ca} (High level), E_{ca} (Very high level), and F_{ca} (No performance determined). These classes combine the test results issued from EN ISO 1716 in addition to EN 60332-1-2 and EN 50399 (ex EN 60332-3) tests.

4.2.2.10 Radiation Resistance

High-temperature cables may be subjected to a high amount of radiation in space and in NPPs. For commercial aircraft, cables can be exposed to considerable amounts of cosmic radiation. Radiations may have a significant effect on the electrical and mechanical properties of the insulating polymer. Cables of K1 and K2 classes in NPPs must be able to work under long-term radiation which is estimated to 25 Mrad for 40 years. For instance, PEEK has been selected as an excellent choice for future nuclear cables due to its excellent radiation resistance property. PEEK can maintain good mechanical and electrical properties after 1100 Mrad irradiation [14]. Since PEKK has a similar chemical structure to that of PEEK, PEKK might be an interesting choice for cables in NPP as it offers higher thermal stability. Several documents of the European Organization for Nuclear Research (CERN) have reported a classification of low- and high-temperature materials with respect to their resistance to radiation as given in Figure 4.13. It can be noticed that PTFE (at the bottom) does not resist high-energy radiation. In practice, fillers can be used sometimes to overcome this issue. As far as aerospace cables are concerned, UV radiation has been determined by laboratory tests as one of the constraints that can enhance the degradation of the insulation even in the case of high radiation-resistant materials such as Kapton. Tests performed by the Air Force Research Lab, with participation and funding from NASA, have confirmed the delamination, shrinkage, and wrinkling of Kapton after being exposed to significant amount of UV radiation.

Other laboratory experiments on FEP material show that vacuum ultraviolet radiation (VUV) may cause a serious degradation of FEP surface and mechanical properties for solar-facing materials in space. X-rays from solar can also have an important degradation effect on the material bulk [71]. On the other hand, ETFE exhibits good resistance to radiation. The Radiation Index (RI) is a measure of the irradiation resistance of insulation as defined in IEC 544 standard. RI represents the logarithm base 10 of the absorbed radiation doses in grays at which the critical property is reduced to half of its initial value under specified test conditions. The critical properties generally chosen are mechanical ones, the flexural strength for thermosets, and deflection for thermoplastics.

4.2.2.11 Chemical Resistance

One of the common characteristics of high-temperature insulation includes chemical bands with high dissociation energy, also known as the chemical bond

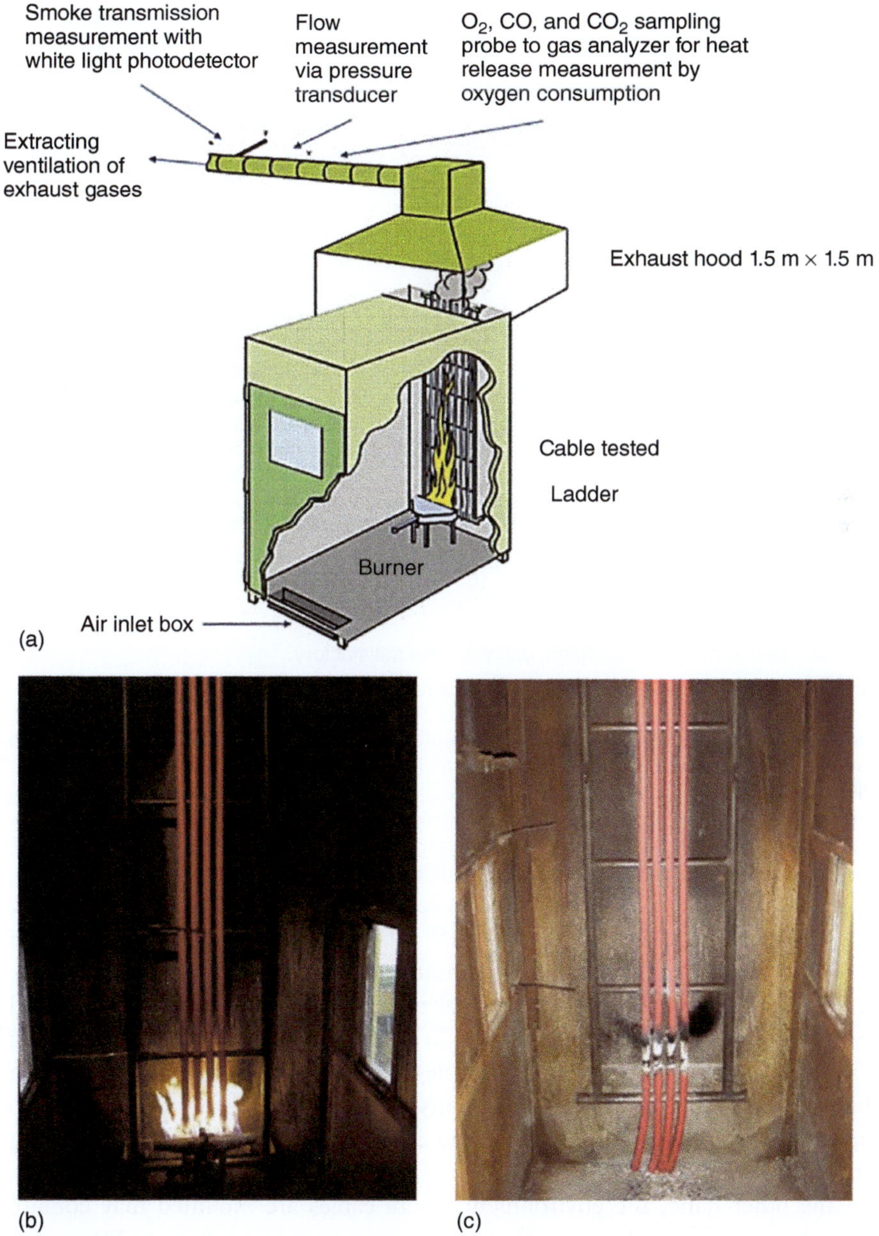

Figure 4.12 (a) Illustration of the large-scale fire test by IEC 60332-1-3, (b) EN 50399 apparatus, during fire test (c) after fire test. Source: From Poutch [69]/with permission from John Wiley & Sons, Inc.

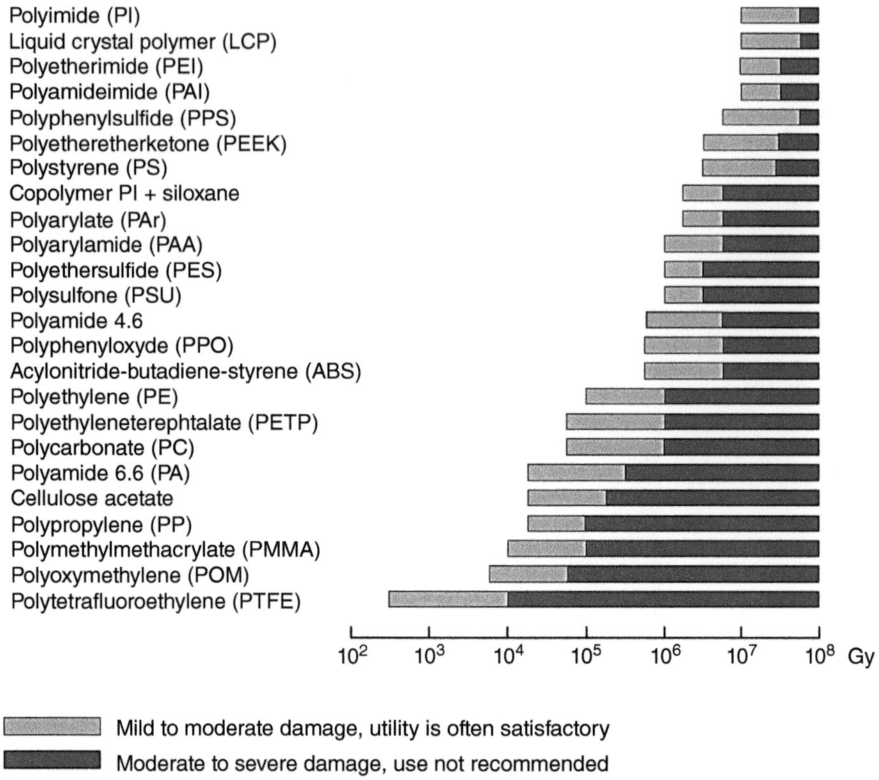

Figure 4.13 CERN Classification of several insulating polymers according to their radiation resistance, Source: Reproduced from Tavlet et al. [70] under CC BY 3.0 licence/CC BY 3.0/Public domain.

strength. Many high-temperature materials contain C=C bonds across their backbones which exhibit high energy of 145 kcal mol^{-1}. Moreover, the bond energy in fluoropolymers (C–F) is 123 kcal mol^{-1} which is much higher than that of low energy bonds like (C–C) and (C–H) ones (83 and 99 kcal mol^{-1} respectively). Good chemical resistance materials can also be characterized by strong inter and intra-molecular interactions (by hydrogen in PAI: –CO–NH–) or conjunctions along the polymer chain (imide rings for example). Such structures allow the insulation to have good thermal and chemical stability. In some cases, exposure of cables to atomic oxygen can lead to mass losses and changes of surface properties resulting from thermos-oxidation [26].

On the other hand, the environment where cables are mounted may contain chemicals and solvents that are able to cause serious damage. Back to Figure 4.9, chemical contamination is one of the three major failure modes of cables in an aircraft. The FAA has listed various types of contaminants in aviation; hydraulic fluid, battery electrolytes, fuel, corrosion-inhibiting compounds, waste system chemicals, cleaning agents, deicing fluids, paint, and soft drinks [28]. Indeed, the chemical stability of the outer insulation in cables should be given special attention in high-temperature cable designs. For example, polyimide is chemically sensitive to moisture, and it has relatively high water absorption. Hence, the predominant process of polyimide deterioration is attributed to the hydrolysis mechanism when water interacts with impurities resulting in the polymer chain scission.

Contaminants do not necessarily have a dangerous effect on the properties of insulation themselves, they may lead to indirect damage such as arcing and chafing. This seems a big threat when it comes to polyimide (Kapton) insulated cables. As mentioned previously, many power cable failures in aircraft involved arcing during 1980s and this occurred because of Kapton uses. This last has a weak arc propagation resistance and the arc-tracking phenomenon can be aggravated in the presence of conductive contaminants. In practice, Kapton tapes are coated on one or both sides with FEP fluoropolymer resin as a moisture barrier and improve its chemical resistance. ETFE can also suffer from moisture when carbonyl difluoride, released during the manufacturing process of extrusion, reacts with moisture and leads to the formation of hydrogen fluoride.

For example, aeronautic cables could be improperly tied in contact with contaminated hydraulic lines. At least three problems might be involved– chemical attack, chafing, and abrasion [28, 72]. High-temperature underground cables used in oil and gas exploration are also susceptible to high levels of contamination. The contaminants consist mainly of hydrocarbons that are capable of getting access through extruded insulating materials into the center of the cable. It is obviously important to avoid inflammable liquids from penetrating to the cable. To overcome this issue, an insulating sheath is usually added as a liquid barrier over the main insulation. In addition to fluoropolymers that have excellent resistance to chemicals, PEEK material is currently one of the most used insulations in cables for the petrochemical industry due to its high tensile strength and excellent chemical and thermal stability. Because of PEEK's high cost, industrial interest is being driven toward other high-temperature materials with similar properties such as PSU.

4.2.2.12 Recapitulation of Important Standards for Cables Testing

In order to fulfil the required specifications and to make an adequate selection of the insulation material, there are many important standards. Such standards specify tests and provide recommendations for cables or material qualifications. At the end of this section, some of the most relevant standards are gathered in Table 4.2, as they can come in helpful for researchers and manufacturers to understand testing and qualification procedures on insulating materials.

4.3 Review of the Most Relevant Electrical Characteristics of High Temperature Insulating Materials

4.3.1 Dielectric Spectroscopy

Dielectric spectroscopy measurement is a widely used tool for evaluating the performance of power cables at high temperatures as it provides information about

Table 4.2 Most relevant standards for measuring different properties of insulating materials and cables.

Life service	
ASTM D2304	Standard test method for thermal endurance of rigid electrical insulating materials
UL746C	Polymeric materials, use in electrical equipment evaluations file review guidelines
Electrical tests	
ASTM D149	Standard test method for dielectric breakdown voltage and dielectric strength of solid electrical insulating materials
ASTM D3755	Standard test method for dielectric breakdown voltage and dielectric strength of solid electrical insulating materials under direct-voltage stress
ASTM D150	Standard test methods for AC loss characteristics and permittivity (dielectric constant) of solid electrical insulation
ASTM D1531	Standard test methods for relative permittivity (dielectric constant) and dissipation factor by fluid displacement procedures
ASTM D257	Standard test methods for DC resistance or conductance of insulating materials
ASTM 2275	Standard test method for voltage endurance of solid electrical insulating materials subjected to partial discharges (corona) on the surface
ASTM D3032	Standard test methods for hookup wire insulation
ASTM D495	Standard test method for high-voltage, low current, dry arc resistance of solid electrical insulation
AS5692B	ARC fault circuit breaker (AFCB), Aircraft, Trip-free single phase and three phase 115 VAC, 400 Hz – constant frequency
AS6019A	ARC fault circuit breaker (AFCB), aircraft, Trip-Free 28 VDC
IEC 60112	Method for the determination of the proof and the comparative tracking indices of solid insulating materials
IEC 60343	Recommended test methods for determining the relative resistance of insulating materials to breakdown by surface discharges
IEC 60853	Calculation of the cyclic and emergency current rating of cables
IEC 60270	High – voltage test techniques – partial discharge measurements
Mechanical tests	
ASTM D882-91	Standard test method for tensile properties of thin plastic sheeting
ASTM D1894	Standard test method for static and kinetic coefficients of friction of plastic film and sheeting
ASTM D1708	Tensile testing for plastics using micro tensile specimens
ASTM D638	Standard test method for tensile properties of plastics
ASTM D790	Flexural test of plastics and composites
ASTM D3028	Standard test method for kinetic coefficient of friction of plastic solids
ASTM D955	Standard test method for measuring shrinkage from Mold dimensions of thermoplastics

Table 4.2 (Continued)

Thermal characterizations	
ASTM C177	Standard test method for steady-state heat flux measurements and thermal transmission properties by means of the guarded-hot-plate apparatus
ASTM D2236	Standard test method for dynamic mechanical properties of plastics by means of a torsional pendulum
ASTM E794-85	Standard test method for melting and crystallization temperatures by thermal analysis
ASTM D5470	Standard test method for thermal transmission properties of thermally conductive electrical insulation materials
ASTM E831-19	Standard test method for linear thermal expansion of solid materials by thermomechanical analysis
ASTM D648	Standard test method for deflection temperature of plastics under flexural load in the edgewise position
IEC 60287	Electric cables calculation of the current rating thermal resistance calculation of thermal resistance
Flammability, smoke and toxicity tests	
ASTM D2863-87	Standard test method for measuring the minimum oxygen concentration to support candle-like combustion of plastics (oxygen index)
ASTM E1354	Standard test method for heat and visible smoke release rates for materials and products using an oxygen consumption calorimeter
ASTM E662	Standard test method for specific optical density of smoke generated by solid materials
BS 6387	Test method for resistance to fire of cables required to maintain circuit integrity under fire conditions
DIN 4102-12	Fire behavior of building materials and elements – fire resistance of electric cable systems required to maintain circuit integrity – requirements and testing
EN 13501-6	Fire classification of construction products and building elements – part 6: classification using data from reaction to fire tests on power, control and communication cables
EN 50200	Method of test for resistance to fire of unprotected small cables for use in emergency circuits
EN 50265-2-1	Common test methods for cables under fire conditions – test for resistance to vertical flame propagation for a single insulated conductor or cable
EN 50399	Common test methods for cables under fire conditions
EN 61034	Measurement of smoke density of cables burning under defined conditions
EN ISO 1716	Heat of combustion (CALORIC BOMB).
EN60754	Test on gases evolved during combustion of materials from cables
IEC 60331	Tests for electric cables under fire conditions – circuit integrity
IEC 60332-1-3	Tests on electric and optical fiber cables under fire conditions

(Continued)

Table 4.2 (Continued)

IEC 754-2	Determination of degree of acidity of gases evolved during the combustion of mater
ISO 4589	Plastics – determination of burning behavior by oxygen index
ISO 5660	Reaction to fire tests heat release, smoke production and mass loss rate
NES 713	Burning smoke toxicity index test chamber apparatus for combustion test
NFPA 271	Standard method of test for heat and visible smoke release rates for materials and products using an oxygen consumption calorimeter
NFPA-258	Recommended practice for determining smoke generation of solid materials
UL 2196	Fire test for circuit integrity of fire-resistive power, instrumentation, control and data cables
UL 94	UL standard for safety tests for flammability of plastic materials for parts in devices
Chemical tests	
ASTM D2303	Standard test method for liquid-contaminant, inclined-plane tracking and erosion of insulating materials
ASTM D570	Standard test method for water absorption of plastics
ASTM D792	Standard test method for density and specific gravity of plastics
BS 6425	Methods of test for gases evolved during combustion of electric cables
IEC 60754	Test on gases evolved during combustion of electric cables
Irradiation tests	
IEC 544	Guide for determining the effects of ionizing radiation on insulating materials
ISO 178	Plastics – determination of flexural properties
All aspects	
Aeronautic and aerospace	
AS50881	Aerospace standard, wiring aerospace vehicle
ASTM F2696	Standard practice for inspection of aircraft electrical wiring systems
EN 3475-100	Aerospace series – cables, electrical, aircraft use – test methods
EN 4681	Aerospace series – cables, electric, general purpose, with conductors in aluminum or copper-clad aluminum
EN3197	Aerospace series – design and installation of aircraft electrical and optical interconnection systems
IPC-D-620	Design and critical process requirements for cable and wiring harnesses
Nuclear	
IEEE 383	Standard for qualifying electric cables and splices for nuclear facilities preview

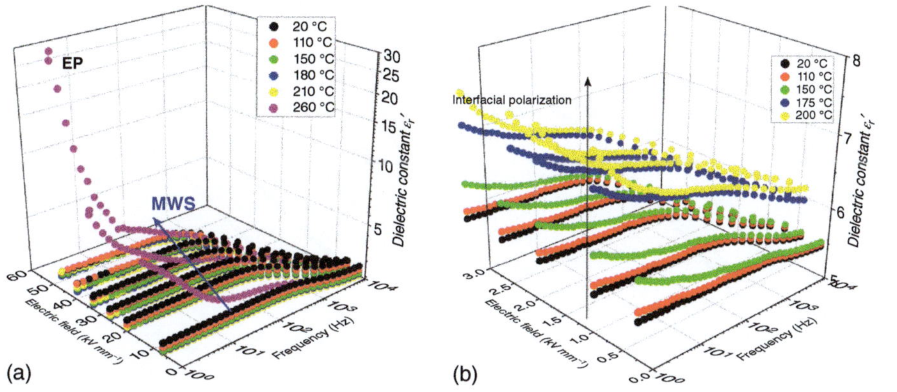

Figure 4.14 Dielectric constant as a function of temperature and electric field– (a) PI/FEP film, (b) PEEK sample. Source: Kemari et al. [73]/IEEE.

the dielectric properties of the insulation material. This information can be used to evaluate the quality and the reliability of the cable insulation, including its dielectric constant, and dissipation factor. Dielectric losses in high-temperature cables can affect their performance by causing an increase in heat generation, leading to a reduction in the life of the cable. High temperatures can also cause a change in the dielectric constant of the insulation material. Therefore, it is important to carefully select materials with low dielectric losses and high thermal stability to ensure a reliable and efficient operation of high-temperature cables.

In the case of high temperature materials, many researches have shown that dielectric constant and AC conductivity reveal significant changes above 200 °C or even below. Examples of spectroscopy measurements on polyimide and PEEK materials are presented in Figure 4.14. High voltage frequency domain dielectric spectroscopy was carried out on pure polyimide (Kapton type) covered on each side by FEP layer and PEEK-based samples. The maximum applied voltage on each sample is 1.4 kV. Besides, the conductivity variations are illustrated in Figure 4.15 for both materials.

A drastic increase in permittivity is clearly observed, mainly at low frequencies. Moreover, the conductivity becomes frequency-independent and this represents the electrical signature of electrode polarization (EP). Such polarization is caused by the buildup of charge carriers excited by heat near the polymer/electrode interface. High electric fields cause electrical conductivity to become more dependent on temperature. Similar trends with temperature increase can be noticed for PEEK material where both dielectric constant and conductivity began to rise above the PEEK's glass temperature (150 °C).

Dielectric properties may also change over time due to various degradation processes, essentially when materials are exposed to high temperatures within their maximum operating one. The physical and chemical processes underlying these changes differ from one material to another. For example, the increase of the dielectric constant of semi-crystalline material can be generally related to a reduction in crystallinity. Chemical changes including oxidation and dehydrofluorination

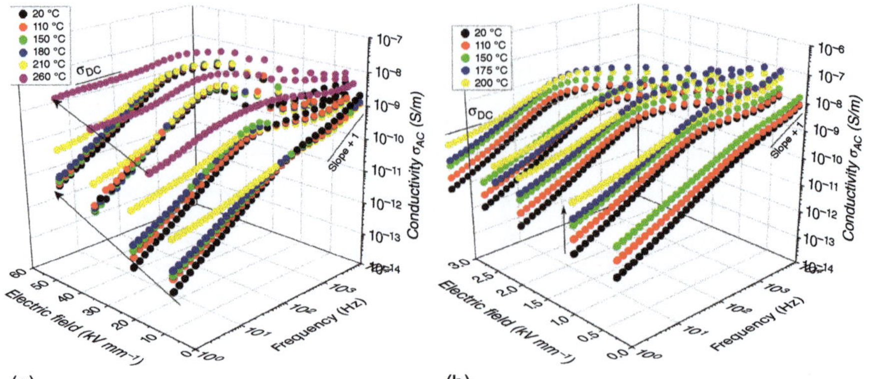

Figure 4.15 AC conductivity as function of temperature and electric field: (a) PI/FEP film, (b) PEEK sample, Source: Reproduced with permission from Kemari et al. [73] IEEE.

due to aging may also contribute to the changes in dielectric properties [74]. Therefore, it is possible to identify any changes in the cable's behavior that may indicate degradation or other issues, by measuring these properties over a range of frequencies and temperatures. This information can also be used to optimize the design and manufacturing process to improve cable performance and extend its life.

4.3.2 Partial Discharges

The increase of voltage levels, combined with high operating temperatures, will increase the risk of partial discharges (PD) occurrence in power cables. PD ignition poses an important concern in aerospace and aeronautic applications where applied voltages exceed 300 V. According to Paschen's law, PD existence is aggravated at high altitude due to low air pressure. For example, the air breakdown voltage of two parallel plates with an air insulation distance of 3 mm at sea level is 10 kV. However, at an altitude of 10 km, discharges will be triggered for the same electrode setup even for an air gap of ~10 mm under the same voltage. This presents the main challenge regarding the compact nature of wiring systems in aircraft as the insulation mass and the volume occupied by cables should be kept as low as possible.

Low pressure related problems could also be encountered inside containment of some small modular reactors (SMRs) where high-temperature cables can face high-temperature environment (250 °C) under low-pressure vacuum less than 70 mbar. Such cables are generally insulated with PEEK or polyimide materials [75]. As illustrated in Figure 4.16, the typical main types of partial discharge (PD) that can take place in unbraided cables are– discharges between the cable and a flat-grounded screen, discharges between cables, and discharges within the cable insulation as a result of voids or cracks, either inherent in the material or from manufacturing processes. As mentioned before, PD might be initiated in wrapped cables within air voids between the conductor and the insulation. For screened

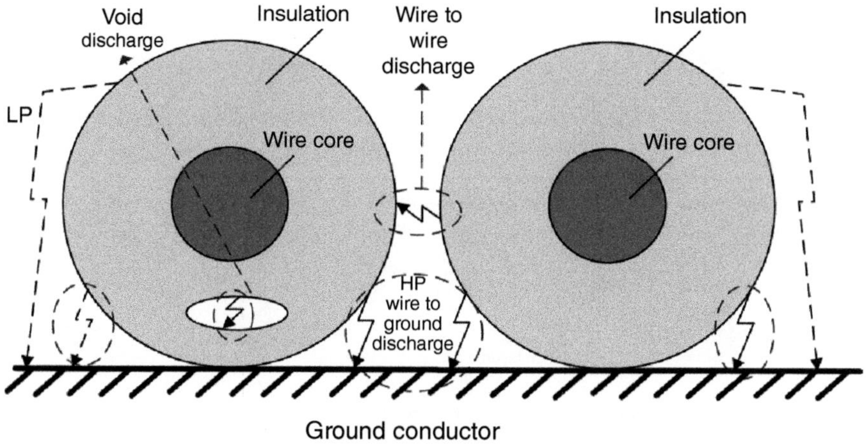

Figure 4.16 Some potential locations of PD in cables according to the pressure level, Source: Reproduced with permission from Christou and Cotton [55] IEEE.

cables with non-tightly installed braids, PD may occur between the insulation surface and the grounded braid [76].

Indeed, simulations of an aeronautic wrapped cable using finite element method (FEM), illustrated in Figure 4.17b,c, highlights hot spots due to the electric field enhancement within air voids between the conductor and the inner insulation layer (Kapton coated with FEP fluoropolymer). These simulations were performed using the cross-section image of the cable in Figure 4.17a, by pursuing a complex procedure in order to create a representative geometry with the exciting defects. Some air voids can also be seen between insulation layers and can be the location of PD as well.

Furthermore, the risk of PD would be significantly higher considering the electrification trend and the harsh environments in which wrapped aeronautical cables may be mounted (high temperature combined with low air pressure due to high altitude) [77, 78]. In addition, increasing insulation thickness is not sufficient to prevent partial discharges without making the cable design oversized [79, 80]. Consequently, the extrusion process has attracted a lot of attention recently by many industries to limit air voids in high temperature cables, mainly for aeronautical and aerospace applications. The triple extrusion by adding inner- and outer semi-conductive layers seems necessary. The development of high temperature materials for high quality extrusion has already been envisaged for future cables used at elevated temperatures.

In the model shown in Figure 4.18a of a cable with a triple extrusion (semiconductor/insulation/semiconductor), the electric field appears to be zero outside the insulation layer thanks to the semiconductor (Figure 4.18b). With the presence of an air void inside the insulation bulk, the electric field is maximal in this cavity (Figure 4.18c). Besides, the triple point at cable terminations, where the insulation, the metal shielding braid, and air join, is needed to be managed by retaining the electric field below the threshold value for PD initiation.

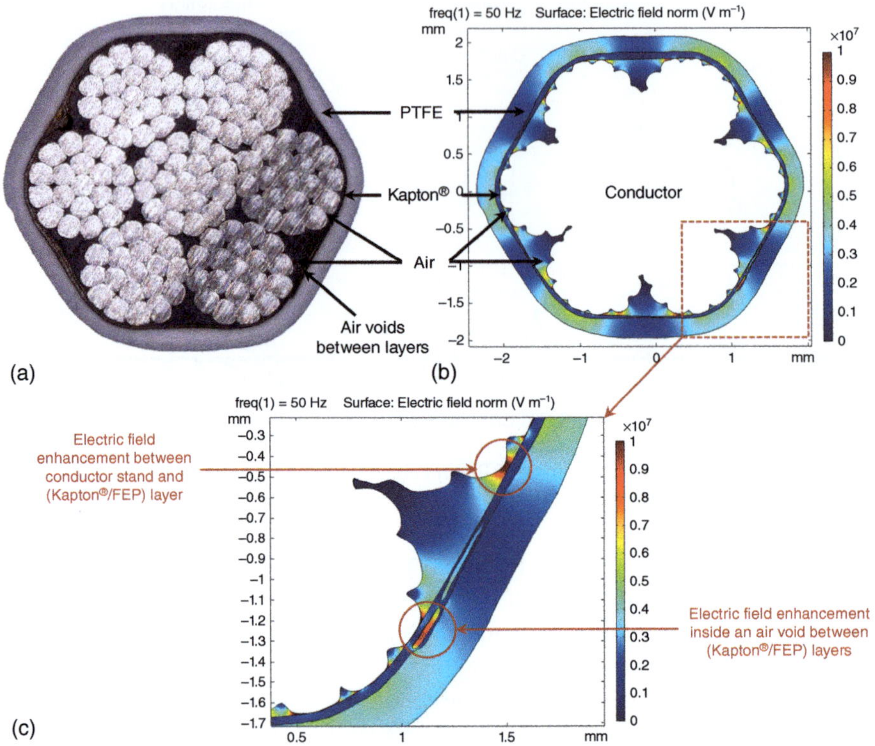

Figure 4.17 (a) Example of a microscopy cross-section of a wrapped (KT) aeronautic cable, (b) electric field cartography of the cable cross-section: voltage applied to the conductor: 1 kV (50 Hz) at 200 °C, (c) Zoom on the electric field distribution and example of weakness points of the tape wrapped cable.

From a material perspective, the choice of the insulation type can have a considerable influence on limiting partial discharge (PD) occurrence in cables. Although not a fundamental parameter of insulation materials themselves, partial discharge inception voltage depends on surface and intrinsic dielectric properties of the insulation (permittivity, surface, and bulk conductivities). It is clearly desirable to have material with low permittivity and electric conductivity.

Moreover, materials will react differently to a specific discharge event depending on those electrical properties and on their chemical nature. When a discharge mechanism occurs, the insulation surface endures several degradation mechanisms (electron/ion bombardment, local heating, chemical reaction), with some of them depending on the material's nature. However, despite the differences between the polymers, their resistance to discharge erosion is not sufficient to enable working under partial discharge activity.

This is all the more true for the townsend's secondary emission coefficient "γ", as several recent researches attempted to estimate a "macroscopic" γ for various applications [81, 82]. It has been found that this coefficient is strongly dependent upon the type of insulating polymer. For example, the estimated γ values at 20 °C

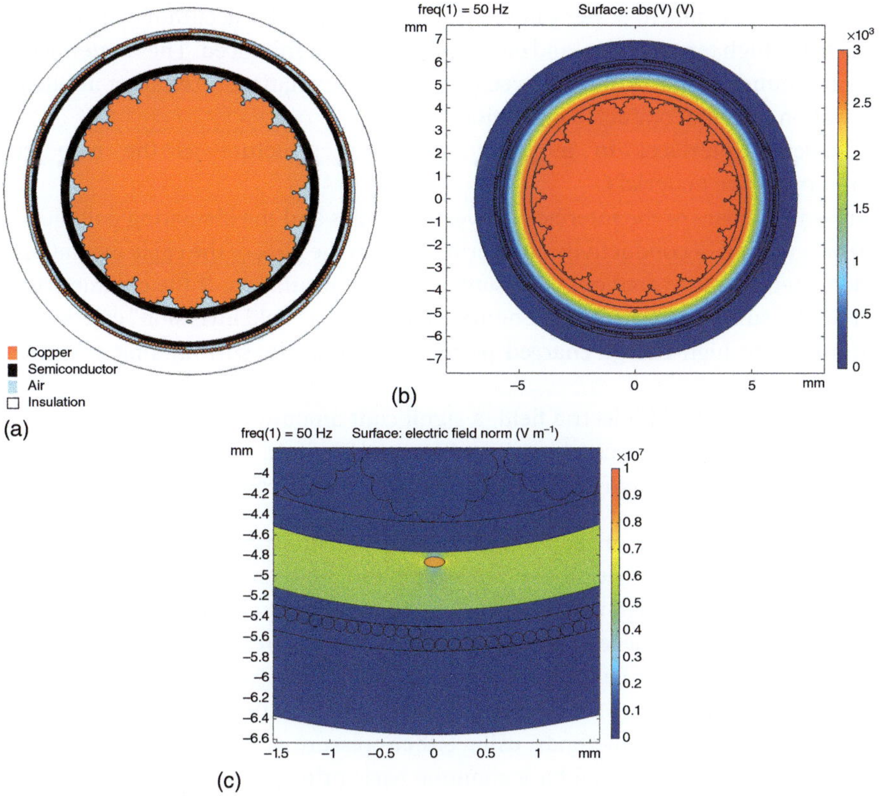

Figure 4.18 (a) Example of a cable model with triple extrusion, (b) electric potential cartography: voltage applied to the conductor: 3 kV (50 Hz) at 200 °C, (c) Zoom on the electric field distribution.

are 6.5×10^{-4} for PI with FEP layer and 10^{-5} for PTFE, which is very low compared to the conventional value reported in literature for metallic electrodes, which is 10^{-2} [82]. According to the relation between Paschen's law and γ coefficient, the lowest is γ the highest is the PDIV. More in-depth investigations are needed to be undertaken regarding the secondary emission coefficient.

4.3.3 Space Charge and DC Conductivity

The space charge property has become a crucial factor in recent years for evaluating the performance of polymeric materials in harsh conditions. Space charge measurements provide valuable insights to cable material manufacturers, enabling them to optimize the insulating material for high-temperature applications and enhance the reliability of polymeric cables [83].

In general, space charge can be built-up for different reasons as follows [13, 83]:

- Bulk impurities, such as cable manufacturing by-products, can act as traps for electrons and ions, which lead to the formation of space charges.

- Interfaces on a macro-scale where space charge can be accumulated and/or injected at high temperature, and electric fields can be enhanced. These interfaces can be found, in particular, between conductor-insulation, insulation-another insulation, or semiconductor-insulation.
- Micro-scale interfaces in a non-homogeneous structure as the case in semi-crystalline materials.
- Nano-mechanism of the interface between particles and the insulating polymer.
- Spatial gradient of conductivity and permittivity in the bulk of the cable insulation when a temperature gradient is present across the insulation. If the conductivity gradient is sufficiently large, a phenomenon known as "field inversion" can occur.
- Exposition to high-energy charged particles exists in the surrounding environment.
- When subjected to AC electric field, a significant amount of charge injection or extraction leads to the recombination of trapped charges, which releases energy to other electrons through radiation and generates high-energy charge carriers.

Space charge phenomenon, including charge origin, transport, accumulation, and dissipation processes, has been identified as a potential trigger for the degradation of cable insulation under high voltage [84]. Accordingly, researchers and manufacturers became aware of the necessity to develop reliable measurement systems to study and evaluate space charge behavior in cables, especially at high temperatures, which is the most aggravating factor. The second half of the twentieth century witnessed the appearance of several methods, using either pressure or temperature as a stimulus to reveal charges. They all have common basic principle consisting in charges excitation at the electrodes, which results in a transient current or voltage, which can be recorded and analyzed. Examples of non-destructive methods using pressure are PEA (Pulsed Electro Acoustic) and LIPP (Laser Induced Pressure Pulse) methods. The LIMM or FLIMM (Focused Laser Impulse Modulation Method) and the Thermal Step Method (TSM) belong to the thermal methods [85].

Today, PEA and high-voltage TSM are the most established methods for cable characterizations. Nevertheless, it must be recalled that the characteristics of high-temperature materials and the structure of some cables make it a complicated task to assess the space charge with such measuring techniques. It is also worth noting that the previous methods have been extensively applied to study XLPE material for HVDC underground and submarine cables.

On the other hand, other approaches might gear widespread industrial interest by studying the polarization phenomena of the insulation. In fact, the relationship between polarization current and space charge is a feedback loop in which the polarization current contributes to the buildup of space charge, which in turn leads to an increase in polarization current. Therefore, polarization/depolarization current (PDC) and return voltage measurements (RVMs) can be very beneficial to determine the threshold field for space charge injection/accumulation and its change with temperature [86]. Choosing the appropriate method to study space charge in high temperature materials depends on many factors such as measurement conditions, sensitivity, material thickness, and cable design.

As long as wrapped cables are concerned, space charge measurements in the tape wrapped cables are more complex than in extruded ones. For instance, the PEA signal is influenced by the multilayer formed due to difference of dielectric properties between the different layers. When associating PTFE and PI, in KT or TKT cables, the charge resolution within PI is further distorted due to the strong diffusivity of PTFE for acoustic waves. In some cases, this issue can be overcome when space charge tests are performed on a design's thick insulating layer to fit to the measurement method, which is not feasible with tapes used in aerospace cables. In KT and TKT cables, the PTFE layer, which is at the outer part of the cable, is in contact with the detection cell, and substantially broadens the signal issued from the PI layers. Therefore, space charge measurements in cables having such structure can be analyzed from the PTFE side only, with indirect effects linked to charge build-up in PI at the interface [87].

Moreover, the electric field distribution in wrapped cable can reveal electric field intensifications at the inter-strand air gaps due to the small air thickness that exists between the conductor strands and the inter insulating layer. Therefore, the electric field in the insulation layers close to these air gaps is directly modified. The electric field intensity increases when the temperature gradient becomes more important. The defects that might take place between insulating layers can lead to high local fields in the fault. The localization of the fault in the insulation layer has been observed and when the fault is far from the conductor of the cable, its internal electric field decreases [41].

Regarding extruded cables, many measurements on underground and submarine HV cables with XLPE material have demonstrated that space charges also accumulate in the vicinity of the insulation surfaces, close to the inner and the outer semi-conductors. Homo-charges (same polarity as the neighboring semi-conductor), or hetero-charges can be caused by charge injection from the semi-conducting surfaces. This space-charge accumulation has led to failures in HVDC extruded cable insulation systems.

The space-charge related field distortion problems can be resolved only by carefully selecting the properties of both the insulation and screen materials [88]. This becomes increasingly important when using high-temperature materials such PFA and other potential candidates for high-temperature cables. Even high-temperature insulation may face interface issues and significant temperature gradients. Further experimentation is necessary to examine space charge behavior in extruded cables made of such materials. From materials perspective, space charge measurements were carried out on PI, PEEK, and some fluoropolymer films with different thicknesses by PEA method at different thermal conditions [89–92]. In Figure 4.19 are presented color maps of the charge density and the electric distribution under 20 kV mm^{-1} at 25 °C and at 50 °C with a thermal gradient across each sample.

As the temperature increases, the maximum trapped charge inside the bulk of ETFE, PEEK, and FEP shows enhancement, while the maximum charge density changes slightly for PI and PTFE. This correlation should be linked to the morphology of these materials. Furthermore, the accumulation of space charge can alter

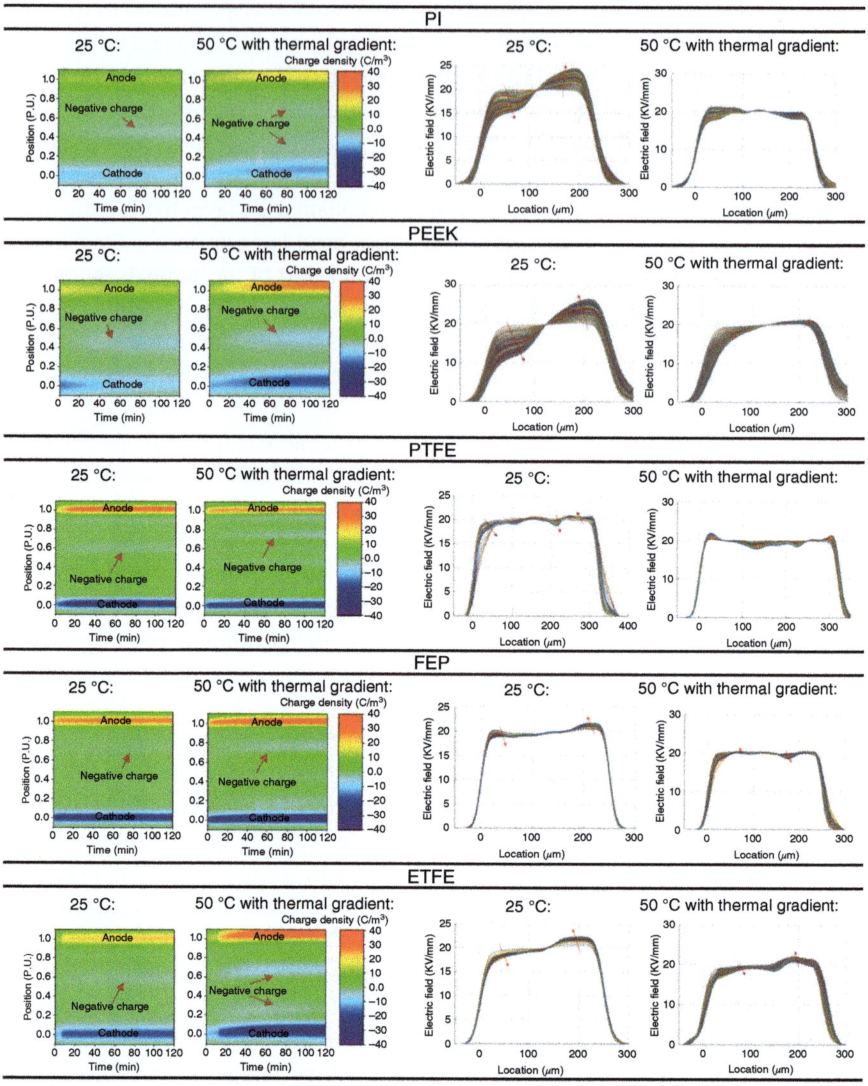

Figure 4.19 Charge density and electric field distributions for different high temperature materials under 20 kV mm^{-1} at 25 and 50 °C with thermal gradient. Left: space charge distributions over time at 25 and 50 °C, Source: Reproduced with permission from Baferani et al. [89] IEEE. Right: Space charge spatial distribution at 25 and 50 °C, Source: Reproduced with permission from Baferani et al. [90] IEEE.

the electric field distribution and potentially cause insulation breakdown. Among the five materials tested, FEP showed the best performance with the lowest accumulated charge in bulk, while ETFE and PEEK exhibited the highest space charge accumulation at 25 and 50 °C with a temperature gradient.

The DC electric conductivity has been determined as well for the above materials via PDC measurements up to 200 °C (Figure 4.20). The thermal activation energy is

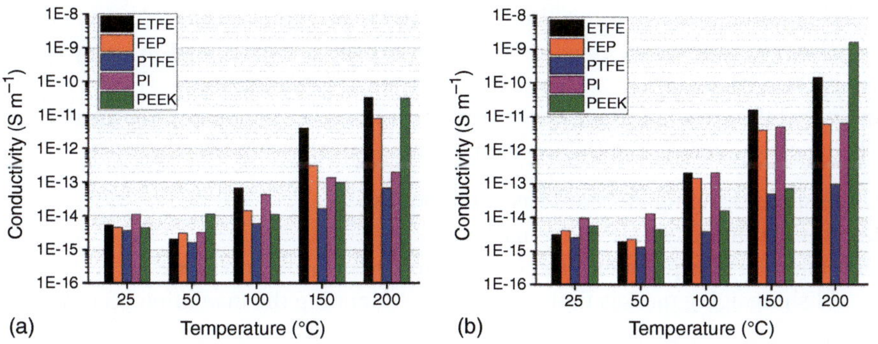

Figure 4.20 DC conductivity of different high-temperature materials versus temperature: (a) under 10 kV mm^{-1}, (b) under 30 kV mm^{-1}, Source: Reproduced with permission from Baferani et al. [89] IEEE.

Table 4.3 Activation energy of different high-temperature materials under 10 and 30 kV mm^{-1}.

Materials	Activation energy (eV)	
	10 kV mm^{-1}	**30 kV mm^{-1}**
ETFE	0.262	0.323
FEP	0.223	0.220
PTFE	0.087	0.109
PI	0.087	0.195
PEEK	0.267	0.378

Source: Reproduced with permission from Baferani et al. [89] IEEE.

also given for every material in Table 4.3. Activation energy reflects the relationship between a material's electrical conductivity and the increase in temperature. The greater the activation energy, the more the electrical conductivity is influenced by temperature changes. It appears that electrical conductivity becomes more temperature dependent under high electric fields.

The activation energy is the lowest for PTFE and PI up to 200 °C. This seems to correlate very well with spectroscopy measurement in low frequency domain. Indeed, as the temperature increases, the maximum trapped charge inside the bulk of ETFE, PEEK, and FEP rises, while the maximum charge density only slightly changes for PI and PTFE. This can be linked to the morphology of these materials. ETFE, FEP, and PEEK are semi-crystalline, so they present more than one phase, making their electrical conductivity temperature dependent. Conversely, PI is highly amorphous, and PTFE is highly crystalline (if it is not sintered).

4.3.4 Aging and Degradation

Degradation of insulating materials has been mentioned several times before in this chapter since it represents an inevitable problem that requires in-depth and serious research. Overall, the major incentives of studying the aging of insulating materials are– safety, reliability, cost-effectiveness, environmental impact, and industry standards of cable systems.

In light of this, we find that most of scientific researches about aging of insulation are being conducted for the following main reasons:

- Understanding aging mechanisms and kinetics.
- Determination of reliable aging markers.
- Development of experimental techniques to assess insulation aging.
- Establishing aging models that might allow estimating the insulation lifetime.

Common failures of power cables are caused by frequent maintenance procedures, repeated bending, and exposure to environmental constraints (high temperature, moisture, and radiation) as well as electrical ones (high electric field, frequency, and transient voltage). The material degradation can also contribute to the issue, and it is not easily visible during regular inspections.

Research work [75] was undertaken on polyimide material under severe constraints representative of those encountered advanced nuclear reactors, i.e. 1000 hours at 250 °C in atmospheric and in low pressure vacuum (70 mbar). A discoloration of polyimide in vacuum environment has been noticed and it was a result of the interaction between the silver-plated copper conductor and the polyimide. Meanwhile, no significant effect on thermal or chemical properties of polyimide was observed, in both atmospheric and vacuum conditions.

On the other hand, different research [24] was carried out on polyimide insulated cables in a bent position at different aging temperatures between 250 and 300 °C for more than 5000 hours. Experimental results have shown that thermal stress had the greatest influence on weight loss and insulation electrical properties, mainly when the cable is subjected to a mechanical strain. Meanwhile, high moisture has the greatest negative effect on the tensile and elongation characteristics of the insulation compared to a simple thermal aging mode in the oven. This is attributed to the mechanism of hydrolytic scission in polyimide. The aromatic polyimide consists of repeating chains formed through the condensation polymerization of dianhydride and diamine. High moisture causes the polymer chain to absorb water through its imperfections, causing the insulation to swell. This swelling allows liquids like water and cleaning fluids to penetrate the insulation [48]. Indeed, hydrolytic degradation represents a major disadvantage of using polyimide. It has also been concluded that cables that are protected from both static (bent) and dynamic (abrasion and contamination) stressors will have a longer lifespan.

In the same study [24], cables with PTFE insulation around polyimide were investigated. As expected, PTFE provided protection for polyimide from external stressors (oxidation and hydrolysis). It is worth noting that this study estimated that the lifetime of cables (temperature = 71 °C, moisture rate = 33%) can be reduced to about half (44 000 hours) when they are bent, in comparison to cables that are not mechanically disturbed (81 000 hours).

The aging mechanism in fluoropolymers is different as they have chemical composition. In general, thermal aging at temperatures higher than 300 °C can promote recrystallization of PTFE, which may affect some dielectric polarizations as mentioned earlier. Similar findings were reported elsewhere on FEP, it was found that thermal exposure of FEP might result in an increase in its crystallinity. Moreover, a combination of X-ray and increased thermal exposures lead to a greater increase in crystallinity compared to heating alone [71].

As long as ETFE is concerned, thermal exposure of ETFE at 160 °C leads to chemical changes, including thermal oxidation and dehydrofluorination, which alters its dielectric properties. Indeed, chain breakage may increase material viscosity and reduce the mobility of charge carriers, hence, decreasing the dielectric constant. On the other hand, polar group formation enhances charge carrier mobility so the dielectric constant rises. Therefore, the dielectric behavior of the material is likely to be determined by the dominant aging mechanism. Such scenario makes it very complicated to correlate certain aging results. According to thermogravimetric analysis by [93], PFA degradation follows straightforward kinetics with an activation energy significantly higher than that of PTFE and PI. It has been suggested that an equivalent mass loss of PTFE at 250 °C can be reached for PFA at a higher temperature of 333 °C.

PEEK can also suffer from thermal aging, mainly at temperatures higher than the glass transition (>150 °C). The PEEK degradation occurs through various mechanisms such as chemical bond breakage, oxidation, cross-linking, and char formation [94]. The temperature is a crucial factor in degradation, while irradiation speeds it up. It has been reported that the mechanical properties (tensile strength and elongation at break) of PEEK may deteriorate faster than the dielectric ones during aging, making the mechanical properties a good marker for monitoring the condition of PEEK-insulated cables. The tensile strength may increase when PEEK is exposed to radiation in addition to heat, hence, PEEK becomes more brittle and loses its flexibility after combined aging. Additionally, dielectric properties measured at low frequency can be a reliable predictive marker of degradation [95, 96]. In Figure 4.21 are presented different examples Fourier-transform infrared spectroscopy (FTIR) analysis on different aged materials to reveal their chemical structure changes.

To sum-up, the aging mechanisms that may occur in high temperature insulation (oxidation, hydrolysis reactions, cross-linking, and chain scission) are strongly dependent upon the chemical and physical composition of the material (nature of chemical bonds, presence of additives) as well as the type of the aging constraints and their severity (temperature, contaminants, mechanical stress, and high electric field). Furthermore, the most common degradation markers in the field of high-temperature cables can be classified in two main groups that are:

- Physical aging markers include changes in physical appearance, such as cracking, discoloration, and weight loss. The changes in mechanical properties, such as tensile strength and elongation at break, are also included.
- Electrical aging markers include changes in dielectric constant, dissipation factor, and conductivity, essentially at low frequencies as well as in DC. An increase in insulation resistance can also be an indicator of aging.

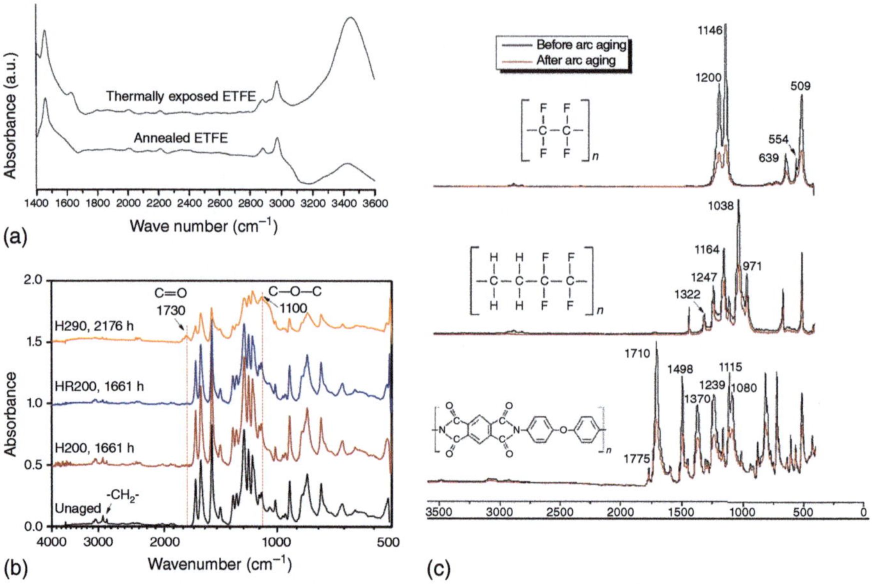

Figure 4.21 Changes in FTIR spectra of different materials aged under different conditions: (a) ETFE before and after thermal aging at 160 °C, Source: Reproduced with permission from Li et al. [74] IEEE, (b) aged PEEK samples under thermal aging (H) and combined aging with heat and gamma rays (HR), Source: Reproduced from Yang et al. [94], with permission from Elsevier, (c) FTIR before and after arc erosion, from top to bottom: PTFE, ETFE, and polyimide, Source: Reproduced with permission from Li et al. [97] IEEE.

4.4 Trends and Outlooks

From the current view, it is not very common to find high-temperature materials as electric insulating in power cables since they are very limited to specific zones in aircraft, NNPs, and oil exploration for instance. However, as demand for higher power densities persist, power cables must meet a number of requirements of operating and environmental conditions which become more challenging (high temperatures, contaminants, vibrations, humidity, low pressure, and radiation). The demand for high-temperature polymers for cable insulation is expected to increase due to the growing demand for high-performance cables in various industries, including aerospace, automotive, and electronics.

The aerospace industry is driving the development of new high-temperature polymers for cable insulation due to trends such as the use of electric and hybrid propulsion systems, which require power cables that can withstand higher voltages and temperatures. Additionally, the development of Ultra High Bypass Ratio (UHBR) aircraft engines, which are designed to be more fuel-efficient and have lower emissions. Therefore, higher temperatures are required rather than traditional turbofan engines. This can result in higher constraints in the engine compartments and surrounding areas, requiring specialized materials, coatings, or insulation to ensure that the cables maintain their performance and do not degrade or fail over time.

Last but not least, future zero-emission aircraft could use fuel cells technology to supply electrical motors. The use of a high-temperature fuel cell is a way to improve the efficiency of such systems. Their operating temperature typically ranges from 600 to 1000 °C, whereas a low-temperature fuel cell operates at temperatures below 100 °C. Therefore, this requires the use of high-temperature resistant metals, ceramics, and coatings for the fuel cell components, including the electrodes, electrolytes, and interconnects. In these applications, high-temperature cables must be able to withstand the high temperatures and thermal cycling associated with the fuel cell system.

Finally, future cables require insulating materials with improved performance and reliability to meet the growing challenges. There is ongoing research in the field of high-temperature polymers for cable insulation, with a focus on developing new materials with even higher temperature resistance, improved mechanical and electrical properties, cost-effectiveness, safety and environmental impact. Some of the current research axis include–

- Development of new high-temperature polymers: Researchers are exploring new chemical structures and compositions to develop high-temperature polymers with improved thermal stability and mechanical properties. For example, the use of cross-linkable PEI to develop high-performance cable insulation with improved temperature resistance. There is also a growing interest in the development of polybenzimidazole (PBI) as a high-temperature polymer for cable insulation.
- Nanocomposites: Nanocomposites are composite materials that contain nanoscale fillers, which can enhance the mechanical, electrical, and thermal properties of polymers. Researchers are exploring the use of nanocomposites in high-temperature polymer cable insulation to improve properties such as thermal conductivity, dielectric strength, and flame resistance.
- Sustainable materials: There is a growing interest in developing high-temperature polymer cable insulation materials that are sustainable and environmentally friendly. Researchers are exploring the use of biodegradable and renewable materials, as well as developing recycling techniques to reduce waste.

For the last point, it is very important to take into account the chemical regulation. Today, for example, PFA polymer has been proposed as a viable option for aeronautic cables. However, polyfluoroalkyl substances (PFASs) have been reported by the European Chemicals Agency REACH (Registration, Evaluation, Authorization, and Restriction of Chemicals) as environmental pollutants and the developed PFA intended for cables is being evaluated to determine if PFA responds to the requirements outlined in REACH.

In addition to the necessity of developing high-temperature materials, design-related issues must be solved in order to achieve the desired reliability of cables with lightweight and high performances. Furthermore, more attention is being given to developing new characterization techniques suitable for high temperature use. Such experimental methods can contribute to the choice of the insulating material(s) for power cables.

Author's Note

Y. Kemari is currently working at the Technology Engineering Center, Capgemini Eng., 4 avenue Didier Daurat, Blagnac, 31700, France.

References

1 Thue, W.A. (2012). Historical perspective of electrical cables. In: *Electrical Power Cable Engineering*, 2e (ed. W.A. Thue), 1–9. Boca Raton: CRC Press.
2 Worzyk, T. (2009). Submarine power cables and their design. In: *Submarine Power Cables Design, Installation, Repair, Environmental Aspect* (ed. T. Worzyk), 10–48. Berlin Heidelberg: Springer-Verlag.
3 Drobny, J.G. (2012). Introduction. In: *Polymers for Electricity and Electronics: Materials, Properties, and Applications* (ed. J.G. Drobny), 1–24. New Jersey: John Wiley & Sons, Inc.
4 Cooke, T. (2021). Coaxial cables. In: *Global Cable Industry: Materials, Markets, Products* (ed. G. Beyer), 331–349. Belgium: Wiley-VCH GmbH.
5 Zuidema, C., Kegerise, W., Fleming, R. et al. (2011). A short history of rubber cables. *IEEE Electrical Insulation Magazine* 27 (4): 45–50.
6 Orton, H. (2013). History of underground power cables. *IEEE Electrical Insulation Magazine* 29 (4): 52–57.
7 Mazzanti, G. (2013). Fundamentals of HVDC cable transmission. In: *Extruded Cables for High-Voltage Direct-Current Transmission* (ed. G. Mazzanti and M. Marzinotto), 11–38. New Jersey: JohnWiley & Sons, Inc.
8 Black, R.M. (1983). *The History of Electric Wires and Cables*. London: P. Peregrinus in association with the Science Museum.
9 Moore, G.F. (1997). Introduction. In: *Electric Cables Handbook*, 3e (ed. G.F. Moore), 1–6. Blackwell Science Ltd.
10 Barber, K. and Alexander, G. (2013). Insulation of electrical cables over the past 50 years. *IEEE Electrical Insulation Magazine* 29 (3): 27–32.
11 Vahedy, V. (2006). Polymer insulated high voltage cables. *IEEE Electrical Insulation Magazine* 22 (3): 13–18.
12 Geusens, T. (2021). Thermoplastics for cables. In: *The Global Cable Industry: Materials, Markets, Products* (ed. G. Beyer), 21–43. Belgium: Wiley-VCH GmbH.
13 Li, C. (2022). Insulating materials for realising carbon neutrality: opportunities, remaining issues and challenges. *High Voltage* 7 (4): 610–632.
14 Liu, F. (2021). Nuclear power station cables. In: *The Global Cable Industry: Materials, Markets, Products* (ed. G. Beyer), 249–289. Belgium: Wiley-VCH.
15 Ebnesajjad, S. (2017). History of polytetrafluoroethylene and expanded PTFE membrane. In: *Expanded PTFE Applications Handbook: Technology, Manufacturing and Applications* (ed. S. Ebnesajjad), 1–7. Chadds Ford, PA: William Andrew Publishing.

16 Dhanumalayan, E. and Joshi, G.M. (2018). Performance properties and applications of polytetrafluoroethylene (PTFE)—a review. *Advanced Composites and Hybrid Materials* 1: 247–268.

17 Li, Q., Yao, F.-Z., and Liu, Y. (2018). High-temperature dielectric materials for electrical energy storage. *Annual Review of Materials Research* 48: 219–243.

18 Lopez, G. (2021). High-performance polymers for aeronautic wires insulation: current uses and future prospects. *Recent Progress in Materials* 3 (1): 1–15.

19 El Idrissi, M. (2013). Development of new high temperature wires and cables for space applications. In: *1st International Symposium Space Passive Component Days* (24-26 September 2013), European Space Components Coordination (ESCC).

20 Ebnesajjad, S. (2021). From fundamentals to applications. In: *Introduction to Fluoropolymers: Materials, Technology, and Applications, 2*e (ed. S. Ebnesajjad), 7–17. William Andrew imprint of Elsevier.

21 Cardoso, V.F., Correia, D.M., Ribeiro, C. et al. (2018). Fluorinated polymers as smart materials for advanced biomedical applications. *Polymers* 10 (2): 161.

22 Ebewele, R.O. (2000). Polymer properties and applications. In: *Polymer Science and Technology, 1*e (ed. R.O. Ebewele), 419–469. Boca Raton: CRC Press.

23 Nezar, A. and Nagu, S. (1999). Dielectrics and electrical insulation: cable insulation. In: *Wiley Encyclopedia of Electrical and Electronics Engineering*. John Wiley & Sons.

24 Kurek, J. (2008). *Aircraft wiring degradation study*. Washington, DC: Office of Aviation Research and Development, Federal Aviation Administration.

25 Furse, F. and Haupt, R. (2001). Down to the wire. *IEEE Spectrum* 38 (2): 34–39.

26 Barry, J.R., Jenkins, D.R., White, D.J. et al. (2003). Maintenance, Material, and Management Inputs : Vehicle and subsystem analysis" in Columbia accident investigation board report, Appendix D.15, pp. 423–427.

27 Brusso, B.C. (2017). History of aircraft wiring arc-fault protection. *IEEE Industry Applications Magazine* 23 (3): 6–11.

28 F. A. Administration (2018). Aircraft electrical wiring interconnect system (EWIS) best practices, job aid. www.faa.gov/training_testing/training/air_training_program/job_aids (accessed 30 November 2022).

29 Portwood, B. and Sadeghi, M. (2001). *"Aircraft Wiring Practices" An Interactive Training and Self-Study Course (25827) - Version 1, Federal Aviation*, 19–24. Administration (FAA).

30 Borghei, M. and Ghassemi, M. (2021). Insulation materials and systems for more- and all-electric aircraft: a review identifying challenges and future research needs. *IEEE Transactions on Transportation Electrification* 7 (3): 1930–1953.

31 Liu, C. and Yan, L. (2014). Development of aircraft used wires for general purpose and a study on the application. *Applied Mechanics and Materials* 551: 214–220.

32 Rickman, S.L. (2018). *Re-Architecting the NASA Wire Derating Approach for Space Flight Applications*. Hampton, Virginia: National Aeronautics and Space Administration (NASA).

33 Nya, B.H., Brombach, J., and Schulz, D. (2012). Benefits of higher voltage levels in aircraft electrical power systems. In: *Electrical Systems for Aircraft, Railway and Ship Propulsion*, 1–5. Bologna, Italy: IEEE.

34 Madonna, V., Giangrande, P., and Galea, M. (2018). Electrical power generation in aircraft: review, challenges and opportunities. *IEEE Transactions on Transportation Electrification* 4 (3): 646–659.

35 Smruti, S., Zhao, X., and Kyprianidis, K. (2020). A review of concepts, benefits, and challenges for future electrical propulsion-based aircraft. *Aerospace* 7 (4).

36 Barzkar, A. and Ghassemi, M. (2022). Components of electrical power systems in more and all-electric aircraft: a review. *IEEE Transactions on Transportation Electrification* 8 (4): 4037–4053.

37 Wu, C. (2022). Rational design of all-organic flexible high-temperature polymer dielectrics. *Matter* 5 (9): 2615–2623.

38 De Leon, A.C.C., Da Silva, Í.G., Pangilinan, K.D. et al. (2021). High performance polymers for oil and gas applications. *Reactive and Functional Polymers* 162: 104878.

39 Mokhtari, M., Archer, E., Bloomfield, N. et al. (2021). A review of electrically conductive poly(etherether ketone) materials. *Polymer International* 70 (8): 1016–1025.

40 Schmidt, J. et al. (2000). *Assessment and Management of Ageing of Major Nuclear Power Plant Components Important to Safety: In-Containment Instrumentation and Control Cables*, vol. I. Vienna, Austria: International Atomic Energy Agency (IAEA).

41 Diaw, E.H.N. (2021). Modeling of physical phenomena involved in high-voltage direct current (HVDC) energy transport cables in aeronautics. PhD Thesis, Université Toulouse-III-Paul-Sabatier.

42 Thue, W.A. (2012). Cable and manufacturing. In: *Electrical Power Cable*, 2e (ed. W.A. Thue), 153. Boca Raton: CRC Press.

43 Worzyk, T. (2009). Manufacturing and testing. In: *Submarine Power Cables Design, Installation, Repair, Environmental Aspects* (ed. T. Worzyk), 127. Berlin/Heidelberg: Springer.

44 Karadjian, M. (2018). Endurance and Dielectric Strength of Electrical Cable Insulation for Aeronautics. PhD Thesis, GeePs - Laboratoire Génie électrique et électronique de Paris, Paris, France.

45 Diaw, E.H.N., Le Roy, S., and Teyssèdre, G. (2020). Current measurements in high performance polymers used in aeronautic cables. *IEEE Transactions on Dielectrics and Electrical Insulation* 27 (6): 2195–2202.

46 Lebey, T., Rumi, A., and Cavallini, A. (2022). Challenges for electrical insulation systems in high voltage aviation applications. *IEEE Electrical Insulation Magazine* 38 (6): 5–11.

47 Driessen, A.B.J.M., Duivenbode, J.v., and Wouters, P.A.A.F. (2018). Partial discharge detection for characterizing cable insulation under low and medium vacuum conditions. *IEEE Transactions on Dielectrics and Electrical Insulation* 25 (1): 306–315.

48 Moffat, B.G., Abraham, E., Desmulliez, M.P.Y. et al. (2008). Failure mechanisms of legacy aircraft wiring and interconnects. *IEEE Transactions on Dielectrics and Electrical Insulation* 15 (3): 808–822.

49 Lafleur, P.G. and Vergnes, B. (2014). Wire coating and cable insulation. In: *Polymer Extrusion*, 1e (ed. P.G. Lafleur and B. Vergnes), 306. Wiley-ISTE.

50 Giles, H.F., Wagner, J.R., and Mount, E.M. (2005). Wire and cable coating. In: *Extrusion: The Definitive Processing Guide and Handbook, 474* (ed. J.R. Wagner Jr., E.M. Mount III, and H.F. Giles Jr.), 469. William Andrew Publishing.

51 IRT Saint Exupéry. (2021). HIVACS Webinar - high voltage aerospace cable systems. https://www.youtube.com/watch?v=4D6ESyT_-AI&t=1390s (accessed 02 February 2023).

52 Wald, D. and Smedberg, A. (2008). Evolution of medium voltage cable technology in Europe. *IEEE Electrical Insulation Magazine* 24 (5): 31–35.

53 Ebnesajjad, S. (2017). Manufacturing polytetrafluoroethylene by emulsion polymerization. In: *Expanded PTFE Applications Handbook: Technology, Manufacturing and Applications* (ed. S. Ebnesajjad), 25–64. Chadds Ford, PA: William Andrew Publishing.

54 Traskos, M. (2019). High voltage impact the aircraft wiring system. Lectromec www.lectromec.com/high-voltage-impact-the-aircraft-wiring-system (accessed 21 October 2022).

55 Christou, I. and Cotton, I. (2010). Methods for partial discharge testing of aerospace cables. In: *IEEE International Symposium on Electrical Insulation*, IEEE, pp. 1–5. San Diego, CA, USA.

56 Saruwatari, M., Tsuji, S., and Fujii, Y. (1991). Extrusion process of polyimide and polyimide pellet used for the process, US5069848A. United States Patent US5069848A, filed 10 May 1991 and issued 3 December 1991.

57 Lizcano, M., Williams, T.S., Shin, E.-S.E. et al. (2022). Aerospace environmental challenges for electrical insulation and recent developments for electrified aircraft. *Materials (Advances in Electric Insulating Materials and Applications)* 15 (22): 8121.

58 Eugene Shin, E.-S., Scheiman, D.A., and Lizcano, M. (2018). Lightweight, durable, and multifunctional electrical insulation material systems for high voltage applications. In: *AIAA/IEEE Electric Aircraft Technologies Symposium (EATS)*, IEEE, pp. 1–21 (12–14 July 2018). Cincinnati, OH, USA.

59 Hergenrother, P. (2003). The use, design, synthesis, and properties of high performance/high temperature polymers. *High Performance Polymers* 15: 3–45.

60 Zhang, X.-M., Liu, J.-G., and Yang, S.-Y. (2016). A review on recent progress of r&d for high-temperature resistant polymer dielectrics and their applications in electrical and electronic insulation. *Reviews on Advanced Materials Science* 46 (1): 22–38.

61 Giles, H.F., Wagner, J.R., and Mount, E.M. (2005). Compounding. In: *Extrusion: The Definitive Processing Guide and Handbook* (ed. J.R. Wagner Jr., E.M. Mount III, and H.F. Giles Jr.,), 425–434. William Andrew Publishing.

62 Huang, X., Jiang, P., and Tanaka, T. (2011). A review of dielectric polymer composites with high thermal conductivity. *IEEE Electrical Insulation Magazine* 27 (4): 8–16.

63 Dai, W., Yu, J., Liu, Z. et al. (2015). Enhanced thermal conductivity and retained electrical insulation for polyimide composites with SiC nanowires grown on graphene hybrid fillers. *Composites Part A: Applied Science and Manufacturing* 76: 73–81.

64 Ouyang, Y., Bai, L., Tian, H. et al. (2022). Recent progress of thermal conductive polymer composites: Al_2O_3 fillers, properties and applications. *Composites Part A: Applied Science and Manufacturing* 152: 106685.

65 Venkateswarlu, G., Sharada, R., and Rao, M.B. (2015). Effect of fillers on di-electric strength of PTFE based composites. *International Journal of Scientific and Research Publications* 5 (9): 1–9.

66 Du, B., Kong, X.X., Li, J., and Xiao, M. (2019). High thermal conductivity insulation and sheathing materials for electric vehicle cable application. *IEEE Transactions on Dielectrics and Electrical Insulation* 26 (4): 1363–1370.

67 Searle, O.B. and Pfeiffer, R.H. (1985). Victrex® poly(ethersulfone) (PES) and Victrex® poly(etheretherketone) (PEEK). *Polymer Engineering and Science* 25 (8): 474–476.

68 Mustafa, E., Afia, R.S.A., and Tamus, Z.Á. (2020). Condition monitoring uncertainties and thermal - radiation multistress accelerated aging tests for nuclear power plant cables: a review. *Periodica Polytechnica Electrical Engineering and Computer Science* 64 (1): 20–32.

69 Poutch, F. (2021). CPR testing of cables. In: *The Global Cable Industry: Materials, Markets, Products* (ed. G. Beyer), 181–212. Belgium: Wiley-VCH.

70 Tavlet, M., Fontaine, A., and Schönbacher, H. (1998). Compilation of radiation damage test data CERN-98-01. CERN. Geneva. Technical Inspection and Safety Commission.

71 Moser, M., Semprimoschnig, C., Van Eesbeek, M., and Pippan, R. (2008). Surface and bulk degradation of Teflon® FEP retrieved from the hubble space telescope solar arrays. *High Performance Polymers* 20 (4, 5): 429–446.

72 Traskos, M. (2011). Basic concepts in understanding aircraft wire degradation.www.lectromec.com/basic-concepts-in-understanding-aircraft-wire-degradation/ (accessed 11 November 2022).

73 Kemari, Y., Belijar, G., Laudebat, L. et al. (2021). Partial discharge inception voltage: precise and robust numerical estimation based on extensive. In: *IEEE Electrical Insulation Conference (EIC)*, IEEE, pp. 264–268 (07–28 June 2021). Denver, CO, USA.

74 Li, L., Bowler, N., Kessler, M.R., and Yoon, S.H. (2010). Dielectric response of PTFE and ETFE wiring insulation to thermal exposure. *IEEE Transactions on Dielectrics and Electrical Insulation* 17 (4): 1234–1241.

75 Ferree, C.R., Ward, P.R., and Toll, T.A. (2022). Condition assessment of cable insulation materials in advanced reactor environments. *Progress in Nuclear Energy* 151: 104341.

76 Christou, I. (2011). Optimisation of high voltage electrical systems for aerospace applications. PhD thesis, The University of Manchester, UK: Faculty of Engineering and Physical Sciences.

77 Borghei, M. and Ghassemi, M. (2020). Investigation of low-pressure condition impact on partial discharge in micro-voids using finite-element analysis. In: *IEEE Energy Conversion Congress and Exposition (ECCE),* IEEE, pp. 3293–3298 (11–15 October 2020). Detroit, MI, USA.

78 Shahsavarian, T., Baferani, M.A., Wu, X. et al. (2019). Partial discharge studies on high-temperature insulation materials for hybrid compulsion systems. In: *IEEE Conference on Electrical Insulation and Dielectric Phenomena (CEIDP),* IEEE, pp. 674–677 (20–23 October 2019). Richland, WA, USA.

79 I. Cotton, R. Gardner, D. Schweickart, D. Grosean and C. Severns, "Design considerations for higher electrical power system voltages in aerospace vehicles," *2016 IEEE International Power Modulator and High Voltage Conference (IPMHVC),* IEEE, San Francisco, CA, USA, pp. 57–61, 2016.

80 Xu, H., Lowndes, R., and Cotton, I. (2021). Power capacity of high voltage cables for future electrical aircraft. In: *IEEE Electrical Insulation Conference (EIC),* IEEE, pp. 177–180 (07–28 June 2021). Denver, CO, USA.

81 P. Collin, D. Malec and Y. Lefevre, "About the relevance of using Paschen's criterion for partial discharges inception voltage (PDIV) estimation when designing the electrical insulation system of inverter fed motors," *2019 IEEE Electrical Insulation Conference (EIC)*, IEEE, Calgary, AB, Canada, pp. 513–516, 2019.

82 Kemari, Y., Van de Steen, C., Belijar, G. et al. (2022). A townsend's secondary ionization coefficient estimation method for partial discharge inception voltage prediction for insulating polymers. In: *IEEE 4th International Conference on Dielectrics (ICD),* IEEE, pp. 226–229 (03–07 July 2022). Palermo, Italy.

83 Fabiani, D., Montanari, G., and Laurent, C. (2007). Polymeric HVDC cable design and space charge accumulation. Part 1: insulation/semicon interface. *IEEE Electrical Insulation Magazine* 23 (6): 11–19.

84 Delpino, S. (2008). Polymeric HVDC cable design and space charge accumulation. Part 2: insulation interfaces. *IEEE Electrical Insulation Magazine* 24 (1): 14–24.

85 Toureille, A., Notingher, P., Castellon, J., and Agnel, S. (2007). Space charges: definition, history, measurement. In: *Dielectric Materials for Electrical Engineering* (ed. J. Martinez-Vega), 119–130. Hoboken: John Wiley & Sons, Inc.

86 Afia, R.S., Mustafa, E., and Tamus, Z.Á. (2022). Condition monitoring of photovoltaic cables based cross-linked polyolefin insulation under combined accelerated aging stresses: electrical and mechanical assessment. *Energy Reports* 8 (1): 1038–1049.

87 Benyoucef, A., Berquez, L., Teyssedre, G., and Aubert, E. (2020). Space charge characterization in insulating materials used in aircraft cables. In: *IEEE International Conference on High Voltage Engineering and Application (ICHVE),* IEEE, pp. 1–4 (6–10 September 2020). Beijing, China.

88 Fabiani, D., Montanari, G., and Laurent, C. (2008). HVDC cable design and space charge accumulation. Part 3: effect of temperature gradient. *IEEE Electrical Insulation Magazine* 24 (2): 5–14.

89 Baferani, M.A., Li, C., Shahsavarian, T. et al. (2021). High temperature insulation materials for DC cable insulation – Part I: space charge and conduction. *IEEE Transactions on Dielectrics and Electrical Insulation* 28 (1): 223–230.

90 Baferani, M.A., Shahsavarian, T., Wu, X. et al. (2019). Study of space charge behavior of insulations for high temperature applications. In: *IEEE Conference on Electrical Insulation and Dielectric Phenomena (CEIDP),* IEEE, pp. 490–494 (20–23 October 2019). Richland, WA, USA.

91 Tanaka, Y., Mima, M., Konishi, S., and Miyake, H. (2019). Space charge accumulation properties in various insulating materials under DC high electric field at high temperature. In: *2nd International Conference on Electrical Materials and Power Equipment (ICEMPE),* IEEE, pp. 22–30 (07–10 April 2019). Guangzhou, China.

92 Perrin, C., Griseri, V., Laurent, C. et al. (2008). Space charge detection in Kapton® and PTFE polymer films by the open pulsed electro-acoustic method. *High Performance Polymers* 20 (4, 5): 535–548.

93 Teyssedre, G., Dinculescu, S., Le Roy, S. et al. (2021) Thermal degradation kinetics of high temperature polymers for aeronautic cables insulation. In: *IEEE International Conference on the Properties and Applications of Dielectric Materials (ICPADM),* IEEE, pp. 294–297 (12–14 July 2021). Johor Bahru, Malaysia.

94 Yang, L., Ohki, Y., Hirai, N., and Hanada, S. (2017). Aging of poly(ether ether ketone) by heat and gamma rays-its degradation mechanism and effects on mechanical, dielectric and thermal properties. *Polymer Degradation and Stability* 142: 117–128.

95 Li, H., Yang, J., Li, X., and Dong, S. (2020). Study on the microstructure of polyether ether ketone films irradiated with 170 keV protons by grazing incidence small angle x-ray scattering (GISAXS) technology. *Polymers (Basel)* 12 (11): 2717.

96 Ding, R., Xu, L., Li, J. et al. (2021). Investigation and life expectancy prediction on poly(ether-ether-ketone) cables for thermo-oxidative aging in containment dome of nuclear power plant. *Polymer Testing* 103: 107362.

97 Li, C., Shahsavarian, T., Baferani, M.A. et al. (2021). High temperature insulation materials for DC cable insulation – Part III: degradation and surface breakdown. *IEEE Transactions on Dielectrics and Electrical Insulation* 28 (1): 240–247.

5

High-Temperature Polymer-Based Dielectrics for Advanced Electronic Packaging

Jie Liu, Peng Li, Jianwei Zhao, and Shuhui Yu

Shenzhen Institute of Advanced Electronic Materials, Shenzhen Institute of Advanced Technology, Chinese Academy of Sciences, Shenzhen 518055, P. R. China

5.1 Introduction

High-temperature polymers (HTPs) are distinguished from typical organic polymers by their superior physical and chemical property stability at high temperatures [1]. In brief, the development of HTPs has experienced three stages up to date. The first stage includes the origin of HTPs and their initial mass production in the 1960s, which was promptly driven by the demand for human space exploration, and VARI (Vacuum Assisted Resin lnfusion) polymers were designed and produced. However, a significant number of the HTPs gradually disappeared in the following years as a result of poor processability or exorbitant cost. The second stage of HTP development occurred between the 1970s and 1980s, at which time the electrical and electronic industries, especially microelectronic industries, experienced great growth. In this stage, for the HTPs invented in the 1960s, only those possessing high thermal stability, decent processability, and relatively low cost survived and were commercialized. The third stage, which is what we are experiencing at present, involves the functionalization of HTPs for applications in advanced electrical (high-speed transportation system, high-power energy generation, storage, and transmission, etc.) and electronic (microelectronic and optoelectronic fabrication) systems. The application of HTPs for electrical and electronic insulation could date back to mid mid-1960s, when synthetic polymers started being applied for insulation purposes [2]. Ever since, HTPs have been increasingly employed as dielectrics in insulation applications. The pressing needs of high-temperature and high-reliability electronics are strongly tied to this emerging trend. Various applications of HTPs in electrics and electronics have been evaluated in literature during the past decades [3].

Nowadays, the manufacturing of integrated circuits (ICs) is one of the most important industries in the world. A major portion of our daily lives revolves around the use of devices that would not be possible without ICs, such as the integrated 5G/satellite 6G communication (Figure 5.1) [4]. Massive research has

High Temperature Polymer Dielectrics: Fundamentals and Applications in Power Equipment, First Edition.
Edited by Jun-Wei Zha and Zhi-Min Dang.

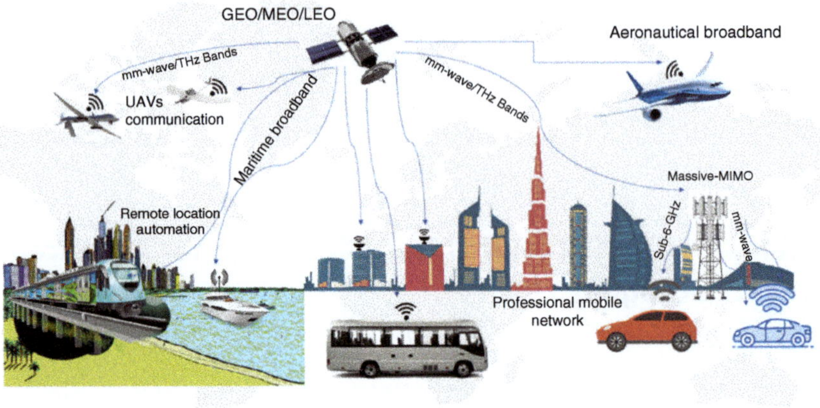

Figure 5.1 Diagram of integrated 5G/satellite 6G communication. Source: Ikram et al. [4]/MDPI/CC BY 4.0.

been conducted in academia as well as in the industry to improve IC's performance and manufacturing processes. These ICs must be enclosed in packages to enable their use in a system. The IC package has many functions, including environmental protection, thermal management, mechanical stability, and electrical connection. Environmental protection is needed to prevent chemical damage that might be detrimental to the IC's performance. Mechanical stability and thermal management are both required to enhance the reliability and service life of the IC by preventing physical damage. Electrical connections must be built in order for the package and the IC to communicate with a larger system on a circuit board [5].

5.1.1 Development of Electronic Packaging Technology

The semiconductor industry has identified five major growth engines (applications), namely: (i) mobile (e.g. smartphones, smartwatches, and wearables, etc.) and portable (e.g. notebooks and cameras); (ii) high-performance computing (HPC), also known as supercomputing, which is capable of processing data and performing complex calculations at high speeds on a supercomputer; (iii) autonomous vehicles (or self-driving cars); (iv) Internet of Things (IoTs), such as smart factories and smart health; (v) big data (for cloud computing) and instant data (for edge computing). Many system-technology drivers, such as artificial intelligence (AI) and fifth-generation (5G technology standards for broadband cellular networks), are boosting the growth of the aforementioned semiconductor applications [6–8]. The robust development of AI and 5G requires the semiconductors to have a faster speed, higher interconnect density, smaller pad pitch, smaller chip size, and higher power dissipation efficiency, all of which present challenges and opportunities for advanced electronic packaging [9]. These package functions are essential to ensure the high performance and reliability of an IC. Requirements for better package efficiency (PE, which is defined as the ratio of the chip area to the package area), as well as the constantly rising performance and component density of ICs, have been the dominant factors facilitating the development of superior packaging technology [10].

The design of these packages has been a long journey. It has evolved from through-hole to surface mount technology, from wire bond to flip-chip, and from dual-inline packaging to chip-scale packaging. Over the past few decades, a wealth of novel forms of packaging technologies have emerged with a collective family name of Advanced Electronic Packaging. Family members of Advanced Electronic Packaging include Fan-in (FI) or Fan-out (FO) Wafer Level Package (WLP), embedded Wafer Level Ball Grid Array (eWLB), Chip Scale Package (CSP), Wafer Level Chip Scale Package (WLCSP), Chip on Wafer (CoW), Wafer on Wafer (WoW), Fan-out Panel Level Package (FOPLP), Integrated Fan-out (InFO), Chip-on-Wafer-on-Substrate (CoWoS), High-Bandwidth Memory (HBM), Hybrid Memory Cube (HMC), Wide Input Output (Wide-IO), Embedded Multi-Die Interconnect Bridge (EMIB), Omni-Directional Interconnect (ODI), 3D IC, SoIC, X-Cube, and more. Advanced Electronic Packaging shares a common feature of being capable of providing higher performance. From dimensional perspectives, the Advanced Electronic Packaging family develops toward two branches, namely horizontal (toward 2D) or vertical (toward 3D) evolution of Advanced Electronic Packaging, with respect to the chip/substrate. 2D Advanced Electronic Packaging heavily relies on the re-distribution layers (RDLs), while 3D Advanced Electronic Packaging mainly makes use of through-silicon vias (TSV), despite indispensable RDLs on certain occasions, for electrical connection between different parts or components in the whole electronic system.

The electronic industries have been successfully coping with the increasing demand for micro-systems with higher functional density and performance involving: (i) downscaling, i.e. increasing the IC functional density by reducing the feature size; and (ii) employing advanced IC packaging and integration techniques such as flip-chip, ball grid array (BGA), CSP, multi-chip module (MCM), system-in-package (SiP), and embedded passives. However, both IC functional density and Advanced Electronic Packaging/integration techniques have reached their practical limits in 2D scale. Only by utilizing the third dimension will the next-generation technologies keep up with the rising functionality and performance demands. Figure 5.2 illustrates the historical progression from 2D to 3D integration. One measure of circuit density is silicon packaging efficiency (SPE), which is the ratio of total silicon area to total area of circuitry. A higher SPE is obtained by combining 2D integration with 3D integration, e.g. embedded passives and stacked module technologies. By reducing the connecting length, 3D packaging design enables a packaging efficiency greater than 100% and enhanced performance. This is possible by stacking chips and electrically connecting the bottom of one chip to another using TSV and the bumping process. 3D integration not only allows a higher density of integration but also offers several other practical advantages over 2D integrations. For example, a conventional circuit board demonstrates a low SPE of 10–50%, while MCMs of the next-generation technologies could enable a higher SPE of 50–70% [11].

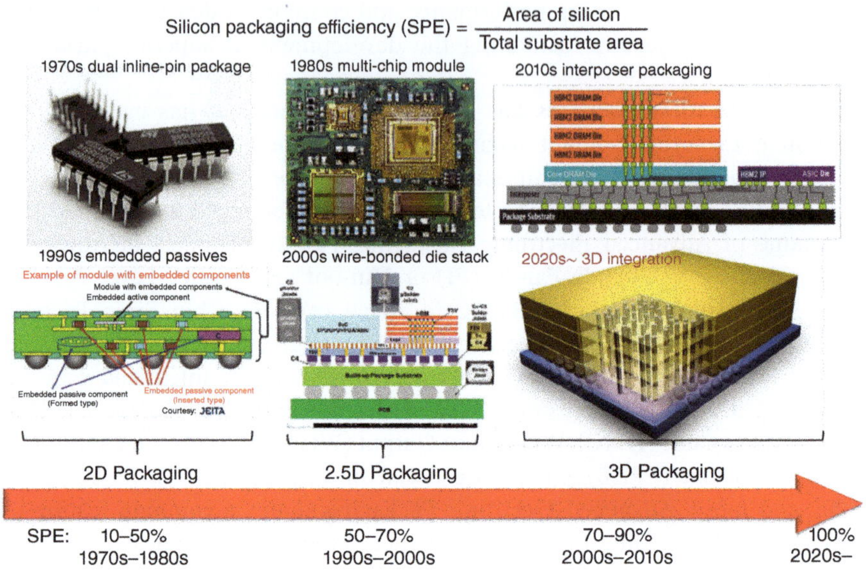

Figure 5.2 Historical progression of integration technologies from 2D to 3D. Source: Zhang et al. [11]/Springer Nature.

The demand for faster technologies continuously grows. The microelectronics industry has been constantly struggling to keep up with this rising demand. Under such circumstances, it is gradually realized that 3D packaging is the technology hotspot fulfilling this demand [12]. 3D packaging is the stacking and integration of multiple ICs. The stacking of chips comes with many advantages that include improved transmission speed, lower power consumption, and higher PE [13]. However, the complex design creates processing and reliability issues [14].

Take a few examples as below: Flip-chip package-on-package (PoP), the most fundamental 3D packaging technology, has already achieved mass production in-memory applications [15]. PoP is a type of 3D package that increases the PE by stacking CSPs on top of one another. Typically, a PoP device is a package with a wire-bonded memory chip affixed to the top of a package with a logic chip and electrically coupled to it. PoP has more flexibility in packaging design, which makes it one of the best options for fabricating memory devices [16, 17]. The SiP design is different from PoP [18]. SiP is a 3D design, bonding a stack of chips together into a single package. An example of this can be seen in Figures 5.3 and 5.4. SiP allows for a reduction in size and weight in comparison with PoP. However, a single defective chip may ruin the SiP device, whereas the PoP design allows the testing of individual packages prior to the 3D integration [19].

5.1.2 Requirement of Polymer-Based Dielectrics for Advanced Electronic Packaging Application

As shown in Figure 5.5, a typical IC chip module consists of many electronic packaging components, such as organic substrates, underfill, epoxy molding

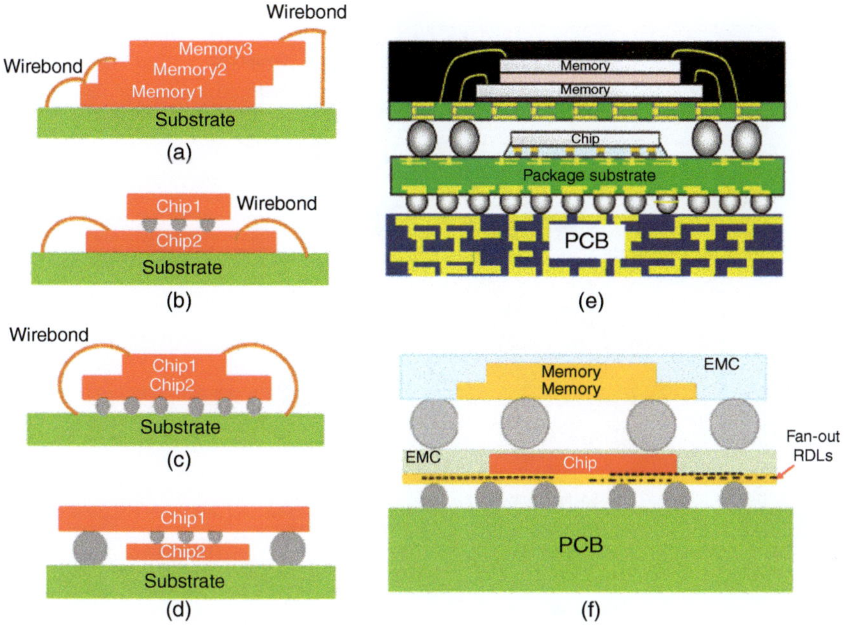

Figure 5.3 Examples of 3D packaging (without TSV). Source: Lau [17]/IEEE. (a) Memory chips stacked with wire bonds (b) Two chips are face-to-face solder bumped flip-chip and then with wire bonds to the next level of interconnect. (c) Two chips have back-to-back bonding; the bottom chip is solder bumped flip-chip to the substrate, and the top chip is with wire bonds to the substrate. (d) Two chips have face-to-face solder bumped bonding, and the top chip is with solderball to the substrate. (e) PoP with flip chip for the application chipsets. (f) PoP with fan-out for the application chipsets.

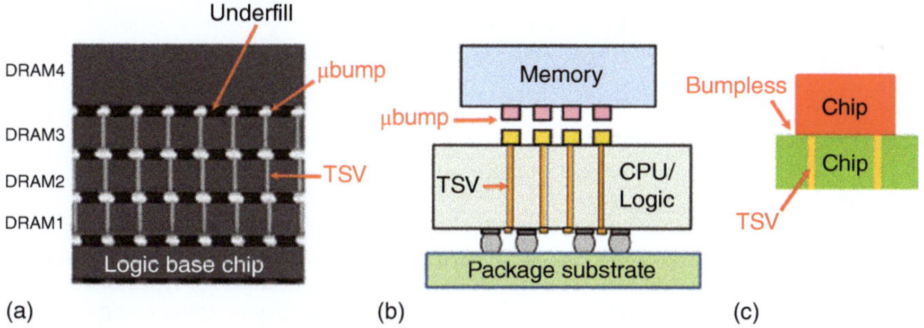

Figure 5.4 Examples on 3-D IC integration. (a) HBM with μbumps and TSVs. (b) CoC with TSVs and μbumps. (c) CoC with TSVs and bumpless . Source: Lau [17]/IEEE.

compound (EMC), photosensitive polyimide (PSPI), and thermal interface material (TIM). Different components require materials with different properties [9, 20–25]. Apart from traditional requirements for dielectrics used in IC packaging, including compatible coefficient of thermal expansion (CTE) and Young's modulus, other thermal and electrical properties, such as low dielectric loss, high thermal conductivity (TC) and stability, and high glass transition temperature (T_g), are also important parameters to be taken into consideration. Low dielectric loss translates

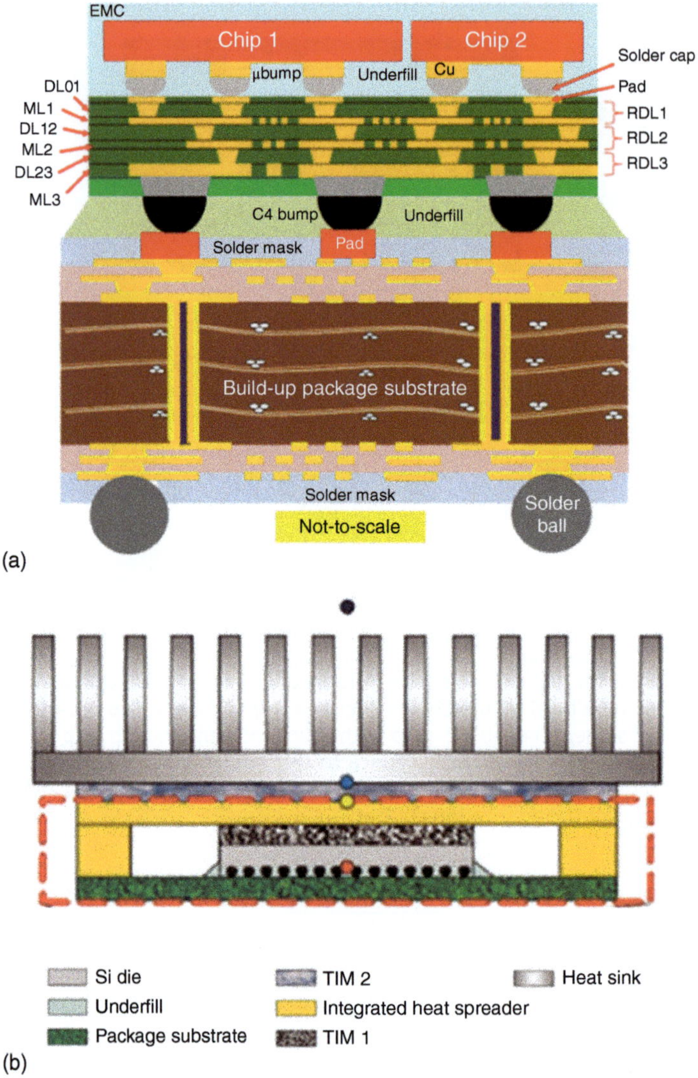

Figure 5.5 (a) Cross section of the heterogeneous integration package. (b) Types of electric packaging materials, including packaging substrates, underfill, epoxy molding compound, and thermal interface materials. Source: Lau et al. [20]/IEEE.

into a low dissipation factor or loss tangent (tan δ) and mitigates dielectric loss in in-package interconnects, feedlines, and antennas, thus preserving high signal transmission efficiency. Integration of higher-density electronic components will generate more thermal sources, which is detrimental to the long-term operation stability of the system and thereby requires ever more efficient and stable heat management packaging materials. TIMs of higher thermal conductivity could promptly dissipate accumulated thermal stress and boost the performance of the system.

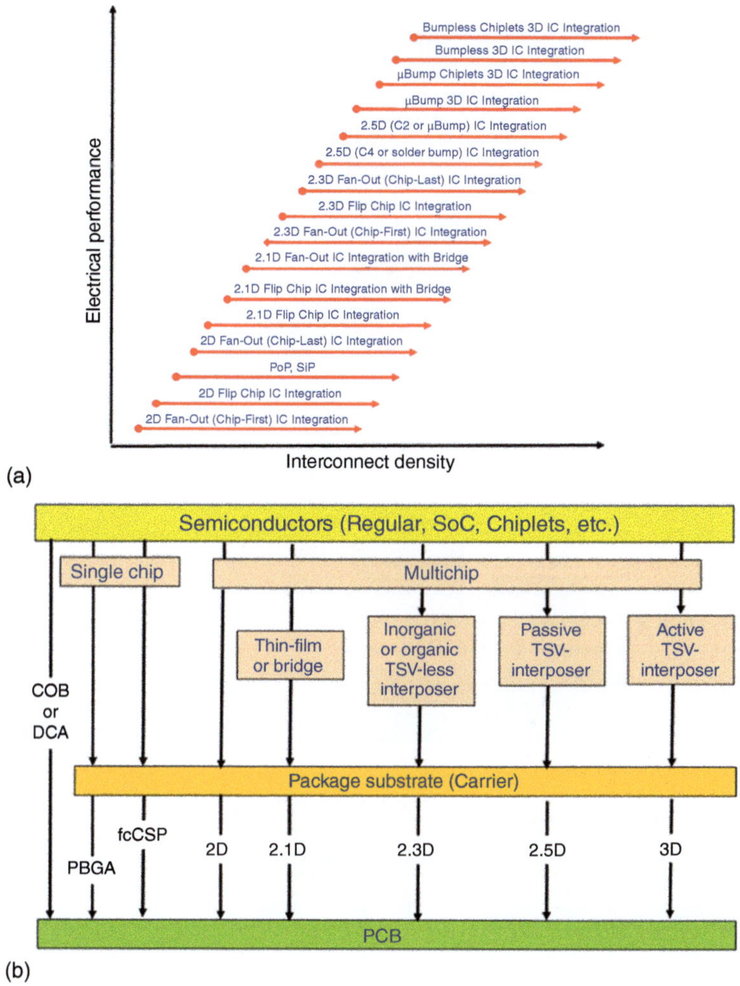

Figure 5.6 (a) Advanced Electronic Packaging ranking according to their density and performance. (b) Groups of Advanced Electronic Packaging from 2D to 3D IC integration. Source: Lau [17]/IEEE.

The 3D chip stacking package design is beneficial, but it also confronts challenges. The increased Joule heating effects observed in chip stacks lead to reliability issues like chip warpage, interfacial cracking, solder bump fracture, device performance, atomic diffusion, etc. [26]. Furthermore, the signal integrity of TSVs is also difficult to manage [27]. Thermal management and electrical issues become more complicated in this design (Figure 5.6).

5.1.2.1 Dielectric Properties

Moore's law, which states that the number of transistors on IC chips doubles every 18 months, was first proposed in 1965 and predicted the rapid development of technology in the field of information technology (Figure 5.7) [30]. Enhancements in

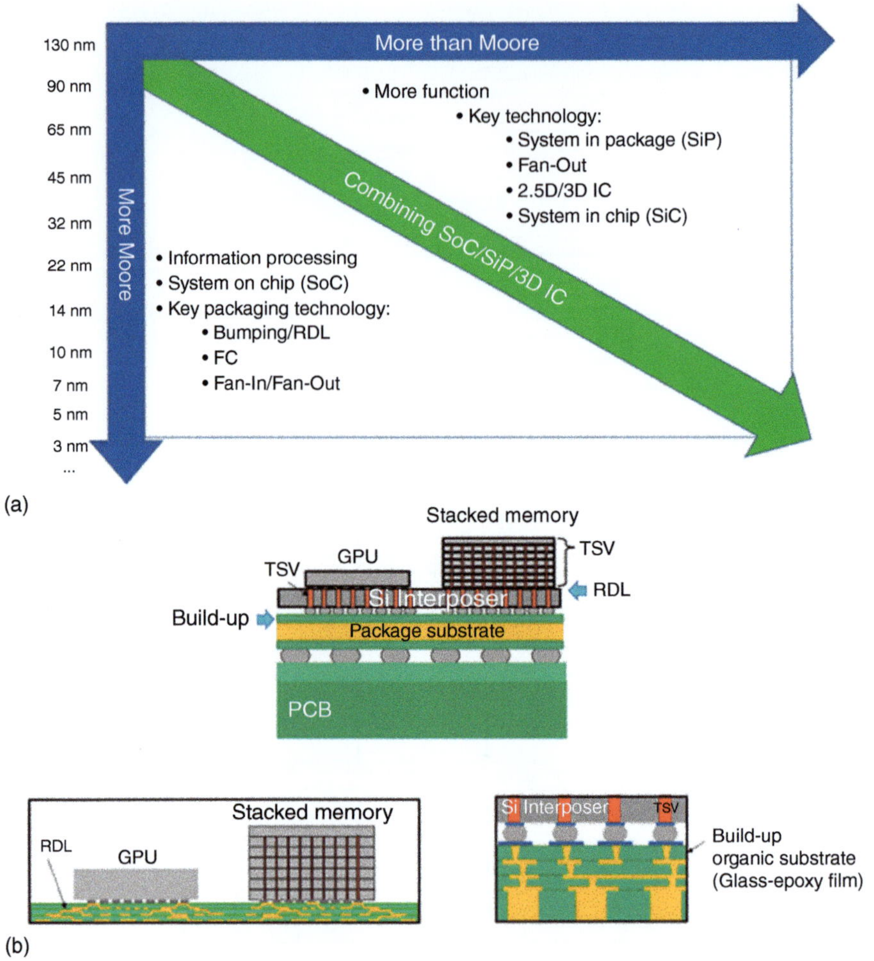

Figure 5.7 (a) More Moore and More Than Moore. (b) GPU Module Package: GPU chip and stacked DRAM chip on RDL (left); Build-up multilayer wiring (right) [28, 29].

resistance and capacitance, as well as the obvious increase in the parasitic effect, become more pronounced when the integration level of IC devices continuously increases and the sizes of the components are pushed ever smaller, which further leads to an increase in power consumption, signal latency, crosstalk, etc. [28, 29].

The calculation of signal lag time is shown in Eq. (5.1),

$$\tau = RC = 2\rho\varepsilon\varepsilon_0\left(\frac{4L^2}{P^2} + \frac{L^2}{T^2}\right) \tag{5.1}$$

where τ is the signal delay time, R is the resistance, C is the capacitance, ρ is the specific resistance of the conductor, ε is the dielectric constant of the insulating material, and ε_0 is the dielectric constant of vacuum ($8.85\times10^{-12}\,\mathrm{F\,m^{-1}}$), L is the length of the conductor, T is the thickness of the conductor, and P is the distance between two conducting lines [31].

The relationship between high-frequency signal transmission speed V, signal transmission loss, and interlayer dielectric resin can be expressed as follows using the signal transmission rate equation:

$$V = \frac{kc}{\sqrt{\varepsilon}} \tag{5.2}$$

where V (m s^{-1}) and c (m s^{-1}) are the signal propagation speed and the speed of light in vacuum, respectively, and k is a constant ($1.38 \times 10^{-23}\ \text{J K}^{-1}$). As can be seen, signal propagation will proceed more quickly the lower the substrate's dielectric constant is. It is worth noting that, in high-frequency circuits, the dielectric loss of the substrate needs to be considered in the equation of signal propagation loss:

$$\alpha = k' f \sqrt{\varepsilon}\ \tan\,\delta \tag{5.3}$$

where α (dB m^{-1}) is the signal transmission loss. $\tan\,\delta$ is the dielectric loss factor of the substrate, and f represents frequency. In general, the attenuation of signal propagation will be mitigated as the substrate owns a small dielectric loss factor. Since the speed of signal propagation and propagation efficiency are significantly impacted by the dielectric properties of copper-clad laminates, the research and development of high-performance substrate materials with low dielectric constant and low dielectric loss is the premise of high-frequency printed circuit boards (PCB) applications [32].

5.1.2.2 Thermal and Thermal–Mechanical Properties

5.1.2.3 Dynamic Thermomechanical Properties

Modulus and glass transition temperature (T_g) are the two key dynamic thermomechanical characteristics of materials used in electric packaging, which characterize the viscoelasticity of polymers and determine the processing conditions and application range of electric packaging materials [33, 34]. Storage modulus (E') and loss modulus (E''), also known as elastic modulus and viscous modulus, refer to the energy stored and lost during the deformation of materials under periodic (sinusoidal) mechanical force. $\tan\,\delta$, also known as loss factor, is the ratio of loss modulus to storage modulus, which is an important parameter to evaluate the interactions between the components and the overall flexibility of composites.

5.1.2.4 Thermal Expansion Coefficient

Due to its higher I/O density, ultra-thin profile, high electrical performance, and low power consumption, FOWLP is one of the most advanced package technologies. However, warpage can be induced by the mismatch in CTE between different kinds of materials, which is a mechanical issue in fan-out wafer level package (FOWLPs) [35, 36].

The CTE of electric packaging materials reflects the fractional change in dimension per degree rise in temperature. The CTE of a finished substrate is a function of the CTE, Young's modulus, and the volume ratio of all the materials involved in the substrate.

A first-level engineering estimation can be expressed by the following equation:

$$\mathrm{CTE} = \frac{\sum_{i-1}^{n} \mathrm{CTE}_{\text{mat-}i} \times E_{\text{mat-}i \times V_{\text{mat-}i}}}{\sum_{i=1}^{n} E_{\text{mat-}i} \times V_{\text{mat-}i}} \quad (5.4)$$

where $\mathrm{CTE}_{\text{mat-}i}$ is the CTE of material-i, $E_{\text{mat-}i}$ is the Young's modulus of material-i, and $V_{\text{mat-}i}$ is the volume ratio of material-i. To achieve a final low CTE close to that of a semiconductor chip as a finished substrate, the CTEs of the component materials must be balanced [37].

The linear CTE is usually measured by thermomechanical analyzer (TMA) and calculated by the following formula:

$$\alpha = \frac{\mathrm{d}L}{\mathrm{d}L^{*}L_0} \quad (5.5)$$

where α is the linear CTE, dL is the size change of material within a certain temperature range, and L_0 is the initial length of material (Figure 5.8). Generally, the CTE of pure polymers is relatively high, which is difficult to meet the requirements. In order to lower the CTE of polymer materials, inorganic particles with low CTE are often introduced. (CTE = 0.5 ppm K^{-1}), Al_2O_3 (CTE = 6.6 ppm K^{-1}), AIN (CTE = 4.5 ppm K^{-1}), BN (CTE = 4.5 ppm K^{-1}), and ZrW_2O_8 (CTE = −8.7 ppm K^{-1}) are the most commonly used inorganic particles added to polymer materials. Specifically, SiO_2 is more widely used because of its low cost and accessibility. For SiO_2/epoxy underfill, CTE is mainly determined by several factors, including the filling amount of SiO_2, the size of SiO_2, and the strength of the interfacial interaction between SiO_2 and epoxy resin.

5.1.2.5 Thermal Conductivity

Thermal management is a necessary function of the package, especially to support the ongoing development of microprocessors. These devices are intended to be smaller and capable of higher power and frequency operation, even exceeding 65 GHz. The increased power density and current levels in these devices may introduce heating problems such as on-chip hotspots and Joule heating. Thermal degradation or catastrophic failure can occur in various ways if the operating temperature is not minimized. Failure rates and gate delay both increase with

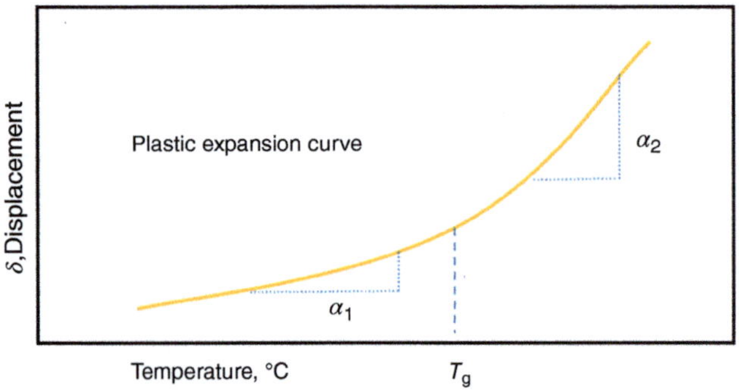

Figure 5.8 Change in coefficient of thermal expansion with temperature.

operating temperature. Therefore, the thermal management ability of the package is essential for product reliability and quality.

The thermal diffusivity, α, is related to the thermal conductivity, k, as shown in Eq. (5.6):

$$\alpha = \frac{k}{\rho \cdot C_p} \tag{5.6}$$

where ρ is the density of the material and C_p is the specific heat capacity. In order to determine the thermal conductivity from the measured thermal diffusivity, it is necessary to know the density and specific heat capacity of the samples. The density is estimated from the weight and dimension of each cylindrical sample [23].

The development of thermally conductive and dielectric polymer composites (TDPCs) technology is one of the critical drivers in upgrading integrated electronic devices. The incorporation of highly thermally conductive fillers into the polymer matrix is to modify the overall thermal conduction properties of the resulting composites. The heat will flow through both the polymer matrix and the inorganic fillers. TDPCs can be used as thermal interface materials (TIMs), hard/flexible substrates, and underfill, widely used in highly ICs, display screens, 5G communication, radar technology, etc.

High temperatures negatively affect the stability and reliability of electronic components, resulting in a reduced service life. Therefore, effective heat dissipation is essential to protect electronic equipment.

5.1.2.6 Other Requirements

Moisture absorption is a key to electric reliability for electric packaging materials. In addition, warpage issues arise when a resin's volume changes significantly after curing or when high internal stress accumulates with curing. As presented in Figure 5.9, there are mainly two types of shrinkage contributing to the volume change in resin, which are curing shrinkage and thermal shrinkage, respectively. To reduce

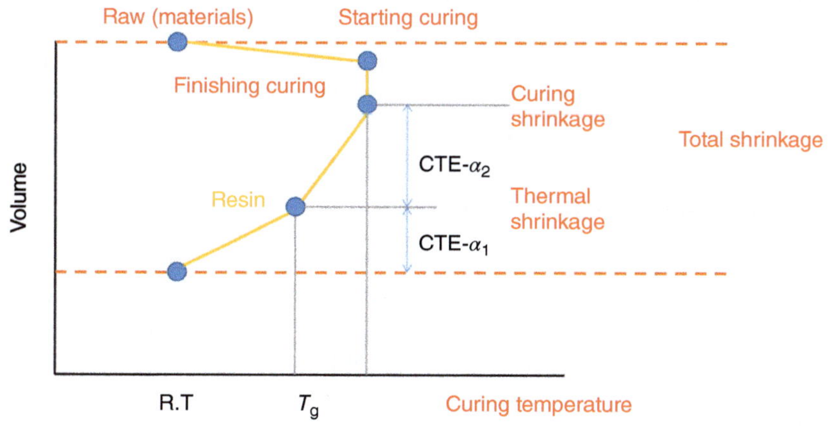

Figure 5.9 Schematic of resin shrinkage after curing. The *x*-axis represents temperature, and the *y*-axis represents the total volume of resin. Source: Ikehira [33]/Elsevier.

shrinkage, it is effective to add a high amount of inorganic filler. Due to the fact that CTE-α_2 tends to be larger than CTE-α_1, increasing the T_g of the cured resin is also important for regulating thermal shrinkage. Residual internal stress after curing, which is the other reason for warpage can be calculated using Eq. (5.7) [33, 37].

$$\sigma = \int E \cdot \alpha dt = \alpha_1 \int_{R.T.}^{T_g} E(t)dt + \alpha_2 \int_{T_g}^{\text{cure-temp.}} E(t)dt \tag{5.7}$$

where σ is the internal stress, E is the modulus, α_1 is the CTE for $<T_g$, and α_2 is the CTE for $>T_g$. Equation (5.7) reveals that the performance with low warpage is optimized when CTE and Young's modulus are low.

5.2 High-Temperature Polymer and Polymer-Based Dielectrics

According to the related mechanism in Section 5.1.2, the rational design of molecular structures of high-performance polymers and delicate recipes of high-performance polymer composites are the keys for them to achieve ideal electric packaging performances.

Many HTPs have been suggested for use as ILD/IMD materials in advanced microchips. Meanwhile, polymer materials have gradually gained lots of research interest due to their inherent low molecular polarity, wide selection of abundant monomers, adjustable functionalization, good processability, and ready hybridization with traditional inorganic materials. Commonly used polymer-based low-k materials mainly include polyaryl ether (PAE, $k \sim 2.6$–3.0), polyimide ($k \sim 2.7$–3.1), parylenes ($k \sim 2.2$–2.6), polytetrafluoroethylene (PTFE, $k \sim 2.0$), and benzocyclobutene (BCB) resin ($k \sim 2.3$–2.8) [38]. In addition, compared with traditional silicon-based inorganic materials and organic polymer materials, hybrid organic–inorganic materials are theoretically ideal candidates for conventional applications in interlayer materials and printed circuit boards. They have attracted lots of research attention recently because of the low dielectric constant of porous inorganic materials and high mechanical properties of organic candidates. In the following sections of 5.2.1 and 5.2.2, the most important HTPs will be reviewed in more detail (Figure 5.10).

5.2.1 High-Temperature Polymer Dielectrics

Moieties of special structures such as fluorine-containing groups, rigid nonplanar largely conjugated structures, and alicyclic structures can effectively reduce the k value of HTPs. Incorporation of such moieties into HTPs relies on rational design and successful synthesis of special monomers [39, 40].

5.2.1.1 Polyimide

Polyimide (PI) is a type of high-performance engineering polymeric material that is constituted of repeated imide unit structure in the main chain. Although PI

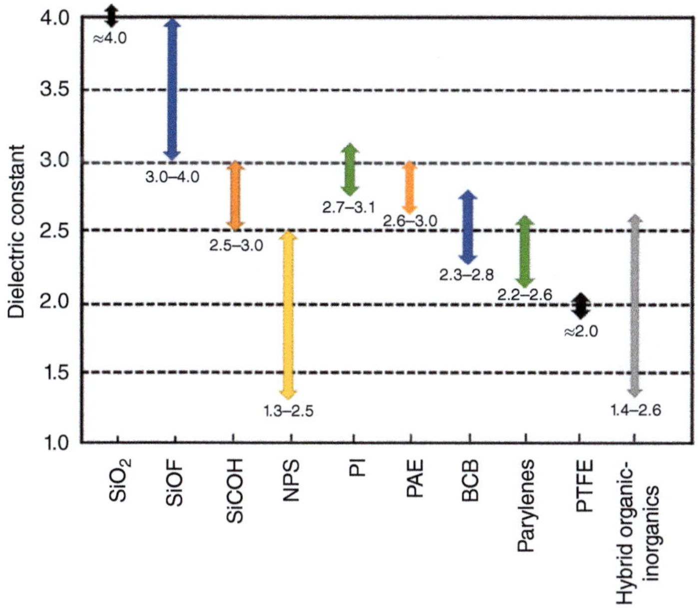

Figure 5.10 Comparison of the dielectric constants of different materials. Source: Adapted from Xie et al. [38].

contains polar groups such as C=O, the carbonyl groups in the conjugated imide ring are symmetrical, so the mobility of free electrons is greatly restricted, leading to excellent electrical insulation properties. Meanwhile, PI also possesses excellent thermal stability, mechanical properties, radiation resistance, low moisture absorption, and chemical stability, so PI film is an ideal choice of material to construct interlayer dielectrics, including RDL [41–43].

With the rapid development of mobile communication, the operation frequency has increased to GHz bands, so PI films with low k and Df at high-frequency (GHz) are urgently demanded. To this end, fluorinated PI has been extensively studied to fulfill dielectric requirements in the GHz region [44].

Among all elements, the electronegativity of fluorine is the largest (~4.0), so the introduction of fluorine can effectively reduce the polarizability of PI. Moreover, fluorine also plays an important role in reducing water absorption of matrix, which enhances the moisture stability of its dielectric properties in high-humidity atmosphere [45, 46]. Thus, the design and synthesis of fluorine-containing monomers has become a research hotspot for the preparation of low-k PI films. As shown in Figure 5.11, the direct substitution of hydrogen with fluorine in the benzene ring of anhydride and amine has become a common method to synthesize fluorine-containing monomers [40].

Compared with the direct fluorination of monomers with elemental fluorine, introducing a $-CF_3$ substituent into the monomers is more common for the fabrication of low-k PI (Figure 5.12) [43]. The effective roles of the $-CF_3$ group for lowering k values originate from the combinatory effect of reduction in polarizability of the

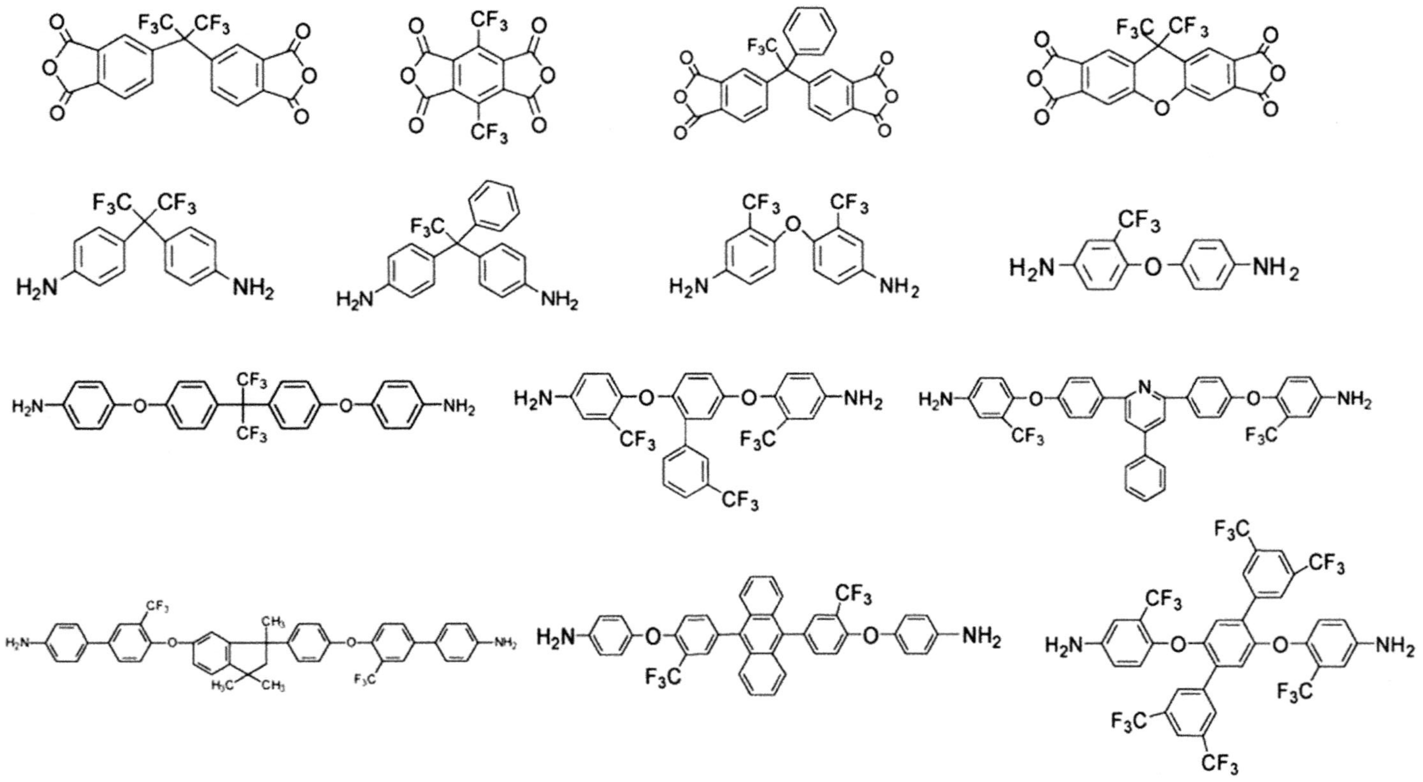

Figure 5.11 Structure of fluorine-containing monomers [40].

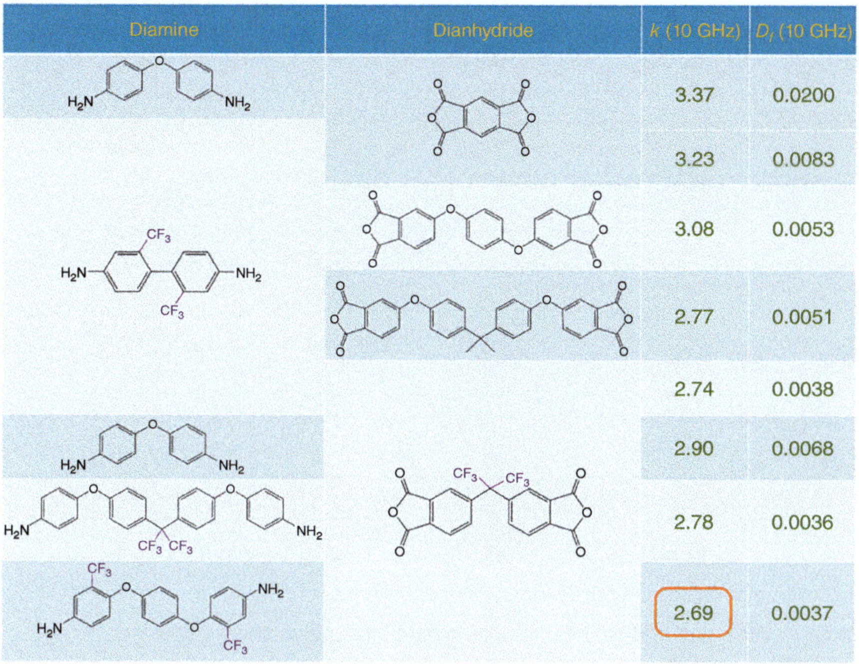

Figure 5.12 Dielectric properties of fluorinated PI. Source: Li et al. [43]/Elsevier.

matrix, increase in fractional free volume (FFV) of the matrix, and decrease in the number of polarized molecules per unit volume [46–49].

Incorporating bulky rigid and nonplanar conjugate structures into monomers is also effective for the synthesis of low-k PI. [50] The bulky rigid and nonplanar conjugated structure possesses large volume and steric hindrance (Figure 5.13) [43]. The introduction of these structures can effectively reduce the packing compactness and the regularity of PI chains. Finally, the comprehensive effects induce the enhancement in FFV and reduction in polarized molecules number per unit volume. Nonplanar polyphenyl structures are usually introduced into fluorinated monomers to further reduce k of PI (Figure 5.14) [43]. The alicyclic structure possesses lower polarizability, and it can also destroy the planarity of molecules and hinder the close packing of polymer chains, contributing to a low k and Df of PI [51–53].

5.2.1.2 Epoxy Resins

Epoxy resins are important players in many electronic packaging scenarios. These resins have a common feature of three-membered oxygen-containing epoxy (oxirane) rings that are incorporated into organic molecules by using either condensation or oxidation reactions. The resin formulations are used in numerous steps in the manufacture of electronic packaging materials including underfill, build-up dielectric films, conductive adhesives, flip-chip encapsulation, die coatings, encapsulation, surface mount placement, and lead bonding adhesives. Epoxy resins are excellent electrical insulators and can protect electrical components from

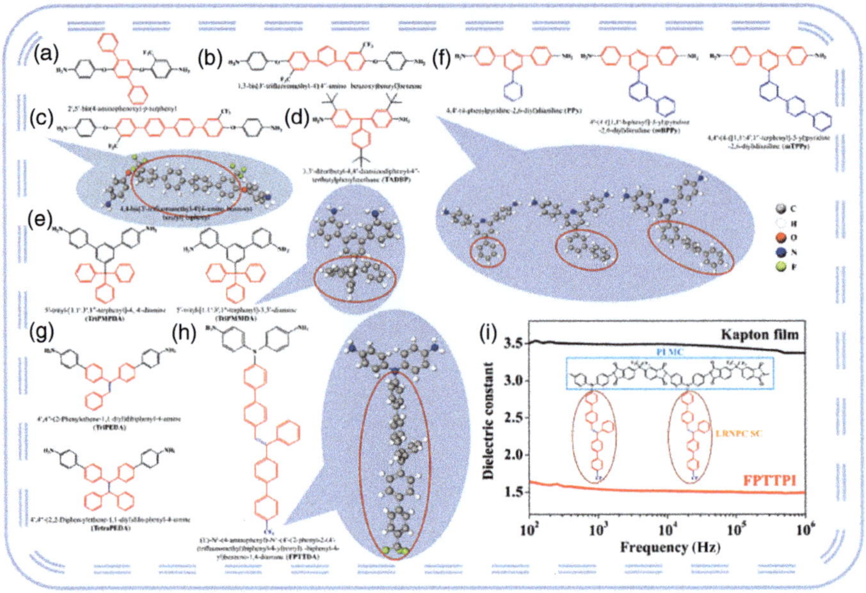

Figure 5.13 Structure of diamines containing nonplanar polybiphenyl. Source: Li et al. [43]/Elsevier.

short-circuiting, dust, and moisture. Due to their excellent electrical properties, mechanical strength, and processability, epoxy resins are widely adapted in the field of electronics [32]. Epoxy resins can be aromatic, cycloaliphatic, aliphatic, monofunctional, or polyfunctional, physically ranging from low-viscosity liquids to high-melting solids. To meet the requirements of advanced electronic packaging, much attention has been focused on developing novel epoxy resins with improved electrical insulating properties as well as enhanced thermal properties. As fluorination has become one of the most popular methods to decrease the dielectric constant of epoxy resins, tremendous research has been concentrated on the investigation of fluorinated epoxy resins.

Among various packaging materials, epoxy-based underfills are often employed in flip-chip packaging. However, widely used capillary underfill materials suffer from low thermal conductivity, and are unable to effectively dissipate the ever-accumulating heat during the operation of next-generation higher-power-density IC chips [54]. Ideal high-performance underfill materials for electronic packaging should have simultaneously high thermal conductivity (>1.0 W mK^{-1}), high electrical resistivity (>10^{12} Ω cm^{-1}), low viscosity (<20 Pa.s at 298 K), proper CTE (25–30 ppm K^{-1}), high T_g (>398 K), low dielectric constant (<4.0 at 298 K and 1 kHz) and reduced dielectric loss factor (<0.005 at 298 K and 1 kHz). The intrinsic thermal conductivity of traditional epoxy resins is low (~0.2 W mK^{-1}), far from meeting the increasing demand for thermal conduction/dissipation capabilities of electrical equipment or electronic components.

The first research paper describing intrinsically thermally conductive liquid crystal epoxy resins was proposed in 2003 by Takezawa et al. They started with

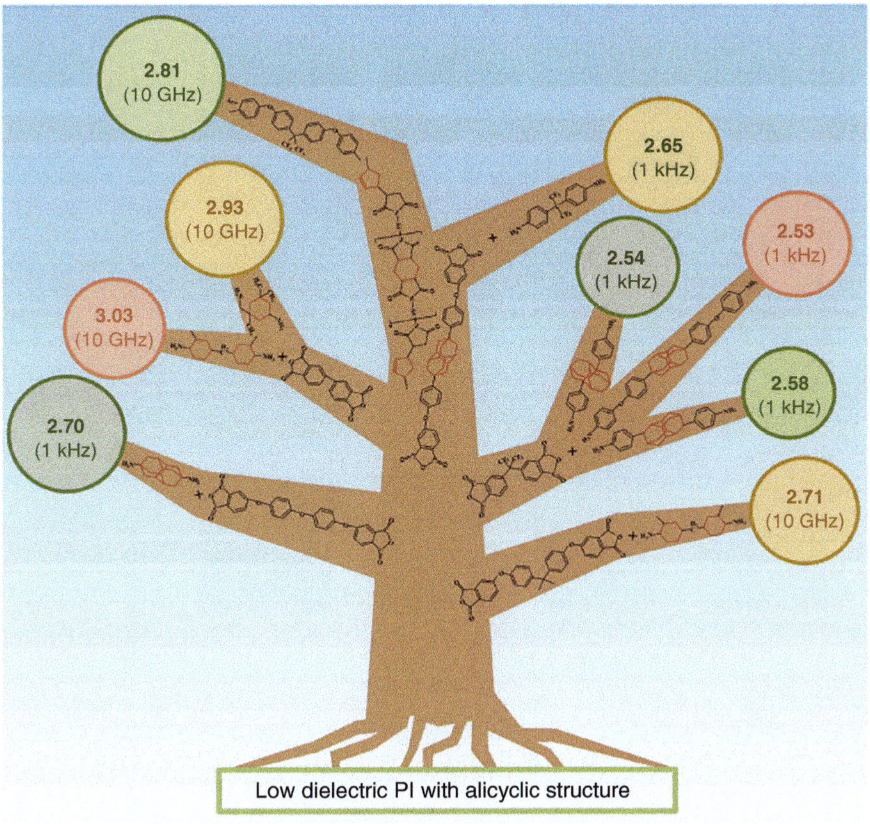

Figure 5.14 Structure of some alicyclic monomers and dielectric constant of the corresponding PI film. Source: Li et al. [43]/Elsevier.

the design of the molecular structure of liquid crystal epoxy monomer and used diphenyl benzoate as the mesogens. By changing the length of the flexible segment on the main chain, different epoxy monomers were prepared where aromatic diamine was used as the curing agent, and the influence of the length of the flexible segment on the main chain on the intrinsic λ was studied. When the number of carbon atoms in the flexible carbon chain on the main chain is 8, 6, and 4, respectively, the intrinsic λ of the liquid crystal epoxy resins reached 0.85, 0.89, and 0.96 $W\,mK^{-1}$, respectively, about five times that of common epoxy resins. It was explained that the highly ordered arrangement of mesogens makes it easy to form the partial crystal-like structure, resulting in microscopic anisotropy of the epoxy resins [55–57] (Figure 5.15).

5.2.1.3 Benzocyclobutene Resins

One of the promising candidates of polymer adhesives for ICs is BCB from Dow Chemical Company [58–61]. BCB is an organic resin with superior electrical,

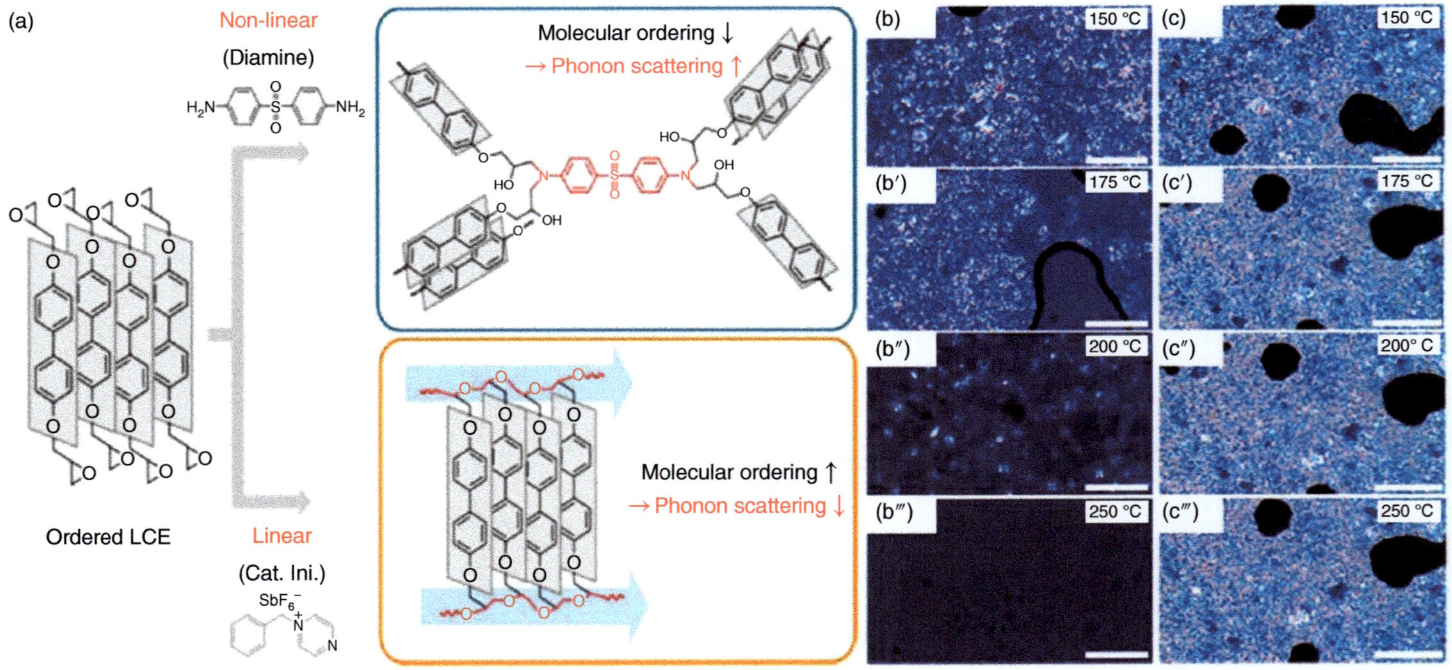

Figure 5.15 Schematic illustration of the different microstructures of liquid crystal epoxy resin obtained using the two curing methods (a), dynamic POM images taken at various temperatures between during heating of the liquid crystal epoxy resins mixed with (b–b‴) DDS and BPH (c–c‴). Source: Ruan et al. [55]/Elsevier.

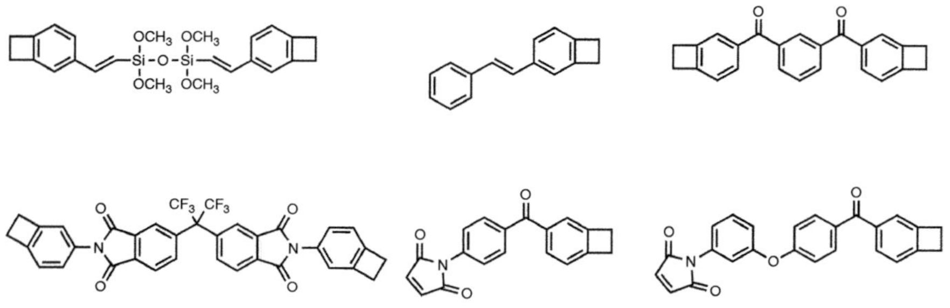

Figure 5.16 Examples of various types of benzocyclobutene monomers. Source: Maier [58]/Elsevier.

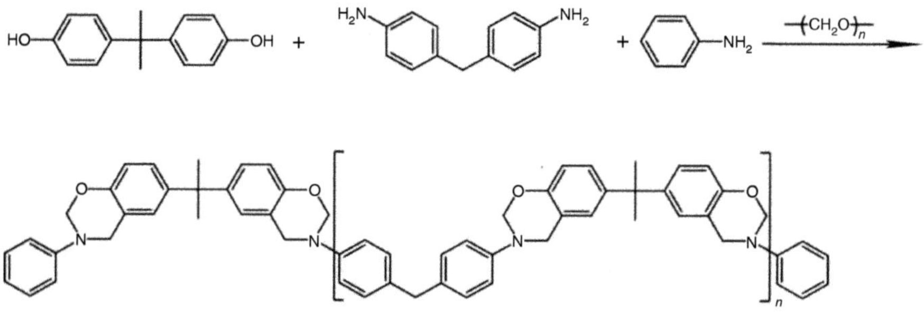

Figure 5.17 Preparation of main-chain benzoxazine copolymer oligomers [64].

mechanical, and thermal properties due to its high resistivity ($10^{19}\,\Omega\,cm^{-1}$), low loss tangent (0.0008–0.002), low permittivity (2.65), low moisture absorption (<0.2%), low out-gassing, low cure temperature (<250 °C), small volume shrinkage during curing (5–6%), and excellent dielectric stability over a wide frequency range [42, 62]. Additionally, BCB as the intermediate bonding material offers the highest bond strength in comparison to other conventional bonding polymers [60, 61, 63] (Figure 5.16).

5.2.1.4 Benzoxazine Resins

Benzoxazine resins and their corresponding polybenzoxazine derivatives are known to have unique properties, including hydrophobicity despite the hydrophilic moieties in each chemical repeat unit, near-zero shrinkage upon polymerization, extremely low surface free energy without using fluorine atoms, low dielectric constants and dielectric loss, high thermal stability and char yield, good chemical resistance, high mechanical strength, and excellent flame retardance [64–68] (Figure 5.17).

5.2.1.5 Polyaryl Ether

PAE is a kind of polymer material formed by the repeated connection of groups including ether and phenyl. Due to its unique molecular structure, it has good

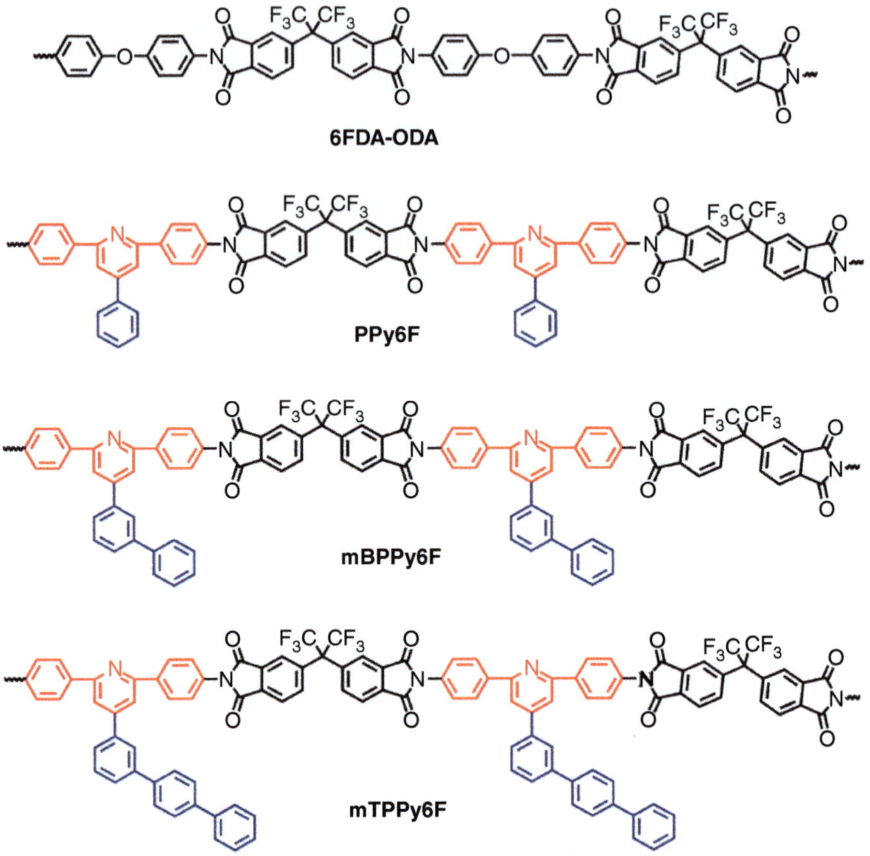

Figure 5.18 "Porous" PI film fabricated by increasing the steric hindrance of PI [73].

resistance to heat, radiation, corrosion, impact, creep deformation, and abrasion, as well as other comprehensive advantages such as nontoxicity, good hydrolysis resistance, flame retardance, excellent dimensional stability, electrical properties, and wide servicing temperature range. It has been widely applied to various fields such as electronics and electrical appliances, aerospace, and transportation. In recent years, PAE materials featuring low dielectric constants and dielectric loss have been developed [69–72].

5.2.1.6 Organic Porous Materials

In addition to abovementioned methods for preparing low-*k* PI film, the enhanced steric hindrance of PI chains can also generate low-*k* "porous" PI film. However, as shown in Figure 5.18, this "porous structure" originates from the "free volume pore" of large steric substituent groups in monomers and is not a true hole. Large steric substituent groups can reduce the tightness of chain and increase the molar volume of PI films, so the number of polarized molecules per unit volume decreases, enabling low *k* values of PI films [38, 43, 73, 74].

Table 5.1 Thermal conductivity of common polymers.

Polymers	Thermal conductivity ($W\,mK^{-1}$)
Polyurethane	0.233
Polymethylmethacrylate	0.19
Polyphenylene sulfide	0.286
Polyetheretherketone	0.18, 0.26
Polyimide	0.2
Epoxy	0.22, 0.148
Polycarbonate	0.19
Polytetraflfluoroethylene	0.32
Polyvinylidene fluoride	0.15, 0.22
Polyvinyl alcohol	0.21
Polydimethylsiloxane	0.18
Nylon-6	0.238
Polyethene	0.35
Polystyrene	0.16
Polypropylene	0.209
Natural rubber	0.10

5.2.2 High-Temperature Polymer-Based Composite Dielectrics

Polymers are widely used in electronic packaging due to their easy processing, lightweight, excellent insulation, and good mechanical properties. However, as electronic devices gradually develop toward high energy density and low signal delay, polymers with high thermal conductivity and low dielectric properties are in large demand. Since polymers have low intrinsic thermal conductivity, thermally conductive fillers need to be added to form polymer composites. However, there is often a trade-off between the thermal conductivity and the dielectric properties of polymer composites [75]. Therefore, this chapter aims to elaborate on how to balance the high thermal conductivity and the low dielectric properties by adding fillers to the polymer matrix to form composite dielectrics.

Table 5.1 lists the thermal conductivity of common polymers in the range of 0.1–0.4 $W\,mK^{-1}$. Its improvement often requires adjustment of the molecular structure to increase the crystallinity of polymers [76, 77]. However, the preparation process is often complicated and costly, and the improvement is limited. In contrast, the preparation of thermally conductive composites by doping thermally conductive fillers has become very convenient and efficient.

5.2.2.1 Inorganic Fillers

A variety of inorganic filler materials have been used to enhance the properties of the polymer matrices. Common fillers include aluminum nitride (AlN), aluminum oxide (Al_2O_3), graphite, graphene, CNTs, boron nitride (BN), SiC, silicon nitride (Si_3N_4), diamond, and metal particles [78–81]. Most of these materials have very low CTE (<10 ppm K^{-1}). AlN has high intrinsic thermal conductivity (TC, 150–220 W mK^{-1}) and high electrical resistivity but low oxidation resistance and is moisture sensitive. AlN has been used in the form of whiskers and particles. BN is a widely studied, highly conductive filler with excellent dielectric properties and low density. Its crystalline form, hexagonal boron nitride (h-BN), is the most stable and is similar to graphite due to its layered structure. Boron nitride nanosheets (BNNSs) and Boron nitride nanotubes (BNNTs) have been widely synthesized and used as encapsulants. The in-plane (IP) CTE of BN has been reported to be positive and negative. The negative CTE is due to the presence of low-frequency bending modes in its phonon spectrum for a certain temperature range (<600 and <2000 K). In addition, the CTE has been shown to increase with the thickness of atomically thin BN. In general, the synthesis of BNNS is associated with high costs, low purity, and low yields. Al_2O_3 is a low-cost filler with a low TC and a high dielectric constant. SiC has a high TC, is thermally stable, but has a high dielectric constant. Silicon nitride (Si_3N_4) has a low CTE, low dielectric constant, and moderate TC. Zinc oxide (ZnO) can be prepared as NPs or tetrapods that have a moderately high TC and low CTE. However, as a semiconductor, it has low resistivity and high dielectric permittivity [82, 83].

Thermally conductive fillers with a low dielectric constant are mainly included in the thermally conductive and insulating inorganic particles, such as h-BN, aluminum oxide (Al_2O_3), and AlN. Among them, h-BN, "white graphite," has a lamella structure, excellent thermal stability, an IP thermal conductivity of 200 W mK^{-1}, and a bandgap of 5.6 eV [84–86]. It has the lowest dielectric permittivity among many inorganic particles and is the most widely used filler in preparing TDPCs. However, considering the interface thermal resistance (ITR), low intrinsic thermal conductivity, and higher dielectric permittivity of inorganic fillers, it is often necessary to modify thermally conductive fillers to expand their high thermal conductivity and low dielectric permittivity [87–89] (Figure 5.19).

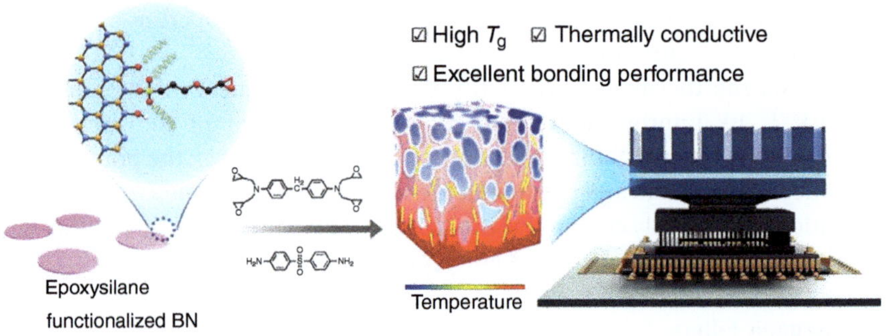

Figure 5.19 Thermally conductive epoxy/boron nitride composites with high glass transition temperatures for thermal interface materials. Source: Yue et al. [90]/Elsevier.

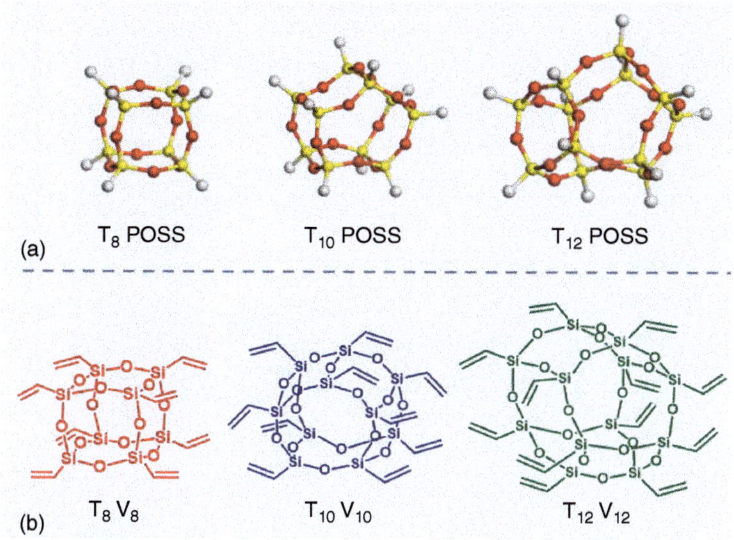

Figure 5.20 (a) Representative 3D structure and (b) chemical structure of POSS. Source: Wang et al. [93]/Elsevier.

5.2.2.2 Inorganic Porous Fillers

Among all porous fillers, POSS, which possesses a three-dimensional cage structure, is the most popular material. POSS has an inorganic framework core composed of Si–O structures with a shape like a cage. Compared with other inorganic porous fillers, POSS has more active functional groups on the surface and is easier to react with organic functional groups, so it is easier to form chemical bonds and homogeneous structure with polymer matrix [38, 43, 91–93] (Figure 5.20).

In addition, hollow glass microsphere (HGM) have been widely used in marine, aerospace, and construction fields because of their lightweight, heat insulation, high electrical insulating strength, high specific strength, and low *k*. Polymer-based composites filled with HGMs are often referred to as syntactic foams [76, 94, 95] (Figure 5.21).

5.2.2.3 Organic Porous Fillers

Metal-organic frameworks (MOFs), as a series of organic–inorganic hybrids, have attracted much attention due to their diverse characteristics, such as permanent porosity, structural diversity, and high surface area [42, 95–98].

In the MOFs family, ZIF-8 is a typical subspecies, which consists of zinc(II) ion clusters linked by 2-methylimidazolate ligands. It possesses high specific surface, porous microstructure, good thermal stability, chemical resistance, and hardness, which make it a kind of promising low-*k* dielectric. Eslava et al. investigated ZIF-8 films to assess their potential application as future insulators in microelectronics. Due to its low polarizability and density, the dielectric constant of the ZIF-8 film was only 2.33 (±0.05) at the frequency of 10^5 Hz. Moreover, it is desirable to fabricate

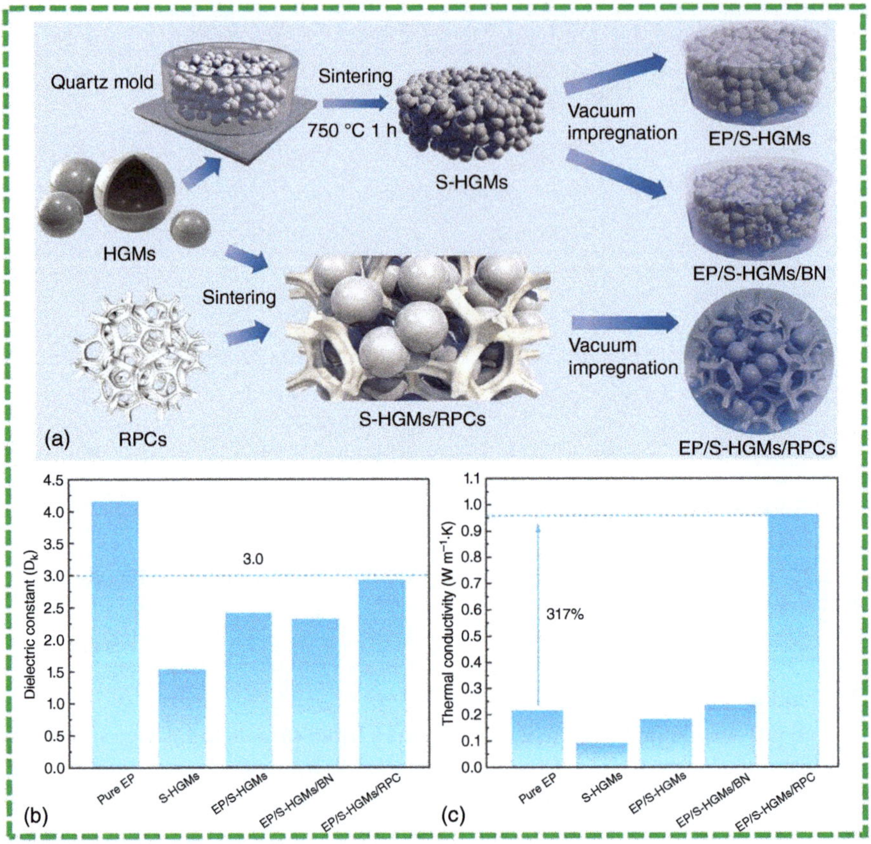

Figure 5.21 (a) Schematic diagram of the preparation of epoxy composites; (b) the dielectric permittivity at 3×10^6 Hz; (c) the thermal conductivity of composites. Source: Li et al. [76]/Elsevier.

polymer/ZIF-8 composite dielectrics to develop practical applications [38, 99, 100] (Figure 5.22).

5.3 Summary and Perspectives

In this chapter, the development and requirements of electronic packaging technology and materials have been reviewed, where polymer dielectrics play very important roles in many aspects. Despite significant advancements in the field of Advanced Electronic Packaging, high-performance polymer dielectrics are still in great demand, especially for applications at high temperatures. Thermal properties such as high thermal stability, compatible CTE, and high thermal conductivity would have to be balanced with mechanical and electrical features for the sake of efficient packaging purposes. HTP dielectrics, added with fillers of required properties whenever necessary, are one of the most important research fields worthy

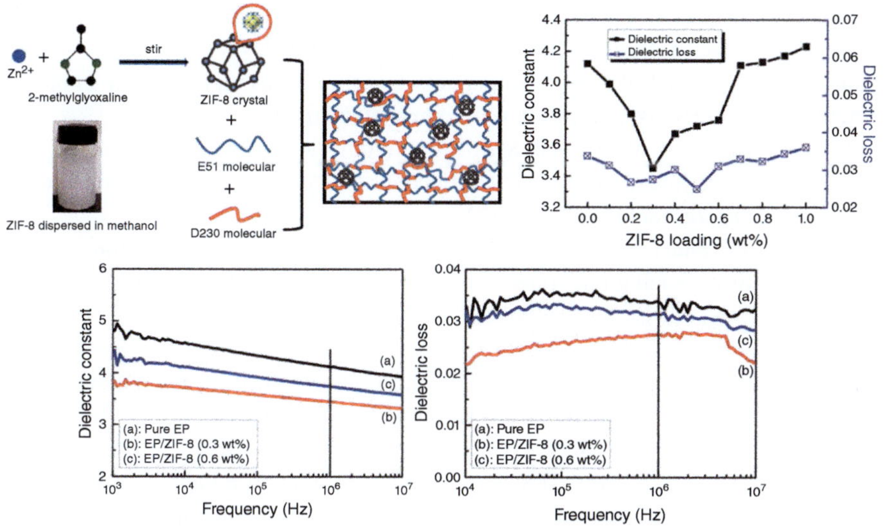

Figure 5.22 Schematic diagram for preparation of the EP/ZIF-8 composites. Source: Zhao et al. [99]/Taylor and Francis.

of devoted efforts. There remains great space for the development of novel HTP dielectrics for advanced electronic packaging. Gradually solving the problems of the current organic–inorganic composite system with intentionally designed structures and integrated functions in the future is promising to prepare next-generation low-*k* materials with excellent comprehensive thermal, mechanical, and electrical performances.

References

1 Hergenrother, P.M. (2016). The use, design, synthesis, and properties of high performance/high temperature polymers: an overview. *High Performance Polymers* 15: 3–45.

2 Mitsui, H. (1996). Progress in Japan in electrical insulation at high temperatures. *IEEE Electrical Insulation Magazine* 12: 16–27.

3 Sillion, B. (1999). Aromatic and heterocyclic polymers - what future? *High Performance Polymers* 11: 417–436.

4 Ikram, M., Sultan, K., Lateef, M.F., and Alqadami, A.S.M. (2022). A road towards 6G communication – a review of 5G antennas, arrays, and wearable devices. *Electronics* 11: 1–29.

5 Shih, M.-K., Huang, Y.-W., and Lin, G.-S. (2022). Next-generation high-density PCB development by fan-out RDL technology. *IEEE Transactions on Device and Materials Reliability* 22: 296–305.

6 Bajracharya, R., Shrestha, R., Ali, R. et al. (2018). LWA in 5G: state-of-the-art architecture, opportunities, and research challenges. *IEEE Communications Magazine* 56: 134–141.

7 Simsek, M., Aijaz, A., Dohler, M. et al. (2016). 5G-enabled tactile internet. *IEEE Journal on Selected Areas in Communications* 34: 460–473.

8 Rao, S.K. and Prasad, R. (2018). Impact of 5G technologies on smart city implementation. *Wireless Personal Communications* 100: 161–176.

9 Watanabe, A.O., Ali, M., Sayeed, S.Y.B. et al. (2021). A review of 5G front-end systems package integration. *IEEE Transactions on Components, Packaging and Manufacturing Technology* 11: 118–133.

10 McGarry, M.P. and Lee, M. (2022). Broadband dielectric properties of integrated circuit packaging materials across the 6G spectrum. *IEEE Transactions on Components, Packaging and Manufacturing Technology* 12: 1575–1582.

11 Zhang, S., Xu, X., Lin, T., and He, P. (2019). Recent advances in nano-materials for packaging of electronic devices. *Journal of Materials Science: Materials in Electronics* 30: 13855–13868.

12 Rao, V.S., Sekhar, V.N., Wee, H.S. et al. (2012). Process and reliability of embedded micro-wafer-level package (EMWLP) using low cure temperature dielectric material. *IEEE Transactions on Components, Packaging and Manufacturing Technology* 2: 13–22.

13 Liu, X., Zhang, W., Hao, D., and Liu, Y. (2022). Cost-effective surface-mount magnetoelectric dipole antenna based on ball grid array packaging technology for 5G millimeter-wave new radio band applications. *IEEE Transactions on Components, Packaging and Manufacturing Technology* 12: 1567–1574.

14 Wu, C.J., Hsieh, M.C., and Chiang, K.N. (2010). Strength evaluation of silicon die for 3D chip stacking packages using ABF as dielectric and barrier layer in through-silicon via. *Microelectronic Engineering* 87: 505–509.

15 Lau, J.H. (2011). Overview and outlook of through-silicon via (TSV) and 3D integrations. *Microelectronics International* 28: 8–22.

16 Ho, D.S.W., Jung, B.Y., Sorono, D.V. et al. (2014). Development of low profile fan out pop solution with embedded passive. In: *2014 IEEE 16th Electronics Packaging Technology Conference (EPTC)*, 3–5 December, Singapore, pp. 597–600.

17 Lau, J.H. (2022). Recent advances and trends in advanced packaging. *IEEE Transactions on Components, Packaging and Manufacturing Technology* 12: 228–252.

18 Hagelauer, A., Wojnowski, M., Pressel, K. et al. (2018). Integrated systems-in-package: heterogeneous integration of millimeter-wave active circuits and passives in fan-out wafer-level packaging technologies. *IEEE Microwave Magazine* 19: 48–56.

19 Lancaster, A. and Keswani, M. (2018). Integrated circuit packaging review with an emphasis on 3D packaging. *Integration* 60: 204–212.

20 Lau, J.H., Chen, G.C.-F., Huang, J.Y.-C. et al. (2021). Hybrid substrate by fan-out RDL-first panel-level packaging. *IEEE Transactions on Components, Packaging and Manufacturing Technology* 11: 1301–1309.

21 Li, Z., Pan, J., Hu, H., and Zhu, H. (2021). Recent advances in new materials for 6G communications. *Advanced Electronic Materials* 8: 1–19.

22 Wen, Y., Chen, C., Ye, Y. et al. (2022). Advances on thermally conductive epoxy-based composites as electronic packaging underfill materials-a review. *Advanced Materials* 34: e2201023.

23 Carias, V., Thompson, J., Myers, P.D. et al. (2015). Development of mold compounds with ultralow coefficient of thermal expansion and high glass transition temperature for fan-out wafer-level packaging. *IEEE Transactions on Components, Packaging and Manufacturing Technology* 5: 921–929.

24 Wan, Y.-J., Li, G., Yao, Y.-M. et al. (2020). Recent advances in polymer-based electronic packaging materials. *Composites Communications* 19: 154–167.

25 Yoon, H., Matteini, P., and Hwang, B. (2022). Review on three-dimensional ceramic filler networking composites for thermal conductive applications. *Journal of Non-Crystalline Solids* 576: 1–27.

26 Yu, C.-F., Huang, Y.-W., Ouyang, T.-Y. et al. (2022). Suppression strategy for process-induced warpage of novel fan-out wafer level packaging. *Microelectronics Reliability* 136: 1–10.

27 Kim, Y., Park, A.-Y., Kao, C.-L. et al. (2016). Prediction of deformation during manufacturing processes of silicon interposer package with TSVs. *Microelectronics Reliability* 65: 234–242.

28 Wang, S., Sun, Y., Sheng, C. et al. (2021). Warpage analysis and prediction of the advanced fan-out technology based on process mechanics. *IEEE Transactions on Components, Packaging and Manufacturing Technology* 11: 2201–2213.

29 Murayama, T., Sato, M., Suzuki, A. et al. (2019). Plasma dry process technology development of glass-epoxy film on the silicon substrate to fabricate RDL for future GPU/AI application. In: *2019 IEEE 69th Electronic Components and Technology Conference (ECTC),* May 25-June 2, Las Vegas, USA, pp. 1865–1869.

30 Edwards, C. (2021). Moore's law. *Communications of the ACM* 64: 12–14.

31 Kahouli, A., Sylvestre, A., Jomni, F. et al. (2012). Dielectric properties of parylene AF4 as low-k material for microelectronic applications. *Thin Solid Films* 520: 2493–2497.

32 Wang, L., Yang, J., Cheng, W. et al. (2021). Progress on polymer composites with low dielectric constant and low dielectric loss for high-frequency signal transmission. *Frontiers in Materials* 8 (8): 1–16.

33 Ikehira, S. (2021). Novel insulation materials suitable for FOWLP and FOPLP. In: *2021 IEEE 71st Electronic Components and Technology Conference (ECTC),* June 1-4, San Diego, California, USA, pp. 729–735.

34 Inamdar, A., Yang, Y.-H., Prisacaru, A. et al. (2021). High temperature aging of epoxy-based molding compound and its effect on mechanical behavior of molded electronic package. *Polymer Degradation and Stability* 188: 1–7.

35 Gao, R., Ma, R., Li, J. et al. (2022). Selection and characterization of photosensitive polyimide for fan-out wafer-level packaging. *IEEE Transactions on Components, Packaging and Manufacturing Technology* 12: 368–374.

36 Tsai, C.H., Liu, S.W., and Chiang, K.N. (2020). Warpage analysis of fan-out panel-level packaging using equivalent CTE. *IEEE Transactions on Device and Materials Reliability* 20: 51–57.

37 Yamanaka, K., Kobayashi, K., Hayashi, K., and Fukui, M. (2010). Materials, processes, and performance of high-wiring density buildup substrate with ultralow-coefficient of thermal expansion. *IEEE Transactions on Components and Packaging Technologies* 33: 453–461.

38 Xie, M., Li, M., Sun, Q. et al. (2022). Research progress on porous low dielectric constant materials. *Materials Science in Semiconductor Processing* 139: 1–13.

39 Guan, X., Ma, Z., Xiang, Z. et al. (2022). Soluble polyarylate with high thermal stability and low-k at high frequency. *Materials Letters* 324: 1–4.

40 Bei, R., Qian, C., Zhang, Y. et al. (2017). Intrinsic low dielectric constant polyimides: relationship between molecular structure and dielectric properties. *Journal of Materials Chemistry C* 5: 12807–12815.

41 Lee, C.-H., Huang, B., See, J. et al. (2022). Study of bondable laser release material using 355 nm energy to facilitate RDL-first and die-first fan-out wafer-level packaging (FOWLP). *IEEE Transactions on Components, Packaging and Manufacturing Technology* 12: 692–699.

42 Zhao, X.-Y. and Liu, H.-J. (2010). Review of polymer materials with low dielectric constant. *Polymer International* 59: 597–606.

43 Li, Y., Sun, G., Zhou, Y. et al. (2022). Progress in low dielectric polyimide film – a review. *Progress in Organic Coatings* 172: 1–19.

44 Han, S., Li, Y., Hao, F. et al. (2021). Ultra-low dielectric constant polyimides: combined efforts of fluorination and micro-branched crosslink structure. *European Polymer Journal* 143: 1–9.

45 Hu, K., Ye, Q., Fan, Y. et al. (2021). Preparation and characterization of organic soluble polyimides with low dielectric constant containing trifluoromethyl for optoelectronic application. *European Polymer Journal* 157: 1–8.

46 Wang, Z., Zhang, M., Han, E. et al. (2020). Structure-property relationship of low dielectric constant polyimide fibers containing fluorine groups. *Polymer* 206: 1–10.

47 Wu, Y., Ji, J., Huang, H. et al. (2021). Facile synthesis of acyloxy-containing fluorene-based Cardo polyimides with high optical transparency, fluorescence and low dielectric constant. *Reactive and Functional Polymers* 166: 1–7.

48 Liu, Y., Zhou, Z., Qu, L. et al. (2017). Exceptionally thermostable and soluble aromatic polyimides with special characteristics: intrinsic ultralow dielectric constant, static random access memory behaviors, transparency and fluorescence. *Materials Chemistry Frontiers* 1: 326–337.

49 Zhao, W., Cao, X., Huang, J. et al. (2022). Construction of micro-branched crosslink fluorinated polyimide with ultra-low dielectric permittivity and enhanced mechanical properties, Express. *Polymer Letters* 16: 142–151.

50 Liu, Y., Qian, C., Qu, L. et al. (2015). A bulk dielectric polymer film with intrinsic ultralow dielectric constant and outstanding comprehensive properties. *Chemistry of Materials* 27: 6543–6549.

51 Popovici, D., Hulubei, C., Cozan, V. et al. (2012). Polyimides containing cycloaliphatic segments for low dielectric material. *High Performance Polymers* 24: 194–199.

52 Wang, Z.-h., Fang, G.-Q., He, J.-J. et al. (2020). Semi-aromatic thermosetting polyimide resins containing alicyclic units for achieving low melt viscosity and low dielectric constant. *Reactive and Functional Polymers* 146: 1–10.

53 Qian, C., Bei, R., Zhu, T. et al. (2019). Facile strategy for intrinsic low-k dielectric polymers: molecular design based on secondary relaxation behavior. *Macromolecules* 52: 4601–4609.

54 Lv, G., Jensen, E., Shan, N. et al. (2021). Effect of aromatic/aliphatic structure and cross-linking density on the thermal conductivity of epoxy resins. *ACS Applied Polymer Materials* 3: 1555–1562.

55 Ruan, K., Zhong, X., Shi, X. et al. (2021). Liquid crystal epoxy resins with high intrinsic thermal conductivities and their composites: a mini-review. *Materials Today Physics* 20: 1–12.

56 Ruan, K., Guo, Y., and Gu, J. (2021). Liquid crystalline polyimide films with high intrinsic thermal conductivities and robust toughness. *Macromolecules* 54: 4934–4944.

57 Guo, Y., Zhou, Y., and Xu, Y. (2021). Engineering polymers with metal-like thermal conductivity—Present status and future perspectives. *Polymer* 233: 1–20.

58 Maier, G. (2001). Low dielectric constant polymers for microelectronics. *Progress in Polymer Science* 26: 3–65.

59 Wu, S., Peng, Q., Deng, Y. et al. (2022). Preparation and performance of low-dielectric benzocyclobutene resins containing dicyclopentadiene. *Journal of Vinyl and Additive Technology* 28: 631–639.

60 Fan, W., Hong, N., Sun, Q. et al. (2022). Thermo-curable and photo-patternable polysiloxanes and polycarbosiloxanes by a facile Piers–Rubinsztajn polycondensation and post-modification. *Polymer Chemistry* 13: 2187–2194.

61 Hou, J., Sun, J., and Fang, Q. (2022). A fluorinated low dielectric polymer at high frequency derived from allylphenol and benzocyclobutene by a facile route. *European Polymer Journal* 163: 1–7.

62 Kong, L., Qi, T., Ren, Z. et al. (2016). High-performance intrinsic low-k polymer via the synergistic effect of its three units: adamantyl, perfluorocyclobutylidene and benzocyclobutene. *RSC Advances* 6: 68560–68567.

63 Gong, Z., Zhang, Y., Guo, X., and Liu, Z. (2018). Wafer-level packaging method for RF MEMS applications using pre-patterned BCB polymer. *Micromachines (Basel)* 9: 1–13.

64 Xu, Q., Zeng, M., Chen, J. et al. (2018). Synthesis, polymerization kinetics, and high-frequency dielectric properties of novel main-chain benzoxazine copolymers. *Reactive and Functional Polymers* 122: 158–166.

65 Dang, P.T.M.T.D., Alexander, M.D. Jr., Grayson, C.J. et al. (2000). Synthesis and characterization of fluorinated benzoxazole polymers with high Tg and low dielectric constant. *Journal of Polymer Science Part A: Polymer Chemistry* 38: 1991–2003.

66 Cai, W., Wen, H., Wang, Z. et al. (2022). Design, synthesis, flame retardancy and dielectric properties of novel aromatic hyperbranched benzoxazine. *Reactive and Functional Polymers* 170: 1–9.

67 Lee, S.H., Kim, K.S., Shim, J.H., and Ahn, C.-H. (2018). High-performance printed circuit board materials based on benzoxazine and epoxy blend system. *Macromolecular Research* 26: 388–393.
68 Yuan, M., Lu, X., Zhao, Y. et al. (2022). Study of two novel siloxane-containing polybenzoxazines with intrinsic low dielectric constant. *Polymer* 245: 1–8.
69 Zhang, Y., Geng, Z., Niu, S. et al. (2020). Preparation and applications of low-dielectric constant poly aryl ether. *Advanced Industrial and Engineering Polymer Research* 3: 175–185.
70 Wang, Z., Shang, Y., Han, X. et al. (2020). Cross-linked fluorinated poly(aryl ether) (C-FPAE) films: preparation strategy, performance study, and low dielectric applications. *Macromolecular Materials and Engineering* 305: 1–7.
71 Wang, Z., Xie, Y., Shang, Y. et al. (2020). Preparation and properties of novel cross-linked fluorinated poly(aryl ether) with low dielectric constant and high thermal stability. *Macromolecular Rapid Communications* 41: e2000100.
72 Wang, Z., Jiang, B., Zhang, Y. et al. (2022). Influence of crosslink density on thermal, mechanical and dielectric properties of cross-linked fluorinated poly(aryl ether)s. *European Polymer Journal* 172: 1–6.
73 Dhara, M.G.B. (2010). Susanta, Fluorinated high-performance polymers: poly(arylene ether)s and aromatic polyimides containing trifluoromethyl groups. *Progress in Polymer Science* 35: 1022–1077.
74 Fang, L., Zhou, J., He, C. et al. (2020). Understanding how intrinsic micro-pores affect the dielectric properties of polymers: an approach to synthesize ultra-low dielectric polymers with bulky tetrahedral units as cores. *Polymer Chemistry* 11: 2674–2680.
75 Nilagiri Balasubramanian, K.B. and Ramesh, T. (2018). Role, effect, and influences of micro and nano-fillers on various properties of polymer matrix composites for microelectronics: a review. *Polymers for Advanced Technologies* 29: 1568–1585.
76 Li, R., Yang, X., Li, J. et al. (2022). Review on polymer composites with high thermal conductivity and low dielectric properties for electronic packaging. *Materials Today Physics* 22: 1–18.
77 Zhou, Y., Liu, F., and Wang, H. (2017). Novel organic-inorganic composites with high thermal conductivity for electronic packaging applications: a key issue review. *Polymer Composites* 38: 803–813.
78 Lokanathan, M., Acharya, P.V., Ouroua, A. et al. (2021). Review of nanocomposite dielectric materials with high thermal conductivity. *Proceedings of the IEEE* 109: 1364–1397.
79 Xu, Y., Wang, X., and Hao, Q. (2021). A mini review on thermally conductive polymers and polymer-based composites. *Composites Communications* 24: 1–11.
80 Hu, Y., Chen, C., Wen, Y. et al. (2021). Novel micro-nano epoxy composites for electronic packaging application: balance of thermal conductivity and processability. *Composites Science and Technology* 209: 1–9.
81 Alim, M.A., Abdullah, M.Z., Aziz, M.S.A. et al. (2021). Recent advances on thermally conductive adhesive in electronic packaging: a review. *Polymers (Basel)* 13: 1–23.

82 Zhang, H., Zhang, X., Fang, Z. et al. (2020). Recent advances in preparation, mechanisms, and applications of thermally conductive polymer composites: a review. *Journal of Composites Science* 4: 1–46.

83 Ge, M., Zhang, J., Zhao, C. et al. (2019). Effect of hexagonal boron nitride on the thermal and dielectric properties of polyphenylene ether resin for high-frequency copper clad laminates. *Materials & Design* 182: 1–9.

84 Liu, Z., Li, J., and Liu, X. (2020). Novel functionalized BN nanosheets/epoxy composites with advanced thermal conductivity and mechanical properties. *ACS Applied Materials Interfaces* 12: 6503–6515.

85 Zou, D., Huang, X., Zhu, Y. et al. (2019). Boron nitride nanosheets endow the traditional dielectric polymer composites with advanced thermal management capability. *Composites Science and Technology* 177: 88–95.

86 He, H., Peng, W., Liu, J. et al. (2022). Microstructured BN composites with internally designed high thermal conductivity paths for 3D electronic packaging. *Advanced Materials* 34: e2205120.

87 Wang, H., Wang, Q., Zhang, Q. et al. (2021). High thermal conductive composite with low dielectric constant and dielectric loss accomplished through flower-like Al_2O_3 coated BNNs for advanced circuit substrate applications. *Composites Science and Technology* 216: 1–9.

88 Yu, X., Bhatti, M.R., Ren, X. et al. (2022). Dielectric polymer composites with ultra-high thermal conductivity and low dielectric loss. *Composites Science and Technology* 229: 1–8.

89 Chen, Y., Zhang, H., Chen, J. et al. (2022). Thermally conductive but electrically insulating polybenzazole nanofiber/boron nitride nanosheets nanocomposite paper for heat dissipation of 5G base stations and transformers. *ACS Nano* 16: 14323–14333.

90 Yue, C.E., Guan, L., Zhang, X. et al. (2021). Thermally conductive epoxy/boron nitride composites with high glass transition temperatures for thermal interface materials. *Materials & Design* 212: 1–9.

91 Huang, C., Li, J., Xie, G. et al. (2019). Low-dielectric constant and low-temperature curable polyimide/POSS nanocomposites. *Macromolecular Materials and Engineering* 304: 1–10.

92 Zhao, H., Zhao, S.-Q., Li, Q. et al. (2020). Fabrication and properties of waterborne thermoplastic polyurethane nanocomposite enhanced by the POSS with low dielectric constants. *Polymer* 209: 1–8.

93 Wang, J., Zhou, D.-L., Lin, X. et al. (2022). Preparation of Low-k Poly(dicyclopentadiene) nanocomposites with excellent comprehensive properties by adding larger POSS. *Chemical Engineering Journal* 439: 1–10.

94 Wu, B., Liu, H., Fu, R. et al. (2021). Epoxy-matrix composite with low dielectric constant and high thermal conductivity fabricated by HGMs/Al_2O_3 co-continuous skeleton. *Journal of Alloys and Compounds* 869: 1–7.

95 Zhou, X., Liu, X., Cui, Z. et al. (2020). Design and development of HMS@ZIF-8/fluorinated polybenzoxazole composite films with excellent low-k performance, mechanical properties and thermal stability. *Journal of Materials Chemistry C* 8: 7476–7484.

96 Eslava, S., Zhang, L., Esconjauregui, S. et al. (2012). Metal-organic framework ZIF-8 films as low-κ dielectrics in microelectronics. *Chemistry of Materials* 25: 27–33.

97 Li, X., Liu, X., Hu, X. et al. (2021). A novel low-dielectric nanocomposite with hydrophobicity property: fluorinated MOFs modified bismaleimide-triazine resin. *Materials Today Communications* 29: 1–7.

98 Burtch, N.C., Baxter, S.J., Heinen, J. et al. (2019). Negative thermal expansion design strategies in a diverse series of metal–organic frameworks. *Advanced Functional Materials* 29: 1–8.

99 Zhao, Y., Chen, G., Zhou, Z., and Li, Q. (2019). Metal-organic frameworks filled epoxy composites: exploring new chances for low-κ polymer dielectrics. *Polymer-Plastics Technology and Materials* 59: 184–194.

100 Liu, C., Mullins, M., Hawkins, S. et al. (2017). Epoxy nanocomposites containing zeolitic imidazolate framework-8. *ACS Applied Materials & Interfaces* 10: 1250–1257.

6

High-Temperature Polymer Dielectrics for Printed Circuit Board

Xu Wang, Xinyu Chen, Junhui Luo, Xin Wang, Yan Chen, and Xiangyang Liu

Sichuan University, College of Polymer Science and Engineering, Chengdu, Sichuan, 610065, PR China

With the development of the electronic, electrical, and information industries, printed circuit board (PCB) has become the main component of most electronic products [1]. As the carrier of electrical connection between components on the insulation substrate, PCB is the support of electronic components. Electronic products manufactured with PCBs have the advantages of high reliability, good consistency, high mechanical strength, lightweight, small volume, and easy standardization. Copper-clad laminate (CCL), as the substrate of PCBs, has developed vigorously in recent decades. As the main part of the PCB, CCL mainly performs the functions of conducting, insulating, and supporting the whole PCB, which determines the mechanical strength, processing performance, heat resistance, dielectric properties, etc., of the PCB. At present, the world's annual output of CCL has reached about 17.8 billion US dollars, which means that the CCL manufacturing industry is a sunrise industry with broad prospects for development. From the perspective of the output value structure of segmented products, FR-4 CCL (glass fiber (GF)-reinforced epoxy resin CCL) and special substrate CCL are the main categories of CCL in the world. The total output of FR-4 CCL accounts for more than half of the total output value. Among them, conventional FR-4 CCL accounts for about 33%, nonhalogenated FR-4 CCL accounts for about 15%, and high-T_g FR-4 CCL accounts for about 9%. In addition, special substrate CCL accounts for about 32% and has become an important class of CCL.

As a leading raw material for CCL as substrate material of PCBs, both the structure and performance of polymer resin play a decisive role in determining the performance of CCL. Here, the constant development of dielectric polymer resins gradually leads CCL to make progress in terms of performance. Along with the fast development of the electronic information industry, upgrading of electronic products and circuit assembly technology makes PCB fabrication technology move toward fine trace, microvia, multiple layers, and high-density tracing, therefore laying higher requirements on CCL's thermal dissipation capability, dimensional stability, and dielectric loss, which thereafter brings forward new demands on the performance of polymer dielectrics. So, polymer dielectrics, especially high-temperature polymer dielectrics, are the main focus of this chapter.

High Temperature Polymer Dielectrics: Fundamentals and Applications in Power Equipment, First Edition.
Edited by Jun-Wei Zha and Zhi-Min Dang.

This chapter starts with the traditional epoxy resins for CCLs and first introduces the most traditional high-temperature polymer resin materials, including epoxy, phenolic resins, and polyimide. Then, we introduce polymer dielectrics suitable for high-frequency communication, such as LCP, PTFE, and so on.

6.1 Epoxy Resin Used for PCB

6.1.1 Structure of Epoxy Resins

Epoxy resin is an organic polymer compound containing two or more epoxy groups in the molecule. It started in the 1930s and began to be industrialized in the late 1940s, after which many new varieties of epoxy resins were developed successively. Now, the annual world production has reached 5.5 million tons. Because epoxy resin and its cured system have a series of valuable properties, they can be used as an adhesive, coating, matrix resin for fiber-reinforced composites, etc. It is widely used in machinery, electrical machinery, chemical industry, aerospace, shipping, construction, and other industrial sectors.

There are many varieties of epoxy resins. According to their molecular structure, they can be roughly divided into five major categories: glycidylamine epoxy resins, aliphatic epoxy resins, cycloaliphatic epoxy resins, glycidylether epoxy resins, and glycidylester epoxy resins. The chemical structure formulas are shown in Figure 6.1.

More than 90% of the production is made up of epoxy resin of the glycidyl ether type, which is made up of bisphenol A and epichlorohydrin condensation. The epoxy resin is usually referred to as epoxy resin. The molecular structure of the epoxy resin produced from bisphenol A and epichlorohydrin is shown in the following Figure 6.2.

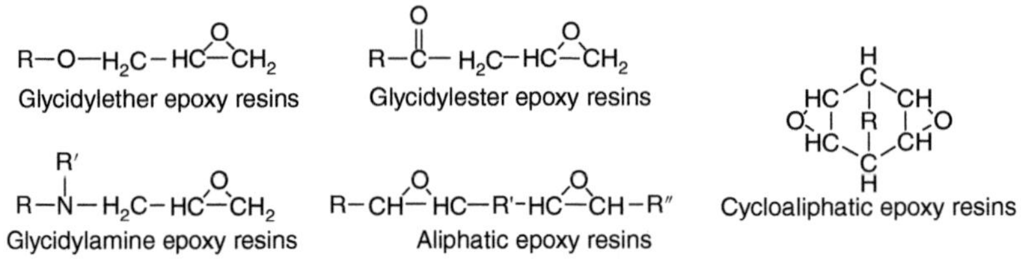

Figure 6.1 Chemical structure formulas of different kinds of epoxy resins.

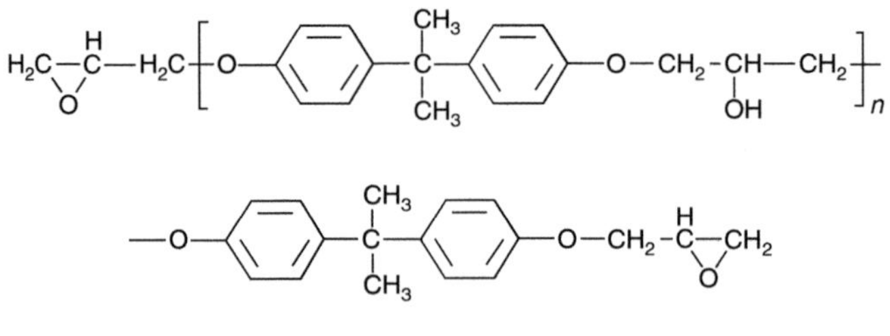

Figure 6.2 The chemical structure formula of bisphenol-A epoxy resins.

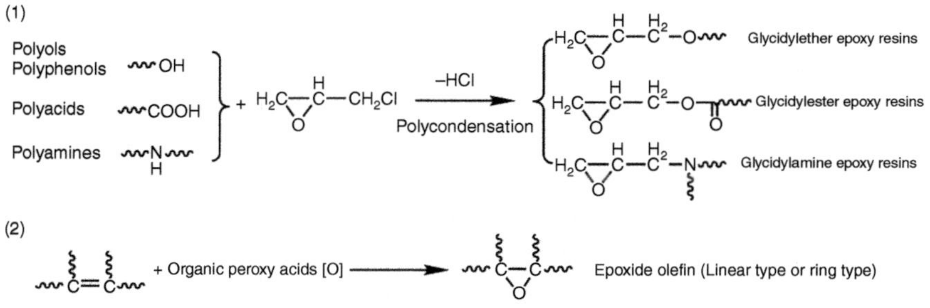

Figure 6.3 Reaction Schematics for Synthesis of Epoxy Resins.

These linear epoxy resins can be divided into three kinds according to the size of their average polymerization degree n. When $n < 2$, it is called low molecular weight epoxy resin, and its softening point is below 50 °C; when $n = 2$–5, it is medium molecular weight epoxy resin, and its softening point is 50–95 °C; when $n > 5$, it is called high-molecular-weight epoxy resin, and its softening point is above 100 °C. Natively, the corresponding performance of different types is also different.

There are two main types of methods for the synthesis of epoxy resins. (i) Epoxy resins are prepared by condensation of compounds containing active hydrogen atoms, such as polyphenols, polyols, poly acids, or polyamines, with compounds containing epoxy groups, such as epichlorohydrin. (ii) Epoxy resins are prepared by epoxidation of the double bonds of chain or cyclic dienes with peroxy acids. The corresponding reaction schematics are shown in Figure 6.3.

In addition to the basic components of epoxy resin, the composition of epoxy plastic also contains the following categories: (i) curing agent; (ii) toughening and other modifiers; (iii) fillers; (iv) diluents; and (v) others.

Among them, the curing agent is an essential addition. Whether for bonding agents, coatings, or castings, it is required to add curing agent; otherwise, the epoxy resin cannot be cured. Under the action of a curing agent, this linear-structure epoxy resin is cured because the epoxy groups are cross-linked with each other. There are always two kinds of epoxy resin curing. (i) cross-linked to a bulk structure by chemical reaction with a curing agent, and the curing agents used are multi-fatty amine, multi-ethylene amine, multi-aromatic amine, multi-acid anhydride, etc.; (ii) cross-linked by polymerization of epoxy group under the action of catalysts such as tertiary amine and Lewis' acid, while the catalyst does not participate in the reaction.

A toughening agent is added to improve its impact strength. There are active toughening agents and inactive toughening agents. Non-active toughening agents are general plasticizers such as dibutyl phthalate, etc., whose dosage is about 5–10% of epoxy resin. Active toughening agents have active groups and participate in the curing reaction, with an obvious toughening effect. Commonly used are epoxidized vegetable oil and multifunctional thermoplastic polyamide resin, whose dosage is 40–80% of epoxy resin. Diluent is added to reduce the viscosity during processing; the dosage is 5–20% of the weight of epoxy resin. Filler is used to improve the performance and reduce the cost. Commonly used fillers include asbestos, GF, mica, and quartz powder. The usage is generally 40% or less.

6.1.2 Properties and Application of Epoxy

Epoxy resins contain unique epoxy groups as well as some reactive and polar groups, such as hydroxyl and ether bonds, and thus have many excellent properties. Compared with other thermosetting resins, epoxy resins have the largest number of types and grades with different properties. There are more types of epoxy resin curing agents, plus many accelerators, modifiers, additives, etc., so that a variety of combinations and groupings can be made to obtain a variety of epoxy curing systems and curing products with excellent performance. They can be adapted to meet the requirements of different performance and process performance. This is incomparable to other thermosetting resins. In summary, epoxy resins have the following outstanding advantages:

① Variety of forms: A variety of resins, curing agents, and modifier systems can be adapted to the requirements of almost all applications on the form, which can range from very low viscosity to high-melting-point solids.
② Easy curing: With a variety of curing agents, epoxy resin systems can be cured in the temperature range of 0–180 °C.
③ Strong adhesion: The presence of polar hydroxyl groups and ether bonds in epoxy resin makes it have high adhesion to various substances. The low shrinkage of epoxy resin during curing also contributes to the formation of a strong, bonding bond with low internal stress.
④ Low shrinkage: The reaction between the epoxy resin and the curing agent used is carried out by direct addition. No water or other volatile by-products are released. At the same time, the epoxy resin has a certain degree of pre-polymerization, which makes it show lower shrinkage during the curing process compared with phenolic resin and unsaturated polyester resin.
⑤ Mechanical properties: After curing, the epoxy resin system has excellent mechanical properties.
⑥ Electrical properties: The cured epoxy resin system has good electrical properties in a wide range of frequencies and temperatures and is an excellent insulating material with high dielectric properties, resistance to surface leakage, and arc resistance.
⑦ Chemical stability performance: The cured epoxy resin system has excellent alkali resistance, acid resistance, and solvent resistance. Its chemical stability depends on the selected resin and curing agent. A proper selection of epoxy resin and curing agent can give it special chemical stability.
⑧ Dimensional stability: The combination of many properties mentioned above makes the cured epoxy resin system have good dimensional stability and durability.
⑨ Mildew resistance: The cured epoxy resin system is resistant to most molds and can be used under harsh tropical conditions.

Nowadays, epoxy resins permeate directly or indirectly into all aspects from daily life products to high-tech fields based on the above advantages. For example, in the field of coatings, it is applied to primers for metal parts such as automobiles, anti-corrosion coatings for industrial equipment, and epoxy resin flooring for construction. In the field of composite materials, epoxy resin is used in the preparation of various containers, storage tanks, pipes, wind power blades, aircraft, composite materials in spacecraft, packaging materials for large-scale integrated circuits, substrate materials for PCBs, insulation materials for generators and other electrical appliances, etc. In the field of adhesives, there are mainly fast-curing, toughness epoxy resin bonding agents, conductive adhesives, optical structural adhesives, automotive repair adhesives, stone adhesives, etc.

6.1.3 Epoxy Resin Used for CCL in PCB

Although CCLs come in numerous types, epoxy CCL accounts for over 70% of all CCLs. As one of the key substrates for electronic devices, the key function of CCL is to provide insulation for traces and electronics. In addition to the performance upgrade requirements due to continuous technological advances, epoxy resins have the following basic requirements: high purity, low humidity, excellent insulating properties, dimensional stability, and mechanical strength.

With the advent of electrification and the information era, electronic components have stepped into an era of high integration and high reliability. This has put forward the requirements for PCBs to develop in the direction of high precision. It not only realizes high-density surface mounting in the X and Y axes but also requires chemical copper deposition and holeization in the Z-axis direction. In addition to single- and double-sided wiring, it also requires layered and flexible wiring. Therefore, it requires a series of CCLs with high performance, so the epoxy CCLs are developing rapidly in terms of varieties, specifications, and output based on their outstanding features.

According to the American Institute of Electrical Engineers, there are epoxy resin GF CCL (G-10, G-11, FR-4, and FR-5), epoxy resin paper base CCL (FR-1 and FR-3), and epoxy resin composite CCL (CEM-1 and CEM-3). Among them, FR-4 epoxy CCLs account for 90% of products in a month. As one of the most representative epoxy glass cloth copper-clad boards, the manufacturing process of FR-4 is shown in Figure 6.4. The resin glue industry should be prepared first, and then the glass fabric or impregnated paper will be impregnated with glue, dried to make pre-impregnated material, and cut into sheets, laminated, and pressed into shape.

Among them, glue solutions of epoxy resin are usually composed of epoxy resin, catalyst, and solvent. Taking FR-4 (EPGC-32F) and FR-5 as examples of the glue solution, the corresponding composition is shown in the following Table 6.1.

Currently, brominated epoxy resins are used in large quantities in the production of CCLs such as FR-4 and FR-5 CCLs, which are mainly used to meet the requirement of flame-retardant performance. Its chemical structure is shown in Figure 6.5. According to the size of the bromine content in resin, brominated epoxy resins are divided into two categories: low bromine epoxy resin (19–21% bromine content) and high bromine-type resin (48–50% bromine content).

With the further improvement of the heat resistance requirement of PCB, the heat resistance level of its based resin of CCLs has been made more stringent. However, the conventional brominated epoxy resin has insufficient molecular

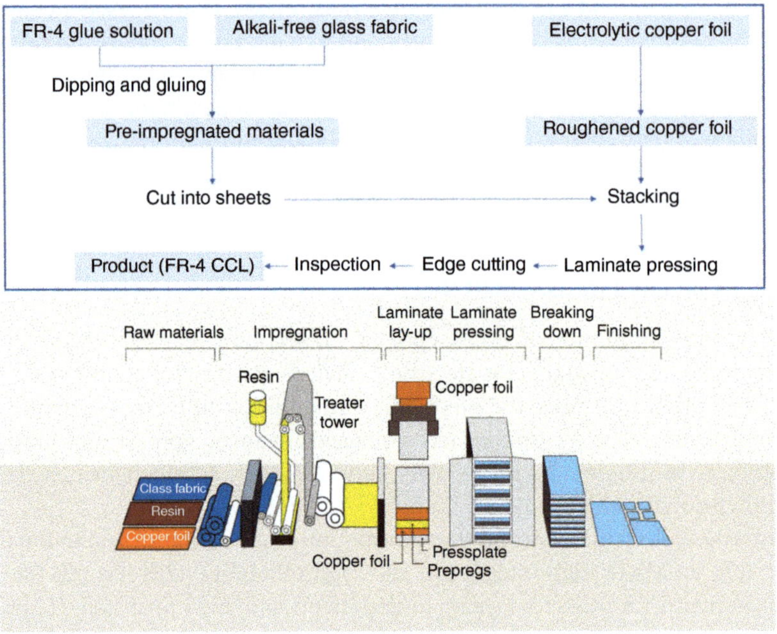

Figure 6.4 Flow chart for the preparation process of FR-4 CCL.

Table 6.1 Compositions of FR-4 and FR-5 glue solutions (unit: mol).

Materials	FR-4	FR-5
Low bromine epoxy	100	49.5
High bromine epoxy	—	10.5
Linear phenolic resin	—	40.0
Dicyandiamide	3.7	3.0
Benzyl dimethylamine (or 2-methylimidazole)	0.2	0.2
Solvent (Acetone etc.)	Adjustable	Adjustable

chain rigidity, and the glass transition temperature (T_g) after curing is 140–150 °C. Thus, the thermal stability performance of the brominated epoxy resin needs to be enhanced. By modifying the brominated epoxy resin with MDI, a special structural functional group, the oxazolidinone group, is introduced into the structure as shown in Figure 6.5; thus, the rigidity of the molecular chain structure is enhanced, and thus the T_g is significantly improved. The thermal stability of the modified epoxy resin meets the requirements of the base resin for CCLs and is widely used in industry.

The increasing T_g temperature can improve products' application temperature, so the research on high-T_g epoxy resin has been going on and has been quite important for CCL development. Therefore, there are many strategies being formed,

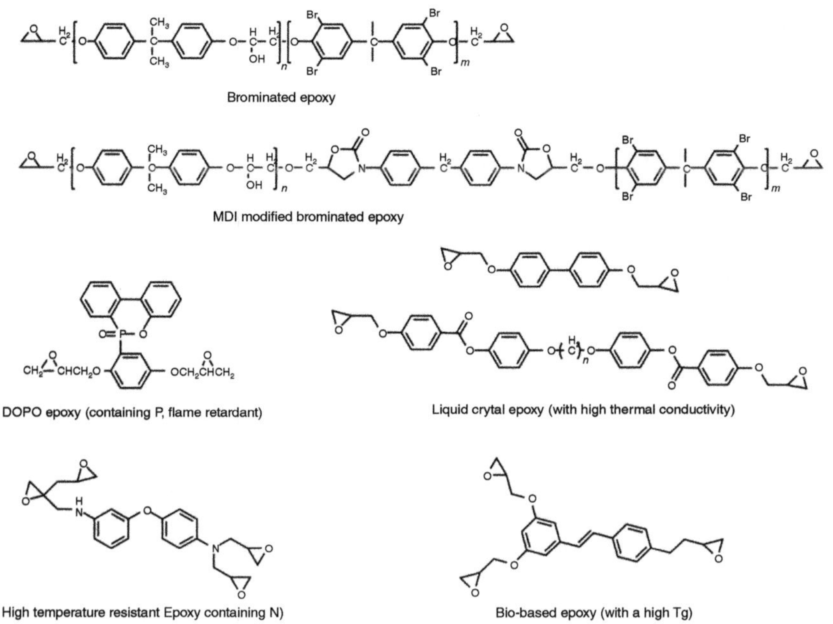

Figure 6.5 Chemical structure of partial epoxy resins.

such as introducing nitrogen-containing heterocyclic structures and synthesizing trifunctional epoxy or tetrafunctional epoxy. The structure of some of the epoxy resins is schematically shown in Figure 6.5.

In addition, it has been reported that brominated fire retardants will release hazardous substances that are bad for the environment and humans. Thus, electronic machines bring forward halogen-free requirements on CCL. Now, the application of N, P, B, Al, etc., is proven to be the feasible method for flame retardance. For example, halogen-free CCL is capable of resisting fire through the introduction of DOPO (9,10-dihydro-9-oxa-10-phosphaphenanthrene 10-oxide) into epoxy molecules. However, there has been a growing awareness of the adverse environmental impact of P and N compounds in recent years, so P, N, and Pb-free refractory epoxy resins will be a leading technology for CCL manufacturing.

As a leading raw material of CCL, epoxy resin also faces more demands of new technology. To cater to the fast development of the electronic information industry, electronic products and circuit assembly have to climb to a new stage, which leads PCB fabrication technology to move toward microvia, fine trace, high-density tracing, and high and multiple layers and brings forward new requirements to CCL in terms of thermal withstanding, low CTE, high dimensional stability, and low dielectric loss. Therefore, many high-performance and multifunctional epoxy resin GF CCLs have been developed, which include high-T_g FR-4 (T_g: 175 °C), lead-free compatible FR-4, high CTI, CAT resistance, high thermal conductivity (Liquid Crystal Epoxy Resin, Figure 6.6), FR-4 used for HDI (high density interconnect) substrate, polyimide modified epoxy, BT modified epoxy, PPO modified epoxy, CE modified epoxy GF CCL, epoxy GF prepreg, and RCC used for multilayer PCBs and BUM PCBs.

6.2 Phenolic Resins Used for PCB

6.2.1 Structure of Phenolic Resins

Phenolic paper-based laminates are the oldest type of material used in the production of PCB substrates [2]. Phenolic resins were one of the first resins synthesized by man, and by controlling the ratio of phenol to aldehyde and the function of phenol, as well as the type of catalyst (acidic or basic), thermoplastic (linear or branched) and thermosetting phenolic (cross-linked) resins can be produced. Thermoplastic phenolic resins require an acid as a catalyst and a phenol-to-aldehyde ratio greater than 1. If there is an excess of formaldehyde, the resulting linear primary polymer is susceptible to cross-linking by formaldehyde. Thermoplastic phenolic resins are rosin-like, friable, soluble, fusible, and soluble in acetone, ether, esters, etc. Thermoplastic phenolic resins do not contain the $-CH_2OH$ group within the molecule, and therefore a curing agent must be added to cure them. Hexamethylenetetramine (HMTA) is commonly used as a curing agent. Thermosetting resins are resins containing further reactive $-CH_2OH$ groups that can be cross-linked and cured under heat or acidic conditions. As the degree of advancement of the polycondensation reaction varies, so do the corresponding resin properties. They can be divided into three stages.

The molecular structure of thermoplastic phenolic resins has the formula (Figure 6.6):

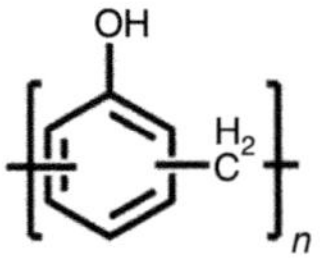

Figure 6.6 Molecular structure of thermoplastic phenolic resins.

The molecular structure of thermosetting phenolic resins is given by the formula (Figure 6.7):

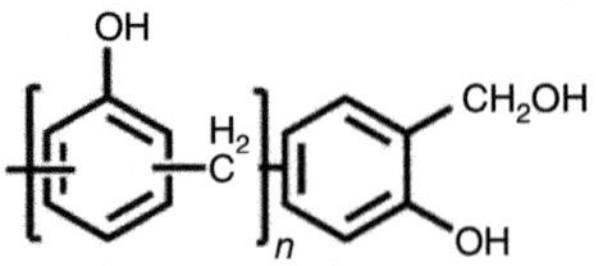

Figure 6.7 Molecular structure of thermosetting phenolic resins.

6.2.2 Synthesis of Phenolic Resins

Depending on the ratio of phenol to aldehyde, the reaction conditions, and the type of catalyst, phenolic resins can generally be divided into two categories. One is linear thermoplastic phenolic resins (Novolak type), which use acid as a catalyst and require the addition of a curing agent during processing, hence the name two-step resins. The other type is thermosetting phenolic resins (Resole type), which are usually synthesized using a base as a catalyst and contain hydroxymethyl reactive groups in the molecular chain that can be further reacted. They are also known as one-step resins.

6.2.2.1 Synthesis of Thermoplastic Phenolic Resins

Condensation Under Strongly Acidic Conditions (pH < 3) Thermoplastic phenolic resins are made by the condensation of formaldehyde with trifunctional phenols (e.g. phenol, m-cresol) or difunctional phenols (e.g. o-cresol, p-cresol) in an acidic medium. If trifunctional phenols are used, the phenol must be in excess (usually a molar ratio of 6 : 5 or 7 : 6 between phenol and aldehyde); otherwise, a thermosetting resin will be produced.

The reaction proceeds as follows (Figure 6.8):

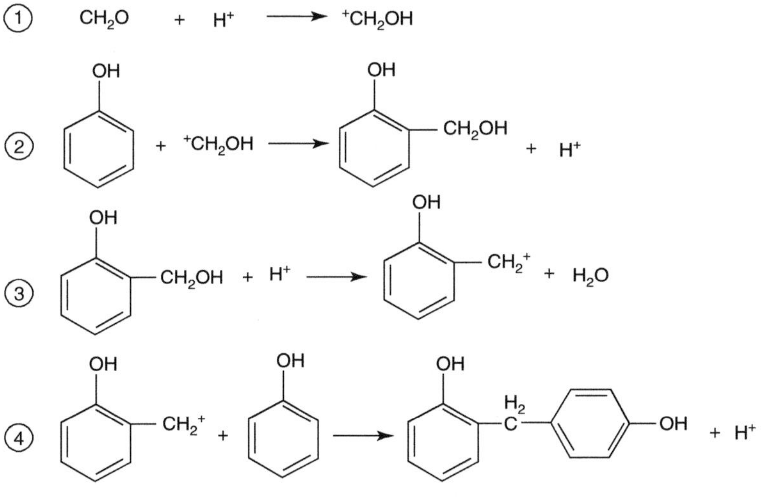

Figure 6.8 Reaction process of thermoplastic phenolic resins.

Reaction equation ② is a side reaction between phenol and formaldehyde; reaction equations ③ and ④ are condensation reactions because the rate of the condensation reaction is greater than the rate of the side reaction, so the end product is a linear resin. As the molecular weight increases, the reaction rate slows down and finally reaches a reversible equilibrium.

The ideal molecular structure of a thermoplastic resin is as follows (Figure 6.9):

Synthesis of High-Adjacent Thermoplastic Phenolic Resins Synthesis of high-adjacent thermoplastic phenolic resins. High-adjacent thermoplastic phenolic resins can be synthesized by the formation of chelates of metal ions in phenolic aldehydes at pH values in the range of 4–7, using certain metal salts as catalysts, mainly with adjacent

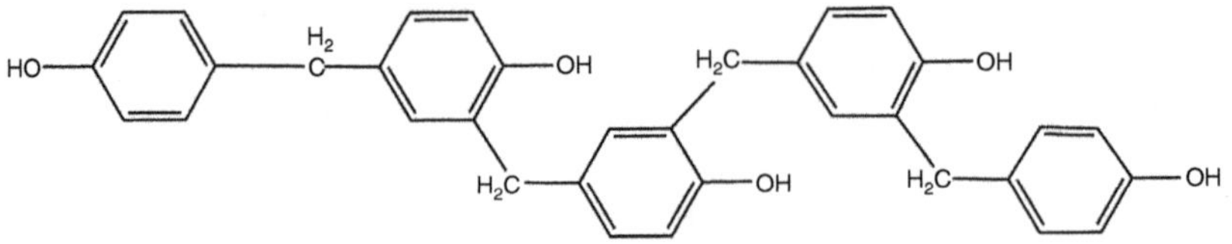

Figure 6.9 Ideal molecular structure of the thermoplastic resin.

linkages. High-adjacent thermoplastic phenolic resins cure two to three times faster than normal thermoplastic phenolic resins and are therefore suitable for fast-curing phenolic moldings.

Curing HMTA is the most widely used curing agent for thermoplastic phenolic resins and decomposes above 100° to form dimethylamine and formaldehyde, which react with the phenolic resin and cause cross-linking. Its structural formula is as follows (Figure 6.10):

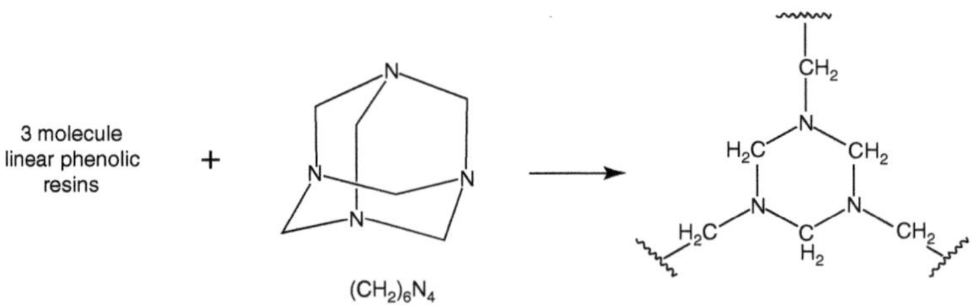

Figure 6.10 Curing process of phenolic resins.

6.2.2.2 Thermosetting Phenolic Resins

Phenol is condensed with excess formaldehyde in an alkaline or acidic medium (generally basic) to produce a meltable, thermosetting phenolic resin.

Addition of Phenol to the Aldehyde The reaction begins with the addition of phenol to the aldehyde, producing a mixture of poly(hydroxymethyl)phenol, monophenols, and polyphenol alcohols. The specific diagram is as follows: (Figure 6.11).

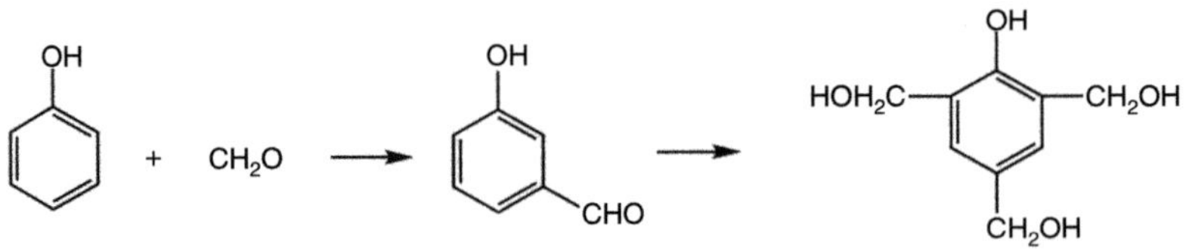

Figure 6.11 Initial reaction of phenol with formaldehyde.

Condensation of Hydroxymethyl Phenol Hydroxymethyl phenol also undergoes the following two possible condensation reactions. Under basic conditions, the products of formula ① are mainly produced. As the rate of the addition reaction is much higher than that of the condensation reaction, the final product obtained has a linear structure. The two reaction routes are shown in Figure 6.12.

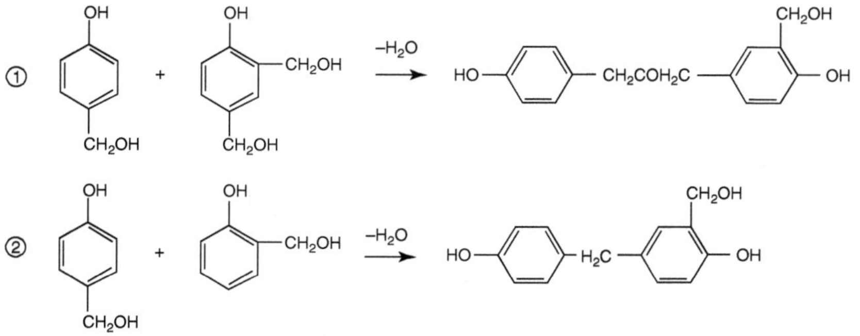

Figure 6.12 Reaction process of thermosetting phenolic resins.

Curing The curing of thermosetting resins does not require a curing agent. Under heated conditions, a condensation cure reaction occurs between the phenolic hydroxyl groups. Depending on the extent to which the condensation reaction occurs, thermosetting resins can be divided into three stages. Gradually, the A-order resin, which is soluble in aqueous solutions of ethanol, acetone, and alkali, changes to an insoluble, non-melting, cross-linked C-order resin.

6.2.3 Properties of Phenolic Resins and Their Application in PCBs

PF is an excellent insulating material due to its cheap price, simple processing and production equipment, low density, high mechanical strength and thermal strength, easy degradation, good mechanical properties, good heat resistance, etc. It is widely used in plastic materials, coatings, adhesives, and foam materials and is an indispensable polymer material for electronic and electrical appliances, machinery, aerospace, and other fields. In order to adapt to the newer and environmentally friendly requirements in the field of high technology, traditional phenolic resins also cannot meet the needs of modern development, and the modification of PF resins has never stopped. The residual polar phenolic hydroxyl groups in PF resins are still present and can absorb moisture in hot and humid environments, leading to a reduction in electrical, physical, and mechanical properties. In addition, the benzene ring in PF resins is too rigid, resulting in greater spatial resistance and reduced free rotation of the molecular chain, making PF resins brittle. Therefore, improving heat resistance and toughness and reducing costs are all goals for the modification of PF resins. In order to improve the electrical insulating properties of phenolic resins in high-temperature and humid environments and to reduce water absorption, the phenolic hydroxyl group must be blocked [3]. There are two general approaches: one is to reduce the number of phenolic hydroxyl groups by causing a polymerization reaction between the phenolic hydroxyl groups and the modifier [4]; the other is to physically blend the phenolic hydroxyl groups by using modifiers with excellent resistance to humidity and heat to separate and surround the phenolic hydroxyl groups [3, 5–7]. In general, the modification of phenolic resins is receiving increasing attention, but the modification process of phenolic resins currently used is more complex, the modifiers used are more costly [8], pollute the environment, etc. Improving the electrical insulation properties of phenolic resins in special environments requires further research.

6.2.3.1 Application of Phenolic Resins in Copper-Clad Laminates

Phenolic CCL is an important branch of phenolic laminate products, but phenolic laminate is also one of the main applications. The current copper laminate substrate (or carrier material) is divided into two main categories: paper-based and glass cloth-based laminates; the former mostly use phenolic resin, due to the traditional phenolic resin made of paper-based copper laminate products being more brittle; water absorption is also on the large side; and electrical insulation performance is not good, so it is suitable for the performance requirements of PCB manufacturing. Later, in order to improve its comprehensive performance, generally using dry oil modification, tung oil-modified phenolic resin made of paper-based copper laminate has good flexibility, low water absorption, and low moisture while still maintaining high electrical insulation performance characteristics. Different from the paper-based laminate, the main resin of the glass cloth-based laminate is epoxy resin, and the role of phenolic resin in this is to cure and modify the epoxy resin.

Phenolic paper-based CCL is the oldest kind of CCL and is commonly used as substrate materials for PCBs. As early as 1909, Dr. L. Baekland dipped fiber paper into phenolic resin and obtained a patent for phenolic insulating material. Nearly a decade later, Formica applied phenolic laminate to radio products for the first time, and in the 1940s, the United States first applied it to the substrate for PCB manufacturing. In the mid-1950s, Wang Tiezhong and others also used phenolic cardboard as the substrate to produce the first PCB in China [9].

Phenolic resins are widely used in the laminate industry and are classified into tung oil-modified phenolic resins (paper-based laminates), linear phenolic resins, and BPA-type phenolic resins (composite substrates, GF cloth laminates), nitrogen and phosphorus-containing phenolic resins (halogen-free flame retardant laminates), benzoxazine and phenotriazine resins (heat-resistant laminates), and epoxy resin-based curing of laminates, depending on their role or structure modifiers [10]. Some of their applications are described in this book.

6.2.3.2 Tung Oil-Modified Phenolic Resin

In order to solve the problem of stiff and brittle phenolic resins for phenolic-paper-based CCL curing and to complement and give it more flexibility, the use of dry oil-modified phenolic resins has long been a mainstream process route. The research on this technical topic, in response to the continuous needs of the actual market, is both an old subject with a long history and a new one in order to adapt to new demands for performance. The manufacturing of tung oil-modified phenolic resin always contains too much unreacted free phenol and aldehyde, causing environmental pollution. This hazard is mainly manifested in two aspects: on the one hand, in the preheating-punching process of the board, it releases foul gas, which is a threat to the health of the operators on site. On the other hand, after the PCB is made, it will be released free out to pollute the environment. Sumitomo

Bakelite Japan [11] achieved a significant reduction in unreacted residues (<2%) of tung oil-modified phenolic resins by using low-molecular-weight phenolics containing low unreacted residues and a “three-step” method in 2011. The patent presents the invention in the process technology for making low molecular weight and low free phenol content phenolic resin, including three elements: ① Reaction of phenol and aldehyde with an organophosphonic acid-type catalyst. ② Control the low composition of water in the reaction system, which should be below 30% (weight ratio) of the total composition of the reaction system, preferably below 20%. ③ The reaction temperature of the addition reaction stage is guaranteed to be above 110 °C (110–200 °C). Chinese scientists engaged in the synthesis of phenolic resins have also done in-depth research in the organophosphonic acid-catalyzed synthesis of phenolic resins, and in the published literature [12], an explanation of the above-patented principle is presented. The reaction system is divided into a catalyst phase containing water and organophosphonic acid and a resin phase containing phenol and phenolic resin. The organophosphonic acid catalyst can make the low-molar-mass components react first because of the different solubility in different molar masses of phenolic resin, phenol, and low-molar-mass components (such as monomers and dimers) dissolve in the aqueous phase at high temperature to react with formaldehyde while the polymerized resin will not be withdrawn from the resin phase to react, and the low-molar-mass components that decompose in the aqueous phase quickly precipitate into the resin phase after the reaction and no longer react. In this way, the reactive substances between the two phases circulate to carry on the reaction, thus increasing the conversion rate, while the content of free phenol in the reaction products is greatly reduced due to the priority reaction of the low-molecular-weight components. This kind of low-molecular-weight phenolic resin has the advantages of low viscosity, low molecular weight, easy control, and reducing the residual amount of unreacted phenol, so that tung oil can react with this low-molecular-weight phenolic resin smoothly and uniformly. Since the mutual solubility of the addition reactants of tung oil and phenolic resin products is excellent, they can also be freely matched with nonpolar aromatic solvents such as toluene.

6.2.3.3 Linear Phenolic Resins

The application of linear phenolic resin in the production of CCLs can effectively improve the structural toughness of lead-free CCLs and avoid the many processing-type problems arising from the brittleness of the material during the use of lead-free CCLs, which are widely promoted and used in the electronics manufacturing industry [13]. Linear phenolic resin toughening mainly uses the rubber particles in it to make it a stress concentration area for inducing silver patterns as well as shear bands, and at the same time, the rubber particles can effectively control the development of silver patterns and make them terminate in time to avoid destructive cracks that can cause damage to the laminate. Toughening methods for linear phenolic resins used in the production of CCLs are divided into two types: nonreactive and reactive. Nonreactive and other toughening methods are one of the methods often used in the manufacturing process of CCL raw materials and are relatively easy to operate. In the production process of CCL, the linear phenolic resin can be heated by the thermoplasticity of linear phenolic resin and fully mixed by physical blending or adding solvent to form the molecular structure or polymer formula alloy of the IPN network, which can effectively improve the properties and toughness of linear phenolic resin. Reactive toughening is mainly based on the basic principles such as chemical reaction to effectively improve the toughness and other properties of linear phenolic resin, connecting the long-chain molecules with good toughness and flexibility to the main chain part of the phenolic resin structure, and connecting it to phenol with the help of the formaldehyde reaction to effectively improve the toughness and other properties of linear phenolic resin. In the process of linear phenolic resin toughening of CCL, many methods are commonly used, such as tung oil modification, nonylphenol modification, allyl bismaleimide modification, and so on.

6.2.3.4 Nitrogen-Containing Phenolic Resins

CCLs as substrate materials for printed circuits must have certain flame-retardant properties to meet the needs of electronic product applications [14]. Early copper laminates were flame retardant by adding halogen-containing (mainly bromine) components to the system, and the flame retardancy of the sheets was generally easily achieved by using brominated epoxy or tetrabromobisphenol A. With the successive implementation of the EU WEEE environmental directive and RoHS standards [15], as well as market competition and human awareness of environmental protection, the requirements for halogen-free substrate materials are increasing, and the current mainstream route for halogen-free flame retardant CCLs is to use phosphorus-based flame retardants for flame retardancy. Phosphorus-based flame retardants are not safe; various intermediates and production processes have a certain degree of toxicity. In addition, phosphorus-containing flame retardants will also reduce the moisture resistance of CCLs, thereby affecting the sheet's resistance to CAF (conductive anode wire) performance. The introduction of the N element into the molecular skeleton of phenolic resin through chemical reaction gives the system an N–P synergistic flame retardant effect, which can improve the flame retardant efficiency on the one hand and make the whole system have a more balanced and comprehensive performance on the other hand. Nitrogen-containing phenolics will simultaneously generate CO_2, NH_3, N_2, NO_2 H_2O, and other nonflammable gases above 300°C, diluting the combustible gases and generating thermal convection to take away part of the heat; and N_2 can capture free radicals and inhibit the chain reaction of free radicals of polymers, which plays the role of scavenging free radicals, thus achieving the effect of flame retardancy [16]. Nitrogenous phenolic resins can be prepared by the joint reaction of nitrogenous compounds such as melamine, aniline, urea, and ammonia with phenol and formaldehyde under certain conditions. Mieczyslaw Maciejewski et al. [17] in Poland used melamine and formaldehyde to synthesize poly(hydroxymethyl)melamine under alkaline conditions, and then reacted with phenol in hydrochloric acid catalysis to obtain nitrogenous phenolic resins with the reaction course given by the following equation (Figure 6.13):

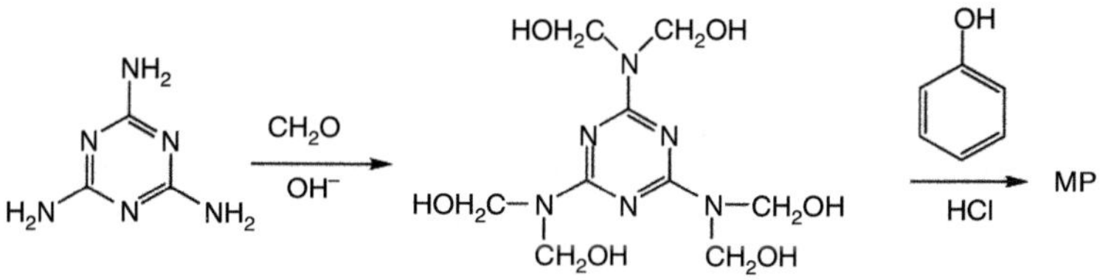

Figure 6.13 Reaction process of nitrogenous phenolic resins.

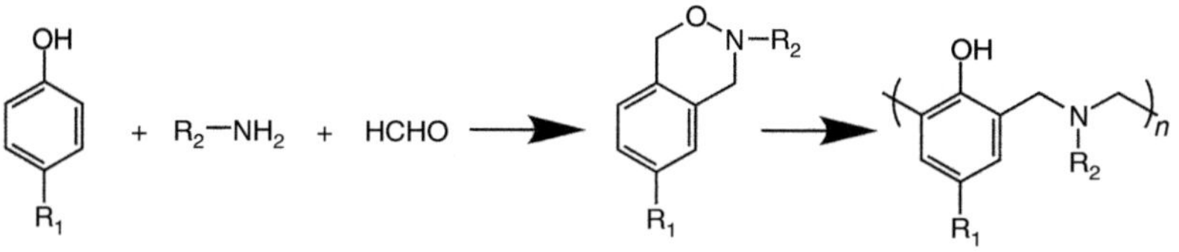

Figure 6.14 Synthetic route of PBZ.

Today, when people pay more and more attention to environmental protection, it is a big trend to develop halogen-free, environmentally friendly CCLs. Nitrogen-containing phenolic resins will play a more and more important role in the development of CCLs.

6.2.3.5 Polybenzoxazine Used for PCB

To further optimize the overall performance of phenolic resin, polybenzoxazine (PBZ) resin was synthesized in 1944 by Holly and Cope. PBZ has received considerable attention in the field of the new generation of high-frequency and high-speed integrated circuits, especially microwave communication system. The specific synthetic reaction route is as follows: (Figure 6.14).

PBZ have a number of unique advantages such as excellent mechanical properties, high char yield, near-zero volumetric shrinkage/expansion upon polymerization, low water absorption, excellent resistance to chemicals and UV light, and high T_g even with rather low cross-link density. In particular, the excellent dielectric property with the relatively low and stable dielectric constant (D_k) value under high frequency makes benzoxazines a good candidate for the next-generation microelectronic industry. Typically, the dielectric constant of PBZ is approximately 3.5 at 1 MHz, which is not satisfied with the requirement of microwave communication. Therefore, the design and preparation of novel low-dielectric benzoxazines and nanocomposites to meet the improved requirements is a stimulating fundamental field that can be a powerful complement to the phenolic resin. For PBZ used at high-frequency, high-speed communication, there are two main principles to reduce the permittivity of benzoxazine resin: one is to increase the free volume inside the molecule; the second is to reduce the polarity of the molecule. At present, the commonly used methods include: (i) introducing low polarity or large volume chemical structures such as aliphatic (ring) chain, fluorine atom, oxazole ring, and dicyclopentadiene through molecular design; (ii) introducing inorganic micro-nanoparticles, such as POSS, mesoporous silica, and GO, to form a porous structure in the system, and improving the dielectric properties of the whole benzoxazine composite system were using the low dielectric constant of air. (iii) Copolymerization with cyanate ester (CE), epoxy resin, polyurethane, and other materials. The polarity of the copolymer is reduced by introducing low polar groups or by consuming hydroxyl groups during the ring-opening polymerization of benzoxazine. In recent years, researchers have introduced low-polarity or large-volume chemical structures such as aliphatic (ring) chain, fluorine atom, oxazole ring, and dicyclopentadiene through molecular design to modify the PBZ.

6.2.4 Prospect

Since the beginning of industrialization in the mid-1950s, the manufacture and application of phenolic resin in CCL substrate have gone through a course of nearly 70 years. Its application has not decreased after 50 or 60 years; on the contrary, a new application market is constantly emerging. For example, the "color TV board" (TV motherboard) based on phenolic-paper-based CCL has become the main application market for this kind of CCL in the past years. In the past two or three years, flat panel TV design with PCB "broken into parts," overcome the "functional excess" of PCB, the pursuit of low-cost design idea prevailed, in the latest listing of flat panel TV PCB, there are multiple pieces of paper-based PCB.

With the development of high-frequency and high-speed communication, especially the improvement of the popularization rate of 5G applications, and the development of miniaturization and lightweight terminal electronics and appliances, CCL is developing in the direction of micro-holes, fine lines, high-density wiring, and high multilayer. Phenolic resin, as one of the main raw materials of CCL, is constantly facing a variety of stringent performance requirements. The intuitive performance is higher T_g and T_d, lower CTE, better high-temperature insulation properties, mechanical property retention, and modulus retention [18]. In addition to modification with other resins, the development of new structure of phenolic resin also plays an important role in improving the performance of high-frequency and high-speed CCLs, such as the linear and oxazine structures mentioned above and the introduction of fluorine, nitrogen, and other elements to modify the dielectric properties, so these aspects will be the focus of technological breakthroughs.

In addition, for PBZ, the introduction of functional groups with low polarity and large volume into PBZ can increase the free volume of polymer materials and reduce the density of dipole moments as well as the dielectric constant, but the reduction of polarity may also lead to a weak bond between the material and the copper foil in PCB. The modification of PBZ with other polymer materials is to reduce the polarity by using the low polarity of the modified material itself or by consuming the polar hydroxyl group produced in the curing process of benzoxazine. However, the compatibility between different polymers must be considered, as well as the adverse effects of other polar groups produced when the consumption of hydroxyl group occurs, such as the introduction of polyurethane, which can consume hydroxyl groups but also produce carbamate polar bonds, thus increasing

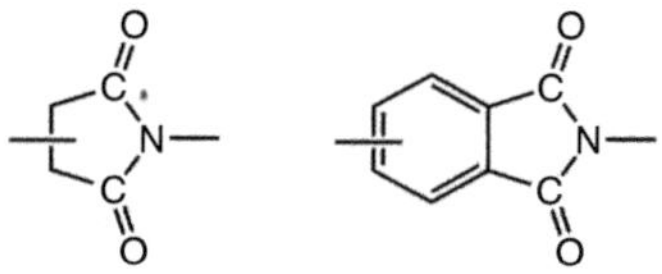

Figure 6.15 Cyclic imide structure and phthalimide structure.

the polarity of the material. The modification of benzoxazine by inorganic fillers mainly uses porous fillers to introduce pore structure into the polymer. Although the dielectric constant can be greatly reduced, it may also lead to defects in the material and reduce its reliability.

6.3 Polyimide Used for PCB

6.3.1 Introduction to Polyimide and its Performance Characteristics

Polyimides (PI) are one of the most important high-performance polymer materials developed in the 1950s to satisfy the demand for heat-resistant, lightweight, and high-strength materials for aerospace. PI is a generic term for a class of high-performance polymers with a characteristic structural unit of an imide ring (Figure 6.15) [19]. Among them, polyimide with a phthalimide group structure is very important in applications, and its structure is shown in Figure 6.15 [20]. The excellent heat resistance, insulation, and mechanical properties make polyimide very promising in many industrial fields. The diverse synthesis routes and processing methods make it a promising product for research and development.

6.3.2 Synthesis Method of PI

PI has a wide range of prospects for a wide range of applications because of the diversity of synthesis, and the choice of method according to the purpose of application. Because of this field variability is difficult to have other polymeric materials. Generally, the PI is prepared from dianhydride and diamine by polycondensation reaction, and the main polymerization processes are one-step synthesis, two-step synthesis, and three-step synthesis [21].

6.3.2.1 One-Step Method

"One-step method" refers to the preparation of high-molecular-weight polyimide solutions by adding diamine and dianhydride monomers together with catalyst in a polar solvent to a reaction vessel, where the polycondensation reaction occurs under high-temperature conditions. This method is a direct synthesis of PI from monomers, and dianhydride or diamine monomers with low activity are usually chosen for monomer selection. The synthesis process of this method is mild compared to thermal treatment conditions, and the key point is the selection of high-boiling point-solvents, such as phenol, m-cresol, p-chlorophenol, and other phenolic solvents, or N-methyl-pyrrolidone (NMP), and *N,N*-dimethylacetamide (DMAC). In the polymerization process, the cyclization reaction will produce a

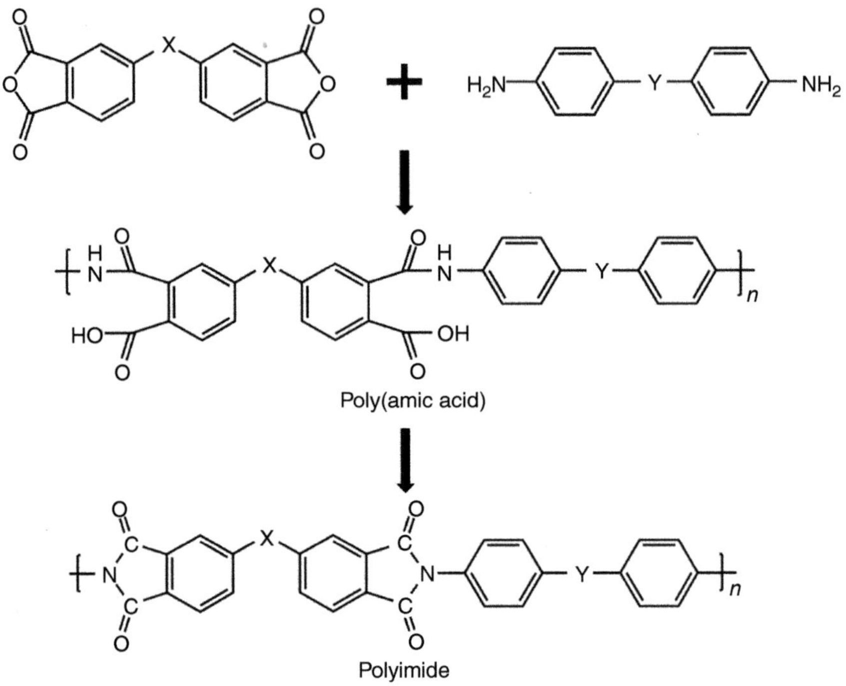

Figure 6.16 Two-step synthesis of PI.

small amount of water, and since water has a side effect on the reaction system, an azeotrope needs to be added to achieve the purpose of removing water from the reaction process [21, 22]. The one-step process is straightforward, but the polymerization temperature is high.

6.3.2.2 Two-Step Method

The "two-step" synthesis route is a classical and representative synthesis method that has been used since the 1960s (as shown in Figure 6.16). The first step is the reaction of diamine and dianhydride in a polar aprotic solvent to obtain polyamide acid (PAA) solution, and the second step is the dehydration and imidization of PAA to form polyimide. The imidization method is subdivided into thermal and chemical imidization. Thermal imidization refers to the thermal dehydration of the PAA solution under a set heating procedure (temperature to 300 °C or more), which has the advantage of simple operation and a high degree of imidization, and the disadvantage that the reaction requires a very high temperature and leaves defects in the material due to the removal of water molecules during the heating process. In contrast, chemical imidization can be obtained by adding a catalyst and dehydrating agent to the PAA solution and reacting at room temperature to obtain polyimide. The advantage is that the reaction can be carried out in solution, which can reduce the influence of water molecules on the material properties and also avoid the occurrence of brittle cracking of the material under thermal imidization.

The disadvantage is that the reaction time is longer and the degree of imidization is lower than that of thermal imidization [19, 21, 23].

Compared with the one-step method, the two-step method is relatively mild in terms of reaction conditions, and polyimide or its precursor PAA is usually prepared at low or room temperature. The reaction process is mainly influenced by the activity of the monomer, and usually, polyimide monomers with high reactivity are applied to prepare polyimide by this method. PAA has good solubility and fluidity and is easy to process, so the two-step method is the most mature polyimide synthesis method [24, 25].

6.3.2.3 Three-Step Method

The three-step method is a relatively new method of synthesis made from polyisocyanimides. Similarly, the first step of the three-step method is also prepared from diamine and dianhydride in a strong polar solvent, after which it is transformed into polyisocyanimide by dehydration cyclization under the action of dehydrating agents such as dicyclohexylcarbodiimide, and finally isomerized into polyisocyanimide under heat treatment catalyzed by acid or base, etc. The structure is stable, solubility is favorable, the glass transition temperature is low, and processability is improved [26]. The advantage of the three-step method in which polyisocimide appears is that the imidization process also does not give off water molecules and no defects such as bubbles appear in the product; therefore, it has received wide attention.

6.3.3 Classification of PI

There are many kinds of polyimide, mainly classified as: non-fusible and insoluble, fusible, and modified polyimide. At the same time, polyimide can be divided into two types: condensation type and polymerization type. The polycondensation type aromatic polyimide is made by the reaction of aromatic diamine and aromatic dianhydride, aromatic tetracarboxylic acid dialkylate, and the addition type polyimide is developed to overcome some disadvantages of the polycondensation type polyimide [27].

6.3.3.1 Non-Fusible and Insoluble PI

The main typical variety of non-fusible PI is poly(homotetramethylene tetramethylene ether imide), which is a condensation product of homotetramethylene dianhydride and 4,4-diaminodiphenyl ether. The synthesis reaction can be divided into two steps: polycondensation and cyclization (imidization). Homotetracarboxylic acid dianhydride and diaminodiphenyl ether are first condensed with wiener acid and then dehydrated and cyclized to polyimide (as shown in Figure 6.17). This type of PI is a thermosetting polymer, that is insoluble and fusible and can usually only be molded into films by the impregnation and salivation methods or molded into products by the molding method [20, 27]. Actually, the DMF solution of PAA produced by the first step of the condensation reaction is molded, and then the product is obtained by the cyclization reaction.

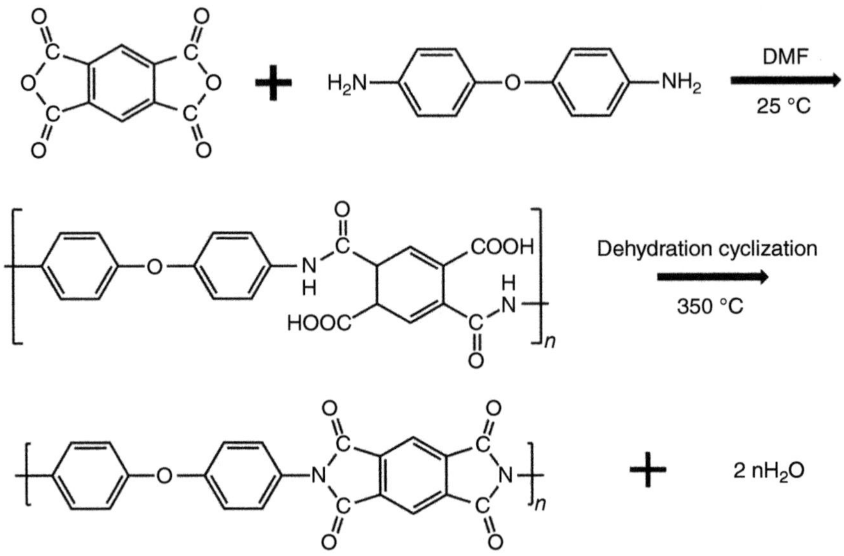

Figure 6.17 Synthesis flow chart of non-fusible PI.

Homogeneous PI has an outstanding temperature resistance. Its long-term use temperature in the air is 260 °C and has excellent radiation resistance and good mechanical properties. But there is a gap in sensitivity, it is not resistant to alkali and strong acids, and it is noncombustible. It can be made into film, reinforced plastic, foam, molded products, coatings, enameled wire, and also fine parts working under special conditions, such as high temperature resistance, high-vacuum self-lubricating bearings, electrical equipment parts used at high temperatures, radiation-resistant products, and parts in contact with liquid nitrogen [28, 29].

6.3.3.2 Fusible PI, Thermoplastic PI

The fusible PI can also be referred to as the thermoplastic PI. Polyetherimide (PEI) is the most representative fusible PI. PEI is a high-performance polymer developed by General Electric as a new species developed to improve the processing properties of non-fusible polyimide made from homophthalic dianhydride under the trade name ULTEM®. The main difference is that instead of homophthalic dianhydride, diphenyl ether tetraacid dianhydride is synthesized from bisphenol A dianhydride and diamine by melt polycondensation [12].

The melting point of PEI is about 380 °C, the glass transition temperature is 270–280 °C, the decomposition temperature is 570–590 °C, the low-temperature resistance reaches −193 °C, and the Young's modulus is about 3.3 GPa. PEI can be processed through molding, injection molding, extrusion, and other methods of molding processing. It can also be secondary processing, such as turning, milling, planning, and grinding. PEI has excellent heat resistance and mechanical properties. Meanwhile, chemically modified PEIs have excellent heat resistance, mechanical properties, chemical stability, and creep resistance, which can be used as wear-resistant materials, dielectric materials, and a substitute for metal materials in some aspects of aerospace engineering [24].

6.3.4 Performance Characteristics of PI

PI is a special engineering plastic with excellent properties, such as stable and excellent physical, chemical, electrical, and mechanical properties, as follows [19, 20, 27, 30]:

(1) Thermal stability: PI has excellent thermal stability; the starting decomposition temperature of aromatic PI is generally around 500 °C. PI synthesized by homophthalic dianhydride and p-phenylenediamine has a thermal decomposition temperature of 600 °C, which is by far one of the highest thermal stabilities of the polymer species. Most thermosetting PI can be used above 300 °C for a long time. At the same time, PI's low-temperature resistance is also very good; in −269 °C liquid helium will not be brittle.

(2) Mechanical properties: PI has excellent mechanical properties; the tensile strength of pure PI is more than 100 MPa, the elastic modulus is generally between 3–4 GPa, and the fiber can reach 200 GPa. According to theoretical calculations, the fiber synthesized by homophthalic dianhydride and p-phenylenediamine can achieve a module of 500 GPa, second only to carbon fiber.

(3) Dielectric properties: PI has good dielectric properties, with a dielectric constant of about 3.4. With the introduction of fluorine or air nano-size dispersion in PI, the dielectric constant can be reduced to about 2.5. The dielectric loss is less than 0.001, and the dielectric strength is 100–300 $KV\,mm^{-1}$. Moreover, these properties can still be maintained at a high level over a wide range of temperature and frequency.

(4) Chemical stability: PI has excellent resistance to water, oil, acid, and organic solvents. Since the five-membered imide ring structure is easily hydrolyzed under strong alkali conditions, PI is not resistant to alkali. This "disadvantage" of polyimide can be exploited to recover raw materials dianhydride and diamine by alkaline hydrolysis.

In addition, PI has a high resistance to irradiation, and its film maintains 90% strength after irradiation at a dose of 5×10^9 rad. PI is a self-extinguishing polymer with a low smoke generation rate and little outgassing in high vacuum.

6.3.5 Application of PI in CCLs

Based on the above advantages, PIs are widely used in CCLs. According to the mechanical rigidity of the laminate, PI-based CCLs can be divided into two main categories: one is a rigid laminate, and the other is a flexible laminate [31].

6.3.5.1 Application of PI in Rigid CCLs

In order to meet the demand for high frequency, high heat resistance, and high reliability printed circuits, copper-clad polyimide glass fabric laminates (hereinafter referred to as PI laminates) have been developed. Bismaleimide has both the heat

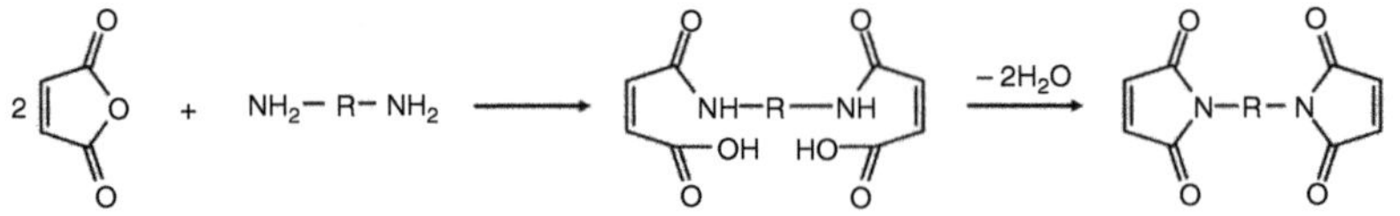

Figure 6.18 Synthesis of Bismaleimide.

resistance of polyimide and the molding process of epoxy resin and is widely used to make PI laminate resin [32, 33]. The specific synthesis route is as follows: (Figure 6.18).

Since the molecular chain of bismaleimide resin has a high rigidity and glass transition temperature, its dipole loss is small in a wide range, and the electrical properties are very good, with the advantages of low dielectric constant, low dielectric loss, high volume resistivity, etc. So, it is widely used in PCBs, and bismaleimide resin can meet the requirements of manufacturing high-speed digital and high-frequency PCB substrates, commonly used to make high-performance multilayer boards [34]. In the early 1980s, Mitsubishi Gas Chemical Company of Japan started to research bismaleimide resins and successfully used them in the manufacture of copper-clad boards. Wang et al. [35] reported a copolymerization system using an epoxy resin modified with bismaleimide and cyanate, and the results showed that the glass transition temperature of CCLs after the post-treatment process could reach more than 190 °C (DSC method).

6.3.5.2 Application of PI in Flexible Copper-Clad Laminate

With the miniaturization, high integration, and multifunctionality of modern electronic products, as well as the acceleration of signal transmission and processing speed, the application of ultrathin PCBs is gradually increasing, and the line width and line spacing of circuit boards are gradually decreasing, and thus the market demand for high-performance flexible printed circuits (FPCs) has increased significantly [36]. FPCs consist of one or more etched conductor layers located on one or both sides of a dielectric film and are widely used in a variety of applications such as smartphones, laptops, wearable devices, and automotive electronics. Flexible copper-clad laminate (FCCL) is a bendable sheet-like composite material made of copper foil laminated on an insulating base film. Its characteristics are mainly manifested in three points: light, thin, and flexible. Flexible copper laminate can be divided into three-layer flexible copper laminate and two-layer flexible copper laminate. As the basic raw material of FPC, the rapid development of FCCL and related products has put forward higher requirements on the heat resistance, stability, and reliability of the insulation substrate [33].

FPC based on PI has a thin thickness, is lightweight, has high dimensional stability, excellent heat resistance, can withstand millions of dynamic bending, and has a high wiring density. It can be arbitrarily moved and scaled according to the spatial layout to achieve three-dimensional assembly, showing the advantages of other types of circuit boards cannot be compared [37]. Polyimide in the flexible copper-clad board is mainly used as an insulating base film. DuPont, in 1962, developed homogeneous polyimide, which began production and commercialization in 1965 and had been dominating the flexible copper-clad board with polyimide film market until 1982. In addition, Kaneka and Ube have developed corresponding products [31].

The bonding mechanism between PI films and copper foils is complex [38]. The bond strength is proposed to depend on the chemical bonding that originates from the interaction between the copper oxide (Cu_2O) and the carbonyl group on the PI imide ring, while the physical interaction between the two surfaces and the possible coordination at the PI/Cu interface are also considered to have an impact on the bond strength. Based on this, Diao [39] et al. improved the bonding performance of FCCL by improving the interfacial wettability, increasing the interfacial physical roughness, enhancing the interfacial reactivity, and improving the bonding ability of the chemical bond between PI film and copper foil.

6.3.5.3 Application of Thermoplastic Polyimide in Double-Sided Copper Laminates

Compared to thermosetting polyimide, thermoplastic polyimide (TPI) is easier to process. In addition, TPI is highly productive and involves essentially no solvent volatilization during its processing, which facilitates the preparation of products with better toughness. At the same time, the advantages in terms of environmental protection and economic benefits have made the preparation of TPI for FCCL applications a new favorite in the research community [28, 37].

The structure of conventional double-sided FCCL consists of three parts: copper foil, PI base film, and adhesive, where the PI base film is bonded to the copper foil by the adhesive. The adhesive is mainly composed of acrylic resin or epoxy resin, which has limited high temperature resistance. When the double-sided FCCL is used in multilayer or dense circuits, it is often accompanied by a large amount of heat generation, which causes the temperature of the laminate to rise sharply. Therefore, the presence of adhesives greatly limits the application of double-sided FCCLs in high-temperature applications [40]. To solve this problem, most use TPI with excellent heat resistance to replace the adhesive. However, TPI does not exist independently but forms a TPI/PI/TPI triple-layer composite film with a PI base film through a double-sided coating process. This composite film consists of three layers: the PI base film in the middle and TPI on the top and bottom, and it can be used to produce double-sided FCCLs for multilayer circuits or dense circuits with high heat generation [41]. The FCCL example and specific structure is shown in the following Figures 6.19 and 6.20).

The application of TPI in the field of FCCLs mainly involves two aspects.

➢ TPI adhesive layer material.

General multilayer flexible substrate material to include the following three steps: the first step is the acrylic or epoxy adhesive on the thermosetting PI; the second step is copper foil bonded to it; the third step is curing of the three molding [42]. However, with the promotion of multilayer circuits and the application of dense circuits, flexible printed circuit boards (FPCB) must meet the requirements of two aspects. On the one hand, in order to reduce waste, FPCB can be used for a long time, so it must have a long service heat resistance; on the other hand, the

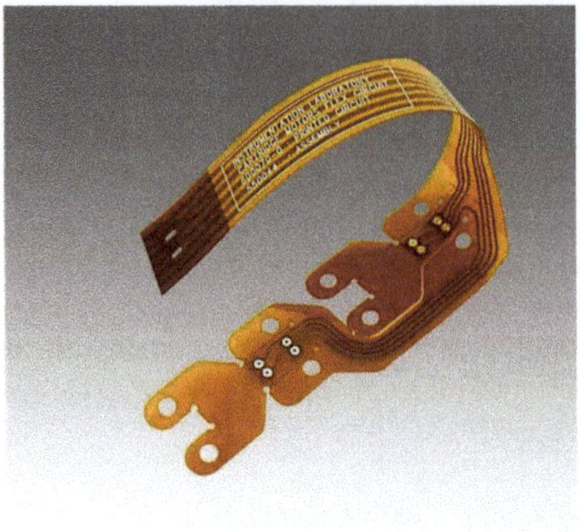

Figure 6.19 Picture of FCCL.

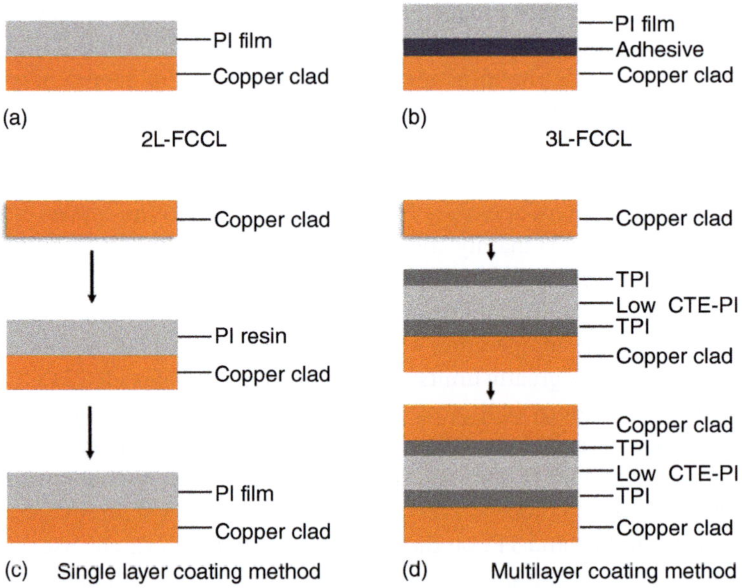

Figure 6.20 Structures and molding techniques of FCCL.

preparation process of FPCB, includes a welding process, and the temperature of welding is very high, so it must be able to withstand the preparation of process of welding heat resistance [43]. However, adhesives such as acrylics or epoxies are not heat-resistant and do not meet the requirements of these two aspects. In order to solve this problem, high-heat-resistant TPI can be used as the adhesive layer material instead of heat-hardening binders such as acrylic or epoxy, which are not heat-resistant, to produce a high-heat-resistant FPCB.

TPI Direct Lamination Base Film In the manufacturing process of FPCB, TPI directly covered with copper foil can also be eliminated by the adhesive heat resistance caused by the adverse effects of poor dimensional stability, known as the glue-free laminate process. FPCBs prepared by the adhesive-free laminate process not only have better heat resistance but also have the advantages of better mechanical properties and dimensional stability [44]. From a technical point of view, the most important technical indicator for PI to be used in FPCs is its coefficient of thermal expansion α. To make dense circuit boards, the coefficient of thermal expansion of the film must be close to the coefficient of thermal expansion α of the copper foil. At present, the α of PI film is generally between $3.0–6.0 \times 10^{-5}$ °C^{-1}, while the α of common copper metal is $1.6 \times 10^{-5}–1.9 \times 10^{-5}$ °C^{-1}. Therefore, the development of PI films with α below 1.6×10^{-5} °C^{-1} is not only a prerequisite for direct lamination but also a necessary way to expand the application field of products [45].

Recently, with the rapid development of smart electronic products such as cell phones and tablet PCs, two-layer adhesive-free copper-clad laminates (2L-FCCL) with PI as the base film have been developed rapidly due to the advantages of high heat resistance and high dimensional stability [46]. The base material of this FCCL is a three-layer laminate film of polyimide with high strength, high modulus, excellent heat resistance, and electrical properties. The PI layer provides the heat resistance, dimensional stability, and mechanical properties required for FCCL, and the TPI layer, which is only 2–4 μm thick, enables efficient bonding of the substrate to the copper foil. At present, there are two main methods for producing FCCL based on TPI/PI/TPI three-layer laminates: one is to achieve simultaneous TPI/PI/TPI three-layer laminates using a three-layer co-flow-delay (extrusion) method, which uses a chemical imidization process, followed by high-temperature lamination with a double-sided laminated copper foil to make FCCL; the other method is to sequentially coat the copper foil with TPI, PI, and TPI precursors, which were subjected to high-temperature imidization and then high-temperature rolled with the other side of the copper foil to obtain adhesive-free double-sided FCCL. Several new structures inspired by this have been designed and proven superior for their applications. Wang et al. [32] prepared a new FPCB (G-FPC) containing a layer of (PI/M-GF/PI) graphene film with a sandwich structure and a maximum thermal conductivity of 739.56 Wm^{-1} K^{-2}. In addition, G-FPC has a good heat dissipation capability and can significantly reduce chip temperature, which has potential applications in future flexible and wearable electronics. Hasegawa and his co-workers at Toho University in Japan have carried out systematic research in the field of basic and applied research on non-fluorinated PI low dielectric films in recent years [5]. They introduced ester bonds into the molecular structure of PI films via dianhydride or diamine monomers and used the low water absorption, high rigidity, high heat resistance, and high planar orientation coefficient (POC) properties of ester bonds to impart excellent dielectric properties at high frequencies to PI films. For example, they have used hydroquinone-bis(trimellitic acid) dianhydride (TAHQ) and bis(4-aminophenyl) terephthalate (BPTP) to prepare PI films. The results of the dielectric performance tests showed that the Dk and Df values of the films were 2.81 and 0.0019, respectively, at 18.3 GHz, while the Dk and Df values of the LCP films were 2.86 and 0.0025, respectively, under the same conditions.

6.3.6 Prospect

With the rapid development of electronics, information industry, and new energy technology, taking into account the complexity of the future application environment of electronic products, the development of PI products with high thermal conductivity, low dielectric, low coefficient of expansion, flame retardancy, and electromagnetic shielding has become an important direction for future research in the field of electronics/microelectronics under low-voltage environment, high-voltage, and ultrahigh-voltage environment. Of course, there are still many issues that require continuous in-depth research, including (i) the preparation of PI materials with a higher grade of temperature resistance, including short-time high temperature resistance and long-term high temperature resistance of two; (ii) improving the molding process performance of high-temperature resin-based composites to reduce costs and broaden the depth and breadth of applications; (iii) accumulating data on the performance of materials in high temperatures and other harsh environments; and (iv) accumulating data on the performance of materials in harsh environments, such as high temperatures, stabilize their performance, and improve safety and reliability.

Simultaneously, the variety of electronic products is becoming more and more diverse and is developing at a high speed toward high density and miniaturization, which require more requirements for the use of FPCB. Thus, the future of the FCCL market will become much brighter. Therefore, the study of TPI in the field of double-sided FCCL applications has also become particularly important and urgent.

6.4 Polymer Materials Used for PCB at High Frequency

The fourth-generation mobile communication technology (4G) uses a frequency range of 700 MHz–3.5 GHz, and the theoretical transmission rate is 150 Mb/s. The most effective way to improve the transmission rate is to apply high-frequency transmission, so the fifth-generation mobile communication technology (5G) came into being. According to the agreement of the 3rd Generation Partnership Program (3GPP) of the International Organization for Standardization, the 5G network mainly uses frequency band 1 (FR1) and frequency band 2 (FR2), in which the range of FR1 frequency band is 410 MHz–7.125 GHz and the FR2 frequency band ranges from 24.25 to 52.6 GHz, which is called the millimeter wave frequency band. Compared with 4G technology, the 5G signal transmission band has high frequency, short wavelength, poor diffraction ability, and fast signal transmission attenuation.

In fact, the transmission rate and transmission stability of electrical signals in PCBs are closely related to the dielectric properties of the latter, where the relationship between signal transmission rate (V_p) and relative dielectric constant (D_k) is shown as follows:

$$V_P = \frac{C}{\sqrt{D_K}}$$

The relationship between signal loss (L), dielectric properties, and frequency of the PCB can be expressed as follows:

$$L = K \times \frac{f}{C} \times D_{\mathrm{f}} \times \sqrt{D_{\mathrm{F}}}$$

where K is a constant, f is the frequency, C is the speed of light, and D_{f} is the dielectric loss, namely the tangent of dielectric loss angle. From the above two relations, it can be seen that the higher the dielectric constant and dielectric loss, the slower the signal transmission rate and the larger the loss. The higher the working frequency, the more obvious the influence of dielectric properties on the transmission loss. Therefore, the higher the communication frequency, the lower the dielectric constant and dielectric loss of the PCB. So as to ensure the rapid transmission of high-frequency signals and reduce the loss in the transmission process, polymer materials used in high-frequency communication usually require a dielectric constant of less than 3 and a dielectric loss of less than 0.005. However, the polymer materials used in traditional PCB are mainly epoxy resin and phenolic resin, whose dielectric constant exceeds 3.5, which makes it difficult to meet the demand for high-frequency transmission. In addition, the heat resistance, moisture absorption, expansion coefficient, and metal/semiconductor adhesion properties of PCBs are also closely related to the polymer substrate. Therefore, the development of high-performance polymer materials with low dielectric constant and low dielectric loss, as well as high heat resistance, low moisture absorption, and dimensional stability, and their application in high-frequency PCBs are of great significance to the development of high-frequency communication technology, which has attracted the attention of many researchers in recent years.

With the development of communication technology toward high-frequency, PCBs based on phenolic resin and epoxy resin are facing obsolescence because their dielectric properties cannot meet the requirements of high-speed and low-loss signal transmission. Therefore, it is crucial to develop polymer materials with low permittivity and low dielectric loss at high frequency for high-frequency communication, as well as good heat resistance, moisture resistance, and dimensional stability. Among them, polytetrafluoroethylene (PTFE), liquid crystal polymer (LCP), CE, and polyphenylene oxide (PPO) have received wide attention due to their low intrinsic dielectric constant and dielectric loss, low moisture absorption rate, and good heat resistance. The application scenarios of some of the materials at HF are listed in the Table 6.2.

6.4.1 Polytetrafluoroethylene (PTFE) Used for PCB

6.4.1.1 Structure and Properties of PTFE

The PTFE resin was first developed by the American chemist Dr. Plunket in 1938 and industrialized in the 1950s. PTFE is a polymer compound made of tetrafluoroethylene by suspension or dispersion polymerization. The CF_2 units in the main chain are arranged in a zigzag shape, since the radius of fluorine atoms is slightly larger than that of carbon atoms, the adjacent CF_2 units cannot follow the trans-cross orientation completely but form a helical twisted chain, and

Table 6.2 Main application areas of high-frequency CCL.

Application	Working frequency (GHz)
Individual receiving base station or satellite launch	13–24
Wireless communication	1–4.8
Automobile, personal receiving satellite	2.4
Satellite frequency demultiplier	2–3
Home receiving satellite	12–14
Global positioning system −40–85 °C	1.57/1.22
Direct broadcast satellite	13
Antenna communication	14
Very small aperture terminal	12–145
Digital microwave system	10–38
Vehicle anti-collision system	75

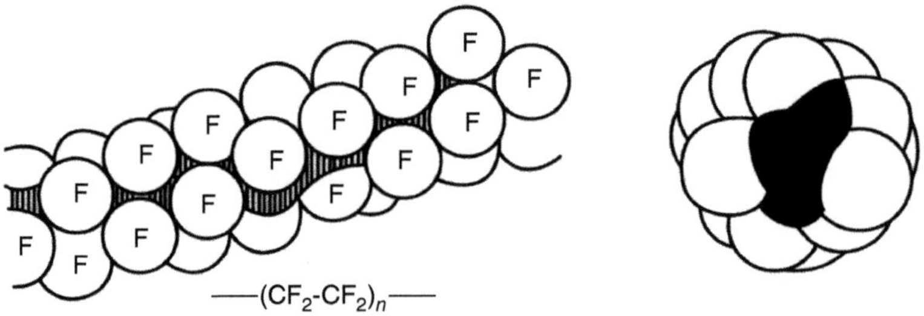

Figure 6.21 Molecular structure of PTFE.

fluorine atoms almost cover the entire surface of the polymer chain. The PTFE topology is shown in Figure 6.21. At the same time, the C—F bond is extremely strong, with a bond energy of up to 460.2 KJ mol^{-1}, much higher than the C—H bond (410 KJ mol^{-1}) and the C—C bond (372 KJ mol^{-1}), which makes PTFE superbly thermally stabile and chemically inert. PTFE has no true glass transition temperature and is resistant to high and low temperatures and corrosion. The dielectric constant of PTFE is 2.1 at a frequency of 10 GHz, and the dielectric constant does not change with frequency and temperature within 10 GHz. At this level, the dielectric loss is 10^{-4} for PTFE. At present, PTFE has the lowest dielectric constant and dielectric loss among all the matrix resins used in high-performance CCL. Because fluorine atoms cover almost the entire surface of the polymer chain, PTFE has strong solubility resistance; even aqua regia cannot dissolve it at room temperature. In addition, PTFE itself is a nonpolar molecule with hydrophobicity; its water absorption is very small. Because of these excellent comprehensive properties, PTFE resin has shown a strong advantage in the field of high-frequency PCB. The special structure of PTFE in its helical conformation gives it many good properties but also the following drawbacks: (i) The mechanical properties are poor. The attraction between PTFE molecules is very small, so its tensile strength and hardness are very low and has no resilience. (ii) The linear expansion coefficient is large. The linear expansion coefficient of PTFE is 1.13×10^{-4}–2.16×10^{-5} at −50–250 °C, this will lead to the cracking of the PTFE composite due to the different size changes in unit temperature of the material after heating. (iii) Poor thermal conductivity. Its thermal conductivity is very low, only 0.20 $W(m \cdot K)^{-1}$, which will suffer thermal expansion, thermal deformation, and thermal fatigue. (iv) Poor radiation and creep resistance. (v) Because of the extremely low surface energy of PTFE and the high symmetry of the molecular structure, the interface bonding of PTFE and other materials is often very poor when they are composited. PTFE also has high melting temperature and crystallinity; in molten state, the viscosity is large and the fluidity is poor, which makes it difficult to process. Besides, PTFE still has some problems such as high prices and insufficient mechanical properties. These shortcomings of PTFE limit its application in the field of high-frequency PCBs to a large extent, so it is necessary to modify PTFE. At present, surface modification and compound modification are mainly used to optimize the properties of PTFE to achieve complementary advantages.

6.4.1.2 PTFE Used for PCB

Because the surface energy of PTFE is very low and the surface wettability is poor, the chemical inertia and the existence of a weak boundary layer seriously hinder the application of the PTFE in the field of CCL. Therefore, the following methods were adopted to modify the surface of PTFE material: (i) Introduce polar groups into the molecular chains on the surface of PTFE. (ii) Improve the surface energy of PTFE. (iii) improve the surface roughness of PTFE. (iv) Eliminate the weak interface layer on the surface of PTFE products. When PTFE is treated with sodium–naphthalene complex, sodium can break the C—F bond and strip some of the fluorine atoms from the surface of PTFE, leaving a carbonized layer and some polar groups on the surface. According to the study of Wake et al. [47] C—H, C=O, C=C, COOH, and other polar groups were introduced into the modified PTFE surface, which raised the surface energy of PTFE and improved the contact angle as well as wetting property, thus enhancing the interlaminar adhesive ability of PFTE when served as CCL. This method is one of the best and most classic methods in PTFE modification. The PTFE CLL structure is shown as follows: (Figure 6.22).

Zhou et al. proposed an O_2/CF_4 plasma treatment to clean the after-drilling PTFE hole, achieving a highly activated surface for following copper plating [48]. The plating results, in contrast with control experiments that were performed without plasma treatment, demonstrate multiple advantages of O_2/CF_4 plasma treatment, including qualified copper plating, excellent thermal shock reliability, and improved wettability. X-ray photoelectron spectroscopy analysis for plasma-treated samples confirms the emergence of hydrophilic groups such as C—O and COO/COF on the PTFE hole wall. These results fully show that plasma treatment of PTFE can change the surface inertness, increase the surface energy, and improve the bonding strength between PTFE and copper foil. In addition, the surface modification methods of

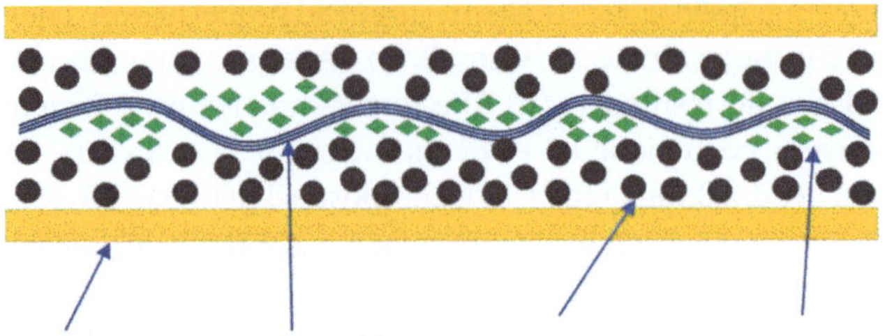

Figure 6.22 Common structure diagram of PTFE CCL.

PTFE also include high-energy radiation, laser radiation, and ion implantation. Filling modification is a simple and effective method that can not only maintain its own advantages but also use the compound effect to overcome the defects of pure PTFE and improve its comprehensive performance. Adding different fillers to PTFE can significantly improve its mechanical properties, hardness, and wear resistance. Compared with pure PTFE, the creep resistance of the product is increased by 1.5–4.5 times at room temperature and 1.5 times at high temperature; the elastic modulus of bending increases by 2–3 times; the hardness increases by 10% to 30%; the wear resistance is improved by 2–3 orders of magnitude; the linear expansion coefficient is reduced by 80%; and the volume resistance and surface resistance are reduced, which can further meet the application in high-frequency and high-speed communication.

Hexagonal boron nitride (h-BN) was incorporated into a PTFE matrix by Johanna [49]. The filler loading in the composite was varied up to a filler volume content of 50%. The incorporation of h-BN improved heat transport significantly, while the loss factor decreased. Fillers with a higher aspect ratio have a greater effect on increasing thermal conductivity. For a filler volume loading of 30%, the in-plane thermal conductivity was up to 14 times higher than the thermal conductivity of the matrix, while the loss factor decreased slightly. The permittivity increased with the increasing filler volume content, while the CTE also decreased. In order to reduce the negative effect of BN on the dielectric constant of PTFE, core–shell structured h-BN-coated hollow glass microsphere (HGM@hBN) particles were prepared using an electrostatic assembly process and incorporated into PTFE [50]. The composites exhibited decreased dielectric constants due to the implantation of air, which is mainly encapsulated in HGM. The hBN layer on the surface of HGM effectively enhanced composite thermal conductivity due to its special core-shell form. The 30 vol% HGM@hBN/PTFE composites had a low dielectric constant of 1.68, low moisture absorption of 0.11%, and thermal conductivity (0.276 W $(mK)^{-1}$ that was improved by 43% compared with that of the 30 vol% HGM/PTFE composites.

According to Wang's work, PTFE/SiO_2 composite substrates were prepared using PTFE resin as matrix, fused silicon dioxide (SiO_2) as filler, and doped with different amounts of GF through surface modification, mechanical mixing, calendering, and hot press sintering [51]. The result showed that composite density decreased with increasing GF content. But water absorption and dielectric loss were just the opposite. As the doping level increased, the tensile and compression modules and dielectric constant first increased and then decreased. When the amount of GF doping was 2%, the PTFE/SiO_2 composites exhibited good density and microwave dielectric properties; the dielectric constant was 2.93 and the dielectric loss was 1.04×10^{-3}.

It is worth noting that due to the lowest dielectric constant of PTFE, polar functional groups generated on its surface when PTFE is modified (such as by chemical impregnation, plasma treatment, and other methods) will often lead to an increase in dielectric constant. At the same time, adding filler with a hollow structure or introducing air is very likely to introduce defects in PTFE and reduce its mechanical properties. Therefore, it is of great significance to introduce new technologies to improve the adhesion property of PTFE, especially for its application in the field of high-frequency and high-speed communication.

6.4.2 Liquid Crystal Polymer (LCP) Used for PCB

6.4.2.1 Structure and Properties of LCP

LCP is an intermediate organic substance whose state is between solid and liquid. Its molecular arrangement has the properties of both solid crystal and liquid, but it is not completely ordered or disordered and will present a certain order on the one-dimensional or two-dimensional scale. It is a new type of polymer material that generally presents liquid crystal properties in the molten state.

LCP has attracted extensive attention as a potential high-performance microwave substrate material for high-frequency PCBs. Applications include flexible microwave circuits for high-end communications, such as highly portable miniature sensing devices, robotics, and wearable systems (Table 6.3).

At present, LCP is mainly synthesized by melting condensation, and the synthesized LCP mainly includes main chain types of polyamides, polyesters, polyethers,

Table 6.3 Basic properties of LCP materials.

Property	Unit	Typical value
dielectric constant (10 GHz)		2.9
dielectric loss (10 GHz)		0.0025
surface resistance	MΩ	1×10^{10}
volume resistivity	MΩ·cm	1×10^{12}
dielectric breakdown strength	KV/cm	1378
water absorption (25 °C, 24 h)	10^{-6}/°C	0.04%
thermal expansivity	X	17
	Y	17
	Z	150
melting temperature	°C	330
fire resistance		V0

Table 6.4 Liquid crystal units commonly used in liquid crystal polymers.

Monomer type	Monomer structure
HO-R-COOH	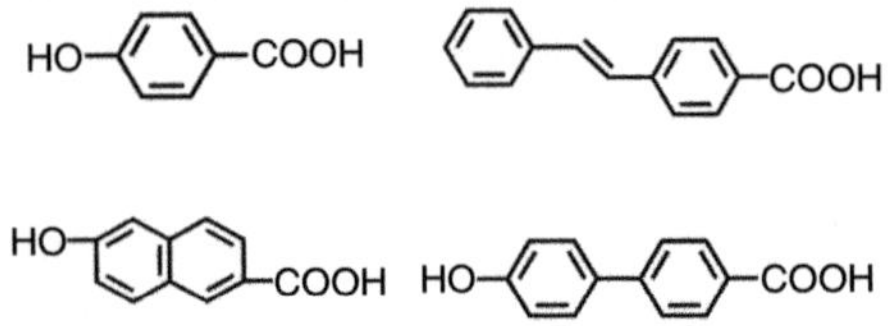
HO-R-OH	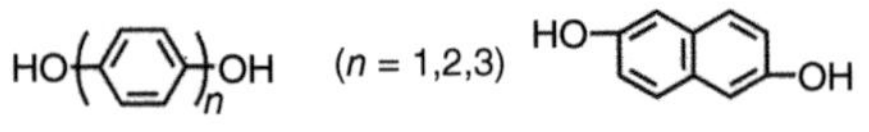
HOOC-R-COOH	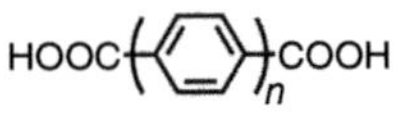 (n = 1,2,3)
H_2N-R-NH_2	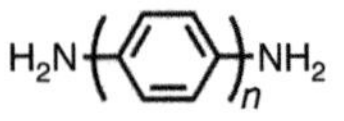 (n = 1,2,3) 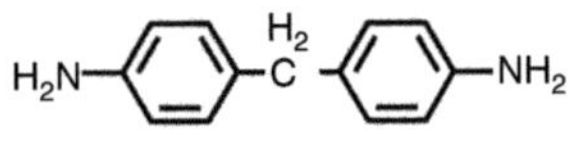
H_2N-R-COOH	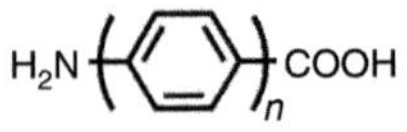 (n = 1,2,3)
HO-R-NH_2	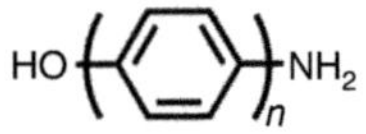 (n = 1,2,3)
Acrylics	$H_2C{=}C(R)$–COOAB
Styrene type	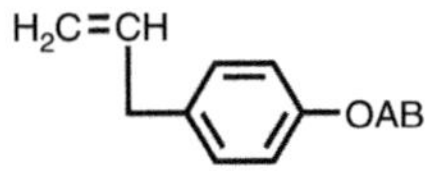
Acrylamide type	$H_2C{=}CH$–CONHAB
Vinyl ether type	$H_2C{=}CH$–COAB

Annotation: A stands for spacer group, and B stands for mesogenic unit.

polythiazoles, and polyimidazoles. Side chain types of polyisocyanate, polyazo, polydimethylsiloxane, polyacrylic acid, and so on (mesogenic units are summarized in Table 6.4); also, there are some other special structure LCP [52]. The first generation of LCPs was polymerized in the 1960s using p-phenylenediamine and terephthalic acid as raw materials; the second generation of thermotropic LCP polymers was aromatic polyester with high strength in the 1980s; and the third generation of thermotropic LCP polymers is currently being developed and applied.

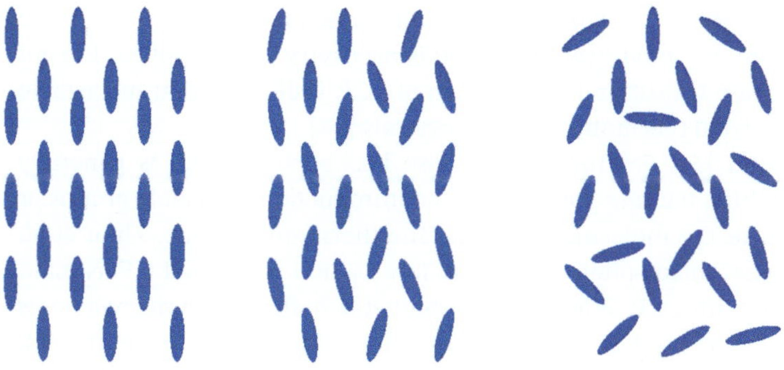

Figure 6.23 Different states of matter.

The synthesized polyamide has an obvious liquid crystal structure in a melting state and belongs to a typical main-chain thermotropic liquid crystal [53]. Compared with other organic polymer materials, LCP has a relatively independent molecular structure and thermal behavior. Its molecules are composed of rigid rod-like macromolecular chains, and after melting or dissolving by solvent, it no longer has most of the properties of solid matter but forms a transitional intermediate phase-liquid crystal state with some properties between solid and liquid. The arrangement of its molecule is between ideal liquid and crystal, showing one-dimensional or two-dimensional long-range order, but still having a certain degree of order in molecular orientation with good anisotropy. This anisotropy of LCP makes it have high strength, high modulus, and self-strengthening performance, outstanding heat resistance, excellent resistance to cold and heat alternating performance, excellent corrosion resistance, flame retardant performance, and process properties. The structures of the liquid, liquid crystal, and solid states of matter are illustrated in Figure 6.23.

6.4.2.2 LCP Used for PCB

LCP has excellent electrical performance and can keep the permittivity close to constant even when used at extremely high frequencies. It can take into account the characteristics of low loss of substrate medium and transmission line so that the products made of LCP are suitable for the substrate manufacturing of millimeter wave equipment. LCP films are highly thermoplastic and have strong morphological plasticity above the glass transition temperature, also, LCP antennas can save space and replace RF coaxial connectors. However, due to the complexity of its manufacturing process, the current yield is not high, and there are few suppliers capable of providing LCP substrate at present. The cost of LCP RF antenna of the same size is about 10–20 times that of a PI antenna, and the antenna with an LCP substrate was first used in Apple's smartphone, iPhone X.

Although LCP has significant advantages in high-frequency communication, there are also some defects, such as not easy to control the fabrication process and the mechanical properties of LCP being poor in the direction at right angles to the shear flow. Therefore, the modification of LCP is necessary when applied. Frosini et al. studied the phase transition phenomenon of LCP, indicating that metastable mesophases exist in solid thermotropic liquid crystals [54].

Studies have found [55, 56] that LCP, whose backbone is rigid, is generally lyotropic and has a high melting point, while the introduction of flexible groups or side chains will make the molecule change to the thermotropic type. Zhou et al. [57] used hollow glass microspheres with very low D_k and reinforced GF as main fillers to compound with LCP. When 10% GF and 20% hollow glass microspheres were added, not only the mechanical properties of the material could be improved, but also the D_k of LCP could be reduced to about 2.7 (1–18) GHz, which can meet the application of high-frequency electronic devices. In order to improve the mechanical properties of LCP materials, Mai et al. [58] prepared polyphenylene sulfide (PPS)/LCP composites with different LCP contents by melting processing. DSC results show that when the mass fraction of LCP in the composite is less than 30%, the heterogeneous nucleation of LCP can increase the crystallization temperature of PPS. With the further increase in LCP content, the crystallization of PPS is inhibited, and the crystallization temperature of composites decreases gradually. For GF-reinforced PPS/LCP composites, with the increase in LCP content, the tensile strength and bending strength of the composites gradually decreased, and the bending elastic modulus gradually increased. The microstructure observation showed that the properties of GF-enhanced PPS/LCP composites were related to the poor interface bonding between PPS/LCP and resin/GF. Further, the ethylene-methyl acrylate-glycidyl methacrylate random terpolymer and epoxy resin were used to improve the interfacial interaction of the composite. The results show that the epoxy resin can significantly improve the mechanical properties of the composite, and the welding-line tensile strength is increased from 31 to 70 MPa.

The existing binding technologies, such as adhesive, lamination, hot pressing, and metallization, are complicated or limited in application scenarios for the bonding properties of LCP materials to copper foils. Since the molecular structure of LCP is generally linear and the polymer chains are not intertwined with each other, the mechanical strength perpendicular to the main chain direction is weak. Therefore, the main way to obtain high mechanical strength LCP film is by hot pressing after adding multiple composites. In this way, the molecular chains need to be oriented in a grid shape. However, the binding strength of LCP and Cu cannot be improved by hot pressing alone. Zou et al. [59] adopted the chemical etching method and used potassium permanganate as the etching agent to produce a rough surface on the surface of the copper layer and improve the binding force between copper and LCP. Redhwan et al. [60] improved the binding force of copper and LCP by treating LCP with high-power O_2 plasma and then treating both LCP and Cu with low-power plasma. Feng et al. [61] found in their study that different power conditions of oxygen plasma treatment had different effects on the peeling strength of the prepared LCP multilayer plates. It is found that when the plasma treatment power is 6 kW, the best peeling strength of the multilayer plate is 8.92 N cm^{-1}. Surface morphology analysis shows that plasma treatment can produce cavities on the surface to increase the area of mechanical bonding. The analysis of the water contact angle showed that the water contact angle decreased and the hydrophilicity improved after plasma treatment. FT-IR analysis showed that plasma treatment had no significant effect on the types of surface groups of LCP, and its effect mainly focused on changing the number of polar groups on the surface, thus enhancing the peeling strength of multilayer plates.

Compared with PI films, LCP films show lower water absorption and CTE and more stable dielectric properties due to their unique molecular structure and aggregation structure but have disadvantages such as poor adhesion to copper foil and poor flexibility, which need to be modified to expand their applications in 5G communications. In addition, there are disadvantages such as the molding process is not easy to control, the physical properties of the product are anisotropic, the mechanical properties of the direction at right angles to the shear flow during molding are poor, and it is easy to fiber, so it needs to be modified to improve the mechanical properties of LCP. On the other hand, LCP raw material, are expensive, which leads to the high price of LCP, which is the most important problem affecting the development of LCP. Of course, with the common production of LCP raw materials, the application of LCP in the field of high-frequency communication will only be further expanded.

6.4.3 Other Resins with Potential in the Field of High-Frequency Communications

6.4.3.1 Cyanate Ester (CE) Used for PCB

CE is usually defined as the binary phenol derivative containing two cyanate (–OCN) functional groups; the general formula is: NCO–Ar–OCN. Because of the existence of –OCN functional group in cyanate resin, under the action of heat and catalyst, the activity of C≡N bond in the molecule is enhanced, and then the C≡N bond is broken and recombined, forming a stable structure of triazine ring between molecules. The reaction route for the preparation of triazine rings and cyanate esters is as follows (Figure 6.24).

The oxygen and nitrogen atoms of the triazine ring in CE are arranged symmetrically and the dipole moment is small; therefore, in a wide range of temperature and frequency, the dielectric constant and dielectric loss of CE are low, and because CE contains a large number of benzene and triazine rings, the cross-linking density is high, the glass transition temperature (T_g) of CE is up to 250 °C, and the performance of heat resistance, moisture resistance, and corrosion resistance is excellent. The cured cyanate resin has the characteristics of low dielectric constant (2.8–3.2), low dielectric loss (0.002–0.008), excellent mechanical properties, low shrinkage, and strong cohesiveness. In general, CE resins have better dielectric properties than PI, a similar processability to epoxy resins, comparable combustion resistance to phenolic resins, and comparable high temperature resistance to bismaleimide resins. Because of these outstanding properties, CE shows a very strong competitiveness in the field of high-frequency PCB. However, CE has the above excellent comprehensive properties, but due to the highly symmetric structure and high crystallinity of

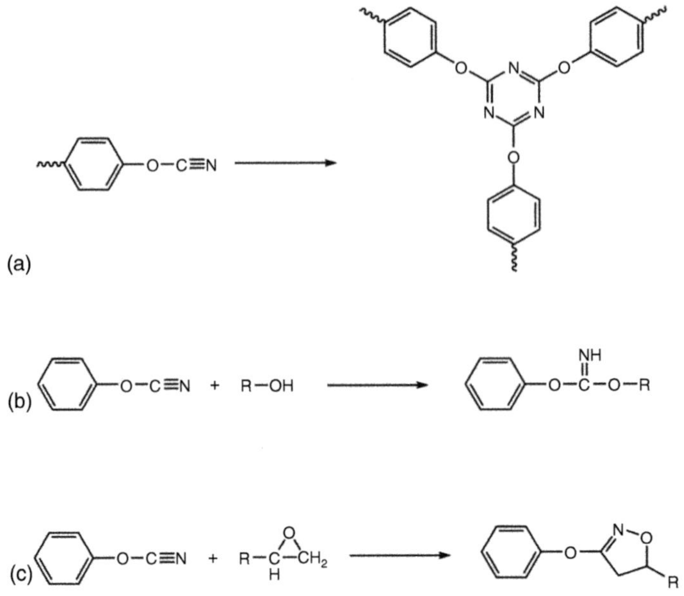

Figure 6.24 (a) Curing mechanism of CE; (b) Mechanism of the reaction between CE and hydroxyls; (c) Mechanism of the reaction between CE and epoxy.

the triazine ring in the molecule, the brittleness of the cured resin is relatively large. Thus, the prepared prepreg has poor coverability and high crosslinking density after polymerization, which limits its application. At present, CE is usually modified by copolymerization or blending with other thermosetting resins, thermoplastic resins, elastomers, and nanoparticles to optimize its performance. The modification of CE by organic polymers usually uses an epoxy group or hydroxyl group in polymer to react with the cyanate group and promote its curing to form a triazine ring, thus reducing the residue of the cyanate group in the material, so as to reduce the dielectric constant and dielectric loss, and give CE better strength and impact resistance.

Cyclodextrin microsphere (M-CDP) with high thermal stability, small size, and suitable content of –OH groups were synthesized by Li et al. [62]. Compared with the CE resin, the M-CDP/CE system has better-curing characteristics, as reflected by significantly decreased curing temperature and reaction enthalpy; moreover, M-CDP/CE resins show stable and lower dielectric constants as well as similar dielectric loss over the whole frequency range. The dielectric constant of M-CDP/CE resin with 5 wt% M-CDP is as low as 2.92. In addition, M-CDP/CE resins with suitable contents of M-CDP have better thermal stability than CE resins. These attractive performances of M-CDP/CE resins are attributed to the different networks induced by the unique structure of M-CDP (Figure 6.25).

New modified CE resins were facilely developed through melt-blending with a phenolphthalein-type polyarylether sulfone (cPES), and the integrated performances, including the reactivity, mechanical, dielectric, thermal, and flame retarding properties, were systematically investigated by Zhao et al. [63]. Compared with CE resin, cPES/CE resins have 1.2–3.2 times higher impact strengths

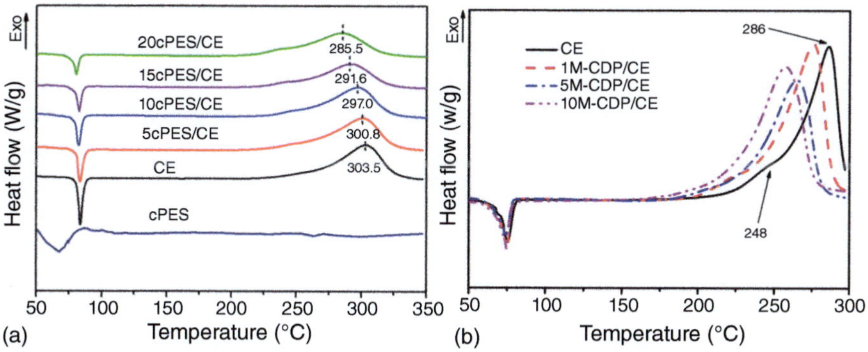

Figure 6.25 (a) DSC curves of CE and cPES/CE prepolymers. Source: Zhao et al. [63]/Royal Society of Chemistry; (b) DSC curves of CE and M-CDP/CE prepolymers. Source: Tang et al. [64]/Royal Society of Chemistry.

according to the concentration of cPES on the basis of maintaining good thermal stability, high modulus, low dielectric constant, and low loss. Through analyzing thermogravimetric kinetics, cone calorimeter tests, and the structures of chars and pyrolysis gases, they proved that the addition of cPES into CE resin greatly increases the difficulty of catching fire, as reflected by the time to ignition (TTI), fire performance index (FPI), and fire growth index (FGI), while once ignited, the cPES/CE resin exhibits a bigger heat release rate and total heat release. However, the addition of excessive cPES will cause a large number of hydroxyl groups to react with cyanate groups and generate N—H, C=N, and C—O bonds, which are more polar than the triazine ring, thus increasing the dielectric constant and dielectric loss of the resin. Tang et al. [64] modified CE with cyclodextrin, a polyhydroxyl cyclic oligosaccharide, whose topological structure is characterized by a large and small opening. Tang et al. first polymerized cyclodextrin with epichlorohydrin to obtain porous microspheres (M-CDP), which contain about 12% hydroxyl groups (mass fraction), and then modified CE to prepare M-CDP/CE composites. DSC characterization results showed that the curing temperature of CE decreased after the addition of M-CDP (Figure 6.25b), indicating that the reactive hydroxyl group in M-CDP structure could catalyze the curing of CE and that the porous structure in M-CDP could reduce the dielectric constant of CE. When the mass fraction of M-CDP is 5%, the dielectric constant of M-CDP/CE decreases from 3.4 to 2.9 at MHz, while the dielectric loss is almost unchanged. The modification of CE with high-toughness polymer containing epoxy or hydroxyl group can promote the curing reaction of the cyanate group and improve the impact resistance of CE; however, the introduction of excessive hydroxyl or epoxy group will also cause side reactions, and the generated oxazole ring or N—H polar bond will increase the dielectric constant and dielectric loss, so only the appropriate amount of addition can better improve the dielectric properties of CE.

In order to further reduce the curing temperature of CE, the multifunctional hybrid (GO/MnMOF) comprised of (GO) and a manganese-based metal-organic framework (MnMOF) was fabricated by Wang et al. [65]. In particular, the addition of only 0.5 wt% of GO/MnMOF can reduce the postcuring temperature of CE from 240 to 220 °C, and the curing temperature continuously decreases as the content of GO/MnMOF increases. Besides, the integrated performance of the modified resin containing 0.5 wt% GO/MnMOF (0.5GO/MnMOF/CE) was much better than that of the original CE resin, including its thermal, mechanical, and dielectric properties, which allows progress on solving the two bottlenecks of thermally resistant thermosetting resins.

Adding inorganic fillers to CE can strengthen and toughen it. The active groups grafted on the filler, such as the hydroxyl group and epoxy group, can promote the curing of the cyanate group and reduce residue. However, the bonds generated by the reaction of the hydroxyl group, epoxy group, and cyanate group are all polar, so excessive fillers will inevitably have a negative impact on the dielectric properties of the material. In addition, the addition of fillers containing transition metals can also enhance the toughness of CE, catalyze the curing reaction of the cyanate group, and reduce the curing temperature. However, whether the residual transition metals in the material will cause the degradation of polymer properties, such as accelerating the aging and decomposition of the material, needs to be further explored.

6.4.3.2 Polyphenylene Oxide (PPO) Used for PCB

PPO, also known as poly(2,6-dimethyl-1,4-phenylene oxide), has a molecular formula shown in the Figure 6.26. It is an amorphous thermoplastic material with high temperature resistance. It was first prepared by Hay of General Electric.

Company in 1959 by oxidative coupling 2, 6-dimethylphenol (DMP) (Figure 6.27). The principle of oxygen coupling is that the monomer containing active hydrogen undergoes dehydrogenation and coupling reactions under the oxidation of the catalyst to generate a straight-chain polymer. The coupling reaction mechanism is shown in the Figures 6.26 and 6.27. In addition, the high density of aromatic rings in PPO makes it difficult to rotate in the molecular chain segment, so PPO has poor liquidity. The methyl group on the benzene ring also increases steric hindrance and makes the structure rigid and hard. The terminal group of PPO is phenolic hydroxyl, so the oxidation resistance is poor, antioxidant or isocyanate can be used to improve the thermal oxygen stability of PPO. Commercialized PPO cannot be directly used in CCL due to its high molecular weight, low processability, low content of active groups, and low degree of cross-linking. Therefore, the modification of PPO has attracted much attention. Therefore, thermoplastic PPO is usually modified into thermosetting when it is used. There are two main ways to modify PPO: one is to introduce cross-linking active groups into the PPO molecular chain and thus realize cross-linking and curing of PPO through the reaction between active groups; Second, another thermosetting material is introduced to form the thermosetting system by blending, which could form an interpenetrating network structure.

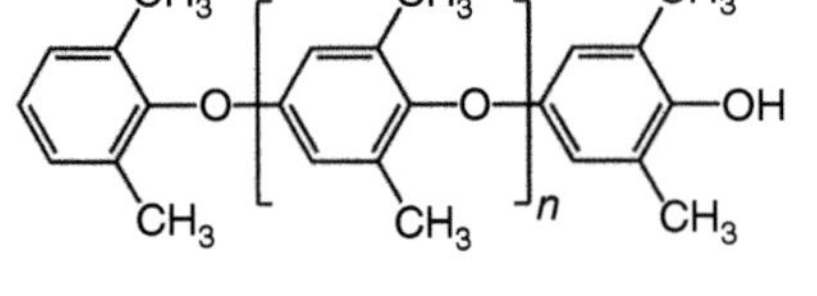

Figure 6.26 The chemical structure of PPO.

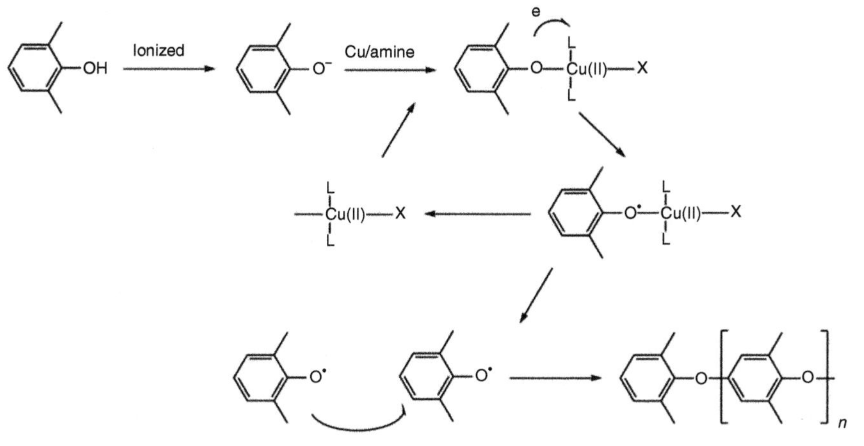

Figure 6.27 Mechanism of oxidative coupling of DMP.

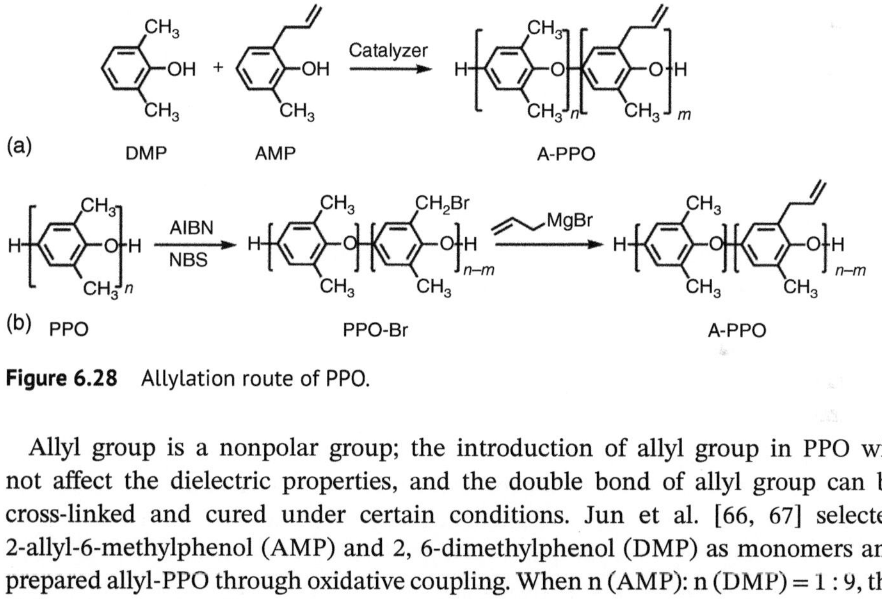

Figure 6.28 Allylation route of PPO.

Allyl group is a nonpolar group; the introduction of allyl group in PPO will not affect the dielectric properties, and the double bond of allyl group can be cross-linked and cured under certain conditions. Jun et al. [66, 67] selected 2-allyl-6-methylphenol (AMP) and 2, 6-dimethylphenol (DMP) as monomers and prepared allyl-PPO through oxidative coupling. When n (AMP): n (DMP) = 1 : 9, the dielectric constant of PPO is 2.4 and the dielectric loss is 0.002 (10 GHz). In addition to selecting the monomer with allyl, the conventional PPO can also be grafted and modified. For example, the methyl group in PPO is easily replaced by halogen to form halogenated hydrocarbon, which can then be coupled with Grignard reagent to graft allyl group [68, 69] (Figure 6.28). Wang et al. [68] used N-bromosuccinimide (NBS) to introduce bromine atoms into PPO, and then reacted the resulting brominated PPO with Grignard reagent to obtain allylated PPO. Finally, allylated PPO was solidified under certain conditions. Fang et al. [70] used bromination reagent to convert methyl group of PPO into bromomethyl group, and the group reacted with hydroxyl group on 4-allyl phenol to obtain PPO containing double bond. Finally, under the action of initiator and heat, the double bond on PPO was cross-linked to form thermosetting PPO (Figure 6.28). After modification, the thermal expansion coefficient of PPO decreased from 155.8×10^{-6} to 23.4×10^{-6} °C^{-1}, and the T_g increased to 270 °C, also, the solvent resistance was significantly improved. Also, end-group modification of PPO can also introduce epoxy group and other reactive active groups. Zheng et al. [71] using epichlorohydrin to seal the end hydroxyl group of PPO, so as to realize the import of epoxy group, making PPO own the properties of epoxy resin to a certain extent. It was then mixed with tetrabromo epoxy resin, and dicyandiamide was added for co-curing reaction. A CCL with a dielectric constant of 3.8 was prepared when GF cloth was used as reinforcing material. Based on the double-ended hydroxyl PPO, SABIC researchers used methacrylate to perform esterification with hydroxyl, introduced a double bond at the end of PPO, and successfully realized the industrial production of thermosetting low molecular weight PPO [72–74], with the trade name SA9000. The PPO modified by double bond can be cured by the crosslinking of double bond and exhibits excellent dielectric properties. The dielectric constant and dielectric loss at 1GHz are 2.5 and 0.002, respectively, which can be directly used to manufacture high-speed CCL or semi-cured sheets.

From the perspective of blending, PPO can blend with epoxy resin to improve the dielectric properties of CCL. However, since PPO and epoxy resin are typical incompatible systems, if directly mixed, the mechanical properties of the substrate will decrease due to the internal inhomogeneity. The main methods to solve the poor compatibility between PPO and epoxy resin are as follows: ① reduce the molecular weight of PPO; ② enhance the reactivity of PPO and epoxy. The redistributive reaction is expected to directly achieve these two purposes, and the redistributive reaction of PPO tends to have higher phenolic hydroxyl content, which can improve the reactivity of PPO with the epoxy resin system. Mo et al. [5] chose phenolic epoxy resin with excellent heat resistance and curing agent 4,4′-diaminodiphenyl sulfone and blended them with low molecular weight PPO to prepare a new resin system as the ideal substrate for high-frequency CCL by virtue of their advantages. The main chain of phenolic epoxy resin is composed of a benzene ring with heat resistance, which can overcome the defect of poor heat resistance of bisphenol epoxy resin. At the same time, the modification of phenolic epoxy resin with low molecular weight PPO can not only overcome the defects of high dielectric constant and dielectric loss of epoxy resin but also improve the compatibility and processability of the system. In addition, PPO was further modified by epoxidation to obtain epoxy-capped PPO, which was then reactively blended with phenolic epoxy resin. The epoxy-terminated PPO can participate in the cross-linking reaction and form a cross-network structure with epoxy resin through chemical interaction, which can not only significantly improve the compatibility of the system but also overcome the poor resistance of PPO to organic solvents.

6.4.4 Prospect of Polymer Resin for PBC at High Frequency

With the development and application of 5G technology, higher requirements have been put forward for the comprehensive performance of polymer materials used in high-frequency PCBs. In addition to the mechanical strength, water resistance, solvent resistance, and other properties of materials suitable for traditional PCB, they also need to own the peculiarity of low dielectric constant and low dielectric loss, high heat resistance, low thermal expansion coefficient, as well as other characteristics. In recent years, researchers have done a lot of work on low-dielectric polymer materials, starting from the structure of the polymer chain, the introduction of large groups, and saturated carbon chain structure. Also, the fluorine atom was introduced to reduce the polarization strength of the polymer materials. Also, in order to reduce the polarization density of polymer materials, the microstructures of polymer materials were thoroughly studied. However, these two methods are the main methods to prepare low-dielectric polymer materials at present, but they have their own disadvantages. Starting from the structure of the polymer chain, it is often necessary to synthesize special raw materials or monomers, which is complicated, and reducing the polarization strength of the polymer will usually cause adverse effects on the adhesion of the material with copper foil. Starting from the macroscopical structure, the introduction of pore structure into the material may lead to a decrease in mechanical properties, and the pore structure is easy to absorb moisture, which then increases the dielectric constant and dielectric loss of the material. Therefore, how to overcome the above problems while reducing the dielectric constant and dielectric loss of the material and improving the heat resistance, thermal expansion, moisture absorption, adhesion with copper foil, and mechanical strength of the material is still an important topic to be studied in depth.

References

1 Cai, J.Q. (2008). Evolvement of materials for PCB. *Printed Circuit Information* 10: 19–28.

2 Huang, F.R. and Jiao, Y.S. (2003). *Phenolic Resin and Its Application*. Beijing: Chemical Industry Press.

3 Dong, R.L. (2001). *Development of High Insulation, High Toughness and Heat Resistant Phenolic Plastics*. Northwestern Polytechnical University.

4 Yang, G. and Tang, C.L. (2003). Phenolic resin modified with aniline derivative. *Journal of Composite Materials* 03: 52–56.

5 Wu, C.Z., Sun, J.S., and He, P.J. (1996). S-155 aromatic alcohol phenolic resin and its properties. *Fiber Reinforced Plastics/Composites* 04: 7–15.

6 Li, C., Ma, Z., Zhang, X. et al. (2016). Silicone-modified phenolic resin: relationships between molecular structure and curing behavior. *Thermochimica Acta* 639: 53–65.

7 Li, C., Wan, J., Pan, Y.T. et al. (2016). Sustainable, biobased silicone with layered double hydroxide hybrid and their application in natural-fiber reinforced phenolic composites with enhanced performance. *ACS Sustainable Chemistry & Engineering* 4 (6): 3113–3121.

8 Wang, H. (1993). Xylene modified phenolic resin. *Contemporary Chemical Industry* 4: 50–53.

9 Tang, L.L., Li, N.N., and Wu, P.X. (2007). *High Performance Phenolic Resin and Its Application Technology*. Beijing: Chemical Industry Press.

10 Tang, L.L., Li, Z.F., and Liu, Y. (2009). Application of phenolic resin in copper clad laminate. In: *Proceedings of the 10th China Copper Sheet Market Technical Seminar*, 251–257.

11 Zhu, D.T. (2012). Research progress of phenolic paper-based copper-clad laminates in recent years(1) – Analysis and synthesis of Japanese patents on phenolic paper-based copper-clad laminates in recent years[J]. *Printed Circuit Information* 3: 10–14.

12 Zhao, J.C., Huang, Z.Q., Wang, L.L., and Yu, D. (2006). Synthesis of novolak by organophosphonic acid. *Plastics Industry* 05: 6–8.

13 Chen, J.G., Meng, F.L., and Liu, M.Z. (2020). Application and research of toughened linear phenolic resin in copper clad laminate. *The Journal of New Industrialization* 10 (08): 135–137.

14 Qin, W.F., Chen, C.H., Liu, J.X., and Wang, L.Y. (2022). Preparation of nitrogen-containing phenolic resin and its application in halogen-free low loss Copper Clad Laminate. *China Academic Journal Electronic Publishing House* 30 (06): 7–11.

15 Xin, Y.J., Tang, J.Q., Zeng, X.P., and Zhou, B. (2013). Research progress of halogen-free and phosphorus-free flame retardant copper clad laminate. *Insulating Materials* 46 (06): 33–40.

16 Diao, Z.Y., Li, Z.F., Liu, Y. et al. (2014). Synthesis of nitrogen-containing phenolic resin and its research progress in copper-clad laminate. *China Academic Journal Electronic Publishing House* 238–243.

17 Maciejewski, M., Kędzierski, M., Bednarek, E. et al. (2002). Highly branched melamine-phenolic novolaks. *Polymer Bulletin* 48 (3): 251–259.

18 Li, Z.F., Liu, Y., Zhu, H.J. et al. (2019). Properties analysis of main resins for rigid copper clad and development progress of 5G material technology in Shengquan. *Copper Cladding Information* 4: 43–48.

19 Meng, X.D. (2006). *Polyimide – Chemistry, Structure and Properties Relations and Materials*, vol. 221, 504. Beijing: Science Press.

20 Edwards, W.M. and Maxwell, R.I. (1955). Polyimides of pyromellitic acid. Google Patents

21 Hergenrother, P.M. (1990). Perspectives in the development of high-temperature polymers. *Angewandte Chemie International Edition in English* 29 (11): 1262–1268.

22 Kuznetsov, A.A. (2000). One-pot polyimide synthesis in carboxylic acid medium. *High Performance Polymers* 12 (3): 445.

23 Ratta, V. (1999). Thermal stability and adhesive properties of novel high performance semicrystalline polyimides.

24 Xie, K., Liu, J., Zhou, H. et al. (2001). Soluble fluoro-polyimides derived from 1, 3-bis (4-amino-2-trifluoromethyl-phenoxy) benzene and dianhydrides. *Polymer* 42 (17): 7267–7274.

25 Rogers, M., Brink, M., Mcgrath, J. et al. (1993). Semicrystalline and amorphous fluorine-containing polyimides. *Polymer* 34 (4): 849–855.

26 Shioda, T., Takamatsu, N., Suzuki, K. et al. (2003). Influence of water sorption on refractive index of fluorinated polyimide. *Polymer* 44 (1): 137–142.

27 Feger, C. (1993). *Advances in Polyimide: Science and Technology*. CRC Press.

28 Yang, T., Zhang, P., Dong, B.T. et al. (2019). Advance in toughening methods of thermosetting polyimide resin matrix composites. *Aeronautical Manufacturing Technology* 62 (10): 66–72.

29 Yang, S.Y., Gao, S.Q., Hu, A. et al. (2000). Progress in high temperature polyimide matrix resins and carbon fiber reinforced composites. *Aerospace Materials and Technology* 30 (1): 1–6.

30 Liu, J.C. (1983). Polyimide film. *New Chemical Materials* 3.

31 Jiang, L.T., Yu, J., Wang, X.D. et al. (2015). Application of thermoplastic polyimide in 2-Layer flexiblecopper clad laminate. *Insulating Materials* 48 (11): 18–21.

32 Wang, Y., Wang, H., Liu, F. et al. (2020). Flexible printed circuit board based on graphene/polyimide composites with excellent thermal conductivity and sandwich structure. *Composites Part A: Applied Science and Manufacturing* 138: 106075.

33 Yang, S.-Y. and Yuan, L.-L. (2018). Advanced polyimide films. *Advanced Polyimide Materials*. Elsevier 1–66.

34 Yan, F.L., Hai, X.Y., Lin, C.Z. et al. (2004). Synthesis and properties of PMR matrix resins of poly (pyrrolone-benzimidazole) s. *High Performance Polymers* 16 (1): 55–68.

35 Wang, J.F., Su, M.S., and Li, W. (2009). Study on bismaleimide resin copper-clad laminate with high T_g. *China Academic Journal Electronic Publishing House* 160–164.

36 Shi, Y.J., Fan, H., and p, Liu S S. (2009). Synthesis and applied research on polyimide used for 2-Layer double-side flexible copper clad laminate. *Chemistry and Adhesion* 31 (2): 5–7.

37 Miki, M., Suzuki, T., and Yamada, Y. (2013). Structure-gas transport property relationship of hyperbranched polyimide-silica hybrid membranes. *Journal of Photopolymer Science and Technology* 26 (3): 319–326.

38 Strunskus, T., Grunze, M., Kochendoerfer, G. et al. (1996). Identification of physical and chemical interaction mechanisms for the metals gold, silver, copper, palladium, chromium, and potassium with polyimide surfaces. *Langmuir* 12 (11): 2712–2725.

39 Diao, E.X., Li, F., Shu, S. et al. (2015). Synthesis and applied research on polyimide used for 2-Layer double-side flexible copper clad laminate. *Insulating Material* 48 (3): 1–7.

40 Zhang, X.J. and Chen, M.C. (1998). Properties, applications and development of polyimides. *Guangzhou Chemistry* 3: 58–66.

41 Wang, K., Gao, S.Q., Zhan, M.S. et al. (2005). Research progress of thermoplastic polyimide. *Chinese Polymer Bulletin* 3: 25–32.

42 Su, G.-X., Wang, J.-Y., Li, G.-H. et al. (2010). Synthesis and properties of soluble copolyimides containing phthalazinone moieties for flexible copper-clad laminates. *High Performance Polymers* 22 (8): 930–944.

43 Tamai, S., Yamaguchi, A., and Ohta, M. (1996). Melt processible polyimides and their chemical structures. *Polymer* 37 (16): 3683–3692.

44 Wise, K.E., Park, C., Siochi, E.J. et al. (2004). Stable dispersion of single wall carbon nanotubes in polyimide: the role of noncovalent interactions. *Chemical Physics Letters* 391 (4–6): 207–211.

45 Shi, Z., Hasegawa, M., Shindo, Y. et al. (2000). Thermo-processable polyimides with high thermo-oxidative stability as derived from oxydiphthalic anhydride and bisphenol A type dianhydride. *High Performance Polymers* 12 (3): 377–394.

46 Gu, X.S. (2017). Manufacturing technology of 2L-FCCL by lamination method. *Printed Circuit Information* 25 (12): 25–26.

47 Wake, W.C. (1977). *Developments in Adhensives-London*. Applied Science Publishers.

48 Zhou, G., Gou, X., Jin, X. et al. (2019). Low fractal dimension modified drilling-hole wall for PTFE high-frequency board copper plating with plasma treatment. *Journal of Applied Polymer Science* 136 (42): 48052–48061.

49 Zimmermann-Ptacek, J., Muggli, M., Wildhack, S. et al. (2018). Thermal, dielectric, and mechanical properties of h-BN-filled PTFE composites. *Journal of Applied Polymer Science* 135 (44): 46859–46867.

50 Pan, C., Kou, K., Zhang, Y. et al. (2018). Investigation of the dielectric and thermal conductive properties of core-shell structured HGM@hBN/PTFE composites. *Materials Science and Engineering: B* 238–239: 61–70.

51 Wang, L.J., Jin, X., Wu, C., and Zhang, L.X. (2021). Effect of glass fiber on properties of PTFE/SiO_2 composite substrate. *Engineering Plastics Application* 49 (01): 130–134.

52 Jiang, L.P. and Wang, J.F. (2000). Research progress of liquid crystal polymers. *Journal of North China Institute of Technology* 04: 328–333.

53 Jin, X.H., Zhang, W.X., Jiao, F. et al. (2005). Synthesis and characterization of thermogenic liquid crystal polyamides. *Journal of Beijing University of Chemical Technology* 01: 100–102.

54 Frosini, V., Marchetti, A., and De Petris, S. (1982). Low temperature phase transition in some thermotropic polyesters quenched from the liquid crystalline state. *Die Makromolekulare Chemie, Rapid Communications* 3 (11): 795–801.

55 Bao, Z., Chen, Y., Cai, R. et al. (1993). Conjugated liquid-crystalline polymers-soluble and fusible poly (phenylenevinylene) by the Heck coupling reaction. *Macromolecules* 26 (20): 5281–5286.

56 Bashir, Z. and Khan, N. (1996). Correlation of observations made by Dsc and hot-stage optical microscopy of the thermal properties of a monotropic liquid-crystal polyester. *Thermochimica Acta* 276: 145–160.

57 Zhou, G.L., Yi, Q.F., Jiang, Z.Q. et al. (2016). Preparation and research of liquid crystal polymer materials with low dielectric constant. *China Plastics Industry* 44 (4): 77–80.

58 Mai, J.H., Yang, S., Long, J.M. et al. (2022). Crystallization behaviors and mechanical properties of GF Reinforced PPS/LCP Composites. *Engineering Plastics Application* 50 (04): 1–13.

59 Zhou, M., Zhang, W., Ding, D. et al. (2012). The effect of pretreatment on adhesive strength of Cu-plated liquid crystal polymer (LCP). *Applied Surface Science* 258 (7): 2643–2647.

60 Redhwan, T.Z., Alam, A.U., Catalano, M. et al. (2017). Direct bonding of copper and liquid crystal polymer. *Materials Letters* 212 (1): 214–217.

61 Feng, H.C., Chen, Y.M., Li, Y.F. et al. (2021). Application of plasma treatment in the combination process of flexible LCP and copper lamination. *Printed Circuit Information* 29 (S2): 214–219.

62 Yuan, L., Liang, G., and Gu, A. (2011). The thermal and dielectric properties of high-performance cyanate ester resins/microcapsules composites. *Polymer Degradation and Stability* 96 (1): 84–90.

63 Zhao, L., Yuan, L., Liang, G. et al. (2015). Novel tough and thermally stable cyanate ester resins with high flame retardancy, low dielectric loss and constant based on a phenolphthalein type polyarylether sulfone. *RSC Advances* 5 (73): 58989–59002.

64 Tang, Y., Yuan, L., Liang, G. et al. (2014). High performance low-k cyanate ester resins with a thermally stable cyclodextrin microsphere. *RSC Advances* 4 (31): 16136–16145.

65 Wang, H., Yuan, L., Liang, G. et al. (2016). Tough and thermally resistant cyanate ester resin with significantly reduced curing temperature and low dielectric loss based on developing an efficient graphene oxide/Mn ion metal-organic framework hybrid. *RSC Advances* 6 (4): 3290–3300.

66 Nunoshige, J., Akahoshi, H., Shibasaki, Y. et al. (2008). Efficient oxidative coupling polymerization for synthesis of thermosetting poly(phenylene ether) copolymer with a low dielectric loss. *Journal of Polymer Science Part A: Polymer Chemistry* 46 (15): 5278–5282.

67 Nunoshige, J., Akahoshi, H., Shibasaki, Y. et al. (2007). Low dielectric loss copolymer obtained from 2,6 dimethylphenol and 2-allyl-6-methylphenol via Cu-catalyzed oxidative coupling polymerization. *Chemistry Letters* 36 (2): 238–239.

68 Wang, Y., Cheng, S., Li, W. et al. (2007). Synthesis and properties of thermosetting modified polyphenylene ether. *Polymer Bulletin* 59 (3): 391–401.

69 Shen, D., Yuan, L., Liang, G. et al. (2019). Thermally resistant photo crosslinked damping poly(phenylene oxide)-fluorosilicone rubber films with broad and high effective damping temperatures. *Journal of Applied Polymer Science* 136 (12): 47231–47238.

70 Wang, Y., Tao, Y., Zhou, J. et al. (2018). Biobased anethole-functionalized poly(phenylene oxides): new low dielectric materials with high T_g and good dimensional stability. *ACS Sustainable Chemistry & Engineering* 6 (7): 9277–9282.

71 Xu, Z.M. (2013). *Preparation and Properties of PPO Modified Epoxy Resin Composite Materials Applied in High Frequency Copper Clad Laminates*. Hangzhou: Zhejiang University.

72 Birsak, J., Chao, S.I., Duffey, B. et al. (2010). Functionalized poly (arylene ether) composition and method. United States Patent, US7781537B2.

73 Chen, C.H., Gu, Z.C., and Tsai, Y.L. (2018). Identification of the reaction mechanism between phenyl methacrylate and epoxy and its application in preparing low-dielectric epoxy thermosets with flexibility. *Polymer* 140: 225–232.

74 Peters, E.N., Jeevanath, M., Fisher, S.M., and Guo, H. (2011). Polyphenylene ether macro monomer: X. Vinyl terminated telechelic macromers. In: *Annual Technical Conference-ANTEC, Conference Proceedings*, vol. 3, 2819–2822.

7

High-Temperature Polymer Dielectrics for New Energy Power Equipment

Meng Xiao, Zhiyuan Zhang, Yuyan Chen, Xiaodan Du, and Boxue Du

Tianjin University, School of Electrical and Information Engineering, Department of Electrical Engineering, 92 Weijin Road, Nankai District, Tianjin 300072, China

7.1 Introduction

In order to promote the revolution of energy production and consumption, and build a low-carbon, clean, safe and efficient energy system, the world will vigorously build a new power system with wind power generation, photovoltaic power generation and other new energy sources [1]. New energy equipment such as high-frequency power transformers and bushings, metallized film capacitors, and insulated gate bipolar transistor (IGBT) devices play an important role in promoting clean energy consumption [2]. In recent years, the voltage level and capacity of new energy power equipment have continuously improved, and the heat problem in electrical equipment become more prominent. In addition, the size and volume of electrical equipment and electronic components are sharply reduced, resulting in difficult heat dissipation of the equipment, thus greatly affecting its accuracy and lifetime, which greatly restricts the further development of integration technology.

Dielectric has the characteristics of excellent electrical insulation, mechanical and fatigue resistance, light-weight, chemical-corrosion resistance, easy processing, and molding. It is widely used in electrical and electronic equipments. Many new energy power equipment have the phenomenon of premature dielectric breakdown, which leads to abnormal damage before the equipment reaches working life as designed. The failure of dielectric is closely related to its own heat resistance. High operating temperature will accelerate the aging of dielectric materials and shorten the service life of the equipment. Moreover, the cases of local overheating caused by various factors eventually developing into insulation damage accidents are also common at home and abroad [3].

In order to ensure the safe and stable operation of new power equipment, higher requirements are put forward for the heat resistance of polymer materials. Many methods can improve the performance of dielectric materials at a high temperature. By introducing rigid groups into molecular chains through molecular structure design, large steric hindrance will limit the movement of molecular chains, thus

High Temperature Polymer Dielectrics: Fundamentals and Applications in Power Equipment, First Edition.
Edited by Jun-Wei Zha and Zhi-Min Dang.

increasing the thermal decomposition temperature. A stable three-dimensional network structure is formed by the crosslinking of polymer molecular chains, which can also effectively improve the thermal stability of polymer molecular chains. In addition, nanocomposites are also the research hotspots of high-temperature polymer [4, 5].

This chapter will take the typical new energy power equipment as the object and combine the characteristics and performance requirements of different equipment, introduce their commonly used insulation materials and modification measures to improve the heat resistance of materials, hoping to provide some references for reducing the volume and prolonging the working life of new energy equipment.

7.2 High-frequency Power Transformer and Dry-type Bushing

Dry-type bushing has good mechanical properties, strong seismic capacity, and excellent dielectric properties. It can effectively avoid secondary damage to the converter station equipment when the insulating medium fails. The secondary inner layer is the dry capacitor core immersed with epoxy resin. As an energy module, high-frequency power transformer is the key to realize the development of power router towards high voltage, compact, high power density and large capacity. In this equipment, there are mainly two types of insulation, namely, the interturn coating film insulation of metal windings and the outer layer of resin-cast insulation to ground. With the increase of operating temperature, the thermal stability of epoxy resin and polyimide needs to be continuously improved.

7.2.1 Modification of Epoxy Resin

Epoxy resin is a low-molecular-weight prepolymer containing two or more epoxy groups composed of carbon, hydrogen, and oxygen atoms. One-component epoxy resins are of little practical value and need to be combined with a suitable curing agent. The molecular skeleton contains organic compounds such as aromatic, lipocyclic, or aliphatic compounds, which can react with curing agents to form a tight cross-linked network to play the function of epoxy resin.

However, at higher ambient temperatures, the cross-linked network of epoxy resins changes, and the movement and migration of molecular chain segments occur, which lead to the decrease of insulating properties. There are many factors affecting the thermal stability of the epoxy resin curing system, mainly including the rigidity of the molecular structure and the density of the cross-linked network. Generally speaking, the stronger the stiffness of the molecular chain segment, the greater the cross-linked density, the better the thermal stability of the curing material. Therefore, in order to enhance the thermal stability of the epoxy resin curing system, the main ways are– change the molecular chain of epoxy resin, develop the curing agent with better thermal stability, and co-mix the epoxy resin with the resin of high heat-resistance or inorganic materials.

7.2.1.1 Change the Molecular Chain

The main structure design of epoxy resin mainly considers two aspects– the introduction of rigid groups into the molecular chain and the development of epoxy resin with multifunctional groups.

Introduce Rigid Groups The introduction of the rigid groups into the epoxy resin skeleton could improve their thermal properties [6]. Due to the increasing stiffness of the molecular chain, the movement of the chain segment in the solidified material system slows down. It increases the tightness of the epoxy resin after curing, increases the bulk density of the curing material, and is beneficial to improve the thermal properties of the curing material.

The biphenyl has a regular structure, which can reduce the free volume in the epoxy curing material system. Meanwhile, it can present the liquid crystal state under certain structural conditions [7, 8], thus giving the epoxy resin a more excellent performance [9]. Previous studies have reported that a completely biologically based epoxy-resin precursor (DGEM) could be synthesized from natural Magnolol by a one-step process [10]. The obtained bio-based epoxy resin is then cured with 4,4-diamino diphenylsulfone (DDS), and the glass transition temperature of DGEM/DDS could reach 279 °C. Bisphenol A epoxy resin could be modified with 4,4-dimethoxymethyl biphenyl by condensation method [11]. The initial decomposition temperature of the biphenyl epoxy resin can reach 335 °C. The bisphenol containing biphenyl and aromatic ester structure could be made from bifenol derivatives and acylation reagents, among which the acylation reagents are prepared from salicylic acid and acetic anhydride [7, 8]. The initial decomposition temperature of the epoxy resin product with DDS is up to 370 °C.

The naphthalene epoxide compounds with planar structure have tighter molecular chain segments and smaller free volume between chain segments, which is beneficial to improve the thermal stability of cured materials. Naphthol and its derivatives are the main raw materials for the synthesis of naphthol series epoxy resins [12, 13]. The structure and properties of the synthesized epoxy resins are diverse [14]. Previous studies have reported that the thermal stability of the two-sided antiposition epoxy resin is better than that of the ipsilateral ortho naphthalene epoxy resin [15].

Fluorene contains a variety of benzene ring structures and has non-polar structural characteristics, which makes the epoxy resin have excellent thermal stability [16, 17]. But the rigidity of fluorenyl is high, which easily leads to the poor toughness and processing ability of epoxy resin. Thus the skeleton structure of epoxy resins usually contains fatty chains or siloxane structures [18]. Previous studies have reported that fluoren-based epoxy resin could be synthesized from bisphenol fluorene and bisphenol fluorene with alkyl ether side chain [19]. The initial decomposition temperature is up to 335 °C. Silane fluorene-based epoxy resin could be synthesized from bisphenol fluorene and dimethyl dichlorosilane [20]. The initial decomposition temperature could reach 347 °C.

The introduction of amamantane structure in the epoxy skeleton improves the thermal properties. Previous studies have reported that amantane is a cage structure composed of three highly symmetrical cyclohexanes. When at least three or more C—C bonds are broken, the cage can be destroyed, and then the initial degradation will occur [21, 22]. Aamantane structure has a reactive active site and the synthesis method is relatively simple.

1,3-2 (4-hydroxyphenyl) amamantane diglycidyl ether (DGEBAD) is a synthetic, low molecular weight amantane epoxy resin. DGEBAD is cured using 4,4′-diamino-diphenyl methane (DDM). The initial dissociation temperature of DGEBAD-DDM solidified substance is higher than that of E51-DDM solidified substance, and its 5% heat loss temperature is about 402 °C. The heat resistant adamantane structure is introduced into the DGEBAD-DDM cured material, so the high-temperature resistance is significantly improved. The glass transition temperature of epoxy resin with diamond alkyl group is 10 °C higher than that of E51 system. The rigid structure of adamantine alkyl group makes the movement between molecular chains difficult, thus increasing the glass transition temperature of the mixed system [23, 24].

Increase the Number of Functionality Developing and synthesizing multi-functional epoxy resin containing three or more epoxy groups is another important way to improve the thermal properties of epoxy resin. The increased number of epoxy groups can promote the cross-linked density of curing material, thus improving the thermal stability of resin curing material. In the study of the multi-functional epoxy resin, the three functional and the four functional epoxy resin occupy the vast majority. *N,N,N′,N′*, -four glyglyceryl diamino diphenyl methane (TGDDM) is the most used, the most comprehensive, the most representative multi-function resin.

At present, most of the synthesized four-functional epoxy resin are similar to the structure of TGDDM. Methyl or halogen atoms are introduced into the symmetric position of TGDDM benzene ring, and the substituted molecular intermediate methylene structures are sulfonyl, ester, ether, cyclohexane or phenyl [25, 26]. The initial decomposition temperature of these four-functional epoxy resins is kept above 300 °C, with good thermal stability. The structure modification of four-function epoxy resin is also one of the key research directions, such as novel N-free tetra-functional epoxy resins with higher thermal stability, mechanical properties, and water resistant than TGDDM without sacrificing the processability [27].

As shown in Figure 7.1a, the distance between the two epoxy groups attached to the same nitrogen atom in TGDDM was only 3.4 Å. Whereas, the epoxy groups attached to the noncoplanar benzene rings of MTEP was far apart, with a distance of 6.6 Å, 2 times that of TGDDM, which made the epoxy groups in MTEP with less constraint. The above results suggested that the MTEP/DDS showed better thermal stability than TGDDM/DDS under both atmospheres, which was ascribed to the satisfactory structural advantage of MTEP relative to TGDDM. First of all, a large number of weak C—N bonds in the structure of TGDDM could trigger thermal degradation leading to the deterioration of the thermal stability for the cured TGDDM/DDS system. Whereas the N-free structure of MTEP could reduce above adverse effect. Secondly, the stable biphenyl structure in MTEP further endowed

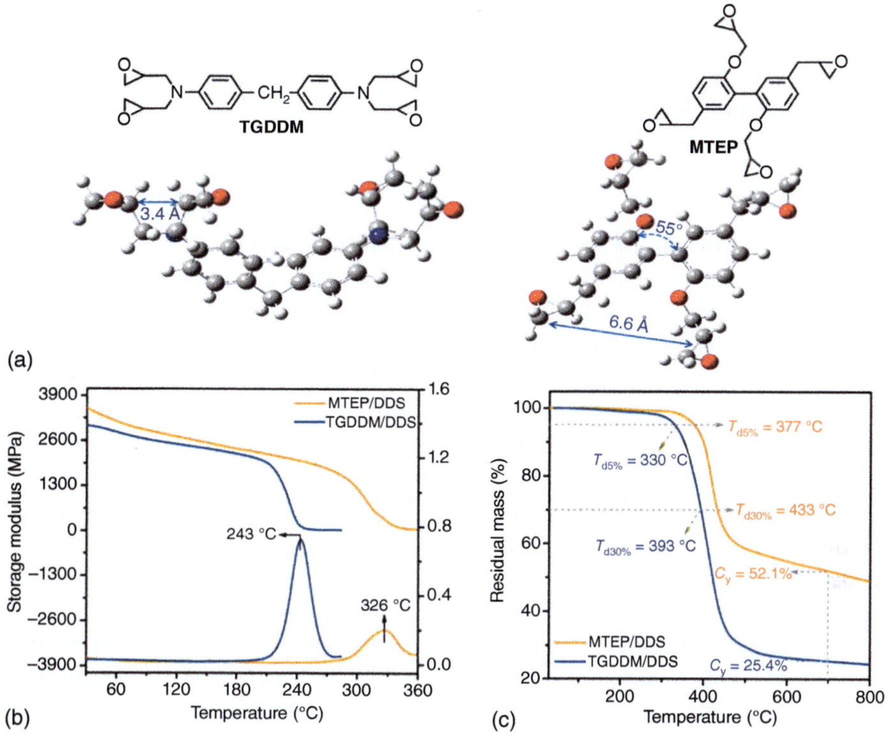

Figure 7.1 (a) Chemical structures and molecular simulation results of TGDDM and MTEP. (b) DMA spectra for storage modulus and tanδ against temperature of MTEP/DDS and TGDDM/DDS. (c) TGA curves of the cured MTEP/DDS and TGDDM/DDS with heating rate of 20 °C/min under N_2 atmosphere. Source: Qi et al. [28]. Reproduced with permission of Elsevier.

the cured MTEP/DDS system with higher thermal stability. Thirdly, thanks to the far-apart epoxide groups in MTEP, the polyfunctional MTEP could get cured more completely, making the cured MTEP/DDS with higher cross-linked density. Generally, increasing the cross-linked density for thermosetting systems is beneficial for enhancing their thermal stability.

As shown in Figure 7.1b,c, after curing with DDS, the cured DGEM/DDS system could achieve a high T_g of 279 °C. Based on that, the MTEP/DDS system reported in this work achieved a further breakthrough in T_g, which was up to 326 °C, 83 °C higher than TGDDM/DDS (243 °C). Thermogravimetric analysis (TGA) was carried out to evaluate the thermal stability of the cured epoxy systems under N_2 and air atmospheres. The results showed that the cured MTEP/DDS exhibited extremely high initial degradation temperature (5% mass loss temperature) under both atmospheres. They were 377 °C under N_2 and 373 °C under air atmosphere, which were both higher than those of TGDDM/DDS.

The structure of three functional epoxy is diversified, obviously different from the four functional epoxy, which can be roughly divided into aromatic ring or aromatic heterocyclic ring and fat chain epoxy resin [29, 30]. Among them, the thermal stability of aromatic ring or aromatic ring epoxy resin is better. In addition, a large number of multifunctional epoxy resins with more than four epoxy groups have also appeared. Previous studies have reported that it could increase the cross-linked density of curing material and give excellent flame-retardant performance of epoxy resin [31, 32].

7.2.1.2 Develop the Curing Agent with Better Thermal Stability

The performance of epoxy curing materials depends on the structural characteristics of epoxy resin itself and the type and quantity of curing agent. Different kinds of curing agents have different reactivity.

First, domestic and foreign researchers have designed and synthesized the curing agent containing heteroatomic structure with organophosphorus [33], imidazole [34], cyanogen [35], triazine ring [36], and cyclic triphosphazene compound [37]. The existence of N—H, O—H or unsaturated bonds (cyanide groups or double bonds) ensures that the system has good thermal stability. The second is the Schiff base metal ion coordination compounds [38–40]. The coordination metal ions are mainly copper, cobalt and nickel. Epoxy curing materials can maintain good thermal stability and excellent mechanical properties. In addition, there are polymer curing agent, generated by polymerization reaction. For example, there are polyphenine nanorods [41] with certain morphological structure, polymethyloxydramine curing agent [42] containing an over branch structure, and phenolic resin [43] generated by using diamine and phenol polymerization. This kind of special structure curing agent provides more ways for epoxy resin curing and enriches the use of products after curing.

The thermal stability of the epoxy resin curing material is also affected by the curing reaction process. Optimizing the curing process can improve the heat resistance of the curing material. Curing process mainly refers to the mass ratio of epoxy resin and curing agent, curing temperature, and curing time. The amount of curing agent determines the cross-linked density of the curing system. Different curing systems have their own optimal reaction temperature, which needs to achieve the appropriate temperature curing reaction to carry out. Generally speaking, small molecule curing agents can be cured at ambient temperature, and macromolecules or polymer curing agents need to react at high temperature. The curing process requires a reaction of enough time at the appropriate temperature to make the group reaction completely form a dense cross-linked structure and play the excellent performance of the curing material.

7.2.1.3 Filling Modification

The resin with higher heat resistance grade is selected for mixed modification. For example, semi-interpenetrating networks could be synthesized by epoxy resin and polyimide. The synthesis process begins with the preparation of polyamide acid, followed by therimide [44]. The prepared copolyimide is mixed with epoxy oligomer and solidified to form a cross-linked network. The results show that the mechanical properties and thermal stability of epoxy resin are obviously improved. Previous studies have reported that resins such as rigid matrix maleimide and cyanate resin can effectively improve the thermal stability of the mixed resin system.

Adding appropriate inorganic nanofillers, such as SiO_2 and TiO_2, can also improve the heat resistance of epoxy resin. Adding inorganic nano-filler to the epoxy resin curing system plays the role of reinforcing and filling effect, reducing the free volume between the molecular chain segments, and could enhance the stability and heat resistance of the curing material [45]. The thermal properties of EP-matrix composites with different filler contents are greatly improved compared to those of raw materials. Findings indicate that organosilanes grafted onto the surface of silica nanoparticles contribute to improving their compatibility and dispersibility within EP matrix, leading to a relatively low level of viscosity of EP-matrix composites, especially under a high nanofiller dosage and improving their dynamic thermomechanical properties and thermal expansion stability as well [46].

However, the preparation process of adding the filler mostly uses the physical mixing mode, which can easily lead to the poor dispersion of the filler in the epoxy resin curing system, forming the agglomeration phenomenon, resulting in the stress concentration in the curing material, reducing the mechanical properties and other properties [47]. The current solution is mainly to modify nanomaterials, connecting with hydroxyl, hydroxy groups, or other groups that contribute to dispersion.

7.3 Modification of Polyimide

Polyimide is a class of polymers containing imine rings on the macromolecular chains. It is formed by the condensation reaction of compounds containing diamine and dianhydride to produce condensers with amide acid structure, followed by cyclarization and dehydration of amide acid.

In particular, aromatic heterocyclic polymers containing phthalimide groups in the molecular backbone of PI play an extremely important role in heat-resistant plastics. The T_g of PI can generally reach over 250 °C, with the starting decomposition temperature above 450 °C.

In the special operation process of the high-frequency power transformer, the corona and partial discharge phenomenon exists in the equipment for a long time, which makes the insulation material stay in a high-temperature environment for a long time. The resulting thermal aging will affect the insulation performance of the high-frequency power transformer. Therefore, both academia and the industry hope to control the performance of polyimide by functional modifications, and develop polyimide with excellent comprehensive performance and special functions to meet special operation requirements. In general, for polymer materials represented by polyimides, the modification methods are mainly divided into three aspects– Make full use of the volume effect, quantum size effect, and macro tunneling effect of nanomaterials. Introduce Rigid Groups into macromolecular chains. Form the cross-linked structure.

7.3.1 Filling Modification

Compared to conventional inorganic fillers, nanocomposite-modified polyimide could achieve molecular-scale binding, with many advantages of both nanoparticles and organic compounds.

The introduction of inorganic components into the PI system will have two opposite effects on the thermal properties of the polymer. On the one hand, due to the high bond energy of the inorganic component, the thermal stability of the inorganic phase itself in the complex is prominent, and the addition of the inorganic component will effectively improve the thermal stability of the polyimide. Inorganic components such as silicon dioxide and titanium dioxide have been widely used in compound modification. The thermal decomposition temperature and the glass temperature improve compared to pure film. Previous studies have reported that when the content of SiO_2 reaches 32 wt%, the thermal decomposition temperature of hybrid film is 30 °C higher than that of pure polyimide, and the glass temperature increases by nearly 10 °C [48, 49]. When the content of TiO_2 is 5 wt%, the thermal decomposition temperature of the composite film is 637.8 °C, which is 40 °C higher than the pure PI film [50].

In addition, by forming hydrogen bonds or other coordination bonds between organic polymer and the inorganic phase, the molecular backbone of PI will become more rigid and the molecular chain movement is blocked. In order to break the molecular chain of PI, the interaction between the organic phase and the inorganic phase must first be overcome, so that more energy is needed to break the molecular chain of PI in the heating process. Coupling agents contain avile inorganic groups and avile organic groups. The introduction of the coupling agent in the nanocomposite modification can disperse the inorganic particles more evenly and make the organic phase more compatible with the inorganic particles. The thermal stability of the hybrid film is further improved [51, 52]. Previous studies have reported that common groups such as hydroxyl and aromatic groups could enhance the two-phase interface force [53–55].

On the other hand, the thermal decomposition temperature of PI is influenced by the structure of its aggregation state. Excessive inorganic particles can cause uneven dispersion and even agglomeration phenomenon, which changes the original network structure of PI and destroys its spatial configuration and aggregation state. Generally, when the inorganic component content is moderate, thermal stability could be improved [56, 57].

TG/DTG curves of hybrid films are shown in Figure 7.2. The temperature corresponding to the maximum decomposition occurring of PI/silica hybrid films, the thermal decomposition temperature at 10 wt% weight loss, T_{max} value, residual of hybrid films at 800 °C under nitrogen (RWN) and airflow (RWA) could be determined [58]. Under nitrogen flow, the PI hybrid films exhibited good thermal stability with thermal decomposition temperatures at 10 wt% weight loss greater than 580 °C, the T_d of API-3 and OPI-2 up to 611 and 607 °C, respectively. All silica doping films demonstrate a higher T_d than the pure PIs. From the DTG curves, maximum decomposition occurring temperatures (T_{max}) increases from 594 to

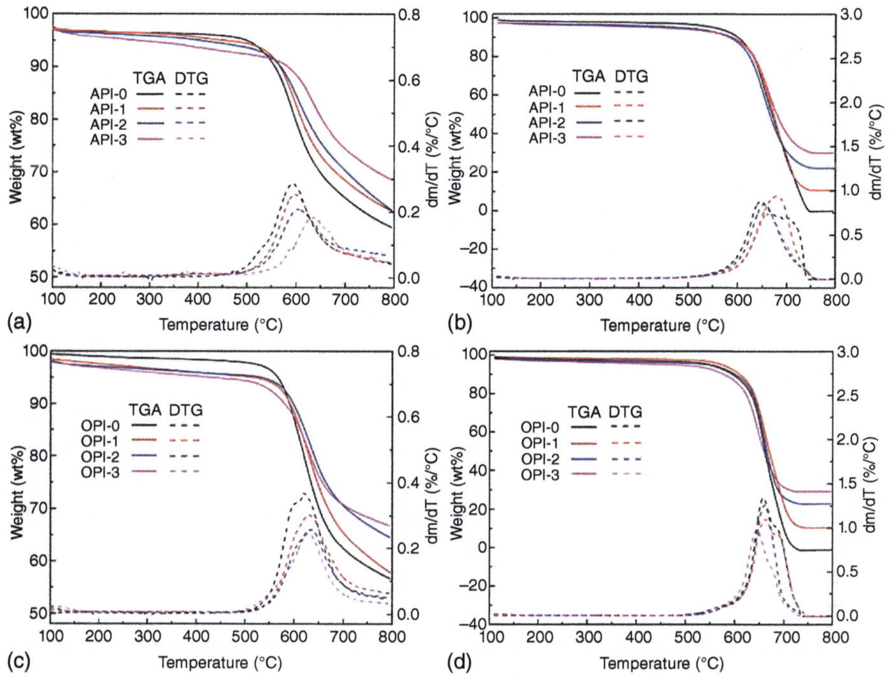

Figure 7.2 TG/DTG curves of PI/silica hybrid films: (a), (b) API system under Nitrogen and Airflow respectively. (c), (d) OPI system under Nitrogen and Airflow respectively. Source: Zheng et al. [58] Reproduced with permission of Elsevier.

635 °C in the API system, from 619 to 629 °C in the OPI system along with an increase in silica content. The intensity of the decomposition rate peak gradually decreased with increasing silica content which indicates that the thermal stability of materials was enhanced efficiently by doping with inorganic silica moiety.

Regarding the peak temperature in the loss modulus curve as T_g, the pure PI of API system possesses a much higher T_g (292 °C) than that of OPI system (271 °C). After the introduction of silica, the T_g of API films increased clearly (about 36 °C). However, the T_g of OPI hybrid films were almost unchanged. According to a previous study, the use of p-APA should increase the rigidity of polymer chains, and thus the potential barriers for rotation would increase, which would lead to a higher T_g of pure PI. Besides, the strong intermolecular force between polyimide chains that contain -NH- group will also contribute to the T_g. The increase of T_g may result from the movement of the polymer chain segment restricted by rigid silica particles that have a strong interaction with the polymer matrix. On the contrary, the poor compatibility between OPI backbone and silica particle is supposed to be responsible for the almost unchanged T_g in OPI films.

7.3.2 Introduce Rigid Groups

The movement of the macromolecular chain segment is mainly the internal rotation of a single bond. The chemical bonds and groups of different structures can change the rigidity and flexibility of the polymer, resulting in different steric resistance affecting the movement of the chain segment. Generally, the greater the rigidity of the macromolecular structure, the higher the thermal decomposition temperature.

As a typical class of molecular chains with relatively electron-rich regions and relatively electron-deficient regions, the electrorich groups between the aromatic polyimide molecular chains will interact with each other through a charge transfer or charge polarization mechanism. This phenomenon caused by the diacid anhydride and the diamine structural units in the long polymer chain is known as the charge complex transfer action. There is π–π interaction between PI molecules. The interaction between PI molecules make the molecular chain accumulate tightly, while the tightly packed molecular structure gives the polymer outstanding heat resistance [59, 60].

As a rigid structure, the ring structure of an aromatic heterocycle is formed by carbon atoms and non-carbon atoms. The non-carbon atoms in the ring are called heteroatom. Conjugate system, ladder, and semi-step chain structure formed by the aromatic ring structure containing heteroatoms such as N and O make the aromatic PI molecular chain very rigid, further improving the heat-resistance and performance. Common heterocyclic structures include imidazole, pyridine, pyrimidine, and triazine. Next is the introduction of different rigid structures [61, 62].

The introduction of the pyrimidine loops into the PI molecules has many advantages. The activation energy of the thermoacylation process could be reduced first. Moreover, due to the large rigidity of the pyrimidine ring structure and the high nitrogen content, the polarity of nitrogen heteroatoms could increase the force between molecular chains and reduce the free activity of the chain, so as to maintain a high thermal stability. The pyrimidine unit also has strong catalytic activity capable of interacting with imide rings.

2,5-Two (4-aminobenzyl) pyrimidine (PRM), in the shape of a rod, is a commonly used diamine monomer in the synthesis of PI-containing pyrimidine. Previous studies have reported that PRM and 4,4′-diaminodiphenyl ether (ODA) could be copolymerized with 4,4′-hexafluoroisopropyl-phthalic acid dianhydride (6FDA) to prepare PI (PI-1) with difunctional structure by a two-step method [63]. PI (PI-2) could be prepared by PRM, p-phenylenediamine (PDA), homophthalic anhydride (PMDA) and biphenyl anhydride (BPDA) by a two-step method [64]. PI fibers (PI-3) could be prepared by PRM and BPDA [65]. Besides PRM, there are many other important structures of pyrimidine units in the preparation of PI. Previous studies have reported that A novel diamine 2- (4-aminophenyl) -5-aminopyrimidine (PAPRM) could be mixed with PDA as a diamine, and the homopolymerization and copolymerization PI (PI-4) containing pyrimidine units could be prepared by two-step thermal amidation with PMDA or BPDA [61]. The PI (PI-5) containing the pyrimidine unit could be prepared by a two-step method with diamine monomer 5,5′-bis [*p*-(4-aminophenoxy) phenoxy] pyrimidine and BPDA. The PI molecular structural unit contains multiple ether bonds, and the combined effect of the polarity of azo atoms in the pyrimidine unit could improve the solubility and flexibility of PI while still maintaining high thermal stability [66].

Table 7.1 Thermal properties of PI-1 to PI-9.

Material	$T_{5\%}$ (°C)	$T_{10\%}$ (°C)	T_d (°C)	T_g (°C)	Reference
PI-1	496–514	525–537	540–550	309–423	65
PI-2	548–588	578–607	—	280–437	66
PI-3	501–525	553–558	587–595	—	67
PI-4	501–525	553–558	587–595	—	68
PI-5	501–525	553–558	587–595	—	69
PI-6	583	614	585	282	71
PI-7	544	604	580	280	71
PI-8	453–483	—	—	226–305	72
PI-9	507–533	—	—	249–313	72

The pyridine ring itself is a rigid symmetric aromatic ring, which can increase the symmetry and aromatic properties of the polymer and improve its thermal and chemical stability [67]. Previous studies have reported that two kinds of PI (PI6, PI-7) films could be prepared by two steps of thermal imimination with two new dianhydride monomers,2,6-bis-4-4-phenylpyridinic anhydride (6FDAPA) and 2,6-bis (3′,4′-dicarboxyphenyl) -4-phenylpyridinic anhydride (DAPA), respectively [68]. The novel diamine monomer containing pyridine heterocyclic and fluorinated groups could be combined with a series of commercial dianhydrides to synthesize a series of PI (PI-8, PI-9) by a two-step therimide process [69]. Table 7.1 lists thermal properties from PI-1 to PI-9.

In addition to the above heterocyclic, heterocyclic structures such as benzimidazole, oxazole, and triazine are also important in the preparation of PI, which are more rigid than the benzene ring. Previous studies have reported that by molecular design, the T_g of PI films containing diimidazole structure could reach to 480 °C, and the coefficient of thermal expansion is between 30 and 350 °C [70, 71]. It changes the supramolecular structure and the morphology between the molecules [72]. PI films containing benzoxazole structure have 5% thermal weight loss temperature in the range of 563 to 570 °C [73]. PI films containing darazine or pyrimidine units have the 10% thermal weight loss temperature in the range of 460 to 505 °C [74].

Large volume alkyls attached near the imide bond could limit the thermal rotation of the macromolecular chain, thus increasing the rigidity of the molecular chain and resulting in elevated T_g. However, the introduction of body side groups can also reduce the degree of molecular interaction and polymer chain accumulation, increase the free volume of polymer, and lead to the reduction of T_g, and this effect is enhanced with the increase of alkyl length [75].

7.3.3 Form the Cross-linked Structure

The cross-linked structure effectively suppresses the thermal agitation of the group and the relaxation behavior of the chain segment so as to ensure that the main body of the molecular chain will not be damaged [76].

Previous studies have reported that the weakening process of hydrogen bonding induces the thermal agitation of other groups in the polyimide backbone, affecting the dimensional thermal stability of PI at high temperatures. The construction of chemical restriction sites by heat-triggered chemical cross-linked reactions can inhibit the weakening and deionization of hydrogen bonds at high temperatures. The thermal properties are greatly improved compared to those of raw materials [77–80].

Figure 7.3a shows the synthesis of PI and crosslinking reaction during thermal imidization. Figure 7.3b depicts the curve of tanδ in DMA in which the temperature at maximal value is regarded as T_g. Compared with BPDA-BOA (T_g is 320.2 °C), T_g value of BPDA/PABZ (PI-0) is 445.4 °C, indicating the remarkable effect of hydrogen bonding on T_g of PI films. The neat PI exhibits T_g of 445 °C. After crosslinking,

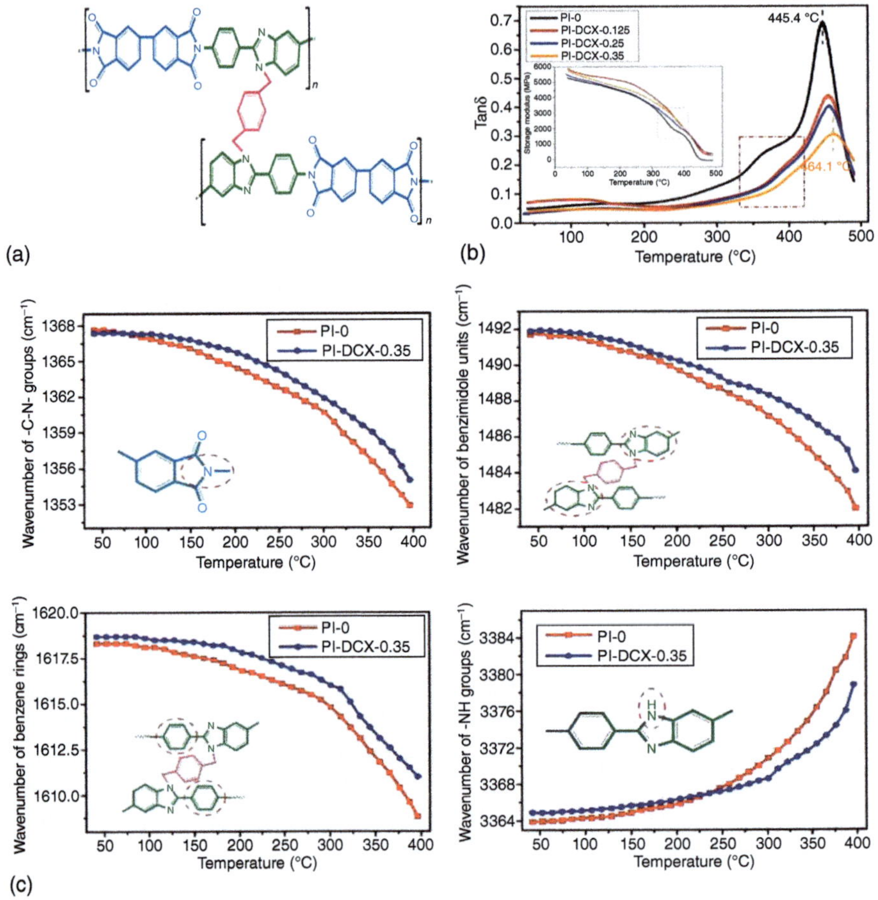

Figure 7.3 (a) Schematic diagram of the synthesis of PI and crosslinking reaction during thermal imidization. (b) Curve of tanδ and storage modulus in DMA of polyimide films with different amount of crosslinker. (c) Dynamic IR spectra and displacement of peaks of different functional groups of PI films in the procession of heating from 40 to 400 °C. Source: Tian et al. [77] Reproduced with permission of Elsevier.

T_g value of PI-DCX-0.35 shifts to a higher temperature (464.1 °C), and tanδ value of α transition reduces from 0.69 to 0.3, indicating that the segment motions of PI were restricted. What's more, it is noteworthy that the value of tanδ in PI-0 sample increases obviously from 300 °C and a small peak (highlighted in a red rectangle) around 370 °C is found, implying an obvious segment relaxation behavior below T_g. Interestingly, with increasing crosslinking degree, the values of tanδ between 300 to 400 °C is decreased and the peak at about 370 °C is weakened simultaneously. This means segment relaxation behavior below T_g is suppressed after crosslinking.

As shown in Figure 7.3c, an increase in temperature results in a red shift of the bands of 1368, 1492, and 1618 cm^{-1}. Peaks in FTIR spectrum shifts to lower wavenumber in the heating process, which implies intensification of group vibration. Below 300 °C, band shift of those peaks is relatively slight, while it becomes obvious above 300 °C, which indicates that drastic group vibration took place above 300 °C. Moreover, there appears a more slight shift for those peaks of PI-DCX-0.35 with increasing temperature, compared with that of PI-0, especially at high temperature. For instance, in the heating process, peak in 1367.6 cm^{-1} gradually shifts to 1352.7 cm^{-1}, and after crosslinking, wave number of PI-DCX-0.35 at 400 °C is 1355 cm^{-1} which is 2.3 cm^{-1} higher than PI-0. This regulation implies that group vibration is obviously restrained after crosslinking.

As for N-H groups at 3357 cm^{-1}, in the heating process, dissociation of hydrogen bond induces peaks to shift to higher wavenumber which compensates shift to lower wavenumber induced by bond force reduction. Below 300 °C, a shift of wavenumber is not obvious while it becomes drastic above 300 °C. This trend of shift of wavenumber reveals that hydrogen bonds are greatly weakened above 300 °C. Interestingly, at 400 °C, wavenumber of -NH groups in PI-DCX-0.35 is 3378.9 cm^{-1} which is 5.3 cm^{-1} lower than that of PI-0 (3384.2 cm^{-1}). This result implies that stronger hydrogen bonding interaction is retained in crosslinked PI at 400 °C in spite of weaker hydrogen bonding interaction at room temperature compared with PI-0.

7.4 High-temperature Resistant Dielectric Material for Capacitor

As an important energy storage component in power and electronic systems, capacitors are widely used in electronic products, electric vehicles, new energy grid connections, oil and gas exploration, aerospace, and other fields [81].

Polymer film capacitor is a kind of capacitor which is wound or laminated with polymer material as dielectric and metal foil or metallized coating as electrode. Compared with other dielectric materials, polymer has the advantages of being lightweight, low cost, having easy processing, and high reliability. Polymer film capacitors have good stability, high breakdown strength, and wide voltage range, which ensures their safe and stable operation. At present, polymer film capacitors have obvious advantages over other types of capacitors in new energy vehicles, grid frequency modulation, wind power generation, and other applications that require high stability and safety [1, 82].

With the vigorous development of new energy power generation, new energy vehicles and other industries, the operating environment temperature of the capacitor is constantly increasing. Relevant research shows that the working temperature of metallized film capacitor reaches up to 105.5 °C when the converter valve is in inverter state, and the working temperature of DC supporting capacitor for electric vehicles reaches 120 °C [83]. Many applications require that thin film capacitor work in high-temperature environment. Contrary to the demand of practical application, the high-temperature electric properties of polymer dielectric materials are generally poor. At high temperatures, the conductivity loss of polymer dielectric materials increases significantly, the breakdown strength of thin film capacitors decreases, and the dielectric loss increases, which seriously reduces the insulation performance of capacitor films, resulting in frequent failures and short lifetime of capacitor.

At present, biaxial-oriented polypropylene (BOPP) has the highest market share among all the dielectric materials for thin film capacitors. However, polypropylene polymer chain is flexible, which tends to curl spontaneously when it is not affected by external force, and its heat resistance is poor. The maximum working temperature of BOPP film in short-term operation is less than 105 °C, and that in long-term operation is less than 85 °C. Therefore, in order to adapt to the widespread high-temperature environment, it is necessary to develop the application of high-temperature resistant materials in capacitors to meet the demand of new energy equipment.

7.4.1 High-temperature Dielectric Polymer

From the structural characteristics, high-temperature resistant polymer dielectric materials usually contain strong intramolecular or intermolecular interactions or high-strength chemical bonds such as hydrogen bonds and conjugated π bonds are easily formed in the molecular structure, as well as the molecular skeleton of aromatic or heterocyclic aromatics. In addition, high crystallinity or cross-linked structure is also beneficial to improve the thermal stability of dielectric polymers.

7.4.1.1 Polytetrafluoroethylene

Among the known polymer bonds, the C—F bond is one of the strongest chemical bonds. The main carbon bond of PTFE macromolecule is covered by fluorine atoms, which prevents the C—C bond from being attacked by active molecules. In addition, the fluorine atoms are large in volume and repel each other, and the whole macromolecular chain is helical, resulting in the fluorine atoms symmetrically being distributed on the macromolecular main chain [84].

The above special structure makes PTFE have excellent properties. PTFE has strong thermal stability, and the material will decompose only when it withstands temperature above 400 °C. Its mechanical and electrical properties remain almost unchanged in the range of −60 to 260 °C. Even at high temperatures, PTFE still maintains excellent chemical corrosion resistance. Because the fluorine atom is symmetrical and evenly distributed, the polytetrafluoroethylene molecule is electrically neutral, and the whole molecule has no polarity, so that its dielectric constant changes little in the power frequency range. Therefore, PTFE has excellent electrical insulation properties and meets the requirements of high-voltage power insulation materials. Due to the high crystalline structure of PTFE, it is poor for the quality of PTFE films to be made by traditional melt extrusion methods, many holes and defects being produced, which limits the application of capacitors to some extent [85].

Considering the practical requirements of production, the modification of polytetrafluoroethylene has attracted much attention, and its modification methods can be roughly divided into three types– filling modification, blending modification, and surface modification.

Filling Modification Filling modification can improve the defects of polymer matrix without destroying the structure of polymer matrix. Generally, adding diversified fillers into PTFE resin, including inorganic nonmetallic fillers and metal fillers, can optimize its performance [86]. Inorganic fillers have high specific modulus, specific strength, and good heat resistance, which are often filled into PTFE matrix to improve the performance of PTFE.

PTFE/ultrafine-glass-fiber (UGF) composites were prepared by cold-pressing sintering method. It can be found that compared with PTFE, the thermal conductivity and heat resistance index of the composites had obviously increased, and the thermal expansion coefficient and high-temperature compression creep were effectively reduced. Therefore, the addition of UGF improved the thermal properties of the composites significantly [87]. In order to realize the miniaturization of PTFE capacitors, some scholars have studied polymer/ceramic composite films, such as PTFE/Sr_2TiMnO_6 (STMO) composite films. The composite material has a higher dielectric constant, which can help the miniaturization of capacitors. Through experimental verification, it is found that its thermal stability has also improved [88].

Blending Modification The basic principle of blending modification is the principle of "similar compatibility." When blending, the solubility and surface tension between the filler and the matrix must be similar [86]. After blending PTFE with other engineering plastics, the obtained composite can integrate the advantages of each component and improve its target performance, and thus, broaden its application fields of PTFE. Commonly blending fillers of PTFE include polyimide (PI), polyphenylene sulfide (PPS), and polymethyl methacrylate (PMMA).

A small amount of aromatic polyimide can make the fluorine-containing polymer sensitive to laser radiation and improve the problem of limiting insulation applications due to PTFE inertia. Therefore, the researchers synthesized polyimide-polytetrafluoroethylene blends, analyzed the thermal behavior of the blends and their components, and found that the blends and their components exhibit excellent thermal stability [89]. In addition, different contents of carbon fiber (CF) and PI were added to the PTFE matrix. It was found that the addition of PI improved the wear resistance of PTFE, and the introduction of CF improved the compactness of PTFE [90].

The poly (phthalazinone ether sulfone ketone) (PPESK)/PTFE blends with different mass ratios were prepared by solution blending, whose thermal properties and dynamic mechanical properties were studied. The experimental results showed that the glass transition temperature (T_g) of PPESK/PTFE composites had significantly increased, and its thermal properties had improved [91].

Surface Modification Physicochemical treatment of PTFE surface can improve the roughness of PTFE surface, and improve the hydrophobicity, nucleophilicity, and antifouling performance of PTFE surface. The copolymer of acrylic acid (AA) and itaconic acid (IA) was used to modify PTFE film by in-situ radiation grafting, and the results showed that its hydrophobicity decreased significantly [92]. PTFE can be easily converted into transition metal-polytetrafluoroethylene hybrid materials by applying the metal oxide atomic layer deposition (ALD) process, and the molecular structure and mechanical properties of PTFE have changed significantly [93].

HBN platelets were surface treated by silane coupling agent KH_{550}, then highly thermal conductive PTFE-based composites with optimal dielectric properties filled with hBN platelets were prepared by the cold pressing and sintering method (Figure 7.5a), then the thermal conductivity and thermal stability of the composites were observed. With the addition of fillers, the thermal conductivity of the composites increased gradually. At a given filler loading, it can be seen that PTFE-based composites with hBN-KH_{550} always exhibits higher thermal conductivity than that with hBN (Figure 7.5b). The thermal degradation behavior of PTFE and hBN-KH_{550}/PTFE composites were investigated by TGA (Figure 7.5c). Both pure PTFE and PTFE composites degrade between 500 and 620 °C, which indicates that the introduction of filler doesn't alter the thermal decomposition mechanism of PTFE matrix. As the filler content increases, the initial degradation temperature of hBN-KH_{550}/PTFE composites slightly transfers to higher temperatures. The experimental results show that compared with the untreated composites, thermal conductivity and thermal stability had significantly improved [94].

7.4.1.2 Polyvinylidene Fluoride

Polyvinylidene fluoride (PVDF) mainly refers to the homopolymer of polyvinylidene fluoride or the copolymer of polyvinylidene fluoride with a small amount of other fluorine-containing vinyl monomers. The molecular chains of PVDF are closely arranged and have strong hydrogen bonds, so its long-term working temperature is −40–150 °C [95]. Otherwise, it has good chemical corrosion resistance, oxidation resistance, and radiation resistance.

PVDF has good heat resistance, suitable for high-temperature working environments. PVDF is a typical ferroelectric polymer with a relative dielectric constant between 9 and 12, which can significantly improve the energy storage density of dielectric materials. It is considered to be the most promising alternative material for new capacitor films [96].

Considering the requirements of heat resistance and high energy storage density of capacitor equipment in industrial practice, researchers need to modify polyvinylidene fluoride, hoping to improve the heat resistance and energy storage density of

PVDF, and develop high-performance polyvinylidene fluoride to meet the operation requirements under special circumstances.

Some researchers use blending or filling modification to improve the heat resistance of PVDF. PVDF/SiO_2 nanofiber membrane was prepared by electrospinning method. The effects of adding nano-SiO_2 particles to PVDF nanofiber membrane on membrane characteristics such as porosity, temperature stability, and mechanical strength were systematically studied. It was found that SiO_2 made PVDF membrane form an electric double layer, which increased porosity and mechanical properties, and stabilized at 150 °C for 30 minutes [97].

With the improvement of people's requirements for energy storage density, the exploration and research of composite organic materials have attracted more and more attention from scientists and researchers, and the research on modification of PVDF to improve its energy storage density is more extensive.

The energy storage density of PVDF can be improved by optimizing the structure of organic polymer. Researchers successfully prepared P(VDF-CTFE) binary copolymer film by adjusting the proportion of VDF to CTFE in order to reduce or eliminate its residual polarization during PVDF copolymerization. The energy storage density of the composite reached 25 J cm^{-3} when the breakdown strength of the composite was 600 MV m^{-1} [98]. In the process of P(VDF-CTFE) copolymerization, Cl atom can be used as the cross-linking point to change the polymer structure, through this crosslinking way to limit the movement of molecular chains, in order to reduce the hysteresis loss [99].

In addition, the methods of solution blending and chemical modification are also widely used. PVDF/PMMA composite film was prepared by solution blending with PVDF as a matrix. The obtained composite material not only retained the high dielectric properties of PVDF, but also reduced the hysteresis loss, which has certain research significance [100]. A series of PVDF/TPU composites with different contents were successfully prepared by blending thermoplastic polyurethane (TPU) with PVDF, which achieved high energy storage density and breakdown strength [101]. Conductive polyaniline (PANI) was embedded into the mixture of P(VDF-CTFE-Tr FE) and copper phthalocyanine (CuPc) oligomer by chemical grafting and physical blending. At 100 Hz, the dielectric constant of the composite reached 516, the loss was only 0.56, and the breakdown strength could reach 28 MV m^{-1} [102]. PVDF-g-PS composite was obtained by grafting polystyrene (PS) into PVDF by electron irradiation. The results show the composite not only keeps the dielectric properties of PVDF matrix, but also improves the overall energy storage performance of the composite [103].

7.4.1.3 Other Materials

Thermostable dielectric polymers can be roughly divided into three categories–aromatic polymers, heteroaromatic polymers, and fluoropolymers. In addition to the above-mentioned PTFE and PVDF, the widely used high-temperature polymer dielectric materials include polyimide (PI), polyetherimide (PEI), polyphenylene sulfide (PPS), polycarbonate (PC), polyaryletherketone (PAEK) and so on [1, 82].

C—F bond in fluoropolymer has high energy, which is not easy to break at high temperatures, so the thermal stability of fluorinated polymers is relatively strong. Its representative polymer is polytetrafluoroethylene.

Aromatic polymer refers to a kind of polymer material whose molecular chain contains only a benzene ring but no other heterocycles. Under the action of π bond in benzene ring, this kind of polymer has higher thermal stability than non-aromatic polymer.

Polycarbonate (PC) is a typical aromatic dielectric polymer [104], which can be used at −60 to 120 °C for a long time. Its glass transition temperature reaches 150 °C and its thermal decomposition temperature is more than 310 °C. In the temperature range of 25–125 °C, the temperature coefficient of capacitance (TCC) of PC is only 3% and its dielectric loss is less than 0.1%, which indicates that PC is a good dielectric material in this temperature range [81]. Polyphenylene sulfide (PPS) has similar dielectric properties to PC [105]. See 4.2.1 for specific properties. Polyethylene terephthalate (PET) is polymerized from terephthalic acid and ethylene glycol monomer. It has excellent physical and mechanical properties in a wide temperature range, and its long-term working temperature can reach 120 °C. Meanwhile, the dielectric constant and dielectric loss of PET will change greatly with change in temperature and frequency [106]. Poly(ethylene 2,6-naphthalate) (PEN) has a similar chemical structure and dielectric response to PET, but the glass transition temperature (T_g) and maximum working temperature of PEN are higher [107]. Polyaryletherketone (PAEK) is a polymer material with ketone group and ether group in the main chain, which is another kind of aromatic polymer dielectric. According to the different order and proportion of ether bond, ketone group and benzene ring in the molecular chain, many different polymers can be formed. The molecular structure of PAEK contains rigid benzene rings, so it has excellent high-temperature properties. Otherwise, in polyaryletherketone series, the higher the content of ketone group, the higher the melting point and glass transition temperature [107].

Heteroaromatic polymer is a kind of polymer with benzene ring and heterocyclic structure in its molecular chain. The interaction between benzene ring and heterocyclic ring makes this kind of dielectric polymer usually have higher thermal stability.

PI and PEI are heteroaromatic polymers containing both benzene and heterocyclic structures in their molecular chains. PI has excellent high-temperature resistance performance, which makes it an ideal choice for high-temperature thin film capacitors, and it is also the only dielectric material for high-temperature capacitors currently used in practice. See 2.2 for its specific performance and application. PEI is a modified version of PI, which is prepared by introducing a flexible ether bond into the main chain of PI, and the processability is improved but thermal stability is affected. The glass transition temperature of PEI is 217–247 °C. PEI' dielectric constant and dissipation factor (DF) are similar to PI's, but PEI is superior to PI in high-temperature energy storage performance. Polybenzimidazole (PBI) has a very high glass transition temperature (435 °C), which is developed into a high-temperature dielectric film material [108]. At present, the application of PBI in capacitive energy storage is still facing some problems, and its dielectric constant and dielectric loss show strong temperature dependence. At 300 °C, the dielectric loss of PBI reaches 50% [109] (Table 7.2).

7.4.2 Nanocomposite Material

Researchers found that the high temperature performance of dielectric materials can be improved by doping nano-fillers into polymer matrix to construct nano-composites. Doping nano-filler with a high aspect ratio in polymer can effectively enhance the mechanical strength and Young's modulus of polymer dielectric, and to some extent, it can prevent the polymer dielectric material from softening when the temperature rises and electro-mechanical breakdown under the action of an external electric field. Moreover, the addition of insulating inorganic nano-filler with a high aspect ratio can also improve the dielectric and insulation properties of high-temperature polymer dielectric. This paper focuses on the influence of nano-filler on the heat resistance of polymer.

At present, boron nitride nanosheets (BNNS) are used as high-temperature dielectric materials, and they are often used as nano-fillers in the process of synthesizing nanocomposites. This is due to the high conductivity of boron nitride, and the selection of BNNS will significantly improve the conductivity of nanocomposites, which is of great significance to improve the heat dissipation of dielectric and avoid thermal collapse. In 2015, c-BCB/BNNS nanocomposites were prepared by crosslinking divinyltetramethyldisiloxane-bis(benzocyclobutene) (BCB) monomer containing BNNS (Figure. 7.4) and verified the stability of its dielectric properties in a wide range of temperature and frequency. It was found that the nanocomposites have excellent energy storage capacity of high-voltage capacitors at high temperatures, so they can be used as the main materials of dielectric films of capacitors at high-temperatures [110].

There are often defects and voids at the interface between filler and polymer matrix, which will seriously damage the mechanical properties of nanocomposites and reduce their electrical breakdown strength. In order to improve the properties of the interface area between BNNS and polymer matrix, a graded nanostructure imitating spider silk was proposed. The long-chain poly (aryl ether sulfone) (DPAES) was compounded with BCB-modified BNNS, namely BNBCB, which effectively improved the compatibility and interface structure of nanocomposites, thus improving the breakdown strength and high-temperature energy storage performance of the materials. The nano-confinement effect between BNNS and DPAES limits the movement of molecular chains and reduces the conductance loss at high-temperature [111].

Alumina nanosheets have excellent characteristics such as strong heat resistance, large specific surface area, and high resistivity. Compared with BNNS, alumina nanosheets have higher dielectric constant and wider band gap. Alumina nanosheets, as nanofillers, can further improve electrical properties of nanocomposites in high temperatures, so they have been widely used in nano-filler modification in recent years. A composite material of PEI and alumina nanoparticles, namely

Table 7.2 Common high-temperature resistant polymers and their heat resistance.

	High temperature resistant material	Structure molecular formula	Glass transition temperature (T_g)	Long-term operating temperature
Aromatic polymers	Polycarbonate (PC)	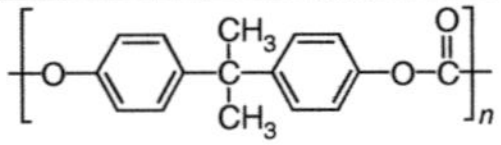	145–150 °C	−60 to 120 °C
	Poly(phenylene sulfide) (PPS)	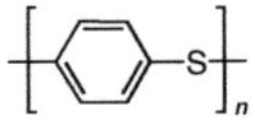	150 °C	260 °C
	Polyethyleneterephthalate (PET)	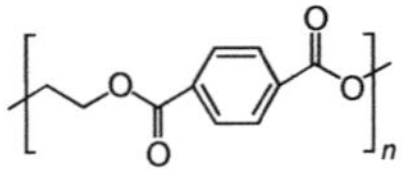	80 °C	120 °C
	Poly(ethylene 2,6-naphthalate) (PEN)	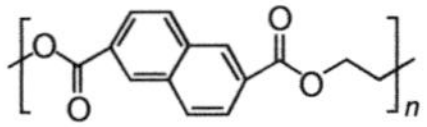	120 °C	120 °C
	Polyetheretherketone (PEEK)	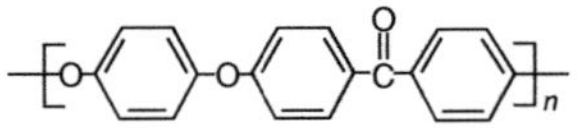	143 °C	200 °C
	Polyetherketoneketone (PEKK)	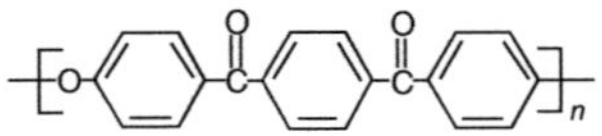	160 °C	250–260 °C
Heteroaromatic polymers	Polyimide (PI)	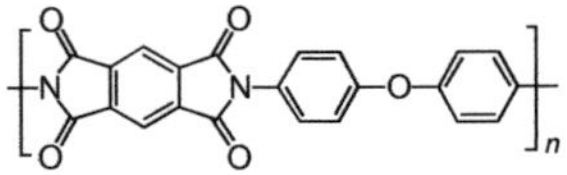	>250 °C	−200 to 300 °C
	Polyetherimide (PEI)	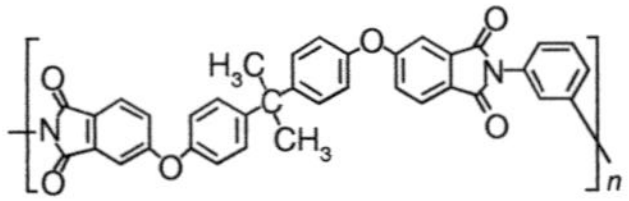	215 °C	160–180 °C
	Polybenzimidazole (PBI)	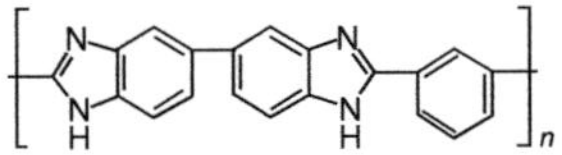	435 °C	310 °C
Fluoropolymer	polytetrafluoroethylene (PTFE)	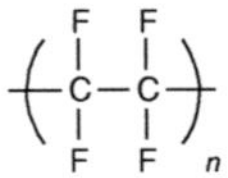	130 °C	−60 to 260 °C

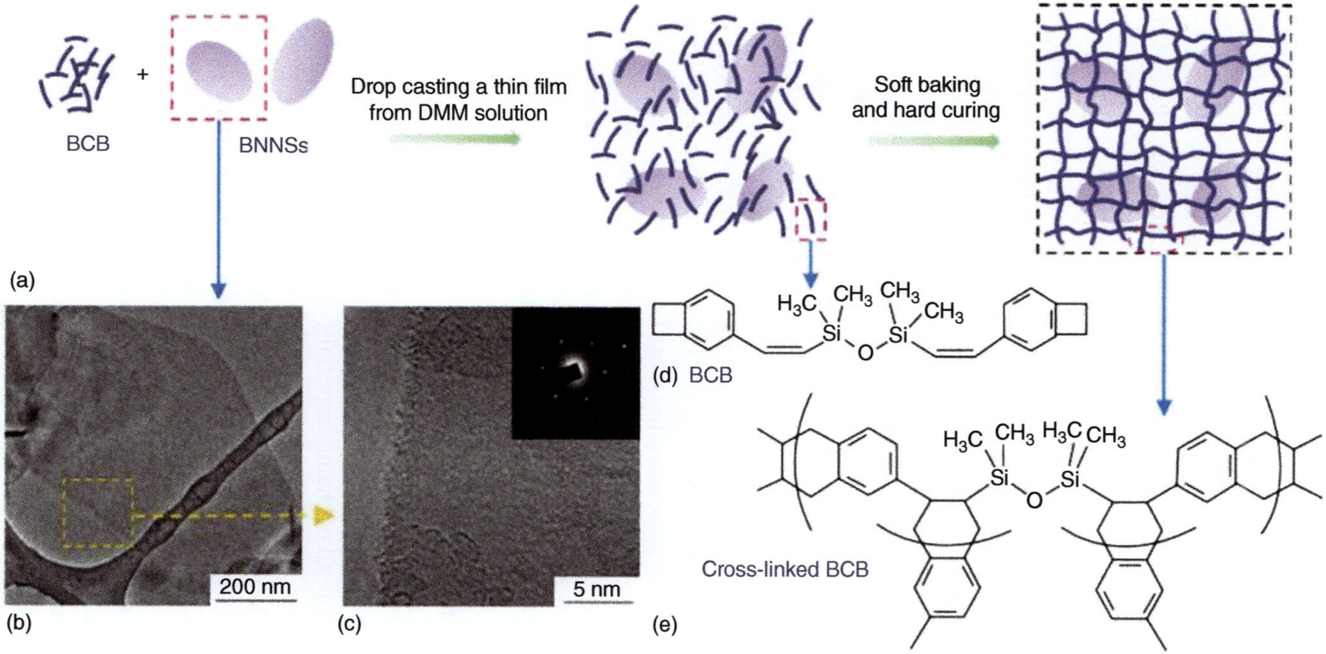

Figure 7.4 (a) Schematic of the preparation of c-BCB/BNNS films; (b, c) Transmission electron microscopy (TEM) images of BNNSs exfoliated from h-BN powders; (c) An electron-diffraction pattern of BNNSs, showing its hexagonal symmetry. (d) Chemical structure of the BCB monomer; (e) The repeating unit of c-BCB. Source: Li et al. [110] Reproduced with permission of Nature.

AO–NPs, was prepared. It has been verified that AO-NPs have a wide bandgap and high breakdown strength, and can reduce the conduction loss at high temperature, showing excellent capacitance performance. The high-temperature applicability of this nanocomposite film makes it a broad prospect in the manufacture of dielectric energy storage devices in harsh environments [112]. In addition, polyesteretherurethane (PEEU)/Al_2O_3 composites were also prepared. It was found that Al_2O_3 nanoparticles with a low doping ratio could reduce the average free path of carriers and increase the deep trap energy level, thus significantly increasing the dielectric constant and inhibiting conductivity loss at high temperatures and high field [113].

By doping wide bandgap inorganic nanoparticles into polymer, the energy storage efficiency of a capacitor can be effectively improved and its heat resistance can also be improved. In order to further improve its energy storage density, researchers try to add nanoparticles with a high dielectric constant into the polymer matrix and propose a sandwich composite structure [114], which can adjust the electric field distribution by the difference of dielectric constants between different layers, and obtain high-temperature dielectric polymer nanocomposites with both high dielectric constant and low dielectric loss. In 2016, researchers tried to mix barium titanate nanoparticles and boron nitride nanosheets into the polymer matrix with high thermal stability at the same time and constructed a sandwich structure: placing the composite material doped with boron nitride nanosheets in the outer layer to block the carriers injected from the electrode, and placing the composite material doped with barium titanate nanoparticles in the inner layer to provide high dielectric constant, thus realizing high energy density. It is found that the energy density, high power density, and remarkable charge-discharge efficiency of the nanocomposite far exceeds those of the existing polymer-based dielectric at 150 °C, and the leakage current and conductivity loss of the nanocomposite are effectively reduced [115].

7.4.3 Crosslinked Polymer

Crosslinking means that the molecular chains of linear polymers are connected by chemical bonds to form a three-dimensional network structure, which will restrict the movement of molecular chains, especially in the case of high temperature and high electric field. Generally, the crosslinking of polymers can be divided into chemical crosslinking and radiation crosslinking. Chemical crosslinking uses organic peroxides, silanes, azides and other crosslinking agents to react, and is often realized by polycondensation and polyaddition. Radiation crosslinking refers to the technical means of using various kinds of radiation to initiate the crosslinking reaction between long chains of polymer macromolecules.

The polypropylene polymer chain is flexible and tends to curl spontaneously without external force, and its heat resistance is poor. Therefore, it is necessary to modify the polypropylene film to realize its application in capacitors at a high temperature. Due to the large free volume in the amorphous region of polypropylene, the flexibility and mobility of molecular chains make it possible to recombine macromolecular free radicals. It is feasible to modify polypropylene by crosslinking.

In view of the broad prospects of cross-linked polypropylene, some scholars have studied this and successfully synthesized cross-linked polypropylene by γ-ray irradiation. It was found that cross-linked polypropylene had compact spherical crystals, and its crystallization performance had improved [116]. It can be seen that the crosslinking reaction improves the breakdown strength and conductivity of polypropylene at high temperatures [117].

The crosslinking of polypropylene blends has also been paid attention to. Some scholars have studied the effect of γ-ray irradiation on the crosslinking of polypropylene and copper (II) acetate monohydrate blends. During γ -ray irradiation, copper (II) could accelerate the formation of crosslinking structure of PP free radicals, which enhanced the thermal stability of the blends. Due to the network structure caused by cross-linking agent and copper (II), the crystallinity and melting behavior of PP blends were inhibited [118].

At present, the research on cross-linked polypropylene mostly focuses on the improvement of foaming property and melt strength, but the research and application of electrical properties are few. Therefore, crosslinked polypropylene can't be effectively applied in the field of capacitors, and attention should be paid to the crosslinking of other high-temperature resistant materials.

The cross-linked PEN film can be prepared by thermal crosslinking [119, 120]. The film shows a decomposition temperature of over 520 °C and a glass transition temperature (T_g) of around 386 °C. And stable dielectric constant and low dielectric loss are observed for this film in the frequency range of 100–200 kHz and in the temperature range of 25–300 °C. This cross-linked PEN film will be a potential candidate as high-performance film capacitor electronic device materials used at high temperatures. Among many crosslinking groups, phenylacetylene group not only shows a wide processing window and relatively good thermal stability, but also provides relatively excellent toughness for the cross-linked polymer. In recent years, the study of introducing phenylacetylene group into polyimide (PI) as crosslinking point has become more extensive. Hyperbranched polyimide can be further prepared by synthesizing dianhydride containing phenylacetylene group, and the formation of crosslinked network makes the polymer have microporous characteristics [121]. A series of phenylethynyl-terminated polyimide was prepared and studied. It was found that the polymer had excellent heat resistance and thermomechanical properties [122]. In order to take advantage of the reaction between amino group and carboxyl group, researchers decided to prepare a new PEI with 2,4,6-triaminopyrimidine (TAP) as crosslinking agent. The cross-linked PEI (c-PEI) films were manufactured via gradient heat treatment (Figure 7.5a). It is seen that the dielectric constants of c-PEI-0, c-PEI-1 and c-PEI-2 gradually increases with elevated temperatures (Figure 7.5b), which can be attributed to enhanced dipole polarization. The sharp increase of the dielectric loss for c-PEI-0, c-PEI-1 and c-PEI-2 at 200 °C can be ascribed to the polarization relaxation and movement of the polymer chains (Figure 7.5c). It is worth noting that the enhancement of the dielectric constant and dielectric loss of c-PEI films with cross-linked networks are smaller than c-PEI-0, indicating that cross-linked networks are conducive to suppressing polarization relaxation at high temperature and improving the energy

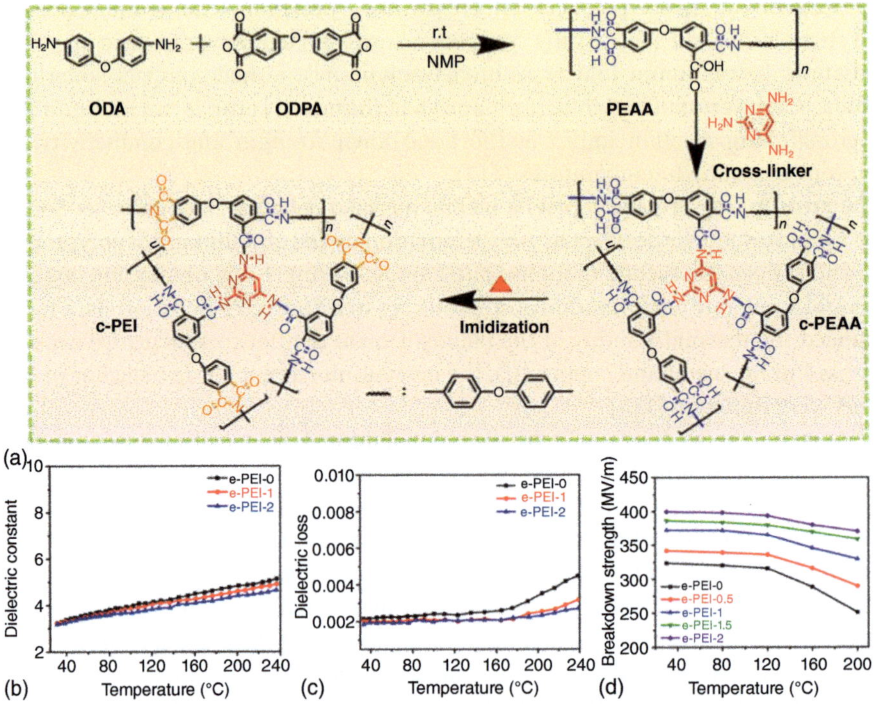

Figure 7.5 (a) The schematic illustration for the fabrication of c-PEI. The red parts represent the cross-linker of TAP, blue parts represent carboxyl and amide bonds, orange parts represent the imide structure in synthesized c-PEI. c-PEI, cross-linked polyetherimide and TAP, 2,4,6-Triaminopyrimidine. (b, c) Temperature dependence of dielectric constant and dielectric loss of c-PEI-0 (black), c-PEI-1 (red), and c-PEI-2 (blue). c-PEI, cross-linked polyetherimide. (d) The temperature dependence of breakdown strengths of c-PEI-0 and c-PEI films. c-PEI, cross-linked polyetherimide. Source: Duan et al. [123] Reproduced with permission of IET.

storage performance of c-PEI films. Otherwise, the temperature dependence of breakdown strengths of c-PEI-0 and c-PEI films can be observed (Figure 7.5d). The declining breakdown strength with rising temperature is due to the leakage current at an elevated temperature. Owing to the effect of cross-linked networks on suppressing migration of space charges and ions, it is clearly seen that the declines in breakdown strength of c-PEI dielectrics are smaller than that of c-PEI-0 [123].

These cross-linked polymer dielectric films with excellent dielectric properties have broad application prospects in developing advanced energy storage applications with high temperature resistance.

7.5 High-temperature Resistant Dielectric Material for IGBT

Insulated gate bipolar transistor (IGBT) is composed of metal-oxide-semiconductor field-effect transistor (MOSFET) and bipolar junction transistor (BJT), which has the advantages of high input impedance, low control power, simple driving circuit, fast switching speed, large on-state current, low forward voltage drop, and low loss [124]. IGBT is a major component in converter systems with DC voltage of 600 V and above, such as AC motors, frequency converters, switching power supplies, lighting circuits, traction drives, and other fields, and is one of the main development directions of power semiconductors in the future.

IGBT is the core device of energy conversion and transmission, commonly known as "CPU" of power electronic devices. As a national strategic emerging industry, it is widely used in rail transit, smart grid, aerospace, electric vehicles, and new energy power equipment. IGBT is the core component in new energy vehicles, which has an important impact on the performance of the vehicle and has been widely used in new energy vehicles. IGBT plays an important role in the field of new energy power generation such as photovoltaic and wind power industries. Photovoltaic power generation needs to be connected to the grid through a photovoltaic inverter, and IGBT is the core component of the photovoltaic inverter.

With the rapid development of electronic equipment towards miniaturization, its working temperature is constantly increasing. Meanwhile, semiconductor devices are sensitive to temperature, and high temperatures may cause abnormal functions of the devices or even burn down and destroy the whole system. Therefore, in order to ensure the normal operation of IGBT modules, new requirements are put forward for the heat dissipation and high-temperature resistance of its packaging materials.

At present, the packaging forms of IGBT devices can be divided into welded packaging and press-pack packaging according to the chip assembly process and installation method. Considering the high-temperature operation of IGBT, its packaging material will generally adopt heat-resistant materials. For welded modules, the typical packaging methods are a combination of silicone gel, epoxy potting glue, and plastic shell, or the combination of silicone gel and plastic shell directly. The stacked materials of press-pack IGBT module include conductive materials, insulating materials, and semiconductor materials. The insulation materials are mainly special engineering plastics with good dimensional stability, high mechanical strength, and good insulation performance, such as polyphenylene sulfide (PPS) and polyetheretherketone (PEEK), which are used for plastic frames in press-pack IGBT packaging. Among them, silicone gel and special engineering plastics are good heat-resistant materials.

7.5.1 Silicone Gels

Silicone gel is a hybrid polymer with inorganic siloxane chain (Si–O–Si) as the main chain and organic carbon-containing group as the side chain. The silicone gel is mostly an elastomer material after curing. The curing process is not exothermic, and there is basically no stress damage to the components, and it can be repaired after curing. Silicone has excellent high and low temperature impact resistance and can be long-term used at 250 °C. Its insulation performance is better than epoxy resin, can withstand voltage more than 10 KV. At present, the silicone gel used for IGBT potting in the market is additive silicone gel.

As the form of IGBT module packaging continues to develop, higher requirements are placed on the silicone gels used for packaging. On the one hand, due to the wide application of power modules such as silicon carbide, the working temperature of IGBT is mainly limited by packaging materials, and the high-temperature resistance of silicone gel can no longer meet the actual needs. On the other hand, the chemical structure and composition of the silicone gels will change to some extent during long-term operation in a high-temperature environment, and the change in structure will cause changes in its related properties, which will have a certain impact on the insulation performance. Ordinary linear polydimethylsiloxane gels become brittle, with a large decrease in mechanical and dielectric properties and even cracking after storage at temperatures exceeding 175 °C for more than 1000 h [125]. Silicone gels harden and become brittle after prolonged exposure at 200 °C or higher, completely losing elasticity and eventually breaking, which is caused by the lack of purity of silicone gels [126]. Considering the above situation, innovative research on silicone gels is necessary to improve their heat resistance properties.

SEMICOSIL 915HT silicone gels can be cured by using UV-activated catalysts, which can shorten the treatment time even at room temperature. After a high-temperature aging test at 210 °C for 2000 h, it was found that the appearance and mechanical properties of SEMICOSIL 915HT silicone gels were basically unchanged, and it had excellent dielectric properties and high-temperature resistance [127]. DOWSIL also developed a high temperature resistant silicone gel for the continuous operation of a new generation of power modules at 200 °C. The silicone gel can withstand temperatures of 215 °C for up to 2000 h [125].

The dielectric properties of silicone gels can also be improved by mixing fillers with high electric strength. For example, the addition of $BaTiO_3$ powder to silicone gel can increase the relative dielectric constant of silicone materials. The study found that the dielectric constant of the composite silicone gel can reach 6.4, and changes with the electric field. This method realizes the adjustability of dielectric constant of silicone gel. The composite organosilicon gels were tested in a 3300 V commercial IGBT module [128].

In addition, the IGBT module also adopts epoxy potting compound. Because of the large internal stress of epoxy resin itself and the large shrinkage in the curing process, it is not common to directly package with epoxy potting glue, which is mostly used in the field of rail transit. Epoxy resin is also a high-temperature resistant material with excellent performance, and its details are shown in Section 7.2.1.

7.5.2 Engineering Plastics

The plastic frame used for press-pack IGBT module packaging must meet high technical requirements. For example, it has high tensile strength and stable mechanical strength in the working temperature range, and can withstand short-term high temperature exceeding 250 °C to be suitable for the welding process of low and medium power modules. In addition, it must have good electrical insulation performance.

7.5.2.1 Polyphenylene Sulfide

Polyphenylene sulfide (PPS) is a thermoplastic resin with phenylthio group in the main chain. It is a special engineering plastic with excellent performance. It has excellent high-temperature stability, flame retardancy, chemical corrosion resistance, good mechanical and electrical properties, and is the chemical corrosion-resistant material second only to PTFE [129]. PPS has strong thermal stability, its glass transition temperature (T_g) is 150 °C, and it can withstand the high temperature of 200–240 °C during continuous use, and it can withstand the high temperature of 270 °C in a short time without damage. The thermal stability of long-term continuous use is better than those of all current engineering plastics. PPS also has a very small dielectric constant, a very low dielectric loss, and excellent electrical insulation. In addition, PPS material can also meet the high comparative tracking index (CTI) value required by IGBT.

The modification of PPS mainly includes copolymerization modification, fiber reinforcement modification, and nano-filling modification, among which fiber reinforcement modification is a common modification method of IGBT.

Copolymerization Modification Copolymerization modification is a chemical modification method, in which several monomers are polymerized together to obtain polymers with special structures and properties. Chemical bonds are produced between the copolymerized components, which enhances the interaction between the components. At present, the common modified materials for copolymerization modification of PPS include polyphenylene sulfide sulfone (PPSS), polyphenylene sulfide ketone (PPSK), polyethersulfone (PES), polysulfone (PSF), and polyphenylene sulfide acyl (PPSA).

The heat resistance of the composite material obtained by copolymerization modification of PPS is significantly enhanced. For example, researchers added different proportions of the third monomer 4,4 dichlorobenzophenone (DCBP) into the synthesis system of polyphenylene sulfide, and synthesize the copolymer PPS/phenylene sulfide ketone (PSK). With an increase of ketone in the copolymer chain, the heat resistance and processability of the PPS/SK copolymer were significantly improved [130].

Fiber Reinforcement Modification In order to improve the ability of plastic frame to bear mechanical stress, fiber reinforced modification is a common modification method for IGBT packaging materials. The fiber materials that can enhance the resin matrix are called fiber reinforced materials. PPS is usually modified by high temperature resistant fibers, such as glass fiber (GF), carbon fiber (CF), and aramid fiber. The existence of fiber material can make PPS and fiber form a good interface adhesion. When external force acts, the matrix can transfer the external force to the fiber through the interface, which improves the mechanical properties of the composite.

Scholars concentrated the recyclable carbon fibers into thermoplastic PPS, and the tensile strength, bending strength, and impact performance of the composite materials have been greatly improved. The crystal content and the thermal stability of injected samples of neat PPS and PPS composites were evaluated by DSC and TGA respectively. The results show that injection molding produces specimens having a degree of crystallinity of about 38 wt%, regardless of the carbon fiber content. Therefore, the increase in mechanical properties should be mainly attributed to the presence of carbon fibers and is not associated to a change in material crystallinity. Thermal analysis shows that the incorporation of recycled carbon fibers is not detrimental to the inherent thermal stability of PPS and the onset of degradation (T_{d5}) remains approximately 515 C for all samples, and the composite still maintains excellent thermal properties [131].

Nano-filling Modification Considering that nanoparticles are prone to the phenomenon of surface agglomeration, the difficulty to be paid attention to is to uniformly disperse the nanomaterials in the matrix when nanoparticles are used to modify PPS. PPS/graphite nanoplates (GNP) nanocomposites were prepared and GNP was functionalized with benzimidazole to improve the compatibility with PPS matrix. The experiment results show that the functionalized GNP is exfoliated and dispersed in PPS matrix, which significantly improves the tensile properties of the composites. Differential scanning calorimetry (DSC) was used to study PPS/functionalized GNP composites. It was found that the crystallization peak temperature (T_c) of PPS/functionalized GNP composites at different cooling rates was always higher than that of pure PPS resin due to the addition of GNP [132]. Otherwise, researchers synthesized PPS/Ti–SiO_2 nanocomposites, and SiO_2 nanoparticles were used to reinforce synthetic rubber and improve the thermal properties of polymer composites. Experimental results show that compared with pure PPS, nano-Ti–SiO_2 reduces the crystal free energy, increases the initial crystallization temperature, and accelerates the crystallization rate. In addition, the crystallization activation energy of PPS/Ti–SiO_2 is higher than that of pure PPS, which means that the crystallization of PPS/Ti–SiO_2 is more dependent on temperature than that of pure PPS [133].

7.5.2.2 Polyetheretherketone

Polyetheretherketone (PEEK) is a thermoplastic with good mechanical wear resistance and insulation properties and is commonly used as a mounting frame for press-pack IGBT.

PEEK is a semi-crystalline polymer. The existence of diphenylketone group is the reason for high thermal stability. Ether linkage provides flexibility, which also gives thermal stability. Therefore, its special molecular structure endows PEEK with excellent properties. For example, PEEK has a glass transition temperature (T_g) of 143 °C and a melting temperature (T_m) of 334 °C, which has good mechanical properties and heat resistance, and can be used continuously at 220 °C.

As the performance of PEEK is difficult to meet the requirements of IGBT operation, PEEK materials need to be modified. The common modification methods are chemical modification, blending modification, and fiber reinforcement modification.

Chemical Modification Chemical modification is to change the molecular structure and regularity of polymers by introducing special functional groups or small molecules, such as changing the proportion of ether ketone groups in the main chain or introducing other groups in the main chain, branching and crosslinking, introducing side chain groups, block copolymerization, and random copolymerization, so as to change its thermodynamic properties.

The surface of PEEK was etched with sulfuric acid, and the bonding strength between modified PEEK surface and resin was studied. It was found that the bonding strength and shear strength between PEEK and resin improved after sulfuric acid etching [134]. Sulfonation reaction is a kind of electrophilic substitution, in which concentrated sulfuric acid connects sulfonate groups to aromatic rings in PEEK to obtain SPEEK. SPEEK with low sulfonation degree designed and synthesized based on PEEK itself has good thermal stability. As a surface modifier, SPEEK was introduced in calcium carbonate /PEEK system to improve the dispersion of calcium carbonate particles in matrix PEEK, reduce the agglomeration of particles, and make the composites show good mechanical properties. The thermal properties of the composites were investigated by DSC. It was found that the glass transition temperature of filled PEEK composites increased, the melting point decreased, and the crystallization peak temperature decreased. The introduction of SPEEK with low sulfonation degree and calcium carbonate improved the thermal properties of the composites [135].

Blending Modification Blending PEEK with organic polymer materials with high glass transition temperature can not only improve the thermal properties of the composites and reduce the production cost, but also greatly affect their mechanical properties.

PEEK and polyetherimide (PEI) were injection molded. The results show that when the PEI content is below 30%, the friction properties of the composites at 20 °C-T_g can be improved [136]. SPEEK/PBI blends were prepared using highly sulfonated poly(aryl ether ether ketone) (SPEEK) and polybenzimidazole (PBI). It was found that with the increase of PBI content, the residual amount of free sulfonic acid group decreased, and the thermal stability of the composites was enhanced [137].

Blending modification improves the comprehensive properties of PEEK materials through the complementary properties of materials, and has the advantages of high efficiency and no pollution.

Fiber Reinforcement Modification The fiber filling system can effectively bear part of the load, and the synergistic effect of fiber and PEEK can better improve the comprehensive properties of the composite. Experiments have proved that glass fiber, carbon fiber, and aramid fiber have good affinity with PEEK, so they are often selected as fillers to reinforce PEEK, which can be made into high-performance composites to improve the strength and working temperature of PEEK resin. In the existing design of rigid press-pack IGBT sub-module, the primary purpose of PEEK frame is to bear mechanical pressure, so in the field of IGBT, the primary modification method of PEEK is glass fiber and carbon fiber reinforcement.

PEEK/GF foam composites were prepared by supercritical carbon dioxide (Sc-CO_2) technology with CO_2 as foaming agent. The results show that the compression resistance and heat resistance of the composites are improved because GF plays a supporting role in PEEK. At the same strain, the compressive strength of the composite foam gradually increased with increase in the GF content. The heat resistance of the composite foams was investigated by measuring their densities before and after heating for 2 h at different temperatures, and the densities of all the heat-treated GF/PEEK composite foams remained unchanged at temperature below 325 °C [138]. PEEK/GF composites were prepared from PEEK and 30 wt% short glass. It was found that the weight loss temperatures of GF reinforced composites were about 75 °C higher than that of pure PEEK, the incorporation of short glass fiber into the PEEK polymer matrix increased thermal stability. And the tensile strength, tensile modulus, flexural strength, and wear resistance of GF/PEEK were higher than those of pure PEEK, which is beneficial for PEEK applications in IGBTs [139].

7.6 Concluding Remarks

This chapter mainly introduces the methods to improve the heat resistance of dielectric materials for high-frequency power transformers and bushings, metallized film capacitors, IGBT devices, and other new energy equipment.

The main methods to improve the thermal stability of epoxy resin for transformers and bushings are changing the molecular chain of epoxy resin, developing high temperature resistant structural curing agent, and filling modification. The structural design could introduce rigid groups into the molecular chain or increase the function of the epoxy resin. Rigid groups can increase the stiffness of molecular chains, and the increase of the functionality could improve the sealing property of the epoxy resin after curing. In terms of curing agent, develop suitable high heat-resistant curing agent, adopt suitable curing process, or optimize the existing curing process flow. In the filling modification, co-mix the epoxy resin with the resin of high heat-resistance or inorganic materials.

The measures to improve the heat resistance of polyimide used in high frequency transformer are mainly divided into two aspects–The first is the nano-composite modified polyimide. Through the thermal stability of inorganic phase itself and the formation of hydrogen bonds or other coordination bonds with organic polymers, the rigidity of the molecular chain is increased, and the thermal movement of the molecular chain is restricted. The other is to change or control the structure of polyimide by molecular design to meet the requirements. The molecular design is mainly carried out by improving the rigidity of molecular chain and improving the interaction between molecular chains. To improve the rigidity of molecular chain, the rigid groups is introduced into the PI to inhibit the rotation and movement of the molecular chain. To improve the interaction between molecular chains, the main method is to improve the cross-linked degree to restrain the thermal vibration of the group and the relaxation behavior of the chain segment.

Thermal stability dielectric polymers of metallized film capacitors can be divided into aromatic polymers, heteroaromatic polymers, and fluoropolymers. Aromatic polymers have high thermal stability due to the action of π bond in benzene ring. The interaction between benzene ring and heterocyclic ring in the heteroaromatic polymer makes it have excellent thermal stability. The thermal stability of fluoropolymer is due to the characteristic that C—F bond is difficult to break at high temperature, whose representative materials are PTFE and PVDF. PTFE modification methods mainly include filling modification, blending modification, and surface modification. The filler modification refers to adding inorganic filler with good heat resistance into PTFE to improve its thermal performance. Blending modification is to blend PTFE with other engineering plastics, and the composite material will integrate the advantages of each component and improve heat resistance. Surface modification of PTFE is often used to enhance hydrophobicity. However, when filling or blending, surface modification of some fillers or blends can be targeted to improve the heat resistance of composites. Besides heat resistance, PVDF has excellent energy storage density, which makes it a potential new capacitor film dielectric material, and modification research of PVDF also focuses on heat resistance and energy storage density. In addition, the high temperature resistance of nanocomposites and crosslinked polymers is also excellent. The three-dimensional network structure formed by polymer cross-linking will limit the movement of molecular chain at high temperature and improve the heat resistance of polymer. The cross-linked polymer dielectric films with excellent dielectric properties have broad application prospects.

Engineering plastics such as PPS and PEEK are commonly used in IGBT press-pack packaging. And modification methods of PPS mainly include copolymerization modification, fiber reinforcement modification, and nano-filling modification. Copolymerization modification refers to the polymerization of PPS with other monomers, which makes the obtained composites have excellent heat resistance. Fiber reinforcement modification is mainly used to improve PPS's ability to bear mechanical stress. When external force acts, the matrix can transfer the external force to the fiber through the interface to improve the mechanical properties of composite materials, which is also the main modification method of engineering plastics in IGBT field. Polyetherketone (PEEK) has good mechanical wear resistance and thermal stability and is a good material for sub-module mounting seat. In addition to the fiber reinforcement modification and blending modification mentioned above, chemical modification is also a common modification method of PEEK, which means that the molecular structure and regularity of PEEK are changed by introducing special functional groups or small molecules, so as to improve its mechanical properties and heat resistance.

References

1 Buttay, C., Planson, D., Allard, B. et al. (2011). State of the art of high temperature power electronics. *Materials Science and Engineering: B* 176 (4): 283–288.

2 Watson, J. and Castro, G. (2015). A review of high-temperature electronics technology and applications. *Journal of Materials Science: Materials in Electronics* 26 (12): 9226–9235.

3 Watson, J. and Castro, G. (2012). High-temperature electronics pose design and reliability challenges. *Analog Dialogue* 46 (7): 3–9.

4 Critchley, J.P., Knight, G.J., and Wright, W.W. (1983). *Heat-Resistant Polymers: Technologically Useful Materials*, 1. New York, USA: Plenum Press.

5 Sroog, C.E. (1996). History of the invention and development of the polyimides. In: *Polyimides: Fundamentals and Applications* (ed. M.K. Ghosh and K.L. Mittal), 1–6. New York, USA: Marcel Dekker, Inc.

6 Zhou, Y.Z., Ren, H., Lu, H.B. et al. (2013). Synthesis of epoxy resin based on biphenyl structure for application in carbon fiber-reinforced composites. *Polymer-Plastics Technology and Engineering* 52: 581–585.

7 Cai, Z.Q., Sun, J., Wang, D. et al. (2007). Studies on curing kinetics of a novel combined liquid crystalline epoxy containing tetramethylbiphenyl and aromatic ester-type mesogenic group with diaminodiphenylsulfone. *Journal of Polymer Science Part A: Polymer Chemistry* 45 (17): 3922–3928.

8 Cai, Z.Q., Sun, J., Zhou, Q. et al. (2007). Synthesis and characterization of a novel liquid-crystalline epoxy resin combining biphenyl and aromatic ester-type mesogenic units. *Journal of Polymer Science Part A: Polymer Chemistry* 45 (5): 727–735.

9 Na, T., Jiang, H., Liu, X. et al. (2018). Preparation and properties of novel fluorinated epoxy resins cured with 4-trifluoromethyl phenylbenzimidazole for application in electronic materials. *European Polymer Journal* 100: 96–102.

10 Qi, Y., Weng, Z., Zhang, K. et al. (2020). Magnolol-based bio-epoxy resin with acceptable glass transition temperature, processability and flame retardancy. *Chemical Engineering Journal* 387: 124115.

11 Zhang, F., Zong, L., Bao, F. et al. (2020). Novel phthalazinone-bearing tetrafunctional epoxy: synthesis, characterization, and their toughening application for TGDDM system. *Polymers for Advanced Technologies* 31 (4): 635–644.

12 Duan, Y.F., Chen, B.L., and Tsai, T.H. (2005). Mechanical properties of copper-clad laminate using composite naphthalene–phenyl-based epoxy as prepreg. *Journal of Applied Polymer Science* 95 (6): 1485–1492.

13 Duan, Y.F., Liu, T.M., Cheng, K.C. et al. (2004). Thermal stability of some naphthalene-and phenyl-based epoxy resins. *Polymer Degradation and Stability* 84 (2): 305–310.

14 Pan, G., Du, Z., Zhang, C. et al. (2007). Synthesis, characterization, and properties of novel novolac epoxy resin containing naphthalene moiety. *Polymer* 48 (13): 3686–3693.

15 Duan, Y.F., Liu, T.M., Cheng, K.C. et al. (2004). Thermal stability of some naphthalene- and phenyl-based epoxy resins. *Polymer Degradation and Stability* 84 (2): 305–310.

16 Liu, W., Qiu, Q., Wu, Y. et al. (2008). Preparation and properties of one epoxy system bearing fluorene moieties. In: *The 5th East-Asian Polymer Conference, Shanghai*, 58–59. Shanghai: East China University of Science and Technology.

17 Zhang, L., Yang, H., and Shi, G. (2010). Curing kinetics study of epoxy resin/hyperbranched poly(amideamine)s system by non-isothermal and isothermal DSC. *e-Polymers* 10 (1): 136.

18 Peng, J.W., Mo, R.H., Liu, Z.F. et al. (2013). Well-defined amphiphilic polymer-Si (100) hybrids via surface-initiated atom transfer radical polymerization. *Advanced Materials Research* 669: 239–245.

19 Liu, W.B., Wang, J., Qiu, Q.H. et al. (2008). Synthesis and characterization of 9, 9-bis (4-hydroxyphenyl)-fluorene catalysed by cation exchanger. *Pigment & Resin Technology* 37 (1): 9–15.

20 Peng, J.W. et al. (2013). Well-defined amphiphilic polymer-Si(100) hybrids via surface-initiated atom transfer radical polymerization. *Advanced Materials Research* 2273 (669): 239–245.

21 Nozaki, K., Watanabe, K., Namiki, T. et al. A new single-layer resist for 193-nm lithography. *Japanese Journal of Applied Physics. Part 2, Letters* 196 (35): 528–530.

22 Llevot, A., Grau, E., Carlotti, S. et al. (2016). From lignin-derived aromatic compounds to novel biobased polymers. *Macromolecular Rapid Communications* 37: 19–28.

23 Lon, J.M., Charles, M.L., and Kurt, N.W. (1997). Poly(ether ether ketone)s and poly(ether slufones)with pendent adamantly groups. *Macromolecules* 30 (19): 5970–5975.

24 Chern, Y.T., Shiue, H.C., and Kao, S.C. (1998). Synthesis and characterization of new polyamicdes containing adamantly and diamantyl moieties in the main chain. *Journal of Polymer Science Part A: Polymer Chemistry* 36 (5): 785–792.

25 Parsania, P.H. and Patel, J.P. (2020). Fabrication and physicochemical properties of glass fabric-multifunctional epoxy resin composite. *Polymer Bulletin* 77: 1667–1679.

26 Duraibabu, D., Ganeshbabu, T., Manjumeena, R. et al. (2014). Unique coating formulation for corrosion and microbial prevention of mild steel. *Progress in Organic Coatings* 77: 657–664.

27 Liu, Q.L. and Huang, Y. (2006). Transport behavior of oxygen and nitrogen through organasiliccon-containing polystyrenes by molecular simulation. *The Journal of Physical Chemistry B* 110 (1737): 5–17382.

28 Qi, Y., Zhihuan, W., Yan, K. et al. (2021). Facile synthesis of bio-based tetra-functional epoxy resin and its potential application as high-performance composite resin matrix. *Composites Part B: Engineering* 214.

29 Cheng, J., Chen, J., and Yang, W.T. (2006). Synthesis and characterization of novel multifunctional epoxy resin. *Chinese Chemical Letters* 18: 469–472.

30 Xie, M.R., Wang, Z.G., and Zhao, Y.F. (2001). Synthesis and properties of a novel, liquid, trifunctional, cycloaliphatic epoxide. *Journal of Polymer Science Part A: Polymer Chemistry* 39: 2799–2804.

31 Zhou, L., Zhang, G., Yang, S. et al. (2019). The synthesis, curing kinetics, thermal properties and flame rertardancy of cyclotriphosphazene-containing multifunctional epoxy resin. *Thermochimica Acta* 680: 178348.

32 Bai, Y.W., Wang, X.D., and Wu, D.Z. (2012). Novel cyclolinear cyclotriphosphazene-linked epoxy resin for halogen-free fire resistance: synthesis, characterization, and flammability characteristics. *Industrial & Engineering Chemistry Research* 51 (46): 15064–15074.

33 Ai, H., Xu, K., Liu, H. et al. (2009). Synthesis, characterization, and curing properties of novel phosphorus-containing naphthyl epoxy systems. *Journal of Applied Polymer Science* 113 (1): 541–546.

34 Fang, W., Jun, X., Jing-Wen, W. et al. (2008). A novel imidazole derivative curing agent for epoxy resin: synthesis, characterization, and cure kinetic. *Journal of Applied Polymer Science* 107 (1): 223–227.

35 Liu, C., Sun, M., Zhang, B. et al. (2019). Diamine-functional bisphthalonitrile: synthesis, characterization and its application in curing epoxy resin. *European Polymer Journal* 121: 109304.

36 Cheng, J., Wang, J., Yang, S. et al. (2019). Benzimidazolyl-substituted cyclotriphosphazene derivative as latent flame-retardant curing agent for one-component epoxy resin system with excellent comprehensive performance. *Composites Part B: Engineering* 177: 107440.

37 Jian, R., Wang, P., Xia, L. et al. (2017). Effect of a novel P/N/S-containing reactive flame retardant on curing behavior, thermal and flame-retardant properties of epoxy resin. *Journal of Analytical and Applied Pyrolysis* 127: 360–368.

38 Chantarasiri, N., Tuntulani, T., Tongraung, P. et al. (2000). New metal-containing epoxy polymers from diglycidyl ether of bisphenol a and tetradentate Schiff base metal complexes. *European Polymer Journal* 36 (4): 695–702.

39 Ristić, I.S., Barta Holló, B., Budinski-Simendić, J. et al. (2015). Synthesis of novel metal-containing epoxy polymers and their structural characterization by means of FT-IR and coupled TG/MS measurements. *Journal of Thermal Analysis and Calorimetry* 119 (2): 1011–1021.

40 Chantarasiri, N., Sutivisedsak, N., and Pouyuan, C. (2001). Thermally stable metal-containing epoxy polymers from an epoxy resin-Schiff base metal complex-maleic anhydride system. *European Polymer Journal* 37 (10): 2031–2038.

41 Jang, J., Bae, J., and Lee, K. (2005). Synthesis and characterization of polyaniline nanorods as curing agent and nanofiller for epoxy matrix composite. *Polymer* 46 (11): 3677–3684.

42 Ma, C., Qiu, S., Yu, B. et al. (2017). Economical and environment-friendly synthesis of a novel hyperbranched poly(aminomethyl phosphine oxide-amine) as co-curing agent for simultaneous improvement of fire safety, glass transition temperature and toughness of epoxy resins. *Chemical Engineering Journal* 322: 618–631.

43 Mustata, F. (2010). Synthesis and characterization of novel methylenedianiline/tert-butylphenol formaldehyde resins as hardeners for epoxy resins. *Materiale Plastice* 47 (2): 146.

44 Qu, C., Li, X., Yang, Z. et al. (2019). Damping, thermal, and mechanical performances of a novel semi-interpenetrating polymer networks based on polyimide/epoxy. *Journal of Applied Polymer Science* 136 (41): 48032.

45 Grishchuk, S., Sorochynska, L., Vorster, O.C. et al. (2013). Structure, thermal, and mechanical properties of DDM-hardened epoxy/benzoxazine hybrids: effects of epoxy resin functionality and ETBN toughening. *Journal of Applied Polymer Science* 127: 5082–5093.

46 Wang, H., Yan, Y., Tian, L. et al. (2022). High content of spherical nanosilica filled epoxy resin master batch with low viscosity and superior thermomechanical performance. *Composites Communications* 36: 101355.

47 Song, X., Zheng, S., Huang, J. et al. (2001). Miscibility and mechanical properties of tetrafunctional epoxy resin/phenolphthalein poly(ether ether ketone)blends. *Journal of Applied Polymer Science* 79: 598–607.

48 Nandi, M. and Sen, A. (1991). Molecular level metal and ceramic/polymer composites. Synthesis of metal and metal oxide containing polyimides and its relevance to polymer metallization. *Proceedings of the Electrochemical Society Symposium on Metallized Plastics* 1: 35–41.

49 Nandi, M., Conklin, J.A., Salvati, L.J. et al. (1991). Molecular level ceramic/polymer composites. 2. Synthesis of polymer-trapped silica and titania nanoclusters. *Chemistry of Materials* 3: 201–206.

50 Shichen, W., Jinghua, Y., and Jialong, L. (2016). Thermal stability of polyimide titanium dioxide nanocomposite film. *Surface Technology* 45 (01): 36–40.

51 Al-Kandary, S., Ali, A.A.M., and Ahmad, Z. (2005). Morphology and thermo-mechanical properties of compatibilized polyimide-silica nanocomposites. *Journal of Applied Polymer Science* 98: 2521–2531.

52 Al-Kandary, S., Ali, A.A.M., and Ahmad, Z. (2006). New polyimide-silica nanocomposites from the sol-gel process using organically-modified silica network structure. *Journal of Materials Science* 41 (10): 2907–2914.

53 Al-Arbash, A., Al-Sagheer, F., Ali, A.A.M. et al. (2006). Thermal and mechanical properties of poly(hydroxy-imide)-silica nanocomposites. *International Journal of Polymeric Materials* 55: 103–120.

54 Rehman, H.U., Schmidt, H., and Ahmad, Z. (2006). Synthesis and characterization of polyimide-silica hybrids: effect of matrix polarity on the mechanical and thermal properties. *Journal of Macromolecular Science, Part A: Pure and Applied Chemistry* 43: 703–717.

55 Lee, A. (2007). Durability characterization of POSS-based polyimides and carbon-fiber composites for air force-related applications. ADA482151.

56 Zhang, M., Zeng, S., Dong, T. et al. (2006). Preparation and characterization of polyimide/silica hybrid films. *Journal of Harbin Institute of Technology (New Series)* 13 (4): 439–443.

57 Liu, L., Liang, B., Wang, W., and Lei, Q. (2006). Preparation of polyimide/inorganic nanoparticle hybrid films by sol–gel method. *Journal of Composite Materials* 40 (23): 2175–2183. https://doi.org/10.1177/0021998306063353.

58 Zheng, L., Jiang-Huai, H., Chen, C. et al. (2018). Preparation and characterization of adenine-based polyimide/nano-silica hybrid films. *European Polymer Journal* 102.

59 Sun, G.H., Dong, G.Q., Kong, L.S. et al. (2018). Robust polyimidenanofibrous membrane with porous-layer-coated morphology by insitu self-bonding and micro-crosslinking for lithium-ion battery separator. *Nanoscale* 10: 22439.

60 Furukawa, N., Yuasa, M., and Kimura, Y. (2015). Structure analysis of a solublepolysiloxane-block-polyimide and kinetic analysis of the solution-imidization of the relevant polygamic acid. *Journal of Polymer Science Part A: Polymer Chemistry* 36 (13): 2237–2245.

61 Xia, A.X., Lu, G.H., Qiu, X.P. et al. (2006). Syntheses and properties of novel polyimides derived from 2-(4-aminophenyl)-5-aminopyrimidine. *Journal of Applied Polymer Science* 102 (6): 5871–5876.

62 (a) Liu, J.M., Lee, T.M., Wen, C.H. et al. (2011). High-performance organic-inorganic hybrid plastic substrate for flexible displays and electronics. *Journal of the Society for Information Display* 19 (1): 63–69. (b) Wang, S.L., Ma, S.Q., He, H.R. et al. (2019). Aromatic polyimides containing pyridine and spirocyclic units: preparation, thermal and gas separation properties. *Polymer* 168: 199–208.

63 Zhang, J., Luo, L.B., Liu, X.Y. et al. (2015). Synthesis and properties of pyrimidine fluor polyimides. *China Plastics Industry* 43 (2): 19–23.

64 Xia, A.X., Lyu, G.H., Qiu, X.P. et al. (2007). Synthesis and characterization of polyimides based on 2,5-bis (4-aminophenyl) pyrimidine. *Acta Polymerica Sinica* 3: 262–266.

65 Gan, F., Dong, J., Tan, W. et al. (2017). Fabrication and characterization of co-polyimide fibers containing pyrimidine units. *Journal of Materials Science* 52 (16): 9895–9906.

66 Peng, X.W., Wu, Q., Ding, Y.C. et al. (2012). Synthesis and properties of polyimides based on 5,5′- bis [*p*-(4-aminophenoxy) phenoxy]pyrimidine. *Polymeric Materials Science and Engineering* 28 (8): 26–29.

67 Liaw, D.J., Chang, F.C., Leung, M.K. et al. (2005). High thermal stability and rigid rod of novel organ soluble polyimides and polyamides based on bulky and noncoplanar naphthalenebiphenyldiamine. *Macromolecules* 38: 4024–4029.

68 Yang, H.X., Liu, J.G., Li, Y.F. et al. (2006). Synthesis and properties of pyridine bridged polyimides. *Acta Polymerica Sinica* 3: 489–495.

69 Guan, Y., Wang, D., Song, G. et al. (2014). *Polymer* 55: 3634–3641.

70 Yue, Z., Cai, Y.B., and Xu, S. (2014). Facile synthesis of a symmetrical diamine containing bis-benzimidazole ring and its thermally stable polyimides. *Journal of Polymer Research* 21 (6): 463–470.

71 Lian, M., Lu, X., and Lu, Q. (2018). Synthesis of superheat-resistant polyimides with high T_g and low coefficient of thermal expansion by introduction of strong intermolecular. *Interactions* 51 (24): 10127–10135.

72 Wang, S., Zhou, H.W., Dang, G.D. et al. (2009). Synthesis and characterization of thermally stable, high-modulus polyimides containing benzimidazole moieties. *Journal of Polymer Science Part A: Polymer Chemistry* 47 (8): 2024–2031.

73 Sun, M., Chang, J., Tian, G.F. et al. (2016). Preparation of high-performance polyimide fibers containing benzimidazole and benzoxazole units. *Journal of Materials Science* 51 (6): 2830–2840.

74 You, N., Nakamura, Y., Suzuki, Y. et al. (2009). Synthesis of highly refractive polyimides derived from 3,6-bis (4-aminophenyleresulfanyl) pyridazine and 4,6-bis(4-aminophenylenesulfanyl) pyrimidine. *Journal of Polymer Science Part A: Polymer Chemistry* 47 (19): 4886–4894.

75 Wang, L., Chang, P., and Cheng, C.L. (2006). Structural effects of pendant groups on thermal and electrical properties of polyimides. *Journal of Applied Polymer Science* 100: 4672–4678.

76 Yu, G., Liu, C., Wang, J. et al. (2010). Heat-resistant aromatic S-triazine-containing ring-chain polymers based on bis(ether nitrile)s: synthesis and properties. *Polymer Degradation and Stability* 95 (12): 2445–2452.

77 Tian, Y.Y., Luo, L.B., Yang, Q.Q. et al. (2020). Construction of stable hydrogen bonds at high temperature for preparation of polyimide films with ultralow coefficient of thermal expansion and high T_g. *Polymer* 188: 122100.

78 Skrovanek, D.J., Howe, S.E., Painter, P.C., and Coleman, M.M. (1985). Hydrogen bonding in polymers: infrared temperature studies of an amorphous polyamide. *Macromolecules* 18 (9): 1676–1683.

79 Skrovanek, D.J., Painter, P.C., and Coleman, M.M. (1986). Hydrogen bonding in polymers: infrared temperature studies of an amorphous polyamide. 19 (3): 699–705.

80 Jiang, Y. et al. (2021). Improving dimensional stability at high temperature and toughness of polyimide films via adjustable entanglement density. *Polymer* 218.

81 Li, Q., Yao, F.Z., Liu, Y. et al. (2018). High-temperature dielectric materials for electrical energy storage. *Annual Review of Materials Research* 48: 219–243.

82 Chu, B., Zhou, X., Ren, K. et al. (2006). A dielectric polymer with high electric energy density and fast discharge speed. *Science* 313 (5785): 334–336.

83 Li, Z., Chen, X., Zhang, C. et al. (2019). High dielectric constant polycarbonate/nylon multilayer films capacitors with self-healing capability. *ACS Applied Polymer Materials* 1 (4): 867–875.

84 Dhanumalayan, E. and Joshi, G.M. (2018). Performance properties and applications of polytetrafluoroethylene (PTFE)–a review. *Advanced Composites and Hybrid Materials* 1 (7): 247–268.

85 Tan, D., Zhang, L.L., Chen, Q. et al. (2014). High-temperature capacitor polymer films. *Journal of Electronic Materials* 43 (12): 4569–4575.

86 Pitenis, A.A., Ewin, J.J., Harris, K.L. et al. (2014). In vacuum tribological behavior of poly tetrafluoro ethylene (PTFE) and alumina nanocomposites: the importance of water for ultralow wear. *Tribology Letters* 53 (1): 189–197.

87 Zhang, Y., Kaichang, K., Shuangcun, Z. et al. (2019). Improving thermal properties of ultrafine-glass-fib reinforced PTFE hybrid composite via surface modification by (3-aminopropyl) triethoxysilane. *Journal of Polymer Research* 26 (9): 1–11.

88 Ashokbabu, A. and Thomas, P. (2019). Dielectric and thermal behavior of PTFE/Sr2TiMnO6 (STMO) composites. In: *2019 IEEE 4th International Conference on Condition Assessment Techniques in Electrical Systems (CATCON)*, 1–4. IEEE.

89 Davis, C.R. and Zimmerman, J.A. (1994). Thermal stability studies of a polyimide–polytetrafluoroethylene blend and its components. *Journal of Applied Polymer Science* 54 (2): 153–162.

90 Chen, B., Wang, J., and Yan, F. (2012). Synergism of carbon fiber and polyimide in polytetrafluoroethylene-based composites: friction and wear behavior under sea water lubrication. *Materials & Design (1980–2015)* 36: 366–371.

91 He, W., Liao, G.X., Jian, X.G. et al. (2006). Dynamic mechanical thermal behaviour of novel poly(phthalazine ether sulfone ketone) and poly (tetrafluoroethylene) blends. *Journal of Polymer Engineering* 26 (6): 617–632.

92 Hidzir, N.M., Lee, Q., Hill, D.J.T. et al. (2015). Grafting of acrylic acid-*co*-itaconic acid onto ePTFE and characterization of water uptake by the graft copolymers. *Journal of Applied Polymer Science* 132 (7): https://doi.org/10.1002/app.41482.

93 Lee, S.M., Ischenko, V., Pippel, E. et al. (2011). An alternative route towards metal–polymer hybrid materials prepared by vapor-phase processing. *Advanced Functional Materials* 21 (16): 3047–3055.

94 Pan, C., Kou, K., Jia, Q. et al. (2017). Improved thermal conductivity and dielectric properties of hBN/PTFE composites via surface treatment by silane coupling agent. *Composites Part B: Engineering* 111: 83–90.

95 Ameduri, B. (2009). From vinylidene fluoride (VDF) to the applications of VDF-containing polymers and copolymers: recent developments and future trends. *Chemical Reviews* 109 (12): 6632–6686.

96 Martins, P., Nunes, J.S., Hungerford, G. et al. (2009). Local variation of the dielectric properties of poly(vinylidene fluoride) during the α-to β-phase transformation. *Physics Letters A* 373 (2): 177–180.

97 Widiyandari, H., Putra, O.A., Purwanto, A. et al. (2021). Synthesis of PVDF/SiO_2 nanofiber membrane using electrospinning method as a Li-ion battery separator. *Materials Today: Proceedings* 44 (3): 245–248.

98 Zhou, X., Chu, B., Neese, B. et al. (2007). Electrical energy density and discharge characteristics of a poly(vinylidene fluoride-chlorotrifluoroethylene) copolymer. *IEEE Transactions on Dielectrics and Electrical Insulation* 14 (5): 1133–1138.

99 Khanchaitit, P., Han, K., Gadinski, M.R. et al. (2013). Ferroelectric polymer networks with high energy density and improved discharged efficiency for dielectric energy storage. *Nature Communications* 4 (1): 2845.

100 Tan, S., Hu, X., Ding, S. et al. (2013). Significantly improving dielectric and energy storage properties via uniaxially stretching crosslinked P(VDF-co-Tr FE) films. *Journal of Materials Chemistry A* 1 (35): 10353.

101 Meng, Q., Li, W., Zheng, Y. et al. (2010). Effect of poly (methyl methacrylate) addition on the dielectric and energy storage properties of poly(vinylidene fluoride). *Journal of Applied Polymer Science* 116 (5): 2674–2684.

102 Wang, J., Wu, C., Liu, R. et al. (2013). P(VDF–Tr FE–CFE)-based percolative composites exhibiting significantly enhanced dielectric properties. *Polymer Bulletin* 70 (4): 1327–1335.

103 Thakur, V.K., Tan, E.J., Lin, M. et al. (2011). Polystyrene grafted polyvinylidenefluoride copolymers with high capacitive performance. *Polymer Chemistry* 2 (9): 2000.

104 Cozens, J. (1959). Development of plastic dielectric capacitors. *IRE Transactions on Component Parts* 6 (2): 114–118.

105 Odian, G. (2004). *Principles of Polymerization*. Wiley.

106 Ho, J.S. and Greenbaum, S.G. (2018). Polymer capacitor dielectrics for high temperature applications. *ACS Applied Materials & Interfaces* 10 (35): 29189–29218.

107 Attwood, T.E., Dawson, P.C., Freeman, J.L. et al. (1981). Synthesis and properties of polyaryletherketones. *Polymer* 22 (8): 1096–1103.

108 Bourbigot, S. and Flambard, X. (2002). Heat resistance and flammability of high performance fibres: a review. *Fire and Materials* 26 (4–5): 155–168.

109 Hammoud A N, Suthar J L. *Characterization of Polybenzimidazole (PBI) Film at High Temperatures*. 1992.

110 Li, Q., Chen, L., Gadinski, M.R. et al. (2015). Flexible high-temperature dielectric materials from polymer nanocomposites. *Nature* 523 (7562): 576–579.

111 Xu, W., Liu, J., Chen, T. et al. (2019). Bioinspired polymer nanocomposites exhibit giant energy density and high efficiency at high temperature. *Small* 15 (28): 1901582.

112 Fan, M., Hu, P., Dan, Z. et al. (2020). Significantly increased energy density and discharge efficiency at high temperature in polyetherimide nanocomposites by a small amount of Al_2O_3 nanoparticles. *Journal of Materials Chemistry A* 8 (46): 24536–24542.

113 Zhang, T., Chen, X., Thakur, Y. et al. (2020). A highly scalable dielectric metamaterial with superior capacitor performance over a broad temperature. *Science Advances* 6 (4): eaax6622.

114 Wang, Y.F., Cui, J., Yuan, Q.B. et al. (2015). Significantly enhanced breakdown strength and energy density in sandwich-structured barium titanate/poly (vinylidene fluoride) nanocomposites. *Advanced Materials* 27 (42): 6658–6663.

115 Li, Q., Liu, F.H., Yang, T.N. et al. (2016). Sandwich-structured polymer nanocomposites with high energy density and great charge-discharge efficiency at elevated temperatures. *Proceedings of the National Academy of Sciences of the United States of America* 113 (36): 9995–10000.

116 Song, Y.N., Xiao, M., and Du, B.X. (2022). Effect of gamma radiation modification on crystallization and breakdown properties of polypropylene. In: *2022 IEEE 4th International Conference on Dielectrics (ICD)*, 721–724. IEEE.

117 Xiao, M., Song, Y.N., and Du, B.X. (2022). Effect of gamma radiation on the high-temperature breakdown strength of polypropylene films for capacitors. In: *2022 IEEE 4th International Conference on Dielectrics (ICD)*, 709–712. IEEE.

118 Yang, T., Cheng, Y., Wu, Y. et al. (2022). Enhanced crosslinking of polypropylene in γ-irradiation via copper (II) doping. *Radiation Physics and Chemistry* 194: 110042.

119 Yang, R., Wei, R., Li, K. et al. (2016). Crosslinked polyarylene ether nitrile film as flexible dielectric materials with ultrahigh thermal stability. *Scientific Reports* 6 (36434) https://doi.org/10.1038/srep36434.

120 Yang, R., Xiao, Q., You, Y. et al. (2016). In situ catalyzed and reinforced high-temperature flexible crosslinked ZnO nano-whisker/polyarylene ether nitriles composite dielectric films. *Polymer Composites* https://doi.org/10.1002/pc.24272.

121 Yao, H.Y., Zhang, N., Shen, K.Z. et al. (2017). From a flexible hyperbranched polyimide to a microporous polyimide network: microporous architecture and carbon dioxide adsorption. *Polymer* 115: 176–183.

122 Yu, P., Wang, Y., Yu, J.R. et al. (2017). Development of novel cardo-containing phenylethynyl-terminated polyimide with high thermal properties. *Polymers for Advanced Technologies* 28: 222–232.

123 Duan, G., Hu, F., Zhang, R. et al. (2022). Preparation of a novel cross-linked polyetherimide with enhanced breakdown strength and high-temperature energy storage performance. *High Voltage*.

124 Majumdar, G. and Minato, T. (2007). Recent and future IGBT evolution. In: *Power Conversion Conference*, 355–359. Nagoya, Japan.

125 Thomas, S., Hiroji, E., and Francois, V. (2018). Silicone gels for continuous operation up to 200 °C in power modules. In: *International Exhibition and Conference for Power Electronics, Intelligent Motion, Renewable Energy and Energy Management*. Nuremberg, Germany: IEEE.

126 Ohara, M. (2018). Development of high temperature silicone gels. In: *PCIM Europe 2018; International Exhibition and Conference for Power Electronics, Intelligent Motion, Renewable Energy and Energy Management*, 1–2. VDE.

127 Li, Z.L., Zhou, H., Rouse, J. et al. (2015). Silicone protection solution for IGBT modules: from dielectric encapsulation to thermal man-agreement. In: *International Exhibition and Conference for Power Electronics, Intelligent Motion, Renewable Energy and Energy Management*. Shanghai, China: VDE.

128 Wang, N., Cotton, I., Robertson, J. et al. (2010). Partial discharge control in a power electronic module using high permittivity non-linear dielectrics. *IEEE Transactions on Dielectrics and Electrical Insulation* 17 (4): 1319–1326.

129 Rahate, A.S., Nemade, K.R., and Waghuley, S.A. (2013). Polyphenylene sulfide (PPS): state of the art and applications. *Reviews in Chemical Engineering* 29 (6): 471–489.

130 Xu, S., Yang, J., Long, S. et al. (2005). Synthesis and characterization of poly (p-phenylene sulfide sulfone/ketone) copolymer. *Polymer Bulletin* 54 (4): 251–261.

131 Stoeffler, K., Andjelic, S., Legros, N. et al. (2013). Polyphenylene sulfide (PPS) composites reinforced with recycled carbon fiber. *Composites Science and Technology* 84 (9): 65–71.

132 Xing, J., Xu, Z., Ruan, F. et al. (2019). No isothermal crystallization kinetics, morphology, and tensile properties of polyphenylene sulfide/functionalized graphite nanoplatelets composites. *High Performance Polymers* 31 (3): 282–293.

133 Lian, D., Dai, J., Zhang, R. et al. (2017). Effect of Ti–SiO_2 nanoparticles on non-isothermal crystallization of polyphenylene sulfide fibers. *Journal of Thermal Analysis and Calorimetry* 129 (1): 377–390.

134 Sproesser, O., Schmidlin, P.R., Uhrenbacher, J. et al. (2014). Effect of sulfuric acid etching of polyetheretherketone on the shear bond strength to resin cements. *The Journal of Adhesive Dentistry* 16 (5): 465–472.

135 Zhou, B., Ji, X., Yu, C. et al. (2003). Dynamic mechanical properties of $CaCO_3$/peek composites. *Polymeric Materials Science and Engineering* 89: 574.

136 Hanchi, J. and Eiss, N.S. Jr., (1994). The tribological behavior of blends of polyetheretherketone (PEEK) and polyetherimide (PEI) at elevated temperatures. *Tribology Transactions* 37 (3): 494–504.

137 Zhang, H., Li, X., Zhao, C. et al. (2008). Composite membranes based on highly sulfonated PEEK and PBI: morphology characteristics and performance. *Journal of Membrane Science* 308 (1–2): 66–74.

138 Liang, L., Jiang, B., Wu, T. et al. (2022). Preparation of glass fiber/poly (ether ether ketone) composite foam with improved compressive strength and heat resistance. *Advanced Engineering Materials* 24 (3): 2100940.

139 Li, E.Z., Guo, W.L., Wang, H.D. et al. (2013). Research on tribological behavior of PEEK and glass fiber reinforced PEEK composite. *Physics Procedia* 50: 453–460.

8

High-Temperature Polymer Dielectrics for Aerospace Electrical Equipment

Daomin Min, Xiaofan Song, Lingyu Yang, Yuanshuo Zhang, Shihang Wang, and Shengtao Li

Xi'an Jiaotong University, State Key Laboratory of Electrical Insulation and Power Equipment, Gaoqiao Street, Xi'an 710049, Shaanxi, China

8.1 Introduction

With the rapid improvement of global industrialization and people's living standards, energy consumption is also increasing, and energy shortage and environmental damage are becoming increasingly serious. Therefore, increasing the development and utilization of renewable energies such as solar energy and reducing the consumption of fossil fuels have become the future development trends. Space-based solar power stations can continuously collect solar energy and convert it into electric energy all day long, and they can supply power to ground power facilities through microwave or laser energy transmission. The development of space-based solar power stations can greatly alleviate the energy crisis while comprehensively driving the progress of aerospace technology. In addition, in order to reduce the dependence on fossil fuels and reduce the emissions of aircraft greenhouse gases and toxic exhaust, all-electric aircraft technology has become the forefront of global challenges and has been selected as the "Top Ten Potential Technologies in 2020" by "Scientific American" [1].

As shown in Figure 8.1, the power and electronic equipment used in satellites and all-electric aircraft include solar panels, power cables, solar array drive assembly (SADA), motors, power electronic devices, capacitors, switchgear, electronic devices, etc. These power and electronic equipment need to use a large number of high-performance insulating materials, such as polyimide (PI), polytetrafluoroethylene (PTFE), polyetheretherketone (PEEK), silicone rubber (SiR), and epoxy resin (EP).

PI is a standard type of imide material and one of the earliest polymer materials used in motor insulation. PI has good high and low temperature resistance and radiation resistance, as well as excellent mechanical properties, dielectric properties, and flame retardancy [4]. PI insulating paint has been widely used in the field of motor insulation due to its excellent, comprehensive performance. Solar sails and solar arrays also have a large demand for PI films. In addition, PI film is also used

High Temperature Polymer Dielectrics: Fundamentals and Applications in Power Equipment, First Edition. Edited by Jun-Wei Zha and Zhi-Min Dang.

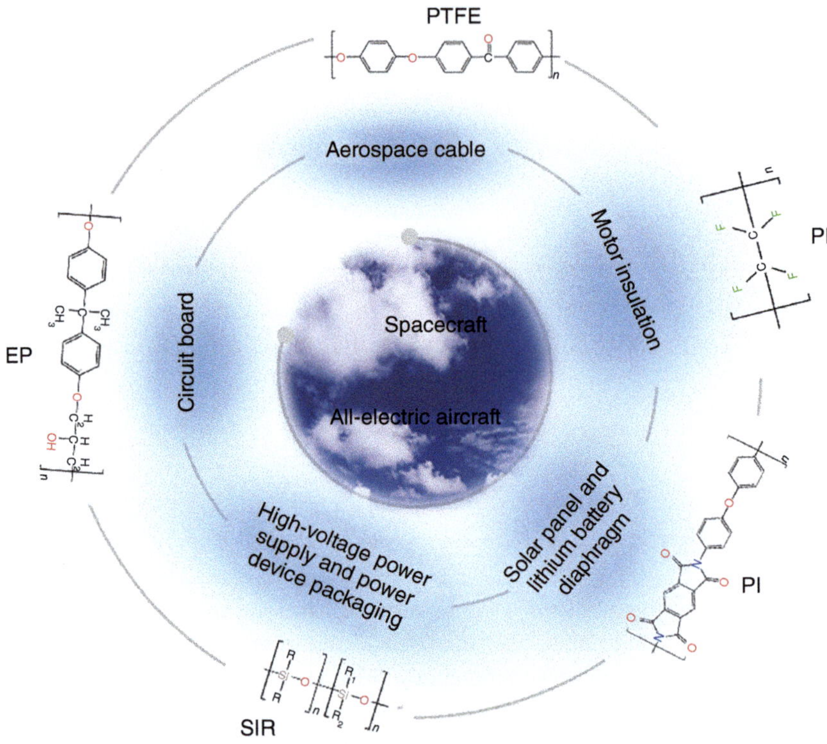

Figure 8.1 Typical power and electronic equipment and commonly used high-temperature-resistant insulating materials on spacecraft and all-electric aircraft. Source: [2, 3]/with permission from Elsevier; /with permission from MDPI/CC BY 4.0.

as a separator material in lithium batteries [5, 6]. DuPont, UBE Corporation, and Kaneka Corporation have developed various Kapton® PI films. It has set the industry standard in terms of high performance, reliability, and durability. Kapton films have unique electrical, thermal, chemical, and mechanical properties and can withstand extreme temperatures, vibration, and other harsh operating environments.

PTFE has excellent thermal properties in addition to its stable chemical and dielectric properties, and its continuous working temperature reaches 260 °C. In addition, PTFE has a low dielectric loss factor, good stability in a wide frequency range, high volume resistivity, and high breakdown strength [1]. Therefore, PTFE is often used as the external insulating material of aerospace cables. Gore Company has been committed to the research of multifunctional polymers such as PTFE and expanded polytetrafluoroethylene (ePTFE) and has developed and manufactured various GORE® Aerospace cables that can provide stable power and data transmission in harsh environments.

PEEK is a kind of semi-crystalline polymer with ether bond and ketone carbonyl group in its molecular structure. It has excellent electrical insulation properties, chemical resistance, radiation resistance, hygrothermal stability, and high strain bending resistance. In addition, the semi-crystalline structure of PEEK enables it to have the hot melting characteristics that other traditional dielectric materials, such as PI, do not have. So, it can be directly thermally bonded with copper wires at high temperatures to obtain insulating products [4]. Therefore, PEEK is often used to replace the traditional PI dielectric in the motor coil manufacturing process. PEEK is also often used as insulation for aerospace ultrahigh-temperature wires. At present, the common brands on the market are TECAPEEK® And Ketron®, which are the most popular advanced engineering plastic materials and are widely used in the aerospace field.

SiR is a semi-organic and semi-inorganic polymer material. The main chain of the molecule is composed of Si-O-Si links. Its bond angle and bond energy are large, and its orientation freedom is high. SiR has excellent high and low temperature resistance, ozone resistance, weather resistance, and electrical insulation. Therefore, SiR is widely used in the packaging of high-voltage power supplies and power electronic devices. According to statistics, a large civil aviation airliner uses an average of more than 500 kg of SiR [7, 8]. Brands such as ELASTOSIL® and WACKER® have produced various types of SiR, which are widely used in aviation and aerospace industries.

EP is an important thermosetting resin that is widely used as the resin matrix of high-performance insulation composites because of its good heat resistance, solvent resistance, chemical corrosion resistance, and excellent electrical and mechanical properties [9]. EP is commonly used to make circuit boards. The CYCOM® EP series polymers researched and produced by SOLWAY can replace engine parts, reduce weight by 50%, and work stably at high temperatures [10]. At present, the temperature resistance of EP is mainly improved by improving its functionality, increasing the rigidity of the main chain, and introducing a large volume of high-temperature-resistant structures such as adamantane or carbon fiber [11].

As shown in Table 8.1, the working environment on spacecraft and all-electric aircraft is relatively harsh, with high temperature, high voltage, and high-energy

Table 8.1 Service Conditions of Satellite and All-Electric Aircraft [12, 13].

Type of aerospace vehicles Service conditions	Satellite (space solar power station)	All-electric aircraft
Temperature	Sunward surface temperature: 100 °C Surface temperature of sunny side: −100 to −200 °C	Skin temperature: 100–350 °C Motor temperature: 180–300 °C
Voltage amplitude	Transmission voltage: 5 kV (in 2030) Transmission voltage: 20 kV (in 2050)	NASA X-57 Li-ion battery output voltage: 400–525 V Battery output voltage of an all-electric drone: 700–1000 V
Voltage frequency	DC	0–400 Hz
High-energy electron irradiation	Energy range: 10 keV–10 MeV Electron flux: 10^7–10^8 e cm^{-2} s^{-1}	—

electron irradiation. This brings great challenges to the insulating materials and insulation structures in power equipment. When a spacecraft flies in space, the side directly exposed to sunlight can produce a high temperature of more than 100 °C, while the temperature of the shady side can be as low as −100 to −200 °C. When the spacecraft reenters the atmosphere, it travels through the dense atmosphere at a high speed and rubs violently with air molecules, making the surface temperature of the spacecraft as high as about 1000 °C. Moreover, with the increase in power generation and current density, the working temperature of an aeromotor can reach 250 °C in the long term, 290 °C in the short term, and 420 °C in the instant [4].

Insulating materials can withstand extremely high voltages while enduring extremely high temperatures. As the power of space-based solar power stations increases, the transmission voltage should also increase accordingly. It is predicted that by 2030, the transmission voltage of space-based solar power stations will be increased to 5 kV, and by 2050, the transmission voltage will be increased to 20 kV [14]. Besides, the battery voltage of NASA X-57 all-electric aircraft can reach 400–525 V, and that of an all-electric drone can reach 700–1000 V. The electric engine of an all-electric aircraft is driven by an electronic control system, and the voltage frequency is in the range of 0–400 Hz. The future all-electric aircraft needs more energy and greater power and may need to use higher voltage and higher frequency. In addition, there are various space irradiations, such as space plasma and high-energy electron, high-energy proton, γ-ray, and X-ray irradiations. The energy of high-energy electrons in space is between 10 keV and 10 MeV, among which those below 50 keV will cause surface charging and unequal charging, and those above 50 keV will cause internal charging of the insulating materials. Space plasma will mainly cause surface charging [12, 13].

Such a working environment puts forward very strict requirements for the high temperature resistance and electrical insulation performance of dielectric materials and insulation structures. Not only in aerospace, with the continuous development of electrical insulation technology toward high temperature, the demand for high-temperature-resistant polymer dielectrics (HTPDs) is increasingly urgent. HTPDs are of great significance for realizing miniaturization, lightweight, and large capacity of electrical equipment and are one of the key development directions for insulating materials in the future [4].

8.2 Challenges of Insulating Materials Under High Temperatures

8.2.1 Substantial Drop in Resistivity Under High Temperatures and Strong Electric Fields

The energy band structures of the crystalline region and the amorphous region in the semi-crystalline polymer are different. Potential barriers will be formed between the crystalline region and the amorphous region for charge carriers, as shown in Figure 8.2a. The charges will absorb energy from the heat bath and jump over the

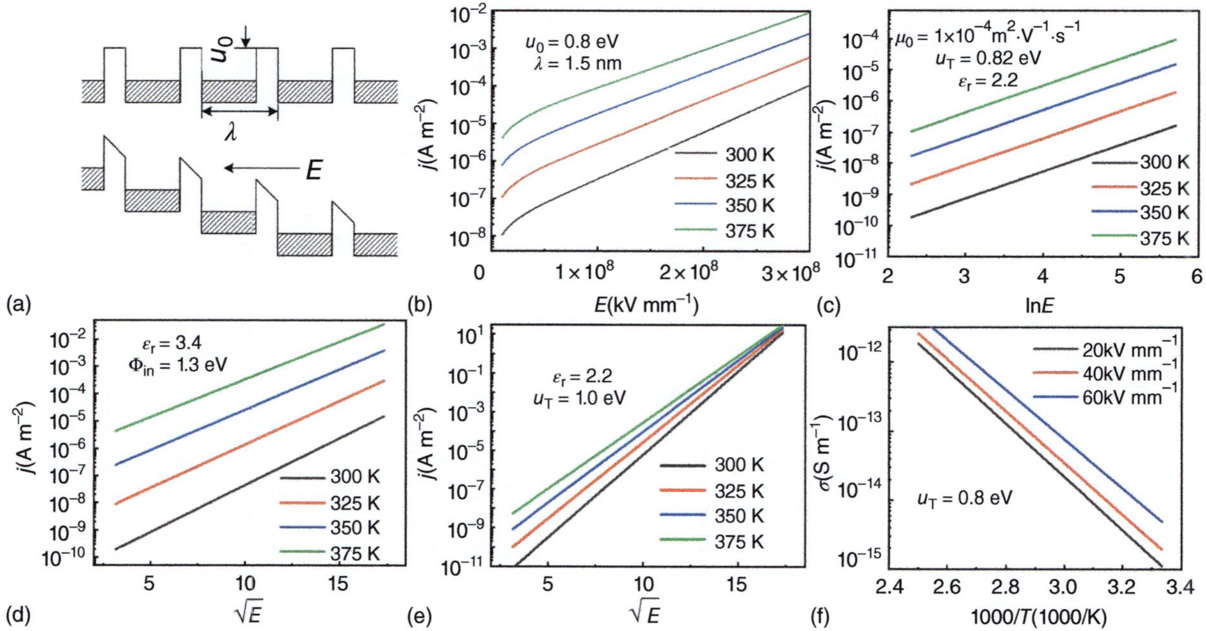

Figure 8.2 (a) Potential barrier diagram of hopping conductance model, (b) current density as a function of the electric field of hopping conductance model, (c) ln*j* as a function of ln*E* of space charge limited current model, (d) ln*j* versus $\sqrt{E}$ of Schottky thermionic emission, (e) ln*j* versus $\sqrt{E}$ of Poole–Frankel conductivity, (f) Arrhenius relationship of conductivity calculated by hopping conductance model with temperature.

potential barrier with a certain probability. When there is an applied electric field, the potential barrier will tilt to one side, and the charges will jump over the potential barrier in a directional manner and form a current. The change of current density j_{hop} of hopping conductance model with temperature T and electric field E satisfies the following equation. [15, 16],

$$j_{hop} = j_0 \exp\left(-\frac{u_0}{k_b T}\right) \sinh\left(\frac{e\lambda E}{2k_b T}\right) \tag{8.1}$$

where e is the elementary charge quantity, j_0 is the exponential prefactor, u_0 is the hopping barrier, λ is the average hopping distance, and k_b is the Boltzmann constant. The relationship between j_{hop} and E at different temperatures is shown in Figure 8.2b.

When the electric field is high, the amount of charge injected into the dielectric from its electrodes is far greater than the amount of charge leaked from the dielectric, and space charges will accumulate in the dielectric. These space charges will change the electric field distribution inside the dielectric, reduce the amount of injected charges, and make the current density inside the dielectric reach balance. Therefore, space charge-limited current, which is a power function of the applied voltage, will be formed. The temperature will affect the current density of the dielectric by affecting the detrapping probability of trapped charges. In a dielectric with a single-level trap, the relationship between space charge-limited current density j_{SCL} and voltage satisfies the following eq. [17],

$$j_{SCL} = \frac{9}{8}\frac{U^2}{d^3}\varepsilon_0\varepsilon_r\mu_0 \exp\left(-\frac{u_T}{k_b T}\right) \tag{8.2}$$

where ε_0 is the vacuum dielectric constant, ε_r is the relative permittivity of the dielectric, U is the applied voltage, d is the dielectric thickness, μ_0 is the carrier mobility of the extended state, and u_T is the trap level. Figure 8.2c shows the relationship between $\ln j_{SCL}$ and $\ln E$ obtained from the space charge-limited current model.

When the potential barrier between the electrode and the dielectric is high, the rate of charge injection from the electrode into the dielectric is less than the rate of charge migration in the dielectric. At this time, the volt-ampere characteristic of the electrode-dielectric system is limited by the charge injection. Under strong electric fields, the potential barrier at the interface between electrode and dielectric will be lowered by the Schottky effect, forming the thermionic emission current. Schottky thermionic emission current density j_{Sch} satisfies the following equation [17],

$$j_{Sch} = AT^2 \exp\left[-\frac{(\Phi_{in} - \sqrt{e^3E/4\pi\varepsilon_0\varepsilon_r})}{k_b T}\right] \tag{8.3}$$

where A is the Richardson constant and Φ_{in} is the charge injection barrier at the interface between electrode and dielectric. Figure 8.2d shows the variation trend of Schottky current density with $\sqrt{E}$.

The similar phenomena of electrode interface potential barrier lowering and thermionic emission current caused by the applied electric field also exist inside the dielectric, which is called the Poole–Frenkel effect. The Poole–Frenkel conductivity σ_{PF} obeys the following equation [17],

$$\sigma_{PF}(E,T) = \sigma_0 \exp\left[-\frac{u_T - \sqrt{e^3E/\pi\varepsilon_0\varepsilon_r}}{k_bT}\right] \tag{8.4}$$

where σ_0 is the exponential prefactor. Figure 8.2e shows the trend of the Poole–Frenkel conductivity with $\sqrt{E}$.

The temperature dependence of electrical conductivity is generally considered to follow the Arrhenius equation, and the equation is as follows [18],

$$\sigma = \sigma_0 \exp\left(-\frac{u_T}{k_bT}\right) \tag{8.5}$$

where σ_0 is the prefactor. Figure 8.2f is the Arrhenius curve for electrical conductivity with increasing temperature.

Through the analysis of the above models, it can be seen that with an increase in the applied electric field, the polymer gradually shows nonohmic conductivity. With the increase in temperature, the electrical conductivity rises sharply, and the conductance loss also increases greatly.

8.2.2 Greatly Increased Dielectric Loss Factor at High Temperatures

Ions produced by ionization in the dielectric at high temperatures will increase, and ions will be transported in the dielectric via hopping. When the applied electric field tilts the potential barrier, the ions will jump directionally. At this time, the number of ions in adjacent positions will be different, forming a dipole moment, that is, hopping polarization, as shown in Figure 8.3a. When the frequency of the applied electric field is close to the hopping polarization frequency, a relatively large dielectric loss will occur [19, 20]. The higher the temperature, the more ions will be generated by ionization, so the dielectric loss caused by hopping polarization will be larger. As shown in Figure 8.3b, the positive and negative charges generated by high-temperature ionization will also accumulate near the electrodes, forming the electrode polarization. This will cause the dielectric constant and dielectric loss factor of the dielectric to increase significantly with the decrease in voltage frequency [21, 22]. In addition, ions and electrons in the dielectric can migrate

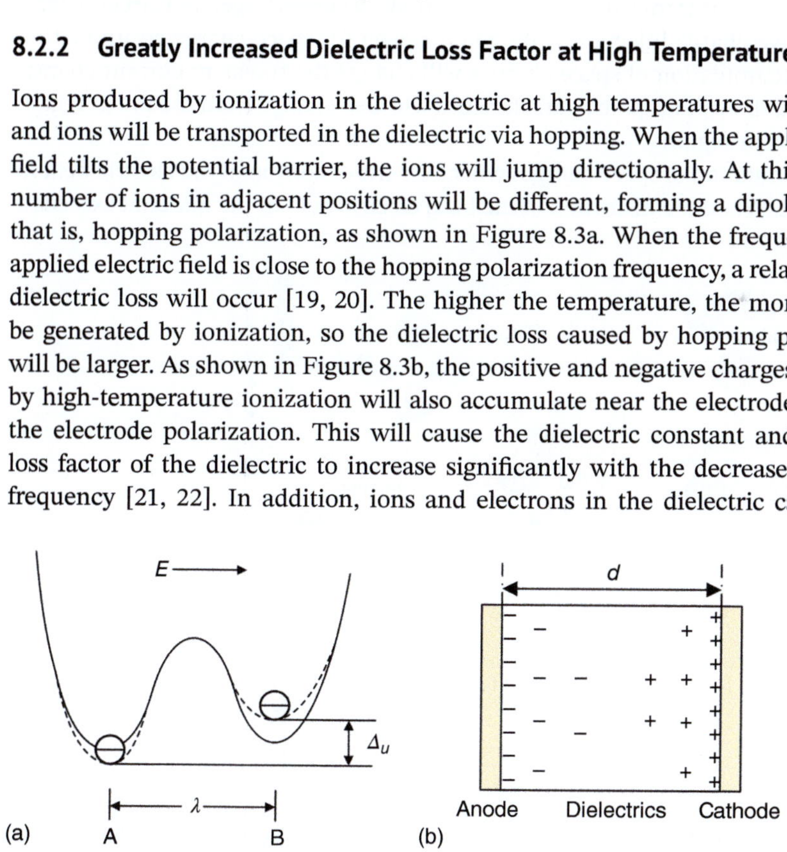

Figure 8.3 Schematic diagram of hopping polarization (a) and electrode polarization (b).

for a long distance to form conductance. At high temperatures, the conductance current increases dramatically, and the corresponding dielectric loss factor and dielectric loss angle tangent tanδ also increase exponentially with the increase in temperature [17].

8.2.3 Space Charge Accumulation and Electric Field Distortion Under High Temperatures and High-DC Electric Fields

As shown in Figure 8.4, under the electric field and heating, the charges in the electrode jump over the potential barrier at the interface between the electrode and the dielectric and are injected into the dielectric through Schottky thermionic emission or field emission. High-energy electron irradiation in space will also deposit charges in the dielectric. The charges injected into the dielectric will migrate or recombine under the electric field. When carriers migrate in the polymer, they can be trapped by deep traps in the material, resulting in a significant decrease in effective mobility. When the amount of charges injected is greater than the amount of charges leaked, it will cause space charge accumulation. With the increase in temperature or electric field, the space charge injection from the electrode becomes larger and larger. Therefore, excessive charge injection is the main reason for space charge accumulation [23]. The accumulation of space charges will change the transient current characteristics, and the volt-ampere characteristics will also change from the linear region to the nonlinear region.

The accumulation of space charges will cause the distortion of the electric field in the material. When the distortion of the local electric field in the material exceeds a

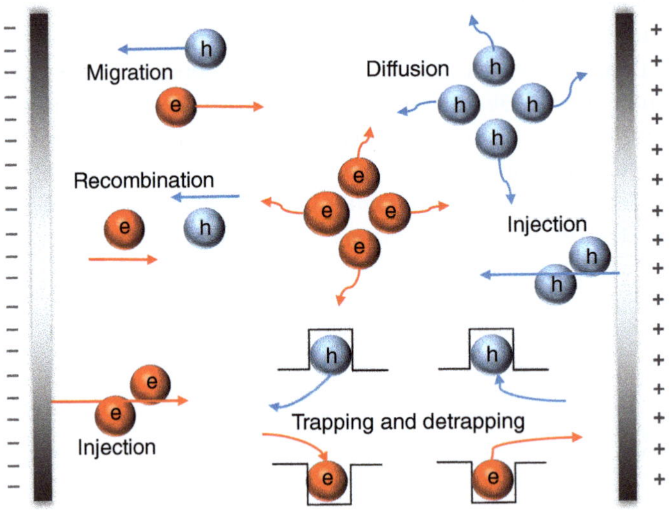

Figure 8.4 Schematic diagram of bipolar charge transport model and space charge distribution characteristics. The electrodes inject positive and negative charges into the dielectric, and they are captured by deep traps during the migration process inside the dielectric, accumulating space charges.

certain strength, the aging of the material at the nanometer level will be caused by electromechanical stress, energy transfer in the charge release process, or current multiplication in the extended states, which will form structural defects. Then, the degradation from nanometer level to micrometer level will occur, which will lead to breakdown or electrostatic discharge inside the material. Ultimately, it affects the efficient, safe, and reliable operation of electrical equipment on spacecraft and all-electric aircraft.

8.2.4 Reduction in Breakdown Strength Under High Temperatures and High-Frequency Voltages

Under high temperatures and high-frequency voltages, the breakdown strength of insulating materials decreases significantly. For solid dielectrics, there is mainly insulation damage caused by thermal breakdown, electrical breakdown, and partial discharge in nonuniform dielectrics.

Power equipment can generate a lot of heat during operation. If the heat is not dissipated in time, it will lead to hot spots in the dielectric. The thermal breakdown is caused by thermal instability in solid dielectrics. For the thermal breakdown model, the thermal breakdown voltage U_b of the solid dielectric can be determined through the Fourier heat conduction law [17],

$$U_b = \left(\frac{8\kappa T_0^2}{\sigma_0 \beta}\right)^{\frac{1}{2}} \frac{\exp(\beta/2T_0)}{d} \tag{8.6}$$

where κ is the thermal conductivity of the dielectric, β is the constant related to the activation energy of electrical conductivity, and T_0 is the ambient temperature. It can be seen that with the increase in ambient temperature, the thermal breakdown voltage of the dielectric will decrease significantly.

When the alternating voltage is applied to the dielectric, the effective resistivity of the solid dielectric is less than that under DC voltage. The effective resistivity of dielectric is inversely proportional to the dielectric constant ε_r, loss tangent $\tan\delta$, and voltage frequency f. The relationship between thermal breakdown strength E_{AC} and frequency can be obtained [17]:

$$E_{AC} = \frac{B}{\sqrt{\varepsilon_r f tg\delta}} \tag{8.7}$$

where B is a constant related to the material. With an increase in voltage frequency, the thermal breakdown voltage of dielectrics decreases rapidly.

The electric breakdown of solid dielectrics is a breakdown phenomenon observed during the breakdown test at a lower temperature and under the condition of strict control, such as the use of electrode devices to eliminate edge effects. The breakdown theory of solid dielectrics is based on collision ionization theory. The breakdown strength E_b obtained by the impact ionization with electron-trapped electron and electron–phonon scatterings is as follows [17]:

$$E_b = C\exp\left(\frac{\Delta u}{2kT_c}\right) \tag{8.8}$$

where C is the constant related to the material, T_c is the lattice temperature, and Δu is the shallow trap level. With the increase in lattice temperature T_c, the breakdown strength decreases. This is also consistent with the experiments showing that the electric breakdown strength of solid dielectric decreases with the rise of dielectric temperature.

8.2.5 More Severe Partial Discharge and Accelerated Aging at High Temperatures and High Frequencies

There may be air gaps in solid dielectrics. As the dielectric constant of solid dielectric is greater than that of gas, the electric field in the air gap will be higher than the average electric field under the applied alternating voltage. The breakdown strength of gas is generally lower than that of solid dielectric. Therefore, under a low electric field, localized breakdowns that do not penetrate the electrodes may occur in the solid dielectric, and they are called partial discharges. Although these partial discharges do not immediately form a penetrating channel, long-term discharge will lead to the gradual expansion of dielectric deterioration and damage, eventually leading to the breakdown of the entire dielectric. Under the applied AC voltage U, the average partial discharge times $\overline{n}$ in half a cycle can be obtained by graphical method as follows [17]:

$$\overline{n} = \frac{2U}{U_0} - 1 \tag{8.9}$$

where U_0 is the initial voltage of partial discharge.

It can be seen from the above equation that the discharge frequency under alternating voltage increases linearly with the applied voltage. If the frequency of the alternating voltage is set to f, the average number of discharges per second N and the average discharge energy per second W are as follows:

$$N = 2f\overline{n} \tag{8.10}$$

$$W = N\omega = 2fQ\left(U - \frac{1}{2}U_0\right) \tag{8.11}$$

where Q is the apparent discharge capacity, and ω is the energy of partial discharges.

The above equations show that the average time and energy of partial discharge are proportional to the voltage frequency. The partial discharges first generally erode the dielectric and change its weight. Then, the localized concentration of the discharge leads to the acceleration of the perforation speed in the local area. On the macroscopic level, it often shows the deterioration of the electrical performance of the dielectric and finally causes breakdown.

For the aging and life prediction of dielectrics, Dissado calculated the characteristic parameter values in the remaining life equation based on aging experiments and previous models. The model can be used to estimate the remaining life of insulating materials. Dissado's universal remaining life model is as follows [24, 25]:

$$L(E,T) = \frac{h}{2kT}\exp\left(\frac{G^{\#} - CE^{4b}/2}{kT}\right)\ln\left(\frac{A^{q} - A^{0}}{A^{q} - A^{*}}\right)\left[\cosh\left(\frac{\Delta - CE^{4b}}{2kT}\right)\right]^{-1} \tag{8.12}$$

where $G^{\#}$ is the activation energy of the material, A^{q} is the fraction of the total that changes when the reversible reaction reaches equilibrium, and A^{0} is the percentage of the total that bonds break before aging. In addition, C and b represent the relationship between the charge stored in the charge center and the applied electric field, and Δ is the free energy difference between reactants and products.

8.2.6 Reduction in Surface Flashover Voltage Under Electron Irradiations and Voltages

The vacuum surface flashover voltage of insulating materials under high-energy electron irradiation is much lower than that without irradiation, which is a technical challenge for the development of high-voltage and high-power spacecraft. The electric field distortion and surface potential increase in insulating materials will be caused by charging under the irradiation of high-energy electrons (Figure 8.5). When the internal distorted electric field reaches the material breakdown threshold or the potential gradient between different materials is high, the discharge or internal breakdown will occur along the material surface. The initial and developing processes of vacuum surface discharge are closely related to the dynamic characteristics of the surface and/or internal charges of the dielectric. Under the conditions of high electron irradiation and high working voltage, there is a relatively concentrated electric field distribution in some positions of insulating materials. The metal electrode at the electric field concentration will emit initial electrons to the insulating materials or vacuum through field emission or thermionic emission.

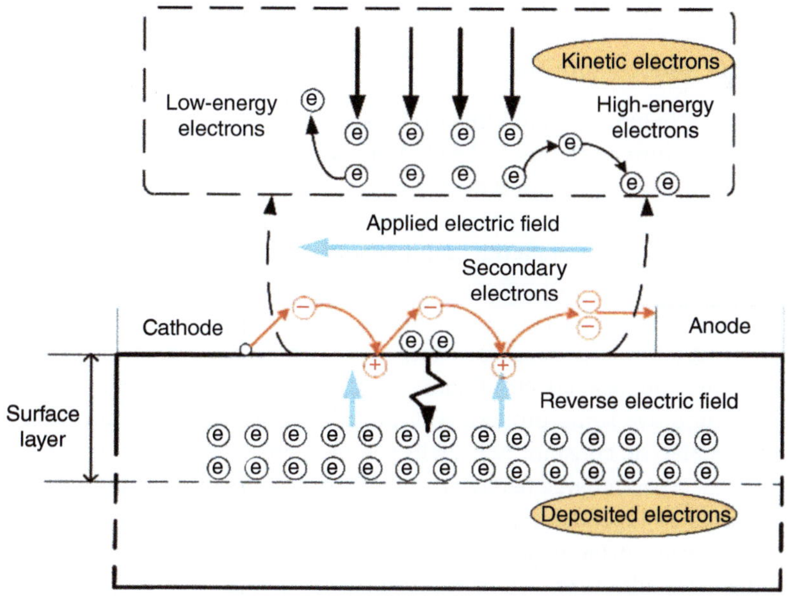

Figure 8.5 Vacuum surface flashover of insulating materials under high-energy electron irradiation and high working voltage. Source: Li et al. [26]. Reproduced with permission of IEEE.

The initial electrons impact the surface of the insulating material or desorbed gas molecules, causing the secondary electrons to multiply and finally form a penetrating conductive channel, which is called surface flashover.

The spacecraft operation failures show that the main discharge form of insulating materials under electron irradiation is the surface flashover. Under 100 V DC voltage and energetic electron irradiation, it may cause surface flashover of insulating materials between electrodes, and the increase in working voltage will further increase the probability of flashover of insulating materials. Therefore, it is necessary to study the suppression methods of surface flashover of insulating materials under high-energy electron irradiation and high working voltage.

8.3 High Temperature Resistant and Strong DC Insulating Polymer Dielectrics

Polymer materials have the advantages of high breakdown strength, low energy loss, high reliability, low cost, light weight, etc., and are ideal strong electrical insulating materials. Various polymers with high thermal stability have been used as high-temperature dielectrics, including PI, PEI, EP, and PTFE. However, due to the sharp increase in leakage current and conduction loss under high electric field and high temperature, it is urgent to design the structure of materials, regulate their resistivity, dielectric loss factor, breakdown strength, partial discharge, aging rate, and life, and prepare high-temperature, strong DC electrical insulating polymer dielectrics.

8.3.1 Polyimide Nanocomposites

The imide group and heterocycle structure of PI have stable properties, which make PI with high temperature resistance (>150 °C), moderate dielectric constant (3.5), low dielectric loss (<1% at 1 KHz), and high glass transition temperature (400 °C). It is an ideal material in the aerospace field. Polymer nanocomposite (PNCs) prepared by doping nano-fillers can integrate the high breakdown strength, ease of processing, and mechanical flexibility of polymers, and the unique thermal, mechanical, and electrical properties of inorganic fillers to improve the comprehensive performance of PNCs. Due to the large specific surface area, high surface activity, and some strange physical and chemical effects of nano-fillers, a small amount of nano-fillers can significantly improve some properties of polymer dielectrics, including electrical resistivity, dielectric constant and dielectric loss, breakdown strength, local discharge, dendritic, space charge behavior, and other dielectric properties.

Ai et al. prepared PI PNCs containing a series of inorganic nano-fillers (Al_2O_3, HfO_2, TiO_2, and boron nitride nanosheets [BNNS]) [15] and studied the electrical properties of the materials. As shown in Figure 8.6a, all PI PNCs exhibit low dielectric loss factors (<0.5%), which are fairly stable in the temperature range of 25 to 200 °C and the frequency range of 100 Hz to 1 MHz. For PNCs operating in a wide range of temperatures and frequencies, the weak dependence of dielectric properties

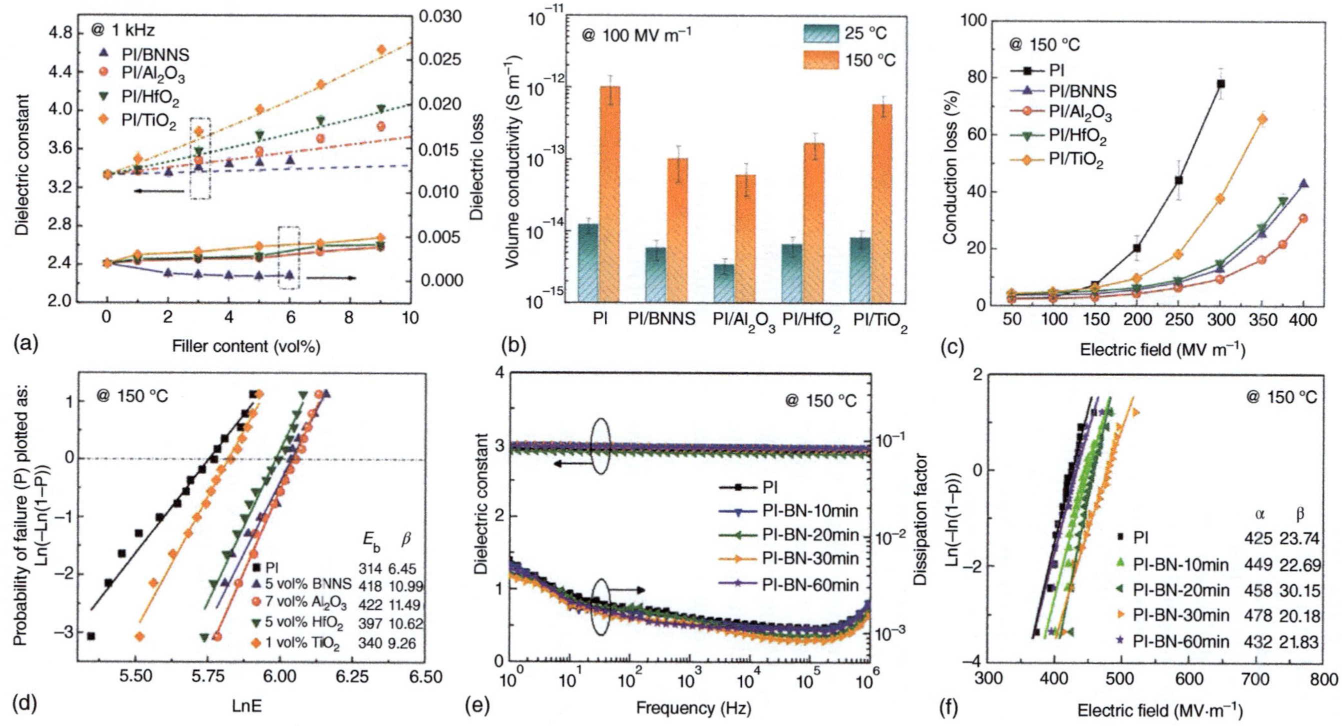

Figure 8.6 (a) The relative permittivity and dielectric loss factor of PI PNCs change with doping amount at 25 °C and 1 kHz, and the dotted line represents the calculated effective relative permittivity; (b) Volume conductivity of PI PNCs at 25 and 150 °C at 100 MV m^{-1}, respectively; (c) At 150 °C, the conductance loss of PI PNCs varies with the applied electric field (the doping content is 5 vol% for BNNS, 7 vol% for Al_2O_3, 5 vol% for HfO_2 and 1 vol% for TiO_2); (d) Weibull breakdown strength of PI PNCs at 150 °C; (e) Dielectric spectrum of PI and PI-BN at 150 °C; (f) Weibull breakdown field strength of PI and PI-BN at 150 °C. Source: [15, 27]. Permission conveyed by Wiley and IEEE.

on temperature and frequency is necessary. From Figure 8.6b, it can be clearly seen that the conductivity of the PI matrix is significantly reduced after the introduction of nano-fillers, which is particularly significant at high temperatures. At 25 °C and 100 MV m^{-1}, compared with pure PI, the volume conductivity of PI PNCs decreased by 33.4%, 47.0%, 53.2%, and 72.6% with the addition of TiO_2, HfO_2, BNNS, and Al_2O_3, respectively. At 150 °C, the corresponding decrease in conductivity is more obvious, which is 41.1%, 83.2%, 89.9%, and 93.9%, respectively. The analysis of experimental data shows that hopping conductance plays a dominant role in PNCs under high electric fields. At 150 °C, the hopping distance (λ) of pure PI is about 1.83 nm, and the λ of PI/TiO_2, PI/HfO_2, PI/BNNS, and PI/Al_2O_3 PNCs decreases to 1.69, 1.41, 1.38, and 1.25 nm, respectively. A reduction in the λ represents a weakening of the strong field effect of carrier mobility, which naturally leads to a reduction in conduction current and a reduction in energy loss. As shown in Figure 8.6c, the conductance loss of PI and PI PNCs increases exponentially with the increase in the applied electric field. After introducing nano-fillers into the PI matrix, the conductivity loss is significantly reduced, especially at high temperatures. The two-parameter Weibull statistical distribution function was used to analyze the breakdown strength (E_b) of PNCs at high temperatures (Figure 8.6d). Compared with pure PI, the E_b of PI PNCs is greatly improved. It shows that adding nano-fillers to polymer matrix can effectively restrain high-temperature leakage current and improve insulation performance. In addition, the increase in the resistivity of PI PNCs is also positively related to the increase in the energy band gap of nano-fillers, which is consistent with the breakdown strengths.

Cheng et al. deposited a thin layer of BN on the surface of PI polymer by magnetron sputtering (a common physical vapor deposition method) [27]. The cross-section of the deposited film shows that the BN layer is compact and uniform without visible defects. The longer the deposition time, the thicker the BN layer is. The BN thicknesses at 10, 20, 30, and 60 minutes deposition times are about 56, 91, 142, and 254 nm, respectively. It can be seen from Figure 8.6e that the relative permittivity of PI-BN film is close to that of pure PI. Because the relative permittivity of BN is similar to that of PI, and the coating is very thin, the overall dielectric constant of PI-BN film cannot be changed. The dielectric loss factor of PI-BN films with different BN thicknesses is slightly lower than that of pure PI. Under the applied electric field, the dielectric loss factors of the samples are very low at high temperatures. Figure 8.6f shows that the E_b of PI-BN films with different BN thicknesses is higher than that of pure PI. It shows that BN coating is beneficial in restraining conduction loss and leakage current, reducing local electric field distortion and space charge accumulation, and improving the breakdown strength.

8.3.2 Polyetherimide Nanocomposites

PEI is the general name of a class of organic polymers that introduce ether bond into the molecular chain by modifying PI. On the one hand, PEI retains the excellent mechanical properties and long-term heat resistance of PI. On the other hand, the

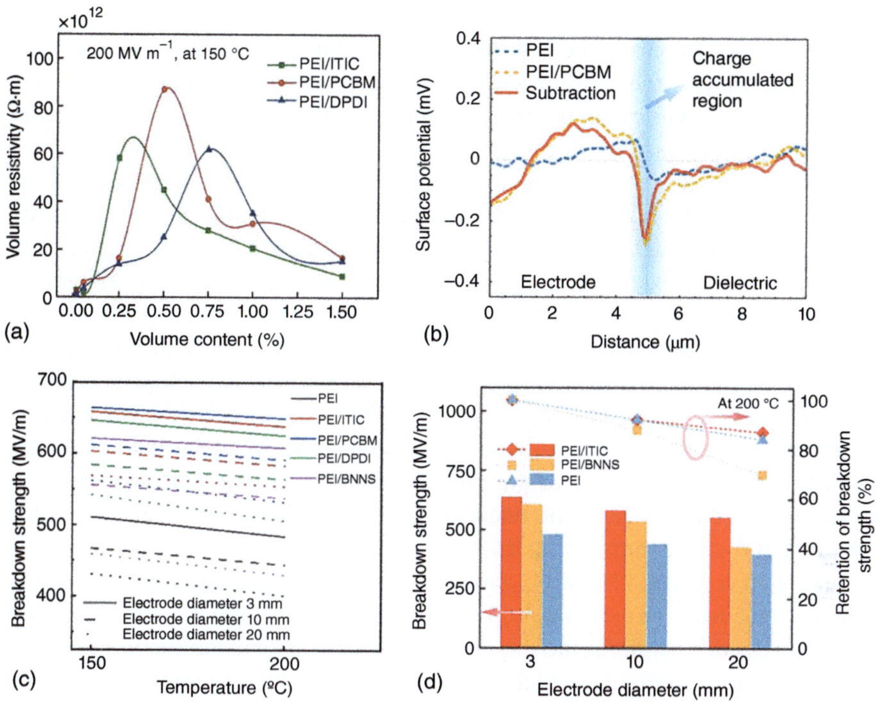

Figure 8.7 (a) The relationship between volume resistivity and filler concentration of PEI PNCs at 200 MV/m and 150 °C; (b) Surface potential profile along the horizontal axis; (c) The Weibull breakdown strengths of PEI, PEI/ITIC (0.25 vol%), PEI/PCBM (0.5 vol%), PEI/DPDI (0.75 vol%), and PEI/BNNS (10 vol%) as a function of temperature between electrodes with diameters of 3, 10, and 20 mm; (d) At 200 °C, the characteristic breakdown strength and retention of PEI, PEI/ITIC (0.25 vol%) and PEI/BNNS (10 vol%) as a function of electrode diameter. Source: Yuan et al. [28]. Permission conveyed by Nature Communications.

existence of ether bond makes PEI have better molding and processing properties, and it is widely used in the aerospace field.

Yuan et al. used PEI as a heat-resistant dielectric polymer matrix and selected three commercial semiconductor particles with thermal stability below 300 °C, namely ITIC, PCBM, and DPDI, as fillers to prepare PNCs [28]. The surface potential was measured by the Kelvin Probe Force Microscope with spatial resolution to study the effect of semiconductor fillers on charge injection and transport. As shown in Figure 8.7a, the resistivity of all organic composites (PEI/ITIC, PEI/PCBM, and PEI/DPDI) under high-temperature environment is significantly higher than that of pure PEI. This shows that the semiconductor particles introduce a deep trap in the composite. At a very low volume concentration, the semiconductor particles with high electron affinity can capture most of the free electrons in the composite. And because the distribution density of capture sites increases, the resistivity increases with the increase in semiconductor concentration. With the further increase in particle concentration, the average distance between adjacent semiconducting particles becomes smaller and smaller, and the electron conduction probability of excited charge carriers becomes significant, which starts to dominate the conductivity. Therefore, the best performance of a semiconductor at a specific concentration may be the result of the competition between the above two factors. A larger band gap will cause the peak to shift to a higher volume concentration because the charge carriers need higher energy to be excited from the valence band to the conduction band, so the probability of electron conduction at the same volume concentration is low. Figure 8.7b is the surface potential profile along the horizontal axis obtained from the surface potential map. The sharp decrease in the number of electrons toward the dielectric side (the increased surface potential) indicates that the injected electrons form a space charge accumulation area near the electrode/dielectric interface and establish a built-in electric field in the opposite direction to the applied electric field. This will inhibit further net electron inflows. The insulation performance was analyzed when the electrode diameter increased at high temperatures (Figure 8.7c,d). When the test electrode diameter increases from 3 to 20 mm, the retention rate of the PEI/ITIC characteristic breakdown strength is 87%, which is equivalent to the retention rate of pure polymer films (83%). This shows that the semiconducting particles have no obvious adverse effect on the film quality. In contrast, PEI/BNNS shows only 70% breakdown strength retention, which may be due to the existence of particle agglomeration and imperfect polymer particle interface. Therefore, the dielectric polymer and semiconducting particles have good mutual solubility, which helps to enhance the advantages of PNCs in forming large-area, high-quality films.

8.3.3 Epoxy Resin Nanocomposites

EP is a kind of organic polymer with more than two epoxy groups in its molecular structure. It can cross-link with various curing agents to form a polymer with a three-dimensional network structure. It has excellent insulating properties, mechanical properties, and chemical stability.

Yang et al. improved the comprehensive properties of EP composites by compounding micron diamond (CBN) and micron alumina (Al_2O_3) [29]. When the Al_2O_3/EP microcomposites are in a uniform electric field, the internal electric fields are distorted at the interfaces between the inorganic filler and the epoxy matrix. The additional electric field generated by the interfacial charges causes the interface electric field distortion. The maximum distortion rate of the electric field at the interface between Al_2O_3 and EP is 31%. In addition, with the increase in applied electric field, the free charges in the composite migrate, and their migration path is mainly along the interface. Due to the difference in the energy band structures of the interfaces, the free charge migration from the matrix to the filler must be greater than the barrier height. In this case, space charge accumulation will occur at the interface, which further distorts the electric field of the interface. Diamond has an energy band structure similar to that of traditional inorganic fillers (Al_2O_3, SiO_2, and BN), and its Fermi level is higher than that of EP. This feature significantly reduces the barrier height of electrons transferred from the matrix to the diamond, which can accelerate the migration of free charges and reduce the accumulation of space charges. Figure 8.8a shows the E_b of the EP composites. With the increase in doping amount, the E_b of CBN/EP composites increases and then slowly decreases, while the E_b of Al_2O_3/EP composites gradually decreases with the increase in doping amount. In addition, the E_b of CBN/EP composites is significantly increased at different temperatures (Figure 8.8b). Figure 8.8d,e shows that with the increase in electric field, the surface charge density and electric field intensity of the two epoxy composites show the same change pattern, but the surface charge density distribution and electric field distribution are different. The surface charge density and electric field intensity of CBN/EP composites are uniformly distributed, while the surface charge density and electric field intensity of Al_2O_3/EP composites are concentrated. In addition, the interface charge density of CBN/EP composites is significantly lower than that of Al_2O_3/EP composites, which indicates that diamond can effectively reduce the distortion of the interface electric field. Figure 8.8c shows that the change rule of dielectric loss of EP composites with frequency and temperature is consistent with that of pure EP, and the dielectric loss of CBN/EP composites at high temperature is lower than that of Al_2O_3/EP composites.

8.3.4 Polytetrafluoroethylene Composites

PTFE is a nonpolar linear polymer that is insoluble in almost all solvents. It has excellent heat and cold resistance, thermal stability up to 260 °C, and can be used for a long time at −180 to 260 °C. Since it was discovered in 1938, it has been widely used in aerospace and other fields. But it is difficult to reduce the thickness of PTFE to 6 μm below. In order to realize the lightweight of power equipment, Luo et al. prepared PTFE film with a thickness of less than 5 μm via a suspension containing PTFE nanoparticles [30]. The heat treatment of the films has been strictly optimized to obtain dense microstructures and smooth surfaces. The particles are about 200 nm (Figure 8.9a). When the heating temperature reaches 330 °C and is close to the melting point of 336 °C, the PTFE particles soften and begin to coalesce with their adjacent particles (Figure 8.9b). After heating at 330 °C for 40 minutes, many adjacent PTFE particles coalesce, and their boundaries are almost invisible. As the heating temperature further rises to 350 °C (Figure 8.9c) and 380 °C (Figure 8.9d) and lasts for 40 minutes, all boundaries among PTFE particles disappear and particles change into interlocking fibers. With the increase in temperature, PTFE particles absorb heat, and their surfaces become soft. The particles tend to close together to form a network and then fully connect with each other to form fibers.

Based on the SEM results, the schematic diagram of the PTFE film formation process is shown in Figure 8.9e, in which the PTFE particles are melted, expanded, and flattened in turn, and finally form a dense film. Figure 8.9f shows that the Weibull breakdown strength of PTFE film (α_b) With the increase in heat treatment time, it first increases and then decreases, which can be attributed to the fact that the extension of heat treatment time contributes to the formation of dense microstructure. But if the heat treatment time is too long, PTFE particles will form a PTFE film with an interlocking fiber structure, and there will be voids and many micropores on the surface. This is also indicated in the surface area difference (S_d),

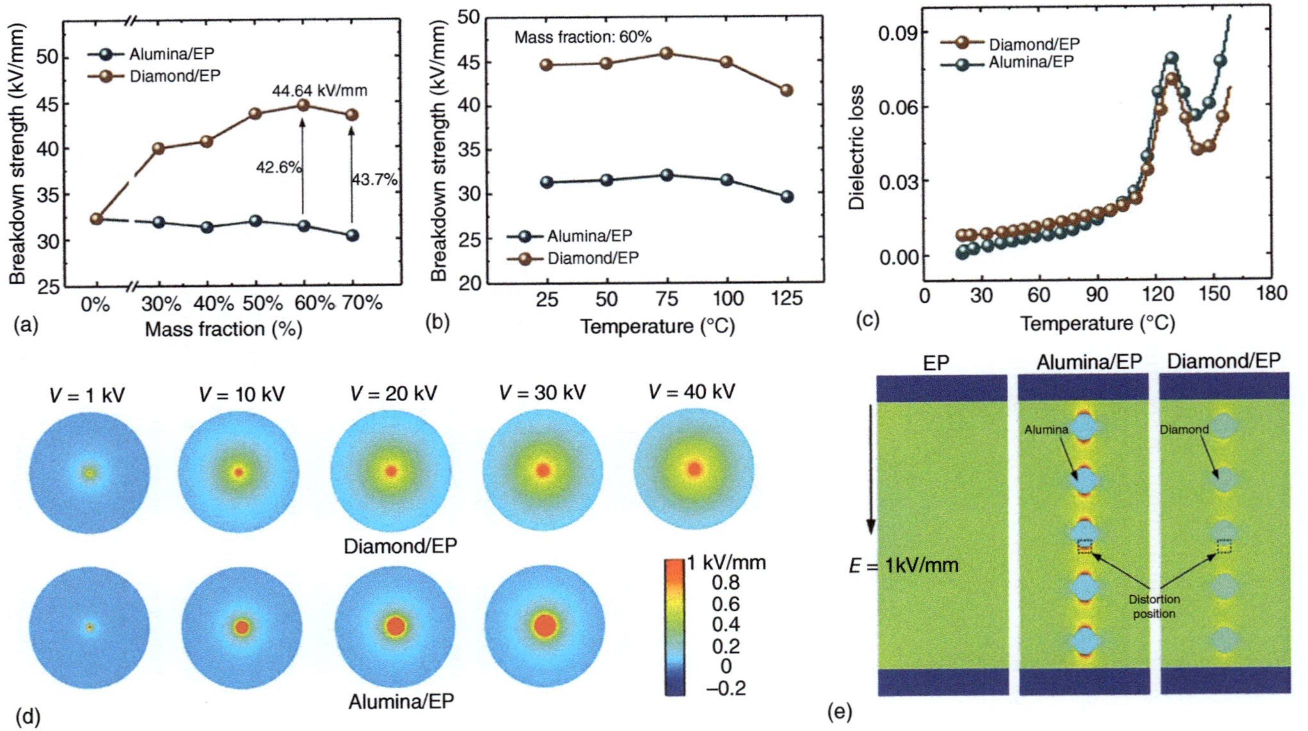

Figure 8.8 Charge density and electric field distributions at the interfaces between Al_2O_3 and matrix (a), breakdown strength versus temperature (b), temperature dependence of dielectric loss (c), surface electric field and electric field distortion (d), and electric field distributions of filler/epoxy interface (e) in EP composites. Source: Yang et al. [29]. Permission conveyed by Elsevier.

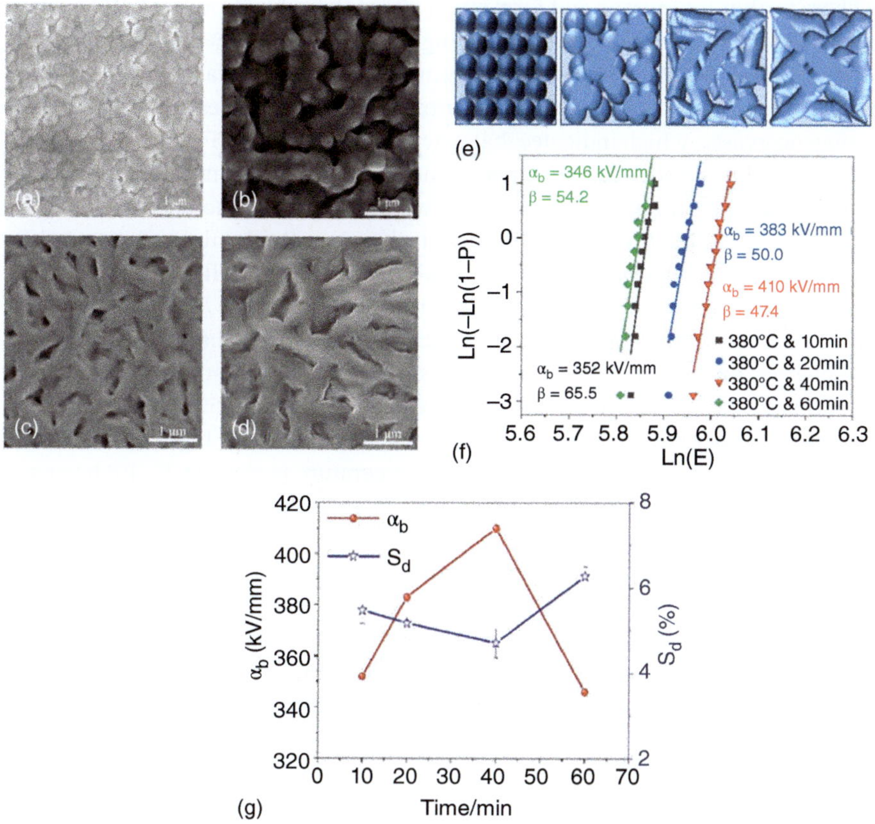

Figure 8.9 Microstructure formation process and breakdown strength measurement of PTFE films. (a) The SEM image of dry PTFE coating on copper foil before heat treatment, (b–d) the SEM images of PTFE films processed at 330, 350, and 380 °C for 40 minutes, (e) structural evolution of PTFE membrane, (f) the Weibull breakdown strength parameter distribution of PTFE films after treatment at 380 °C for different times, and (g) the characteristic breakdown strength and surface area difference versus treatment time. Source: Luo et al. [30]/with permission from Elsevier.

reflecting the surface flatness (Figure 8.9g). Space charges tend to accumulate at the interface between PTFE and electrodes, and more space charges accumulate at the interface with higher S_d. This means that under the same external electric field, the films with higher S_d are easier to break down. In addition, the Weibull breakdown strength of PTFE film at 25, 150, and 200 °C shows that the breakdown strength obtained in the second and third tests is almost the same as the original value regardless of the temperature, which means that the prepared PTFE film has excellent self-healing ability.

In order to reduce the surface roughness of PTFE film, Luo et al. used EP to fill the depression on the surface of PTFE film with its good fluidity and high dielectric constant [31]. There is an interlocking fiber structure on the surface of the initial PTFE film. During the heat treatment, the PTFE particles connect to form fibers;

the fibers further interweave and form small pits at the submicron level. PTFE films were impregnated with solutions containing 0.25, 0.5, and 1 wt% EP. There are pits on the surface of P-0.25%E film, and no pits are found on the surfaces of P-0.5%E and P-1%E. The results show that as the concentration of EP increases, α_b first increases and then decreases, which indicates that when the pits on the PTFE surface are filled with EP, the film surface becomes smooth and the charges accumulated at the interface decrease, thus improving the breakdown strength. But the decrease in α_b of P-1%E film means that the low α_b of EP coating becomes the dominant factor. A suitable EP coating is an important factor in obtaining high α_b. β_s, reflects the dispersion of the E_b, and the higher the β_s, the smaller the dispersion of the E_b. Compared with pure PTFE, the surface of EP-PTFE film becomes flatter, and the negative effect of space charges on the breakdown strength is reduced, which causes the E_b dispersion of the test to decrease and the β_s to increase. The α_b of P-0.5%E films decreases faster with the increase in temperature, but the α_b value is still higher than that of PTFE film until 150 °C. When the ambient temperature is close to T_g, EP becomes soft and does not have breakdown resistance. Therefore, selecting a polymer with a higher T_g to coat the surface defects of the PTFE film can further improve α_b at high temperatures. Using finite element modeling, the electric field strength at the bottom of the PTFE film pit is much higher, and the sharp increase in local electric field value may be the key reason for polymer breakdown. EP coating makes the connection between electrode and polymer smooth, so EP-PTFE film will have higher breakdown strength. As the dielectric loss factor of EP is relatively high (0.01 at 1 kHz), the dielectric loss factor of EP-PTFE film shows a small increase compared with PTFE film at high temperatures.

8.4 High-temperature-Resistant Polymer Dielectrics with Strong Nonlinear Conductivity

High-voltage DC power equipment in space-based power stations deposits charges under high-energy electron irradiation. The discharge under the effect of deposited charges and working voltage will accelerate the aging and deterioration of insulation materials, even leading to insulation failure, which seriously threatens the safe operation and service life of spacecraft. Nonlinear composite dielectrics have the functions of self-homogenizing the distribution of electric fields, inhibiting the accumulation of surface or space charges, and improving their ability to resist internal discharge of dielectrics, thereby improving the working voltage of equipment.

8.4.1 Charge Accumulation and Electric Field Distortion in Polymers Under High Electric Fields

Space charges are usually considered as the trigger mechanism for insulation degradation. The accumulation of space charges will strongly affect the electrical performance of devices. Wei et al. studied the space charge distribution and electric field distortion characteristics of PI films with a thickness of 8 μm under different

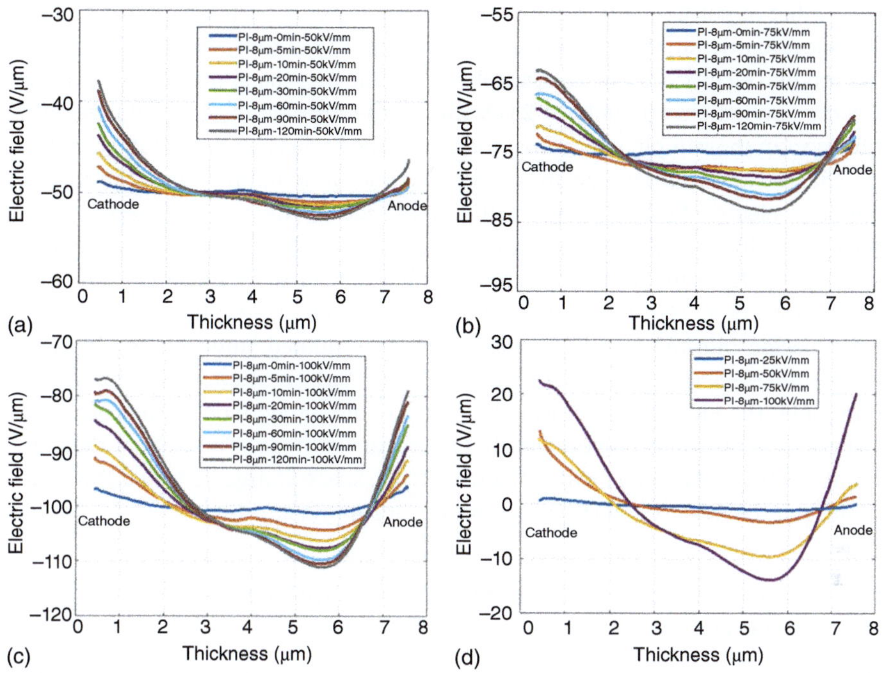

Figure 8.10 Internal electric field distributions under external voltages applied for 2 hours at 50 kV mm^{-1} and 313 K (a), for 2 hours at 75 kV mm^{-1} and 313 K (b), and for 2 hours at 100 kV mm^{-1} and 313 K (c), and after removing the applied voltages at 313 K (d). Source: Wei et al. [32]. Permission conveyed of IEEE.

DC electric fields (from 25 to 100 kV mm^{-1}) at 313 K for 2 hours [32]. When the electric field is less than 25 kV mm^{-1}, the distortion factor of the electric field inside the PI film is very small and can be ignored. Under an electric field of 50 kV mm^{-1}, the electric field distortion inside the sample is obvious near the cathode, and the electric field distortion near the anode changes relatively little within 30 minutes (Figure 8.10a). It shows that the negative charges are easier to inject and accumulate than the positive charges at a lower electric field. With the passage of time, the internally distorted electric field gradually appears negative peak, and the mobility and migration of negative charges are higher than those of positive charges. The amplitude of the negative electric field peak gradually increases, while the electric field enhancement rate near the cathode and anode of the sample decreases significantly, which indicates that the accumulation of bipolar space charges increases. Compared with the results under the electric fields of 75 and 50 kV mm^{-1} (Figure 8.10b), the change trend of the electric field is similar, but the change rate is higher, which indicates that a higher injection rate and higher charge migration rate lead to the enhancement of space charge accumulation. Under the 100 kV mm^{-1} electric field, the electric field distortion near the cathode and anode of the sample is obvious, and the negative peak also appears earlier (Figure 8.10c), which indicates that after the voltage is applied, a large number of positive and negative charges are injected from the electrodes in a short time and migrate to the inside of the film. Figure 8.10d shows the distribution of the residual electric field in the sample after removing the applied voltage. Even under the applied electric field of 25 kV mm^{-1}, the residual electric field can still be observed in the sample after removing the applied voltage, which indicates that space charges are accumulated in the film. At the same temperature, the higher the applied electric field, the higher the amplitude of the residual electric field formed in the film, and the greater the cumulative space charge density. The applied electric field reduces the metal-dielectric interface barrier and reduces the difficulty of charge injection. Therefore, under a higher external electric field, the charges can be injected into the film more easily and accumulated, and the amplitude of the distortion electric field inside the film is larger. With the increase in the applied electric field, the electric fields near the cathode and anode tend to increase, and the negative peak value moves slightly toward the cathode. This shows that the external electric field has a great influence on the space charges injected from the cathode, and the negative charges are easier to accumulate.

8.4.2 Charge Accumulation Induced by High-Energy Electron Irradiation and Working Voltage

High-energy electrons can penetrate the aluminum shield and deposit inside the insulation layer, leading to the accumulation of space charges and the establishment of high electric field. When the local electric field exceeds a certain threshold, discharge may occur, leading to deterioration or even failure of the insulation performance of the spacecraft. Wang et al. established a deep dielectric charging model for the SADA under the coupling effect of electron irradiation and operating voltage and studied the effects of electron flux enhancement, electron direction, and operating voltage on the dielectric charging characteristics [33]. Figure 8.11a shows the schematic diagram of deep dielectric charging model of SADA structure under the coupling effect of electron irradiation and operating voltage.

Bipolar charge transport processes include charge injection, migration, capture, detrapping, and recombination. There are two types of charge injection processes. First, high-energy electrons penetrate the aluminum shield and deposit in the interior of the insulating material. Second, under high DC voltage, the charge on the metal electrode is continuously injected into the insulating material through Schottky thermionic emission or field emission. Two types of charges injected into insulating materials will migrate under the drive of an electric field. The conductivity formed by charge migration in insulating materials under the action of only an electric field is called dark conductivity. As high-energy electrons radiate into insulating materials, they will also transfer energy to target atoms, forming radiation-induced conductivity that is much larger than dark conductivity. Therefore, under the joint action of high-energy electrons and working voltage, both radiation-induced conductivity in the radiation area of high-energy electrons and dark conductivity in the non-radiation area need to be considered in the insulating materials. The processes of charge deposition and injection, as well as conductance,

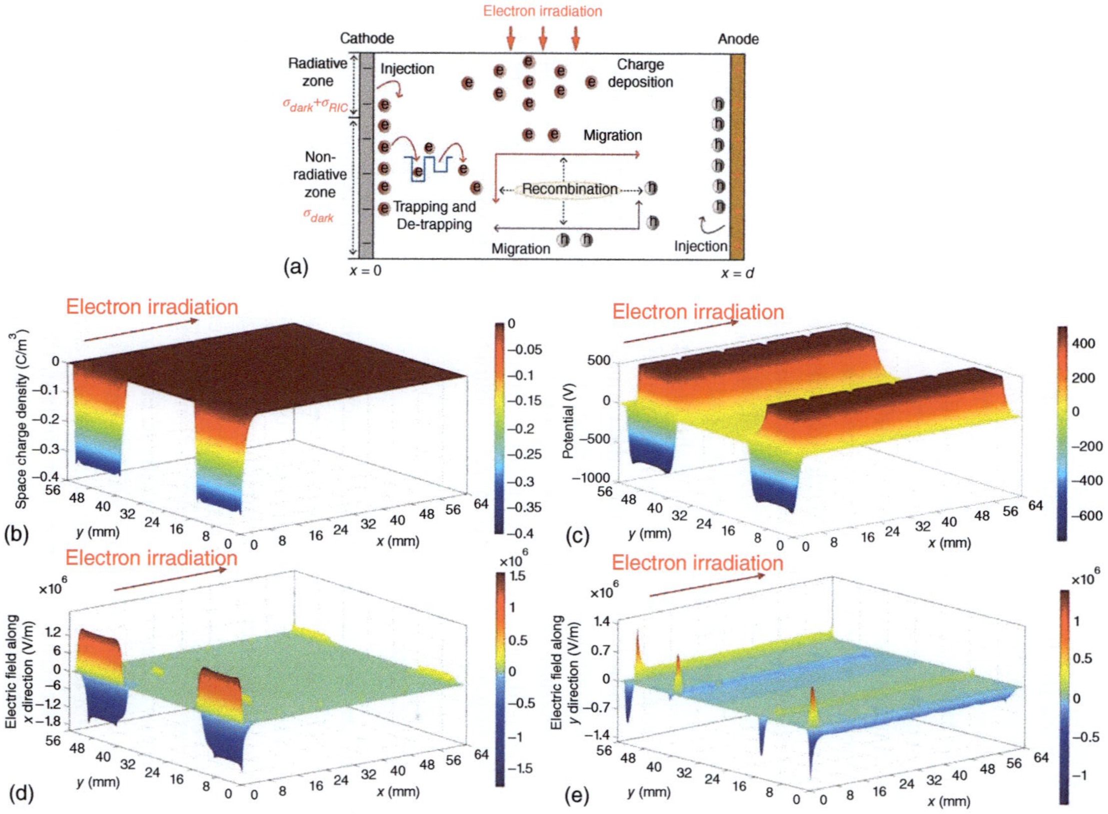

Figure 8.11 (a) Schematic diagram of deep dielectric charging in SADA structure. Under the coupling effect of typical electron irradiation and 500 V working voltage, the space charge distribution (b), potential distribution (c), electric field distribution along the x direction (d), and electric field distribution along the y direction (e) in the SADA structure. Source: Wang et al. [33]. Permission conveyed of IEEE.

will determine the level of space charge accumulation and the degree of electric field distortion in SADA.

Figure 8.11b–e shows the distribution of space charge, potential, and electric field in SADA structure under the coupling effect of typical electron irradiation along x and y directions and 500 V working voltage. Figure 8.11b shows that the charges deposited in the materials are mainly contributed by the irradiated electrons. After the electron beam penetrates the aluminum shielding layer, it is deposited in the materials near the irradiation side and gradually decreases along the x direction. Figure 8.11c shows that the maximum positive potential appears on the high-voltage electrode with the applied working voltage, while the maximum negative potential appears in the outermost PI near the irradiation surface due to the deposition of high-energy electrons. Since the deposited charges in the materials mainly come from the irradiation electrons, the electric field distribution in the irradiation area along the incident direction of the electron beam is similar to that when the electron beam irradiation acts alone. With the decrease in deposited electrons, the electric field distribution in the non-radiation region is mainly determined by the applied voltage. Figure 8.11d,e shows that the maximum electric field in the x direction appears in the PI near the irradiation side, and the maximum electric field in the y direction is located at the corner where the PI on the irradiation side contacts the electrode. In addition, when the irradiation condition is a typical FLUMIC spectrum, the maximum electric field under the coupling effect increases linearly with the increase in operating voltage. With the working voltage increasing to 9 kV, the maximum electric field in SADA is kept below $10^7\ \mathrm{V\,m^{-1}}$. When the electron flux increases to 100 and 1000 times the quiet electron irradiation, the electron irradiation plays a leading role in the deep charging characteristics of SADA.

8.4.3 Nonlinear Conductivity

The PI composites with core–shell heterostructure fillers have good nonlinear conductivity, which is conducive to improving the working voltage and safety reliability of spacecraft [34]. As shown in Figure 8.12, under a high electric field, the PI composites with core–shell heterostructure fillers have higher nonlinear coefficient and threshold electric field than traditional nonlinear dielectrics. The nonlinear index of composites increases with the increase in filler content. The conductivity increases with the increase in content. The threshold electric field decreases with increasing content. In addition, due to the difference in the interface barrier, the different types of fillers will lead to different energy levels and densities of deep and shallow traps. With the increase in the filler content, the density of deep and shallow traps increases, the energy level decreases, the conductivity and nonlinear coefficient increase, and the threshold electric field decreases. The conductance of PI composites with core–shell heterostructure fillers at low field strengths originates from thermally excited electron conductivity, while the nonlinear conductivity at high field strengths is due to the tunneling effect. Therefore, by changing the type and content of core–shell heterostructure fillers, the interface characteristics can be adjusted, thus changing the nonlinear conductivity of PI composites.

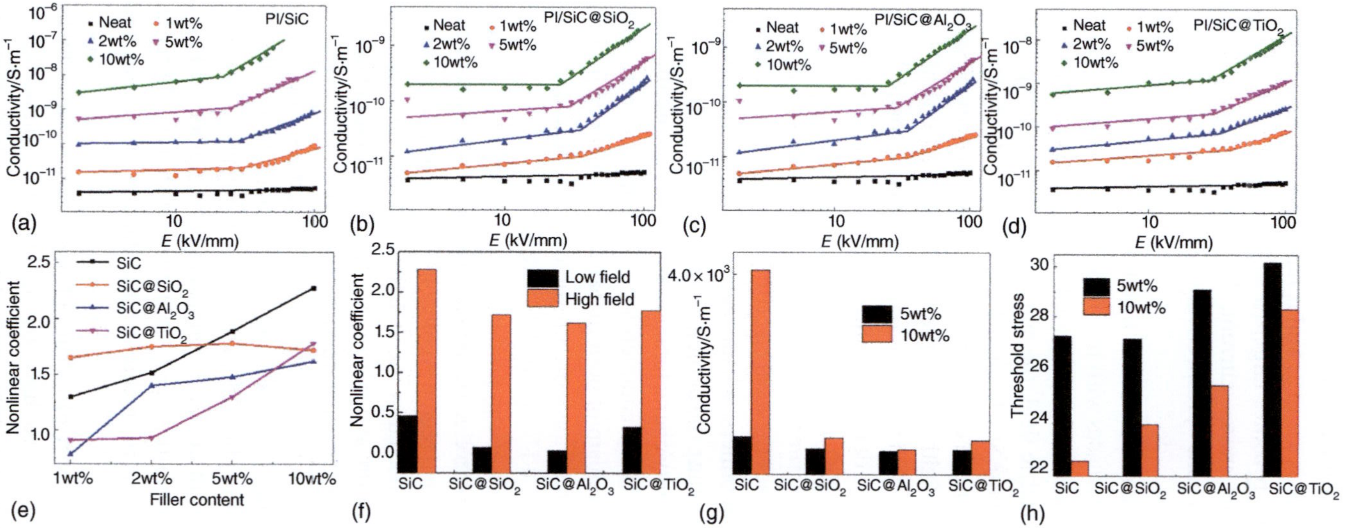

Figure 8.12 Electrical conductivity versus externally applied electric field of PI/SiC (a), PI/SiC@SiO_2 (b), PI/SiC@Al_2O_3 (c), and PI/SiC@TiO_2 (d) composites in double logarithmic coordinates. (e) Nonlinear coefficients as a function of filler content for different PI composites at high electric fields. Comparisons of nonlinear coefficients at low and high electric fields (f), conductivity (g), and threshold electric fields (h) of different PI composites. Source: Shusai [34]/Xi'an.

Ndongo et al. studied the nonlinear conductivity characteristics of PI and PTFE when they are used as aviation cables [35]. At different temperatures, the current densities of PI and PTFE have two slopes. Under low electric fields, the current density rises slowly, the slope is about 1, and it has ohmic characteristics. Under a high electric field, the slope becomes larger (3.36 for PI and 5.8 for PTEF), which corresponds to space charge-limited current. The threshold electric fields of PI and PTFE are 40 and 100 $kV\,mm^{-1}$, respectively, which have little correlation with temperature.

Sun et al. studied the conductivity of SiR by adding nanoparticles [36]. The conductivity of SiC/SiR PNCs increases with the increase in doping content. In the low electric field region, the conductivity of pure SiR is higher than that of composites. With the increase in the electric field, the conductivity of the composite gradually increases, exceeding that of pure SiR. Under a low electric field, SiC destroys the original conductive channel of the material, making the conductivity of SiC/SiR PNCs decrease. When the electric field rises to the threshold electric field, the interface between grains is conductive, and the conductivity increases rapidly. The higher SiC content increases the dependence of the conductivity of the composite on the electric field. The conductivity of the SiO_2/SiR composite decreases first and then increases with the increase in doping content. Excessive doping content of SiO_2 will lead to agglomeration, which increases the local electric field and conductivity.

EP is doped with SiC nanofillers, which regulate the nonlinear conductivity of SiC/EP PNCs [37]. By controlling the doping content, SiC-filled EP composites have different nonlinear conductivity characteristics. Li et al. doped multilayer graphene (GO) in EP to study its nonlinear conductivity [38]. With the increase in electric field, the conductance mechanism of both materials has changed from the Ohmic regime to the SCL regime. This transition threshold electric field decreases with the increase in temperature. The addition of GO increases the conductivity activation energy of EP, and the activation energy gradually decreases with the increase in electric field, leading to a decrease in electric field threshold. The nonlinear conductivity of GO/EP PNCs in the high-field region is controlled by the interface barrier, and the conductivity mechanism is affected by the SCL theory and tunneling effect.

Yang et al. added micron CBN and micron Al_2O_3 into EP to study its conductivity [29]. As shown in Figure 8.13, the conductivity of CBN/EP composites is slightly higher than that of Al_2O_3/EP composites. With the increase in electric field, the conductivity of EP composites increases, and the conductivity of CBN/EP composites increases more significantly but still maintains a strong insulation performance. The nonlinear coefficient (β) of CBN/EP composites is 3.9, higher than the nonlinear coefficient of Al_2O_3/EP composites ($\beta = 3.4$). This shows that the charge in CBN/EP composites is more likely to migrate and leak out dangerous charges under strong electric fields.

Modulation of the conductivity properties of polymer dielectrics by micro- and nano-doping makes the materials have strong insulating properties under low electric fields and excellent nonlinear conductivity under high electric fields. At rated voltage, polymer composites have good insulation performance, which can ensure the normal operation of power equipment. Under the combined action of

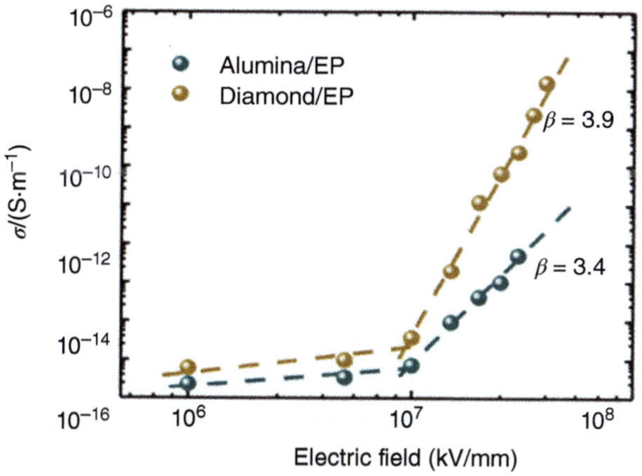

Figure 8.13 Electrical conductivity as a function of the electric field of epoxy composites. Source: Li et al. [37]. Permission conveyed of IEEE.

high-energy electron radiation and working voltage, space charges will accumulate in the insulation material, which will lead to the concentration of the electric field in some areas. If the insulating material does not have excellent nonlinear conductivity, the space charge accumulated in the insulating material will not be released in time. When the space charges accumulate to a certain extent, the local electric field may be high enough to cause discharge and finally affect the failure of the electrical equipment on the spacecraft. If composite insulating materials with excellent nonlinear conductivity are used, when a certain amount of space charges accumulates inside the insulating material and the electric field reaches the threshold electric field, the high conductivity path will be triggered, the conductivity will increase significantly, and the dangerous charges will be discharged in time. Therefore, selecting composite insulation materials with excellent nonlinear conductivity and optimizing the insulation structure of electrical equipment will greatly improve the safety and reliability of spacecraft under the combined action of high-energy electron irradiation and working voltage.

8.5 High-Temperature-Resistant Polymer Dielectrics Under the Coupling of Electron Irradiation and High Voltage

In space, due to the loss of the protection of the geomagnetic field, spacecraft directly receive various forms of space irradiation, such as high-energy electron, proton, solar cosmic ray, and galactic cosmic ray irradiations [39, 40]. Especially for deep space exploration missions with stellar radiation belts, such as Jupiter exploration, the radiation conditions are even worse [41]. Spacecraft materials must not only withstand extreme conditions, including extreme high and low temperatures, but also be subjected to the irradiation of various charged particles in space, resulting in surface and deep charging, which eventually leads to electrostatic discharge of dielectric materials [42, 43]. One of the most common discharge phenomena in spacecraft is vacuum DC surface discharge of insulating materials under electron irradiation [44].

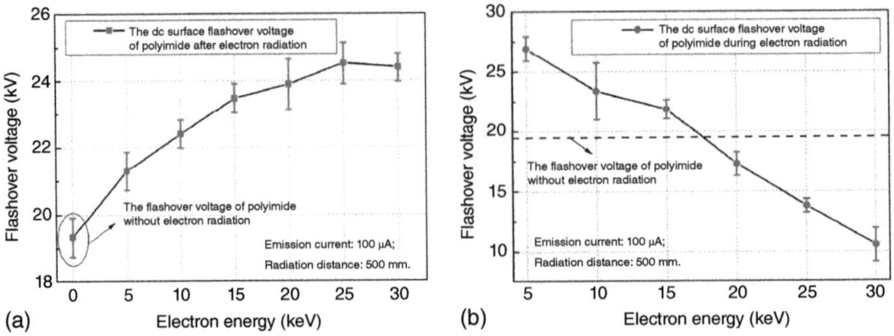

Figure 8.14 The PI vacuum DC surface discharge voltage changed with electron beam energy after electron beam radiation (a) and during electron beam radiation (b). Source: Li et al. [26]. Reproduced with permission of IEEE.

8.5.1 Mechanism of Vacuum DC Surface Discharge Under the Coupling of Electron Beam Irradiation and High Voltage

Generally, the process of vacuum DC surface discharge is divided into three stages: initiation, development, and breakdown. There is a general agreement that flashover begins with field emission or thermally assisted field emission electrons at the cathode triple junction (CTJ) where the insulating materials, cathode, and vacuum are closely combined. The breakdown occurs in the desorption gas layer or solid surface layer, forming a penetrating conductive channel [45, 46]. In the development stage of the surface flashover process, the secondary electron emission avalanche (SEEA) model has been recognized as the dominant theory [47–50]. The model emphasizes that the initial electron emission along the CTJ and the secondary electron emission on the surface of the materials have a great influence on the surface flashover.

Figure 8.14 shows the schematic diagram of surface flashover under the combined action of electron irradiation and operating voltage. High-energy electrons and low-energy electrons refer to incident electrons, while moving electrons refer to electrons flowing to the dielectric surface. In a space radiation environment, the effects of electron irradiation and working voltage should be considered simultaneously for vacuum DC surface discharge. They include the following physical processes: the interaction between incident electrons and dielectrics; the initial electron emission and secondary electron transport under operating voltage; and the combined effect of electron irradiation and operating voltage.

8.5.2 Effect of Electron Beam Irradiation on Vacuum DC Surface Discharge

The vacuum DC surface discharge characteristics of PI after and during electron beam irradiation were studied [26, 51]. As shown in Figure 8.14a, after electron beam irradiation, the discharge voltage of PI is significantly higher than that before electron beam irradiation, and with the increase in electron beam energy, the discharge voltage has an upward trend. This is related to the high electric field formed by the electrons deposited on the dielectric surface. On the one hand, it reduces the electric field at the CTJ. On the other hand, it can repel the secondary electrons, thus increasing the discharge voltage. Compared with the vacuum DC surface discharge voltage of PI after electron beam irradiation, the changing trend of discharge voltage during electron beam irradiation is obviously different.

As shown in Figure 8.14b, during electron beam irradiation, the vacuum DC surface discharge voltage drops sharply with the increase in electron beam energy. In order to simulate the space environment, a working voltage was applied after the electron radiation source had been turned on. As a result, the surfaces of the materials were deposited with electrons before a working voltage was applied. Low-energy electrons will be repelled by these deposited electrons. Therefore, the discharge voltage of PI during electron beam irradiation is greater than that without electron beam radiation. The deposited electrons are limited in their ability to repel electron beams with higher energy, so high-energy electrons have a possible impact on the material surface. These high-energy electrons will act as the initial electrons in the surface discharge. The initial electrons collide with the surface atoms of the material to ionize them, which promotes the generation and multiplication of secondary electrons. Also, the initial electrons collide with the desorbed gas molecules on the material surface, causing them to ionize and multiply electrons. Due to a large number of electron ionizations and multiplications, the material surface discharge voltage was significantly reduced.

8.5.3 Influence of Incident Electron Beam Characteristics on Vacuum DC Surface Discharge

The influence of incident electrons on the vacuum DC surface discharge of PI during electron beam radiation was analyzed by changing the electron beam energy, incident angle, and beam current density [45]. In this case, the incident angle is the angle between the initial direction of the electron beam and the positive direction of the applied electric field. As shown in Figure 8.15a, with the increase in electron beam energy, the vacuum DC surface discharge voltage of PI decreases significantly. When the incident angle changes, the vacuum DC surface discharge voltage of PI varies with the electron energy. As shown in Figure 8.15b, when the electron beam energies are 5 and 15 keV, as the incident angle increases, the vacuum DC surface discharge voltage of PI basically continues to rise. When the electron beam energy is 35 keV, with the increase in the incident angle, the vacuum DC surface discharge voltage of PI decreases first and then increases, reaching its minimum value at 90°. When the incident angle increases, the low-energy electron beam is accelerated by the applied electric field, which tilts the electron beam toward the anode. Deposited electrons gradually increase in importance, so the vacuum DC surface discharge voltage increases. This is because when the electron beam energy is high, there is enough energy to overcome the effect of the applied electric field, strike the surface of the insulating materials, and increase the initial number of electrons, resulting in a decrease in the vacuum DC surface discharge voltage. Figure 8.15c shows that under the continuous vertical radiation of a 5 keV electron beam, the vacuum DC

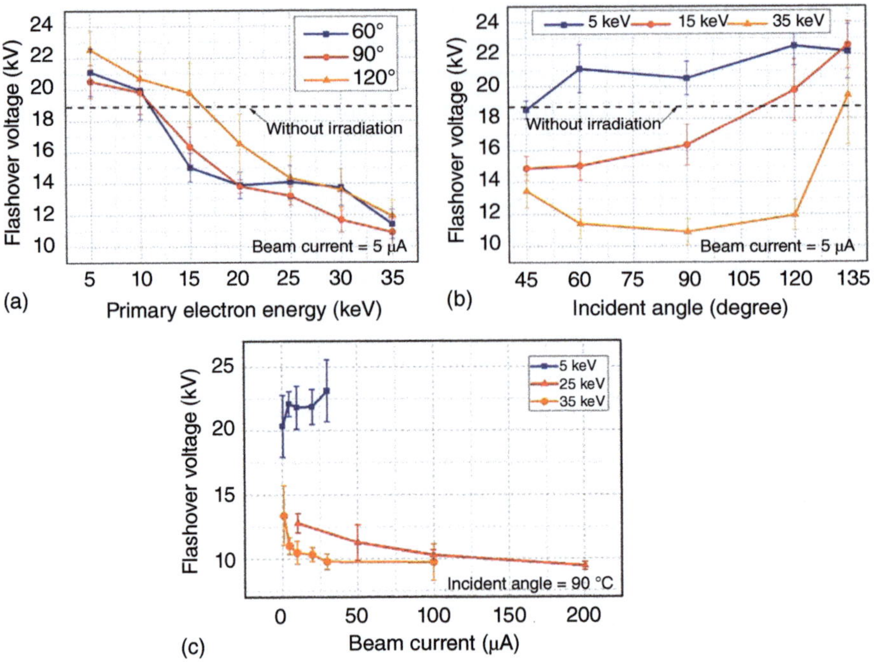

Figure 8.15 Variation of surface discharge voltage with electron beam kinetic energy (a), incident angle (b), and electron beam current density (c). Source: Li et al. [45]. Reproduced with permission of IEEE.

surface discharge voltage of PI increases slightly with the increase in beam current density. At this time, the increase in beam density and deposition electrons leads to the decrease in the electric field at the CTJ and the increase in the vacuum DC surface discharge voltage. For 25 and 35 keV electron beams, with the increase in beam current density, the vacuum DC surface discharge voltage of PI decreases. When beam current density increases, more electron–hole pairs are generated on the surface of the materials due to the energy absorbed by the surface, which changes the charge distribution and transport characteristics of the surface, making the surface flashover easier to occur.

8.5.4 Influence of Insulation Distance and Electrode Height on Surface Discharge

It is possible to modify the external electric field distribution by changing the electrode insulation structure, thus altering the incident electron radiation trajectory. Therefore, a specific electrode system with adjustable insulation distance and electrode height was used to study the influence of electrode parameters on dielectric surface discharge under electron beam irradiation [51]. Figure 8.16a shows the size and shape of the finger electrode. Under three different conditions, Figure 8.16b illustrates the relationship between the vacuum DC surface discharge voltage and the electrode gap distance after, during, and without the electron

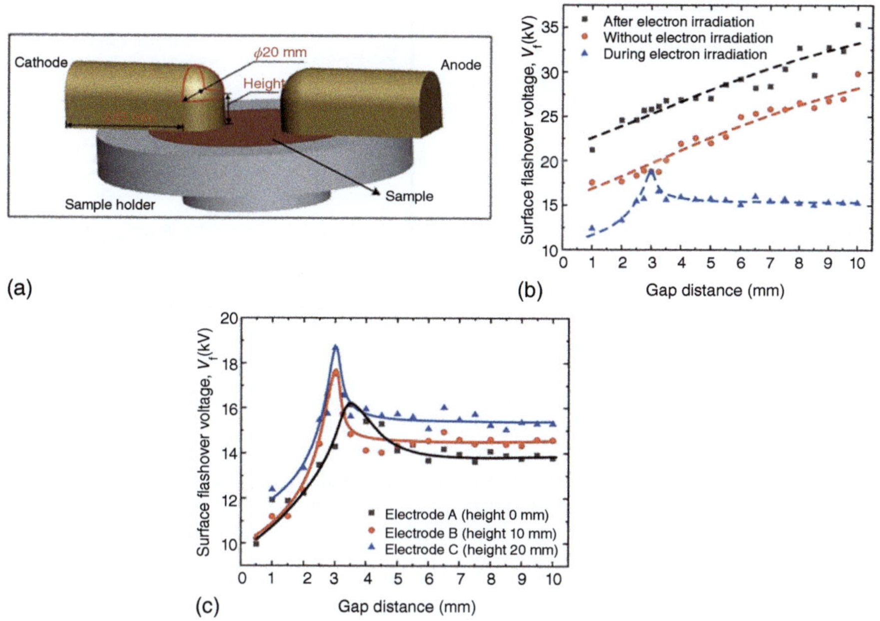

Figure 8.16 (a) Dimensions and shapes of electrodes. Surface discharge voltage during electron irradiation at different gap distances (b) and different electrode heights (c). The electron beam vertically radiates the sample, the electron beam energy is 30 keV, the beam current density is 10 μA, and the temperature of the vacuum test chamber is room temperature. Source: Wang et al. [51]. Permission conveyed through IOP Publishing, Ltd.

beam irradiation. Discharge voltage is the highest after electron irradiation, then without electron irradiation, and the lowest during electron irradiation at different gap distances. The trend of discharge voltage after electron irradiation is basically the same as that without irradiation, and differences of about 5 kV are generally stable between them. During electron irradiation, the surface discharge voltage first increases, then decreases, and finally tends to be stable with the increase in electrode spacing. With the increase in the gap distance, the area of the incident electron bombardment on the surface of the material becomes larger, and the effect of the incident electron is enhanced. Further increasing the electrode gap distance does not change the effect of incident electrons. Therefore, the surface discharge voltage presents a stable trend. Figure 8.16c shows that regardless of electrode height, discharge voltage changes similarly during electron irradiation with gap distance, rising initially, then falling, then stabilizing. With an increase in electrode height, voltage values become more stable and maximum. This is because, as the electrode height increases, the incident electrons gradually deflect until they cannot reach the material surface.

By constructing a three-layer $BaTiO_3$/PI PNCs, the interlayer interfaces and traps can be introduced to regulate the distribution of electrons deposited on the surface of the dielectric. The influence of the characteristics of insulating materials on the surface discharge of insulating dielectric under electron beam radiation was

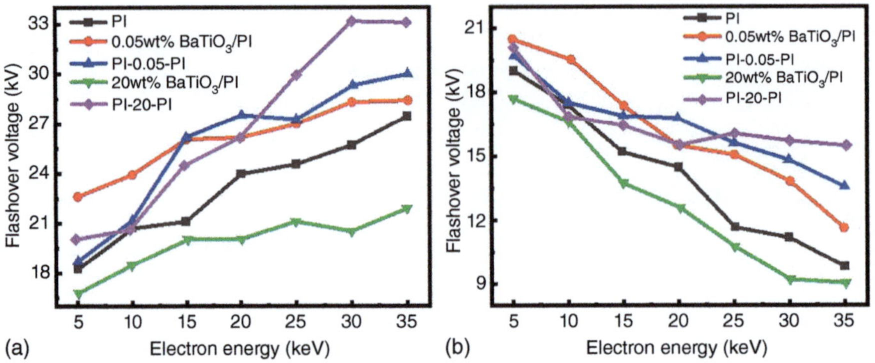

Figure 8.17 Comparisons of the surface discharge voltage of the single-layer and three-layer composites after the vertical irradiation of the electron beam (a) and during the radiation (b). PI-0.05-PI and PI-20-PI are three-layer composites, the left and right sides are pure PI, and in the middle are PI composites doped with $BaTiO_3$ nanoparticles with different contents. Finger electrodes without increased height are used, with an electrode spacing of 4 mm and an electron beam current density of 5 μA. The sample is at room temperature. Source: Wang [46]/Xi'an Jiaotong University.

explored [46]. Figure 8.17 shows the changing trend of the vacuum DC surface discharge voltage of single-layer and three-layer composites with the radiation energy of electron beam after and during irradiation. According to Figure 8.17a, all five samples have a positive correlation between the surface discharge voltage and the electron beam energy after irradiation. After the irradiation of energy ranging from 5 to 35 keV, the discharge voltage of single-layer pure PI is higher than those of single-layer 20 wt%-$BaTiO_3$/PI and lower than that of single-layer 0.05 wt%-$BaTiO_3$/PI. When the electron beam energy exceeds 15 keV, the surface discharge voltages of the two three-layer composites are higher than that of the single-layer pure PI. During the electron beam radiation, as shown in Figure 8.17b, for all five samples, the surface discharge voltage decreases gradually with the increase in electron beam energy. Increased electron beam energy enhances the effect of incident electrons. Similarly, during the continuous irradiation of an electron beam with energy ranging from 5 to 35 keV, five samples showed similar patterns to those after irradiation. Based on the experimental results, the three-layer composites exhibit better vacuum DC surface discharge performance than the single-layer PI both after and during the electron beam radiation. Its role is related to the design of different three-layer nanocomposites and the energy irradiated by the electron beam.

8.6 High Temperature Resistant and High-Frequency Strong Insulating Polymer Dielectrics

Power electronic transformers are convenient and efficient ways of converting electric energy, and they can be applied to spacecraft and all-electric aircraft [52].

The important insulating materials in power electronic transformers are EP and PI. However, unlike traditional transformers, the insulation materials in power electronic transformers must withstand the effects of high-frequency and high-voltage bipolar square waves, as well as the high temperature generated by high-frequency bipolar square waves [53]. PI is also a commonly used insulating material in motors, which also need to withstand the coupling effect of high-frequency voltages and high temperatures [54]. Therefore, it is of great significance to develop strong electrical insulation polymer dielectrics withstanding high temperature and high voltage frequency for high-performance power electronic transformers and motors.

8.6.1 Electrical Tree Characteristics of Epoxy Resin Under Bipolar Square Wave Voltage

Du et al. summarized that the initiation rate and number of electrical branches of EP increase with the increase in temperature under pulse voltage, and its growth rate presents a minimum near the glass transition temperature [55]. Zhang et al. studied the electrical tree characteristics in EP under bipolar square wave voltage in the frequency range of 50 Hz–20 kHz, as shown in Figure 8.18 [56]. From Figure 8.18a, it can be found that at 50 Hz–2 kHz, the growth time of electrical branches decreases

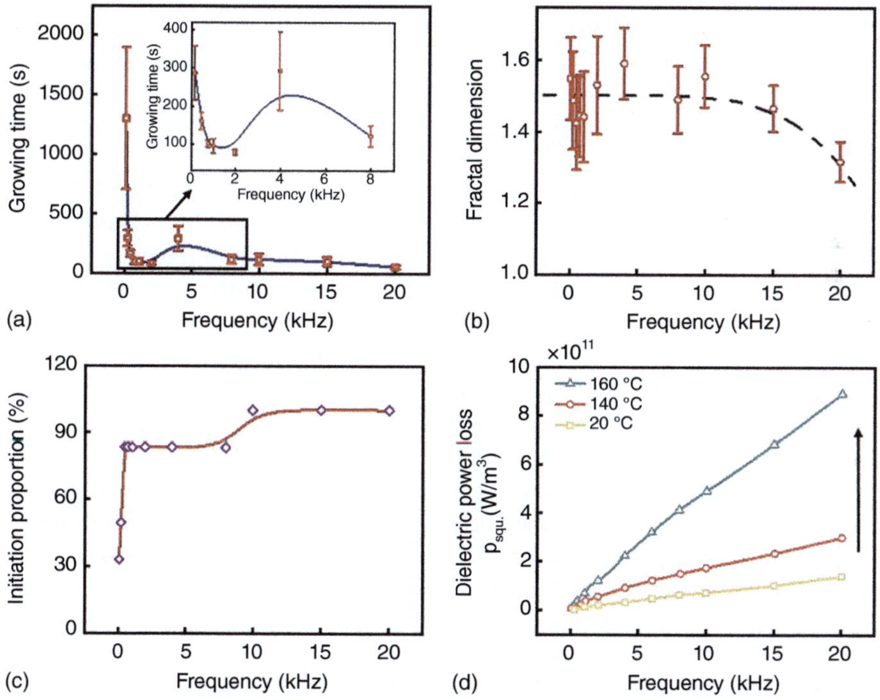

Figure 8.18 Variation of growth time (a), fractal dimension (b), and initiation rate (c) of electrical tree with voltage and frequency. (d) Variation of dielectric power loss with frequency under bipolar square wave electric field at different temperatures. Source: Zhang et al. [56]. Reproduced with permission of IEEE.

rapidly. The growth time increases slightly at 2–4 kHz and then decreases with increasing frequency. As shown in Figure 8.18b, when the frequency increases from 4 to 20 kHz, the number of trunks decreases, which corresponds to the rapid growth of electrical trees. Therefore, the fractal dimension is used to represent the density of electrical trees. With the increase in frequency, the number of trunks increases while the number of electrical tree channels decreases, leading to a decrease in fractal dimension. Although fractal dimension can reflect electrical tree growth time, it does not completely correspond to it. The initiation of an electrical tree branch is defined as the tree length can grow to 10 μm after the voltage is applied for 10 minutes. Figure 8.18c shows that voltage frequency increases the initiation rate of electrical trees. This is related to the charge injection/extraction and dielectric power loss of EP under the bipolar square wave electric field. Figure 8.18d shows the dielectric power loss at 20, 140, and 160 °C under bipolar square wave voltage with frequency from 50 Hz to 20 kHz. It can be seen that the dielectric power loss increases with the increase in frequency, and higher temperatures result in higher dielectric power losses. When partial discharges occur in the electrical tree channel, the EP will lose more dielectric power due to an increase in temperature. When the dielectric power loss increases, more weak points will be formed near the electrical tree channels, resulting in partial discharges and a larger number of branches. Therefore, due to the mutual promotion of dielectric power loss and partial discharge, the initiation rate of the electrical tree under bipolar square wave voltage increases with the increase in frequency.

8.6.2 Micro/Nano-Doped Epoxy Resin Composites

Figure 8.19 shows the space charge behavior of pure EP and micro-nano AlN-doped EP composites (MN) at 80 °C under the 40 kV mm^{-1} electric field with different polarity and different frequencies [53]. Under all test conditions, the average charge density of pure EP samples is higher than that of MN samples. The negative electric field with higher frequency and the positive electric field with lower frequency can accelerate the space charge accumulation in the sample. In addition, the residual charge in the MN sample is more stable than that in pure EP during depolarization. By using micro- and nano-fillers, the space charge injection barrier and the charge transport characteristics based on Schottky emission can be adjusted, preventing the accumulation of charged energy during transport. In addition, the overlap of the interaction region introduced by the fillers can increase the charge transport path and reduce the influence of frequency.

Figure 8.20 shows the electric field distribution behavior of different EP samples under 40 kV mm^{-1} electric fields with different applied voltage polarity and frequency at different temperatures [53]. At different temperatures, the electric field distortion of pure EP is obviously higher than that of the MN sample. In addition, the negative voltage has more influence on the electric field distortion in EP than the positive voltage. For pure EP, the electric field distortion increases with the increase in test temperature. The electric fields in the MN sample remain stable even at high temperatures.

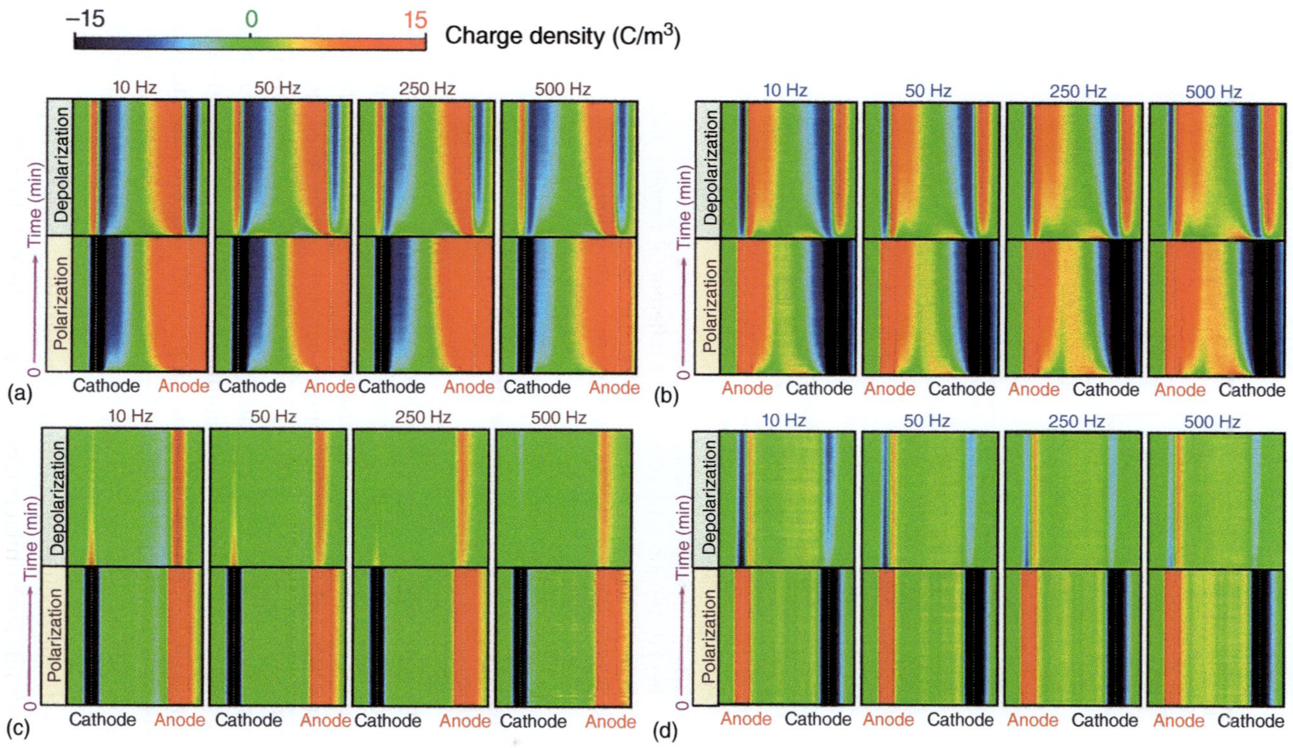

Figure 8.19 Space charge behavior of pure EP and MN samples under an electric field at 80 °C, polarization and depolarization time are 30 minutes respectively. The space charge distributions of pure EP samples under $+40\,kV\,mm^{-1}$ electric fields of different frequencies (a) and under $-40\,kV\,mm^{-1}$ electric fields of different frequencies (b). The space charge distributions of the MN sample under $+40\,kV\,mm^{-1}$ electric fields of different frequencies (c) and under $-40\,kV\,mm^{-1}$ electric fields of different frequencies (d). Source: Dai et al. [53]. Reproduced with permission of IEEE.

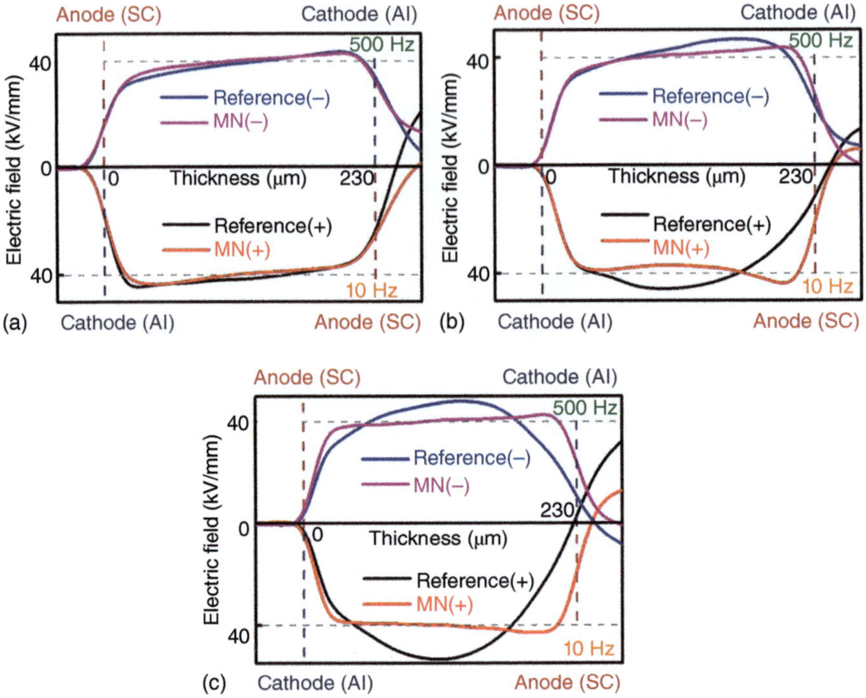

Figure 8.20 Electric field distribution characteristics in different EP samples under a 40 kV mm^{-1} electric field with different polarities and frequencies at 25 °C, (a) 50 °C (b) and 80 °C (c), respectively. Source: Dai et al. [53]. Reproduced with permission of IEEE.

8.6.3 Corona Resistance Life of Polyimide Modified by Nano-Doped Multilayer Structure

Ultraviolet rays, high-energy electrons, and highly active chemical groups generated by corona discharge can accelerate the degradation of insulating materials [54]. Therefore, the anti-corona performance of insulating materials plays a key role in their lifespan. Figure 8.21a shows the relationship between corona resistance lives of aged and unaged samples of PI and PI/Al_2O_3 composites and temperature [57]. It has been shown that nano Al_2O_3 significantly improves the corona resistance life of materials and weakens their temperature dependence. The lifetime of PI/Al_2O_3 nanocomposites decreased with increasing temperature but still improved compared to the pure PI samples.

For bipolar pulse voltage, when the polarity of the pulse voltage reverses rapidly, the direction of the electric field generated by the internal space charges almost remains unchanged. Due to different attenuation times, the inverse field generated by square waves with different frequencies (such as 1 and 10 kHz) after discharge is also different. Inverse field intensity at 1 μs square wave is greater than at 10 μs square wave. Therefore, it can be observed in Figure 8.21c that the film under the square-wave voltage with a rising time of 1 μs has a higher discharge rate. With the

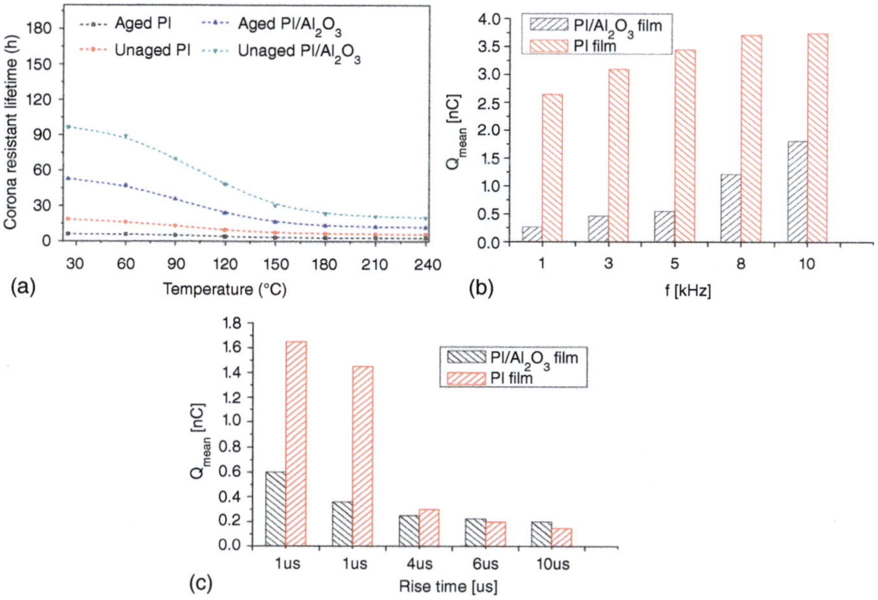

Figure 8.21 Corona resistance lifetime as a function of temperature (a), mean discharge quantities at different frequencies (b), and mean discharge quantities under different rise time square waves (c) for PI and PI/Al_2O_3 films. Source: Akram et al. [57]. Reproduced with permission of IEEE.

increase in frequency and the decrease in rise time, the average discharge capacities of both the modified and pure PI films are increased, but the average discharge capacity of the pure PI film is always greater than that of the PI/Al_2O_3 film, as shown in Figure 8.21b,c. This is because the PI/Al_2O_3 film is a three-layer film with an outer layer of nano-doped film, and its surface conductivity is higher than that of pure PI film. Therefore, the residual charges of PI/Al_2O_3 films after discharge are less than those of pure PI films.

8.6.4 Characteristics of Polyimide Modified by Molecular Structure

Huang et al. grafted the phenyl thioether group into the main chains of PI and synthesized PIs with different percentages of phenyl thioether (PSI) by thermal imidization of cross-polymerization [58]. Figure 8.22a shows the dielectric loss tangent (tanδ) of the PSI membrane prepared in the phenyl thioether content range of 0–100%. In all samples, with the increase in frequency, the tanδ increases, while with the increase in phenyl thioether content, the tanδ decreases first and then increases. The working range of high-frequency power supply equipment is between 1 and 100 kHz. Within this range, the PSI40 sample containing 40% phenyl thioether had the lowest tanδ. When the portion of phenyl thioether disrupts intermolecular charge transfer and molecular stacking, dipole steering receives less binding, and the tanδ is lower. In the case where the thioether content is similar

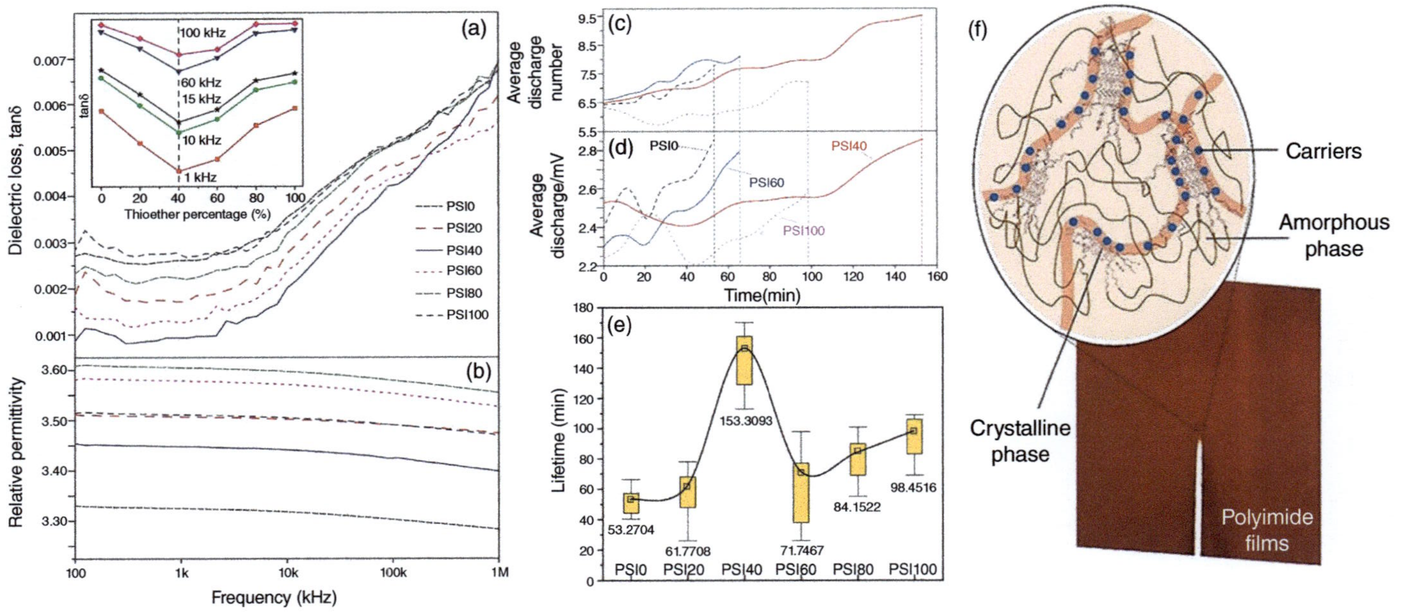

Figure 8.22 Variation of dielectric loss tangent (a) and relative permittivity (b) of different PSI films with frequency. The average discharge number (c) and average discharge amplitude (d) of different PSI films as a function of time. (e) The lifetimes of different PSI films under high-frequency surface discharge. (f) Schematic diagram of charge carrier migration in the PSI films. Source: Huang et al. [59]. Reproduced with permission from IOP Publishing.

to, but less than, the ether content, the smallest crystal zone is formed because of the weakest molecular accumulation. Therefore, PSI40 has the lowest tanδ. As shown in Figure 8.22b, the relative permittivity of all samples decreases slightly with increasing frequency. Because the S element is easier to give electrons than the O element and the dipole moment increases, the relative permittivity of all PSI films is higher than that of pure PI films.

Discharge signals were collected continuously from the time of applying voltage to the time of flashover on the film to analyze the surface discharge behavior of PSI films. In general, with the development of surface discharge, the discharge amplitude of each sample shows an increasing trend, as shown in Figure 8.22c,d. During the first half of the discharge process, the amplitudes of all samples fluctuate, showing a trend of increasing and then decreasing. After reaching the local minimum value, the average amplitude increases until flashover. According to Figure 8.22e, all PSI films show longer life than pure PI films when discharging along the surface at high frequency. This is because charge carriers can migrate through the crystal amorphous phase interface on the PSI film, as shown in Figure 8.22f. So, the energy generated by carrier movement can be slowly dissipated without affecting the molecular structure, and the life can be increased [59].

References

1 Benyoucef, A., Berquez, L., Teyssedre, G. et al. (2020). Space charge characterization in insulating materials used in aircraft cables. In: *IEEE International Conference on High Voltage Engineering and Application (ICHVE)*, 2020. Beijing, China: IEEE.

2 Mankins, J. (2009). New directions for space solar power. *Acta Astronautica* 65 (1–2): 146–156.

3 Lizcano, M., Williams, T., Shin, E. et al. (2022). Aerospace environmental challenges for electrical insulation and recent developments for electrified aircraft. *Materials* 15 (22): 8121.

4 Liu, J., Zhang, X., Tian, F. et al. (2017). Recent progress of research and development for high-temperature resistant polymer dielectrics. *Transactions of China Electrotechnical Society* 32 (16): 14–24.

5 He, X., Han, X., Qin, G. et al. (2016). Research progress of polyimide diaphragm material for lithium ion battery. *China Plastics Industry* 44 (01): 1–15.

6 Sheng, D., Wang, H., and Ying, Z. (2013). Application of thermoplastic polyimide composite in aerospace field. *Plastic* 42 (4): 46–48.

7 Wang, L., Yue, Z., Sun, Q. et al. (2022). Properties of high temperature resistant silicone rubber with double curing system. *Journal of Aeronautical Materials* 42 (5): 135–141.

8 Ou, Z., Gao, F., Zhu, L. et al. (2021). Study on aluminum nitride/addition-cure liquid silicone rubber composite for high-voltage power encapsulation. *PLoS One* 16 (6): e0252619.

9 Wu, Y. and Zhang, J. (2022). Preparation and properties of high temperature and high humidity resistant epoxy resin composites for petroleum logging instruments. *Insulating Materials* 55 (8): 28–31.

10 Genc, G., Sarikas, A., Kesen, U. et al. (2020). Luffa/epoxy composites: electrical properties for PCB application. *IEEE Transactions on Components, Packaging and Manufacturing Technology* 10 (6): 933–940.

11 Han, W., Zhang, F., Wang, J. et al. (2021). Synthesis and properties of high temperature resistant epoxy resin containing carborane. *Polymer Materials Science and Engineering* 37 (11): 22–28.

12 Li, S. (2015). *Principle of Spacecraft Electrification-Interaction between Spacecraft and Space Plasma*. Beijing: Science Press.

13 Garrett, H. and Whittlesey, A. (2011). *Guide to Mitigating Spacecraft Charging Effects*. Pasadena: CA: Wiley.

14 Toyoda, K. (2018). High-voltage power generation on spacecraft. *Electrical Engineering in Japan* 203 (4): 25–30.

15 Ai, D., Li, H., Zhou, Y. et al. (2020). Tuning nanofillers in in situ prepared polyimide nanocomposites for high-temperature capacitive energy storage. *Advanced Energy Materials* 10 (16): 1903881.

16 Yin, Z. (2003). *Dielectric Physics*. Beijing: Science Press.

17 Jin, W. (1997). *Dielectric Physics*. Beijing: China Machine Press.

18 Li, G. (2017). *Charge Deposition of Dielectrics under Electron Radiation and its Effect on Surface Flashover in Vacuum*. Xi'an Jiaotong University.

19 Chen, J. and Liu, Z. (1982). *Dielectric Physics*. Beijing: China Machine Press.

20 Min, D., Li, S., Hirai, N. et al. (2016). Dielectric spectroscopic analysis of degradation in ethylene-propylene-diene copolymer. *IEEE Transactions on Dielectrics and Electrical Insulation* 23 (6): 3620–3630.

21 Baferani, M., Shahsavarian, T., Wu, X. et al. (2019, 2019). Study of space charge behavior of insulations for high temperature applications. In: *IEEE Conference on Electrical Insulation and Dielectric Phenomena*, Richland, WA, USA.

22 Min, D., Yan, C., Huang, Y. et al. (2017). Dielectric and carrier transport properties of silicone rubber degraded by gamma irradiation. *Polymers (Basel)* 9 (10): 533.

23 Min, D., Cui, H., Wang, W. et al. (2019). The coupling effect of interfacial traps and molecular motion on the electrical breakdown in polyethylene nanocomposites. *Composites Science and Technology* 184: 107873.

24 Dissado, L., Thabet, A., and Dodd, S. (2010). Simulation of DC electrical ageing in insulating polymer films. *IEEE Transactions on Dielectrics and Electrical Insulation* 17 (3): 890–897.

25 Dissado, L. and Thabet, A. (2008). Simulation of electrical ageing in insulating polymers using a quantitative physical model. *Journal of Physics D: Applied Physics* 41 (8): 085412.

26 Li, G., Li, S., Pan, S. et al. (2016). Effect of electron irradiation on DC surface flashover of polyimide in vacuum. *IEEE Transactions on Dielectrics and Electrical Insulation* 23 (3): 1846–1853.

27 Cheng, S., Zhou, Y., Hu, J. et al. (2020). Polyimide films coated by magnetron sputtered boron nitride for high-temperature capacitor dielectrics. *IEEE Transactions on Dielectrics and Electrical Insulation* 27 (2): 498–503.

28 Yuan, C., Zhou, Y., Zhu, Y. et al. (2020). Polymer/molecular semiconductor all-organic composites for high-temperature dielectric energy storage. *Nature Communications* 11 (1): 3919.

29 Yang, K., Chen, W., Zhao, Y. et al. (2022). Enhancing dielectric strength of thermally conductive epoxy composites by preventing interfacial charge accumulation using micron-sized diamond. *Composites Science and Technology* 221: 109178.

30 Luo, S., Yu, J., Ansari, T. et al. (2020). Elaborately fabricated polytetrafluoroethylene film exhibiting superior high-temperature energy storage performance. *Applied Materials Today* 21: 100882.

31 Luo, S., Ansari, T., Yu, J. et al. (2021). Enhancement of dielectric breakdown strength and energy storage of all-polymer films by surface flattening. *Chemical Engineering Journal* 412: 128476.

32 Wei, Q., Zheng, F., and Zhang, Y. (2022). Measurement of electric field distribution in thin polyimide film. In: *2022 IEEE 4th International Conference on Dielectrics (ICD)*, 518–521.

33 Wang, X., Min, D., Pan, S. et al. (2021). Coupling effect of electron irradiation and operating voltage on the deep dielectric charging characteristics of solar array drive assembly. *IEEE Transactions on Nuclear Science* 68 (7): 1399–1406.

34 Shusai, Z. (2022). *Investigation on Mechanism of PI Nonlinear Conduction Characteristics with Effect of Core -Shell Heterostructure*. Xi'an Jiaotong University.

35 Ndongo, D., Roy, S., Teyssèdre, G. et al. (2020). Current measurements in high performance polymers used in aeronautic cables. *IEEE Transactions on Dielectrics and Electrical Insulation* 27 (6): 2195–2202.

36 Sun, L., Zhang, P., Li, Z. et al. (2019). Silicone rubber nonlinear composites for HVDC cable terminals. *High Voltage Engineering* 45 (5): 1654–1665.

37 Li, K., Zhang, B., Li, X. et al. (2021). Electric field mitigation in high-voltage high-power IGBT modules using nonlinear conductivity composites. *IEEE Transactions on Components, Packaging and Manufacturing Technology* 11 (11): 1844–1855.

38 Li, H., Wang, C., Liu, L. et al. (2018). Experimental study on nonlinear conduction mechanism of epoxy-based nanocomposite with low content of graphene filler. *High Voltage Engineering* 44 (3): 812–820.

39 Pan, S. (2020). *Investigation on Vacuum DC Surface Flashover Mechanism of Polyimide under Electron Beam Irradiation*. Xi'an Jiaotong University.

40 Zhou, Y., Shen, Z., Qi, Z. et al. (2021). Demand for high performance materials in development of China's aerospace science and technology. *Journal of Materials Engineering* 49 (11): 41–50.

41 Shen, Z., Ouyang, X., and Gao, H. (2021). Demand for aerospace materials and technology for China's deep space exploration. *Aerospace Material Technology* 51 (5): 1–14.

42 Zhang, Z., Miao, J., Wang, X. et al. (2014). Surface flashover characteristics of polyimide during electron beam irradiation under DC voltage in vacuum. *High Voltage Engineering* 40 (1): 117–123.
43 Lai, S. (2003). A critical overview on spacecraft charging mitigation methods. *IEEE Transactions on Plasma Science* 31 (6): 1118–1124.
44 Lai, S. (2012). *Fundamentals of Spacecraft Charging: Spacecraft Interactions with Space Plasmas*. Princeton, New Jersey: Princeton University Press.
45 Li, S., Pan, S., Li, G. et al. (2017). Influence of electron beam irradiation on DC surface flashover of polyimide in vacuum. *IEEE Transactions on Dielectrics and Electrical Insulation* 24 (2): 1288–1294.
46 Wang, X. (2022). *Modulation of Surface Flashover Properties in Vacuum for Spacecraft Components Based on Trajectory of Kinetic Electrons and Distribution of Deposited Electrons*. Xi'an Jiaotong University.
47 Harris, J., Caporaso, G., Blackfield, D. et al. (2007). Displacement current and surface flashover. *Applied Physics Letters* 91: 121504.
48 Pillai, S. and Hackam, R. (1982). Surface flashover of solid dielectric in vacuum. *Journal of Applied Physics* 53 (4): 2983–2987.
49 Neuber, A., Butcher, M., Hatfield, L. et al. (1999). Electric current in dc surface flashover in vacuum. *Journal of Applied Physics* 85 (6): 3084–3091.
50 Sun, J., Song, S., Zheng, J. et al. (2022). A review on surface flashover phenomena at DC voltage in vacuum and compressed gas. *IEEE Transactions on Dielectrics and Electrical Insulation* 29 (1): 1–14.
51 Wang, X., Zheng, S., Li, Z. et al. (2022). Radiation Electron trajectory modulated DC surface flashover of polyimide in vacuum. *Journal of Physics D: Applied Physics* 55 (20): 205201.
52 Magambo, J., Bakri, R., Margueron, X. et al. (2017). Planar magnetic components in more electric aircraft: review of technology and key parameters for DC-DC power electronic converter. *IEEE Transactions on Transportation Electrification* 3 (4): 831–842.
53 Dai, C., Tanaka, Y., Miyake, H. et al. (2022). Space-charge characteristics in epoxy composites under square pulse wave of different polarities with various frequencies at various temperatures. *IEEE Transactions on Dielectrics and Electrical Insulation* 29 (1): 137–144.
54 Mayoux, C. (2000). Degradation of insulating materials under electrical stress. *IEEE Transactions on Dielectrics and Electrical Insulation* 7 (5): 590–601.
55 Du, B., Xue, J., Su, J. et al. (2017). Effects of ambient temperature on electrical tree in epoxy resin under repetitive pulse voltage. *IEEE Transactions on Dielectrics and Electrical Insulation* 24 (3): 1527–1536.
56 Zhang, C., Wang, S., Fu, H. et al. (2021). Effect of bipolar square wave voltage with varied frequencies on electrical tree growth in epoxy resin. *IEEE Transactions on Dielectrics and Electrical Insulation* 28 (3): 806–814.
57 Akram, S., Gao, G., Liu, Y. et al. (2015). Degradation mechanism of Al_2O_3 nano filled polyimide film due to surface discharge under square impulse voltage. *IEEE Transactions on Dielectrics and Electrical Insulation* 22 (6): 3341–3349.

58 Huang, X., Wang, J., Li, Q. et al. (2019). Impact of the phenyl thioether contents on the high frequency dielectric loss characteristics of the modified polyimide films. *Surface and Coatings Technology* 360: 205–212.

59 Huang, X., Li, Q., Ding, Z. et al. (2019). The high frequency surface discharge behaviours of modified polyimide films: the effects of phenyl thioether in the backbone. *Materials Research Express* 6 (8): 085082.

9

Smart Polymer Dielectrics

Xiaoyan Huang, Lu Han, Zhiwen Huang, and Qi Li

State Key Laboratory of Power System, Department of Electrical Engineering, Tsinghua University, Beijing, 100084, China

9.1 Introduction

Traditional dielectric materials are developed to be durable in order to extend their lifetime, but material degradation is unavoidable, especially in complex conditions and severe surroundings [1, 2]. Nature inspires us to find new ways to overcome this difficulty. Many creatures in nature can live in hard environments not just because of their ability to fight predators and withstand injuries but also because such species have developed numerous behaviors and functions to adapt to environments, detect dangers, and repair injuries. Chameleons typically alter their skin colors to hide and avoid detection by predators. Animals usually feel pain when they are exposed to harmful stimuli, which can activate their defense system in order to prevent additional injury. Organisms can possess self-healing capabilities after being damaged, leading to restoration of lost functions [3–5]. Similarly, in the field of electrical insulation, we can develop bionic dielectric materials with autonomous functions to avoid, diagnose, and even repair electrical degradation, which is beneficial to electrical equipment innovation and energy utilization.

There is a growing need for power equipment with more dependable performance and a longer service time with the development of ultra-high voltage power transmission technology. Electrical device robustness is dependent on insulating materials [6]. Therefore, any deterioration greatly raises the danger of equipment and system failure. For example, uneven electric field distribution of cable terminations, wall bushings, and other insulating parts in high-voltage devices tends to bring great difficulties to the insulation design. Traditional techniques of electric devices such as stress cones of terminals and capacitor plates of bushings are no longer applicable at higher voltage levels. Moreover, space charge accumulation, overheating, mechanical stress overload, ultraviolet irradiation, partial discharge, a microdefect, and other potential factors may eventually result in a dielectric breakdown, which is challenging to detect insulation defects in power apparatus. Furthermore, permanent deterioration of the unrecoverable solid insulation continuously operating in severe conditions will inevitably lead to catastrophic failure [7–9].

High Temperature Polymer Dielectrics: Fundamentals and Applications in Power Equipment, First Edition.
Edited by Jun-Wei Zha and Zhi-Min Dang.

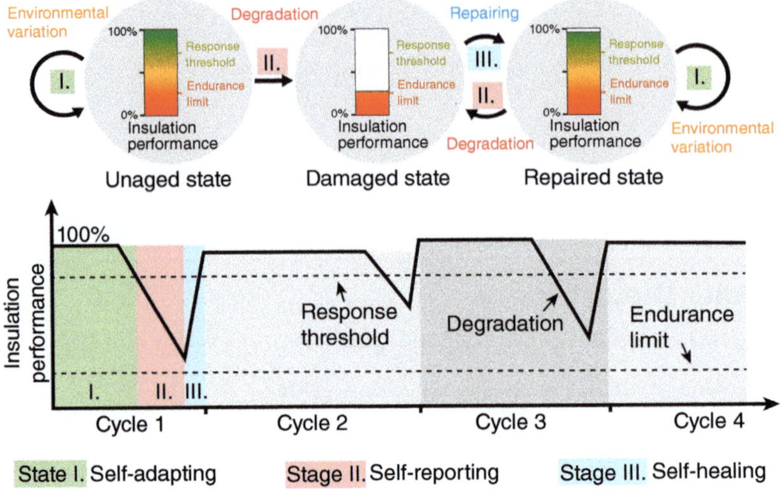

Figure 9.1 Life cycle control of biomimetic smart materials for next-generation electrical insulation.

Designing and manufacturing smart materials with self-adaptive, self-reporting, and self-healing functions is a brand-new solution to uneven distribution of electric fields, insulation defect monitoring, and electrical damage recovery in electrical devices [10]. Due to the nonlinear evolution of electrical parameters with the change of external field strength, self-adaptive dielectrics (SADs) exhibit excellent abilities to uniform local strong electric fields and quickly release dense charges, avoiding insulation failure caused by excessive concentration of field strength. Self-reporting dielectrics (SRDs) can display color changes in response to local performance degradation or detrimental surroundings, so the risk can be detected with the naked eye or with optical monitoring facilities to realize early warning of catastrophic failures. Self-healing dielectrics prolong material life by repairing internal damage after a solid insulating defect has formed and restoring electrical performance. Lifecycle control of insulating materials can be accomplished by rationally incorporating self-adaptive, self-reporting, and self-healing functions into dielectric polymers (Figure 9.1). At the stage of self-adapting, the electrical parameters of the smart dielectrics are modified corresponding to environmental conditions to preserve performance and minimize degradation. When environmental deterioration surpasses the capacity of the smart dielectric, performance degrades, and the material enters the self-reporting stage. At the stage of self-reporting, unacceptable deterioration causes changes in the appearance of the smart dielectrics and initiates a self-healing procedure. At the stage of self-healing, the damage is repaired and the performance is restored, so the material returns to the self-adapting stage. Theoretically, the preceding cycle can continue indefinitely until the material fully loses its autonomous functions. Compared with traditional dielectrics, smart dielectrics increase robustness and reduce manufacturing costs as well.

In this review, research on self-adaptive, self-reporting, and self-healing dielectric polymers in the field of electrical insulation and inspiring studies about bioinspired polymers with autonomous functions in other fields are introduced.

9.2 Self-Adaptive Dielectrics

Some intelligent creatures in nature can self-adapt to different colors in order to blend into the background and avoid predators. Similarly, SADs can adapt their electrical parameters to the spatial electric field to avoid insulation failure due to local intensified field strength. SADs typically have nonlinear conductivity (E) or permittivity (E) that increases rapidly with the local electric field when it exceeds a certain value, as illustrated in Figure 9.2a [10–12]. This nonlinear electrical property endows SADs with the ability to self-adaptively grade electric fields or release unwanted charges, the mechanism of which can be generalized as follows: when SADs work in moderated field strength, they behave just like common dielectrics with relatively low conductivity (10^{-14} S m^{-1}) and permittivity (a little higher than the polymer matrix); and when local field strength tends to exceed the acceptable range, SADs exhibit much larger σ or ε [13–15]. According to the law of current continuity shown in Eq. (9.1), larger σ or ε means lower field strength E under a continuous current density J, which can effectively suppress E to a safe value.

$$J(E) = \sigma(E)^{*}\,\mathrm{E} + \varepsilon(E)^{*}\,\frac{\mathrm{d}E}{\mathrm{d}t} \tag{9.1}$$

Furthermore, when overvoltage or discharge surges occur, SADs must have higher nonlinear conductivity to quickly release the impulse energy and protect the equipment. Simulations and experiments have shown that the field grading or stress control performance of SADs is far better and more stable than traditional techniques, which rely mostly on the optimization of the electrode configurations of the apparatus. The advantage of field grading by SADs can be illustrated in Figure 9.2b. In conventional approaches, material electrical property and electrode structure directly determine the spatial electric field distribution in power apparatus. Assuming the E-field as the control target, this approach is an open-loop control without any feed-back mechanism. In this case, the control performance is not satisfying, and it is rather sensitive to parameter disturbance. For instance, a little discrepancy in the geometry of the field grading structure with a fixed design can considerably affect the field grading effect, and temperature change or

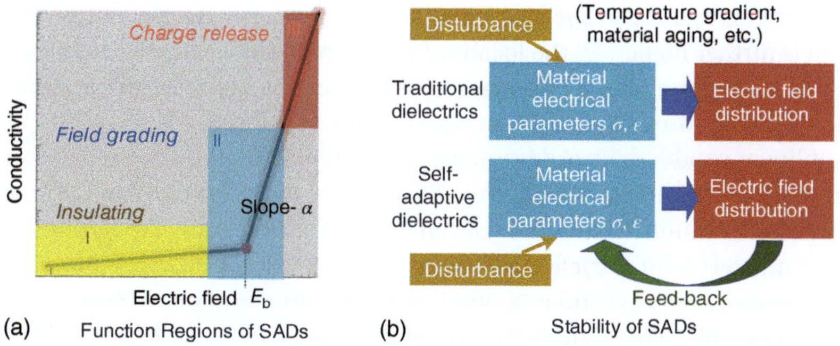

Figure 9.2 Application mechanism of self-adaptive dielectrics. (a) Typical nonlinear $J(E)$ characteristics of SADs and their function regions. (b) Schematic diagram of the field grading mechanism of SADs. Source: Huang et al. [10]/IEEE/CC BY 3.0.

material aging process can also lead to a much different E-field distribution than expected. As a result, traditional approaches have encountered great challenges in the application of power apparatus with increasingly higher voltage levels and more uneven electric fields. On the contrary, field grading and stress control by SADs are typical closed-loop controls. On one hand, the material parameters affect the electric field distribution; on the other hand, the electric field also feeds back to the material parameter, which self-adaptively changes with it. In this case, an effective and stable field grading effect can be obtained by SADs regardless of the geometry discrepancy, different temperatures, or material aging.

For SADs, the most important parameters for their performance are the switching field E_b and the nonlinear coefficient α. E_b is defined as the switching point above which the material begins to exhibit a rocketing conductivity (permittivity) [16, 17]. It should be adapted to specific applications, as different power apparatuses or electronic devices with different voltage levels have different requirements for the electric field. Thus, E_b of SADs is expected to be flexibly tuned according to practical cases, which may range from 0.1 to 5 kV mm^{-1}. How to choose an appropriate E_b for different applications is also an important issue, which is elaborated in the application part. In terms of the nonlinear coefficient α, it is the slope of the $J(E)$ curve (in logarithmic coordinates) and reflects the degree of effect of the electric field on material parameters. The greater the α, the better the field grading performance will be. Ideal SADs are of an infinity α with vertical switching behavior. In practical cases, an α above 10 is high enough for field grading and stress control [18].

The above illustration shows that the key function of SADs is realized by a nonlinear conductivity or permittivity that rapidly switches with the electric field. For the former, many high-field conduction effects such as tuning or Schottky emission can result in nonlinear conductivity. However, these effects are not stable enough or may lead to puncture of the insulation matrix. Furthermore, the nonlinearity due to these mechanisms is rather low, and the switching field cannot be flexibly controlled [19–21]. Therefore, most SADs with nonlinear conductivity are based on the mechanism of the double-Schottky barrier, whose height Φ_B is tuned by the localized states Q_i and applied electric field. Under relatively low electric field, the barrier height Φ_B only decreases a little with E, and the localized state Q_i even increases with E. The increased Q_i can stabilize the barrier and prevent Φ_B from fast decay. At this time, the conduction barrier is relatively high and it has good insulation properties. When the field strength is further higher, all the localized states are fulfilled, and the stabilization effect of Q_i is of minor effect. This leads to a fast decrease of the barrier Φ_B, and the current J flowing through the barrier significantly enhances, which results in excellent nonlinear conductivity. Additionally, the conduction relaxation effect can also lead to nonlinear effective permittivity. Thus, SADs based on this mechanism usually exhibit both nonlinear $\sigma(E)$ and $\varepsilon(E)$. In terms of SADs featured as nonlinear permittivity, materials with ferroelectric properties may exhibit a field-dependent $\varepsilon(E)$. Nevertheless, both the coefficient and the maximum ε of these materials are not satisfying and can hardly be applied for field grading. Above all, the nonlinear mechanism of most current SADs relies on the double-Schottky barrier effect, and their difference lies in various fabrication and connection approaches to the barrier.

For SADs that are applied in power apparatus or electronic devices, their major material system is the same as traditional polymer dielectrics for good compatibility, like cross-linked polyethylene (XLPE) for cables, silicone rubber for insulators, epoxy paper for bushings, polydimethylsiloxane (PDMS) for device packaging, etc. [22, 23]. As polymer dielectrics themselves do not possess desired field-dependent electrical properties, doping of functional fillers is performed to endow them with nonlinear behaviors. Thus, SADs are also called nonlinear composites. The most used functional (nonlinearly conducting) fillers are SiC and ZnO microvaristors, which are raw materials for surge arrestors widely used in power systems. Therefore, the microvaristor/polymer composite SADs are expected to possess excellent nonlinear $J(E)$ characteristics of arrestors. It has been elaborated that the nonlinear mechanism of SADs lies on the double Schottky barrier in the above part. There are also reports that the composites obtained by the graphene oxide or graphite nanosheets and insulation matrix have nonlinear conductivity, which may come from the energy barrier formed on the surface of such fillers. The foundation of this kind of material is not comprehensive enough, and it has yet to be used in engineering practice. For SADs with SiC fillers (denoted as SiC composites below; SADs with other fillers are the same), the Schottky barriers are formed over the interfaces between contacted SiC fillers. In this case, the barrier height of the interfaces is relatively low, which means a relatively low α is also easily affected by filler concentrations and temperature. But for ZnO microvaristor fillers, the Schottky barrier lies on the grain boundary inside each filler particle. Thus, each of the barrier heights is high and the nonlinear properties of the composites are much more stable than SiC. Additionally, the switching field E_b of ZnO composites can be flexibly tuned according to specific requirements in the range of 0.3–2 kV/mm by controlling the grain size in the fillers, the filler diameter, as well as the filler concentrations [24]. As a result, ZnO composites have attracted wide attention in the past decades, and their performance has been well studied. However, the diameter of SiC particles is usually several micrometers, but that of ZnO microvaristors is around tens of micrometers, which makes them easily precipitate in the polymer matrix, and their combination with the matrix is rather loose. Thus, for epoxy-based SADs and other applications that require good mechanical properties, SiC fillers are preferred.

In summary, modern SADs are pursuing stable and high-level nonlinearity, flexible tuned switching field, as well as the light load of nano-scaled fillers, which have little effect on the original property of the polymer matrix. Advanced fabrication, modification, and assembling technology on low-dimensional particles is expected to endow SADs with further enhanced performances.

Up to now, conventional SADs with SiC, ZnO, and other semiconductors as fillers have drawn great attention from academia and industry. Numerous researches report the potential application of such materials in power systems, such as outdoor insulators, high-voltage direct current (HVDC) spacers, bushings, and power electronics, as verified by experiments and simulations [25–29]. A design of outdoor insulation with a nonlinear layer to achieve a homogenous electric distribution is shown in Figure 9.3(a). After the optimization simulation of geometry and parameters of the SADs layer, they prepared a 220 kV prototype and electric field distribution measurements were performed by an optical electric field sensor at 10 reference

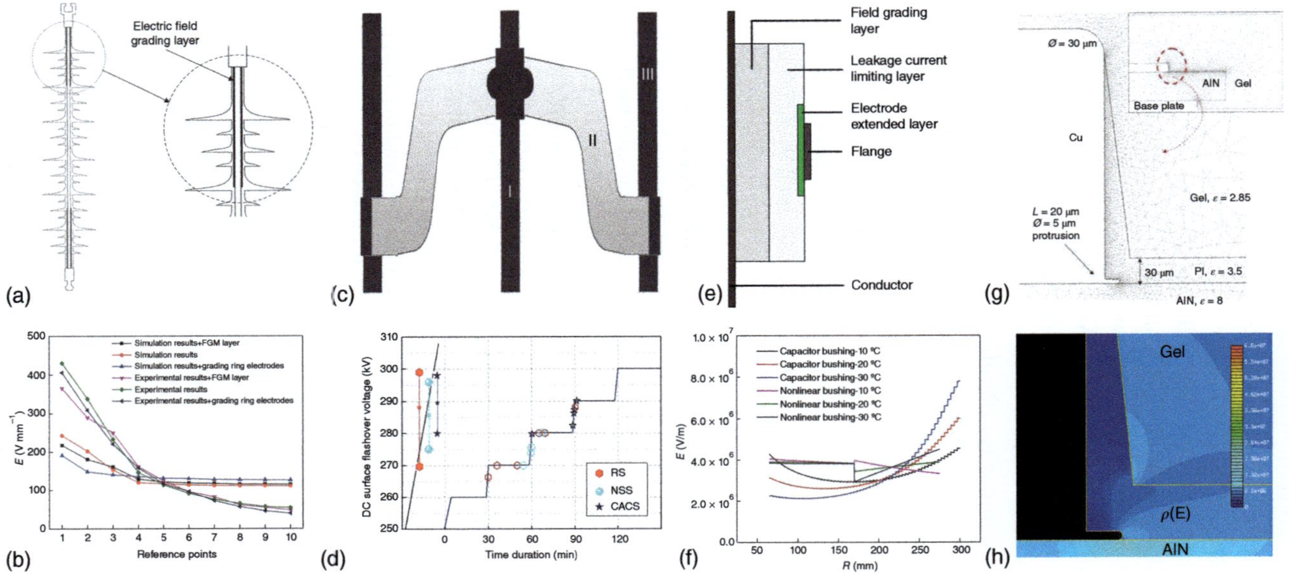

Figure 9.3 Examples of power apparatus and devices based on SADs. (a) Schematic of outdoor insulators with SADs and (b) electric field distribution from experimental and simulation. (c) Geometries of the CACS spacers and (d) DC surface flashover test results. (e) Structural diagram of adaptive bushing and (f) the simulated radial electric field distribution. (g) Geometry and mesh used for the simulations of IGBT with SAD materials coated on AlN substrate and (h) the simulated electric field. Source: Huang et al. [10]/IEEE/CC BY 3.0.

points. As shown in Figure 9.3(b), the maximum electric field and nonuniformity of electric field distribution of the traditional insulator are greater than that of the insulator with a nonlinear layer. A similar application occurs in HVDC spacers, as shown in Figure 9.3(c)–(d), where the DC flashover voltage increases significantly when nonlinear materials placed on the surface in coated form. Figures 9.3(e)–(h) show the application cases of nonlinear materials in bushings and power electronics. Nonlinear materials, rather than coatings, act as a volumetric structure that tailors the electric field in space inside electrical equipment, effectively avoiding internal discharge and material breakdown caused by an excessive local electric field. However, most research mainly focused on the field grading effect of ZnO composites under normal conditions, their performance in complicated operation conditions should be verified. Both Joule heat generated by the conducting portion and variations in ambient temperature can result in temporal and spatial nonuniformity of temperature distribution throughout the device in real insulating applications. Furthermore, under transient overvoltage, the conduction routes in SADs can generate a lot of heat, resulting in a small hot spot in the composites. The nonlinear electrical behavior and field grading impact of SADs may be considerably influenced under these conditions. To assess the dependability of SADs in high-voltage equipment, the nonlinear conducting behavior of SADs at different temperatures or under more complex electrical–thermal stress must be obtained. In terms of time-dependent nonlinear conducting behavior, previous study results reveal that the influence of temperature on ZnO composites with different insulating matrixes (such as silicone rubber, EPDM, and HDPE) might be extremely different [30]. However, there has yet to be a comparison of different matrixes with the same filler characteristic, as well as the effect of the filler on the temperature-dependent conductivity of the composites. Furthermore, few relevant studies have addressed the nonlinear electrical feature of SADs under thermal stimulation such as heat cycle or transient overvoltage heating. As a result, a thorough investigation of how insulating matrix and filler parameters impact the temperature-dependent nonlinear conductivity of ZnO composites is required. Also, research into the electrical response of the composites under various thermal excitations should be performed.

For the nonlinear conducting behavior of ZnO composites under a wide temperature range, DC V–I characteristics of the composite samples are measured in the chamber under temperatures of −30 to 120 °C. At each temperature point, the samples were placed for at least an hour before measurement for temperature stabilization throughout the sample. To investigate the stability of ZnO composites to large temperature changes, samples were heated to 100 °C for two hours and then naturally cooled to room temperature. The U–I characteristics of the same sample before and after the heat cycle are compared.

The nonlinear conductivity of ZnO/silicon rubber (SiR) composites and ZnO/EPDM composites with 37 and 45 vol% filler volume fractions was evaluated at different temperatures to investigate the effect of temperature on SADs. The switching field E_b of the ZnO composites is defined as the applied electric field when the current density reaches 3 μA cm^{-2}, which corresponds to a total current of about 10 μA flowing through the composite sample. The measurement is performed by raising the temperature of the closed container containing the adaptive material samples for 12 hours, waiting for the internal temperature of the adaptive material to stabilize, and then measuring its nonlinear volt–ampere curve.

It should be noted first that for SADs with ZnO microvaristor fillers, sufficient conduction routes comprised entirely of the fillers are often required for stable nonlinear parameters [32, 33]. When the filler concentration falls below a threshold of roughly 33 vol% in the percolation process of functional composites, the conduction routes comprise both fillers and an insulating matrix. In this scenario, E_b is mostly determined by the breakdown strength of the polymer matrix and is quite unstable, with considerable differences across samples, making it unsuitable for field grading. When the filler concentration exceeds the percolation threshold, the conduction routes are completely filled with fillers. In this situation, the polymer matrix's properties have little impact on the material's nonlinear conducting behavior, which is driven by the fillers and their contact surfaces. This ensures that the composite's nonlinear performance is steady. As a result, this work focuses on SADs with relatively high filler concentrations over the threshold, such as 37 and 45 vol%, with the analysis focusing on the properties of conduction routes. Because the 37 vol% ZnO/EPDM complex is prone to thermal breakdown once the test current exceeds 5 μA cm^{-2}, only a nonlinear conductivity curve with a maximum current density of about 3 μA cm^{-2} is provided, and the complex's nonlinearity coefficient at this volume fraction cannot be calculated, as shown in Table 9.1. The effect of temperature on the nonlinear conductance characteristics of adaptable materials may be seen

Table 9.1 The threshold electric field E_b (V mm^{-1}) and nonlinearity coefficient α of the ZnO composites at different temperatures T (°C).

(a) 45 vol% ZnO/SiR composites									
T	−30	−10	0	20	40	60	80	100	120
E_b	993	885	840	725	750	753	785	838	873
α	13.6	10.5	9.2	8.8	8.8	8.3	8.1	8.1	8.1
(b) 37 vol% ZnO/SiR composites									
T	−30	−10	0	20	40	60	80	100	120
E_b	1322	1194	1103	991	1031	1119	1284	1450	/
α	13.3	12.1	9.9	8.1	8.3	7.7	8.1	8.8	/
(c) 45 vol% ZnO/EPDM composites									
T	−30	−10	0	20	40	60	80	100	120
E_b	1288	1163	1090	975	1000	958	950	925	918
α	14.1	13.7	12.6	9.0	8.8	8.6	8.4	7.8	7.7
(d) 37 vol% ZnO/EPDM composites									
T	−30	−10	0	20	40	60	80	100	120
E_b	2038	1863	1781	1756	1763	1884	1891	1944	1972

Source: Yang et al. [31]/IEEE.

in two aspects: First, temperature influences the conductivity properties of grain boundaries within ZnO microvaristors [31].

The single-grain boundary nonlinear conductance characteristics at different temperatures can be calculated using the relationship between the conductance current flowing through the grain boundary and the grain boundary voltage and temperature T as Eq. (9.2) [34, 35].

$$\mathrm{J} = J_0{}^{*} \exp\left[-\frac{E_b - \beta' F^{\frac{1}{2}}}{kT}\right] \tag{9.2}$$

where J, E_b, F, k, and T are the current density, the Schottky barrier height, the field intensity, the Boltzmann's constant, and the temperature, respectively. And J_0, β' are the constants. With rising temperatures, the density of current passing through grain boundaries in the same electric field increases noticeably as shown in Figure 9.4. It follows that the voltage-sensitive voltage of the single-grain boundary decreases with increasing temperature. The same conclusion also applies to the complex, which is made up of an infinite number of conductive nonlinear grain boundaries. In other words, as the temperature rises, the compound's pressure-sensitive electric field weakens. Secondly, temperature also affects the conduction path and contact resistance between the nonlinear filler particles. As the temperature increases, the composite insulation matrix expands, which makes the contact between the packers more relaxed, or the force between the packers is lower, and the loose contact increases the contact resistance so that the threshold field strength of the compound increases and the nonlinearity coefficient decreases. In addition, the expansion of the complex may also cause changes in the conduction path. Since the coefficient of expansion of materials such as silicone rubber is much higher than that of fillers,

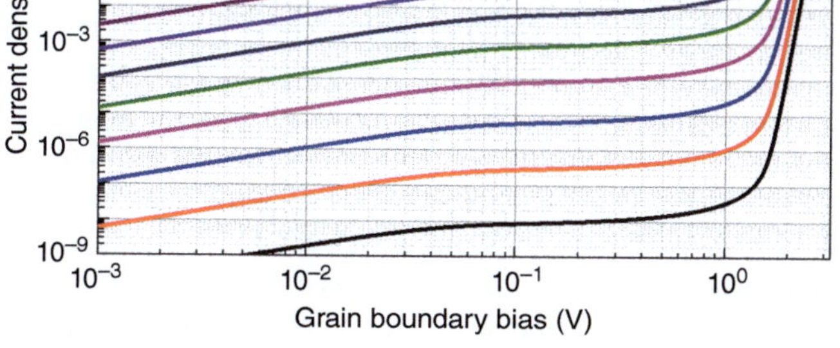

Figure 9.4 Calculated nonlinear conductivity curves of single grain boundaries at different temperatures. Source: Reproduced with permission from Yang et al. [31]. Copyright 2023, IEEE.

the conduction path composed of fillers will be interrupted when the compound expands, creating a silicone rubber gap between the fillers, which further makes the threshold electric field higher. The effect of the expansion of the composite on the contact resistance shows that the temperature increase will cause the compound to expand, resulting in an increase in its pressure-sensitive electric field. The above two factors make the increase in temperature have opposite effects on the nonlinear conductivity characteristics of the compound, and the nonlinear conductivity characteristics of the composites at any temperature are the result of the balance of the above two factors. If the former factor dominates, the threshold electric field decreases with the increase in temperature; if the latter factor dominates, the threshold electric field increases with the increase in temperature. Therefore, the effect of temperature on the pressure-sensitive electric field of the compound will be affected by many factors, including the case when the filler constitutes a conductive channel, including the thermal expansion of the matrix itself, and the effect of different temperature ranges on temperature will also be affected. In general, the high temperature (120 °C) that may be reached as a new functional material oriented to the internal electrical overstress for power and electronic equipment does not directly cause the filler to fail but rather poses a threat to the matrix filler (such as XLPE and SiR), which in turn affects its electrical properties through the complicated effects discussed above.

In terms of heat cycle stability, Figure 9.6 shows nonlinear $J(E)$ characteristics of 45 and 37 vol% ZnO/SR composites before and after a heat cycle (heating to 100 °C and subsequently cooling to room temperature). Samples 1–4 in the figure are distinct measured samples from the same set of 45% ZnO/SR composite and 37% composite. It clearly indicates that the nonlinear curves shift to the right following the heat cycle, indicating a greater switching field. The reason for this is that the composites' expansion during heating and contraction may be symmetric, while cooling may not be. Some conduction routes are broken during expansion and can return following shrinking, as seen in Figure 9.5. Figure 9.5a shows the initial state of the material before heating. In Figure 9.5c, due to the shrinkage of the insulation matrix during the cooling process, fillers separated by silicone rubber gaps are in contact again, and the conduction path 2 has recovered. But in the other possible process of Figure 9.5b to Figure 9.5d, during the shrinkage process, the separated fillers fail to form a thorough conduction path, which is still blocked by silicone rubber gaps. Under this circumstance, the number of conduction paths decrease, which means a larger switching field. This process of expansion and shrinkage is nearly random: if more conduction paths recover after shrinkage, the variation of E_b is small, as shown in the case of samples 2 and 4 in Figure 9.6a; otherwise, the variation of E_b is relatively large, as shown in the case of samples 1 and 3 in Figure 9.6a. For ZnO composites with lower filler concentration, there are fewer conduction paths, and the effect of the heat cycle may be more significant, shown as in the case of sample 2 in Figure 9.6b. The average variation of E_b is about 7% for 45 vol% ZnO/SR composites and 11% for 37 vol% composites. This indicates that a short-term but significant temperature increase (caused by overload, etc.) is also likely to considerably affect the nonlinear behavior of SADs under working conditions.

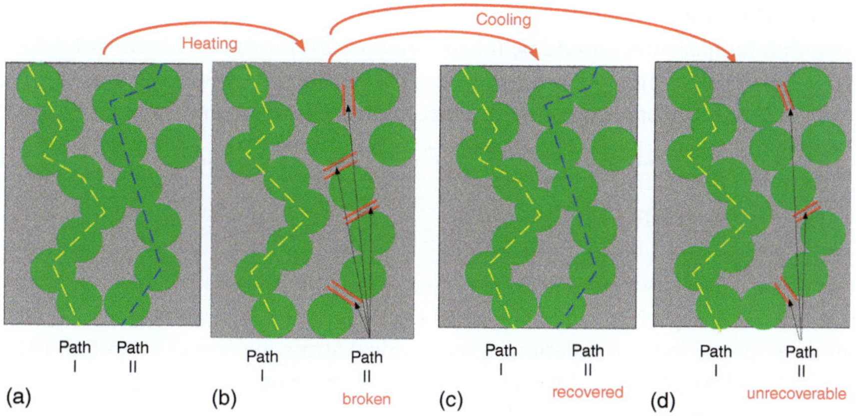

Figure 9.5 Schematic diagram of conduction paths within the compound as a function of temperature. Source: Reproduced with permission from Yang et al. [31]. Copyright 2023, IEEE.

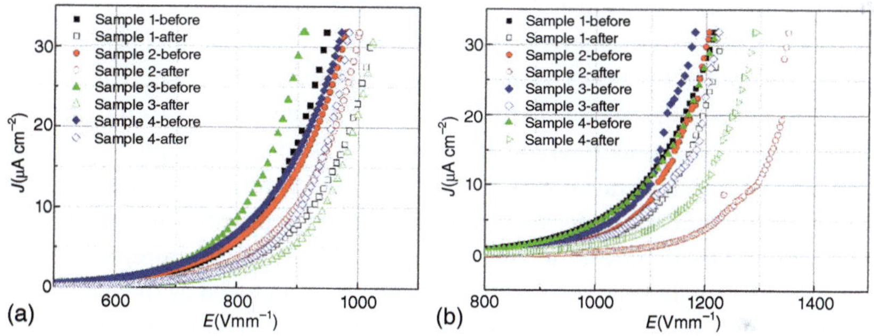

Figure 9.6 Nonlinear *J*(*E*) characteristics of several ZnO/SR samples before and after a heat cycle. (Samples 1–4 are different measured samples in the same group.) (a) 45 vol% ZnO/SiR composites, (b) 37 vol% ZnO/SR composites. Source: Reproduced with permission from Yang et al. [31]. Copyright 2023, IEEE.

The nonlinear V–I properties of SADs can change significantly with temperature. The temperature-dependent nonlinear conducting behavior of SADs is caused by the influence of nonlinear conducting filler and insulation matrix expansion. Temperature range, filler concentration, and insulating matrix can all influence the temperature dependency of nonlinear conductivity. When the filler concentration is low or the thermal expansion rate of the insulating matrix (such as silicone rubber) is quite high, the switching field of the composites can change dramatically with temperature. Furthermore, because of the possibility of asymmetric thermal expansion and shrinkage, a short-term but large temperature increase is anticipated to have a major impact on the nonlinear behavior of SADs. Besides, transitory high electrical stress can cause a local hot spot in the material due to Joule heat, resulting in filler degradation and thermal runaway of the polymer matrix. This is another crucial component influencing SAD thermal stability. Thus, in field grading design, the operation temperature should be taken into account for the nonlinear behavior of SADs. When selecting a good filler or insulation matrix property for SADs, the effect on SAD stability of temperature change should be considered.

9.3 Self-Reporting Dielectrics

Self-reporting materials can convey information about performance or the environment by changing physical or chemical characteristics in response to external stimuli such as UV light, temperature, mechanical stress, electrical potential, pH, and ions [20, 36]. Many natural phenomena inspire the production of self-reporting materials. For example, the animal's skin bleeds and produces pain after injury, which can help animals perceive and avoid danger. Self-reporting materials have mainly been used to monitor mechanical stress, microcracks, and metal corrosion recently [37–39], but the utilization of self-reporting insulating materials, to the best of our knowledge, in the field of electrical insulation has not been reported. Since the damage to solid insulation is irreversible, the accumulated degradation increases the likelihood of electrical failure when the power apparatus is used in harsh surroundings for a lengthy period. Moreover, even a microinsulation flaw anywhere in highly intense electric fields will cause a failure and exhibit no visible warning, so the unpredictable insulation breakdown makes monitoring and prewarning of material performance exceedingly intractable. Compared to detection by external devices, designing SRDs with functionalities to offer a visual alarm to possible threats in solid insulation is a simpler and more efficient solution to addressing the aforementioned challenges. Degradation in solid insulation can be easily recognized by the naked eye or optical observing equipment as SRD synthesis and utility technology advances, and the robustness of transmission lines and power apparatus will be highly improved.

Stimuli-response materials and sensing materials, which are analogous to the concept of self-reporting materials, are more broadly and intensely investigated in the field of fundamental materials science. Until now, numerous smart materials have been designed to sense and detect a wide range of physical, chemical, or biological stimuli. The most common reaction of these materials, when they are activated by external stimuli, is stimuli-responsive chromism, such as photochromism, thermochromism, mechanochromism, hydrochromism, solvatochromism, and electrochromism. These smart materials can be widely used in aerospace, automotive, civil engineering, electrical engineering, microelectronics, printing, chemical sensors, bioimaging, and drug delivery, which may spur the development and implementation of SRDs in the electrical insulation field. In this section, the underlying notions and key principles of extensively researched common smart polymers that contribute to the manufacture of SRDs are introduced.

Self-reporting materials based on photochromic compounds, conjugated polymers, and microencapsulated systems are described below. Photochromic compounds are organic molecules isomerizing between two chemical forms that have different physical or chemical properties when exposed to UV light. They are typically incorporated into polymeric matrices physically or chemically to create a variety of self-reporting materials that can react to different stimuli. Conjugated polymers, such as polydiacetylenes (PDAs), have a distinctive backbone made of alternating single and double bonds, which gives them useful and unusual optical and electrical features. PDAs present a wide range of macroscope colors because of varying levels of backbone distortion brought on by interactions between side chains and external stimuli. By modifying diacetylene monomers and designing composite materials, a variety of highly visualized, adaptable, and selective self-reporting materials can be developed. Microencapsulated systems are separate material systems made up of indicator solutions for visualizing degradation in bulk polymers or on their surfaces and encapsulating containers for sealing off indicators from their surroundings. The negative interactions between microencapsulated indicators and polymeric matrices are minimized due to the presence of encapsulating containers. This kind of material is frequently used to identify metal corrosion and polymer wear.

9.3.1 Self-Reporting Materials Based on Photochromic Compounds

The physical and chemical characteristics of photochromic compounds, such as spiropyran (SP), spirooxazine, azobenzene, diarylethenes, and their derivatives, alter as a result of photo-induced isomerization. Due to the strong chemical activity of particular groups or bonds, these compounds also exhibit variations in visible light absorption or fluorescence emission in response to other stimuli, such as polarity, pH, and temperature [40–43]. SP and its derivatives are the most widely investigated photochromic compounds, which are constitutive of indoline and chromene moieties. The chromene moiety exhibits a ring-closed structure, the two moieties' planes are orthogonal in the inactivated state, and the SP molecule is hydrophobic, nonpolar, and colorless. The molecule isomerizes to a hydrophilic, polar, and colored merocyanine (MC) structure with a conjugated planar zwitterionic structure and undergoes the UV light-induced bond cleavage of the spiro C–O. Additionally, when exposed to heat or visible light, the ring-opened MC molecule transforms into the ring-closed SP form (Figure 9.7). Similar to SP, spirooxazine and its derivatives possess a two-part structure and a stimuli-chromic property, which exhibit better responsiveness to a variety of external stimuli, a higher level of fatigue resistance, and a longer lifespan compared to SP. When azobenzene and its derivatives are heated or subjected to UV/visible light irradiation, the azo group undergoes trans-cis isomerization. Aside from changing the color of azobenzene, the difference in chemical structure also has a substantial impact on the physicochemical properties of the media in which it is present. A common kind of diarylethene derivative is dithienylethene. Under alternate UV and visible light irradiation, a reversible cycloaddition process and photochromic phenomena happen. Photoswitchable conductive materials

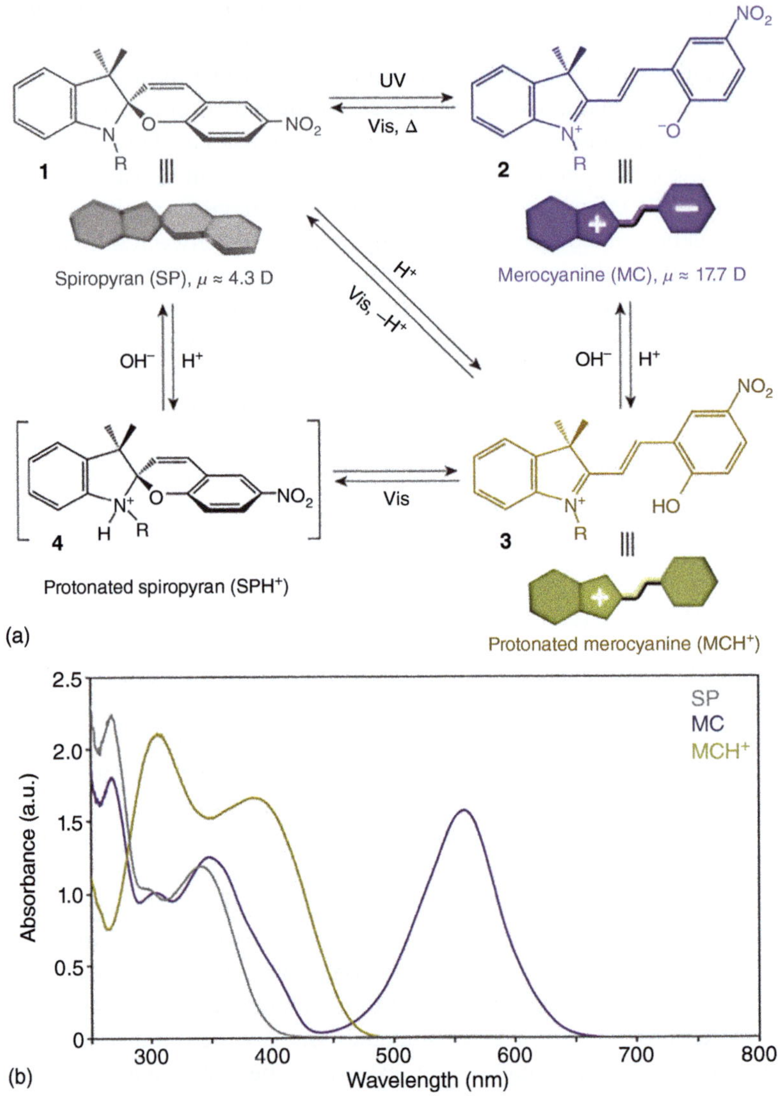

Figure 9.7 Photochromism and acidochromism of spiropyran. (a) Reversible transformations between the four states: spiropyran (SP) **1**, merocyanine (MC) **2**, protonated merocyanine (MCH^+) **3**, and protonated spiropyran (SPH^+) **4**. (b) UV–vis spectra of the parent spiropyran before (gray) and after (purple) UV irradiation, and after the addition of HCl (yellow). Source: Klajn [40]/Royal society of chemistry/CC BY 3.0.

containing dithienylethenes have practical application potential in the commercial field. In comparison to its ring-opened state, activated dithienylenthene has better conductivity and stronger visible light absorption. Using techniques including doping, copolymerization, cross-linking, and grafting, photochromic chemicals may be physically and chemically incorporated into polymeric matrices to produce a variety of self-reporting materials with different stimuli-responsive behaviors.

The most fundamental property of photochromic substances is photochromism. In the course of preparation, it is necessary to prevent the chromophore that is sensitive to the surrounding qualities from being destroyed. The rational design and synthesis of photo-responsive polymers are anticipated to increase their photoswitchability, reversibility, photostability, and photofatigue resistance. In order to create surface-attached, light-switchable polymer networks, Schenderlein et al. [44] chemically integrated a terpolymer made up of SP moieties, benzophenone groups, and dimethylacrylamide units onto planar silicon and glass surfaces. The UV illumination's intensity and duration had an impact on the photochromic surface's UV/vis absorption peak and maximum wavelengths (max). As a consequence, the color or visible light absorption spectrum of the surface might be used to estimate the strength of unknown UV irradiation.

Numerous polymers are subjected to mechanical stress in engineering applications, including tension, compression, and shear. Therefore, self-sensing of material wear, deterioration, and failure contributes to the stability and robustness of mechanical systems. To detect the degradation and failure of polymers brought on by force, Davis et al. [45] devised a self-reporting approach. Through the use of SP as a mechanophore, polymers were created, and each of the molecule's two moieties was independently linked to a polymeric chain. The colorless molecules isomerized to the colored MC form as a result of the force-induced deformation of the polymeric matrix, which increased the tensile stress of the SP molecules and caused a bond cleavage of the spiro C–O. A hue change was then noticed at the location of the deformation or degradation, and as the stress on the matrices increased, the intensity of it intensified until the material fractured. SP mechanophores have been effectively used in a variety of glassy (poly(methyl methacrylate), polystyrene, polycaprolactone, etc.) and rubbery (polymethacrylate, polyurethane, PDMS) polymeric matrices. Additionally, using diarylbibenzofuranone (DABBF) as a mechanophore, Kosuge et al. [46] created a new force-induced self-reporting polymer and thoroughly examined its mechanochromic properties. Mechanical strain caused the DABBF group's core C—C bond to rupture, resulting in the generation of stable blue radicals and a change in the material's hue from light yellow to blue. Using the two forms of mechanophore discussed above, they also created a multicolor mechanochromic polymer/silica composite that displayed several color states in response to various intensities of external stimuli [47] (Figure 9.8).

The photochromic compounds are combined with electroactive matrices, such as conductive polymers, gold nanoparticles, and electrolyte solutions, to create the electrochromic materials. Two different oxazine derivatives were studied by Zhu et al. [48] for their electrochromic properties in both solutions and indium tin oxide (ITO) devices. The opening reaction of the oxazine occurred when the chromophore was stimulated by electrical stimulation, changing the color from yellow to blue. As the applied electrical potential changed between 0 and 2.1 V, they also examined the

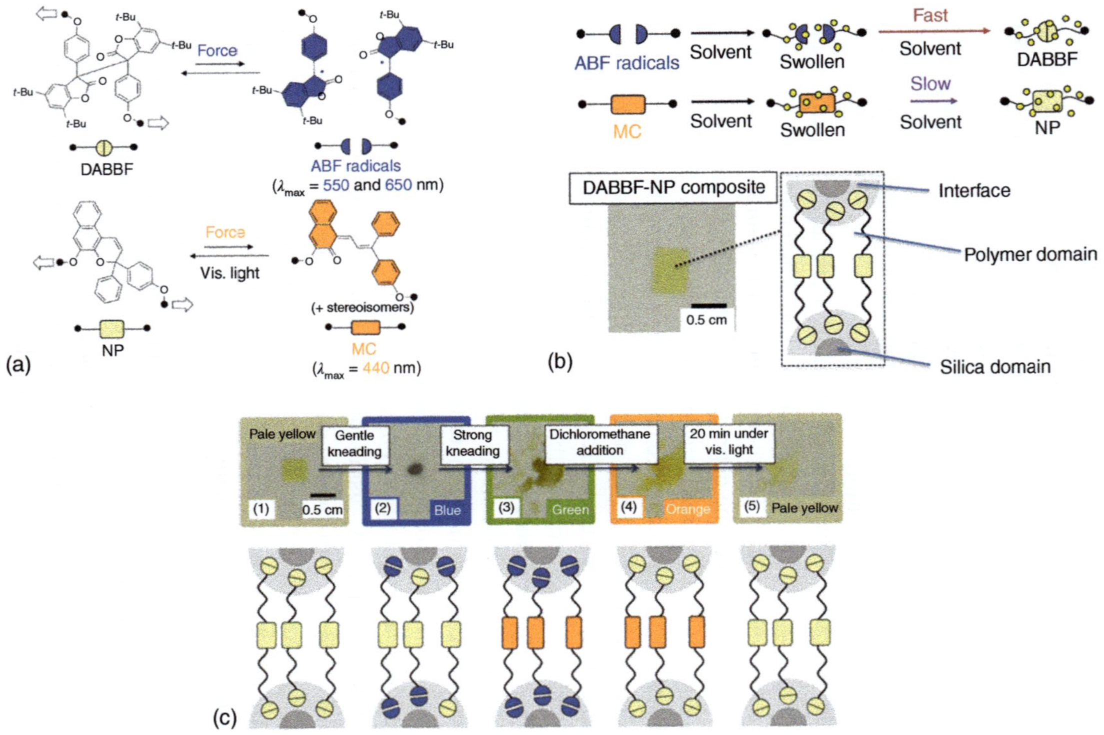

Figure 9.8 Multicolor mechanochromism of a polymer/silica composite with dual distinct mechanophores. (a) Chemical structures of two mechanophores and their activated states. (b) Illustration of the two mechanophores' locations in the sample and their reversible mechanochromic behaviors. (c) Optical photographs and plausible illustrations showing mechanophores' activation states in the pristine, gently kneaded, strongly kneaded, and dichloromethane-added samples, and the sample left under visible light for 20 minutes after dichloromethane addition. Source: Kosuge et al. [47]/with permission from American Chemical Society.

material's UV/vis absorption spectrum. Their results showed that the more voltage supplied, the more intensely visible light was absorbed and the more noticeable the color fluctuation was. The material has the potential to be utilized to measure the magnitude of an unknown voltage applied to it based on the qualities listed above.

The surrounding media have an impact on SP's photochromic characteristics. To be more exact, the polarity of the photo-triggered zwitterionic MC molecules is affected by the interactions with the solvents or media, leading to the presentation of a range of hues. The name of this phenomenon is SP photoswitchable solvatochromism. In several solvents with variable polarity, Rosario et al. [49] investigated the solvatochromism of glass substrates treated with SP. The surface on which the SP chromophore was covalently integrated showed a hypsochromic shift when the polarity of the solvents was increased in both the UV/vis absorption and fluorescence emission maxims. A negative photochromic phenomenon was seen in high-polarity solvents with hydrogen bonds because the ring-opened molecules were stabilized by the hydrogen bonds and the dipole–dipole interactions between the zwitterionic MCs and the solvents. Based on the aforementioned method, self-reporting materials that can distinguish between different organic solvents and detect the polarity of their surroundings may be created. By integrating SP-functionalized polymer brushes into fused silica microcapillaries, Florea et al. [50] created a sensing material that can detect solvents with various polarities. As a result of solvents of various polarity running through the microcapillaries continually, the SP chromophores that had previously undergone UV irradiation to activate them to MC isomers presented a variety of optical and spectral responses. Further examination of the hue and spectra allowed for simple identification of the unidentified solvent type.

According to the acidochromic process, the ring-opened MC isomer of SP occurs in protonic or acidic environments as yellow protonated MCH^+ but transforms back into purple MC under basic or aprotic conditions. SP must be activated by UV light in order to prevent it from isomerizing back to the ring-closed state and sustain the chemical activity of the acidochromophore in nonpolar liquids. Abdollahi et al. [51] chemically incorporated latex particles containing SP with cellulose to create a unique photoswitchable pH-sensing material. The substance displayed a range of colors in response to solvents with various pH values.

The ability of the zwitterionic MC isomer to couple with different ions and produce distinct colors is referred to as SP's ionochromic feature. While the electron-deficient indolium ring is sensitive to anions, its electron-rich phenolate moiety is simple to combine with cations. By covalently attaching the copolymer of spiropyran methacrylate derivative (SPMA) and methyl methacrylate (MMA) to the surface of a glass substrate, Fries et al. [52] were able to create a photoswitchable ionochromic material. With numerous divalent metal cations, the surface displayed a distinctive optical and spectral response. Shiraishi et al. [53] also showed that the cyanide ion (CN^-) in aqueous solutions may be detected using nitrospiropyran.

9.3.2 Self-Reporting Materials Based on Conjugated Polymers

PDA, polythiophene, and polypyrrole are typical conjugated polymers, which are organic macromolecules having a backbone made up of alternating single and double bonds. Attractive optical and electrical features are produced by their π-electron delocalization, which is caused by the overlap of p-orbitals. Excellent stimuli-responsive PDA provides a lot of promise to give polymeric dielectrics the capacity to self-report. Many different physical, chemical, and biological stimuli, including temperature, mechanical stress, current, organic solvent, ions, and biomolecules, cause it to change color. The conformational distortion of the PDA's backbone increases the side chains' freedom of motion and unpredictability when it is stimulated by outside stimuli. The UV/vis absorption peak experiences a hypsochromic shift from 650 to 550 nm, which causes the material to change color from blue to red. This is due to fewer coplanar side chains and a shorter p-conjugation length [54].

The most essential stimulus-responsive quality of PDA material is thermochromism. The visible light absorption intensity of PDAs increases continuously as the ambient temperature rises when the ambient temperature is higher than the activated temperature of the PDAs. As a result, the color of the materials exhibits a series of intermediate states that are visible to the unaided eye [55]. To manufacture PDAs with diverse activation temperatures, reversibility, and thermostability, numerous molecular modification approaches are offered. The thermochromic behaviors of PDA materials may be impacted by variations in head groups, alkyl side chains, and the location of conjugate diacetylene. As a result, it is simple to create chemosensors based on PDAs to monitor equipment's internal overheating and ambient temperature in a range of situations. By adding artificial nacre to montmorillonite substrates and adding the PDA thermochromophore to the composite, Peng et al. [56] created a sturdy thermochromic composite material. By successively raising the temperature from 20 to 70 °C, the material's apparent absorption peak progressively moves from 620 to 530 nm, changing the hue from purple to red (Figure 9.9).

Solvatochromism results from the interaction of the pendant groups with the organic solvent molecules when PDA is exposed to them, which causes the backbone to bend. The degree of backbone deformation and color fluctuation depends on the polarity of the solvents rather than on particular organic molecules, and this behavior is often nonspecific and connected to the polarity of the environment. Because of this, it is impossible to tell apart various solvents with similar polarities by looking at the color or spectrum of PDA sensors. Similar to thermochromism, modifying the headgroups can change the solvatochromic behaviors of PDA in an identical solvent. As a result, preparing a group of PDAs with different solvatochromic behaviors is an effective strategy for identifying solvents with approximate polarities. Park et al. [57] deposited four diverse PDAs onto conventional paper by an inkjet printer to form a detect array, which exhibits various color patterns ("fingerprint") in response to 11 organic solvents. Comparing the color response of the PDA array with the fingerprint of each solvent, an unknown solvent could be established.

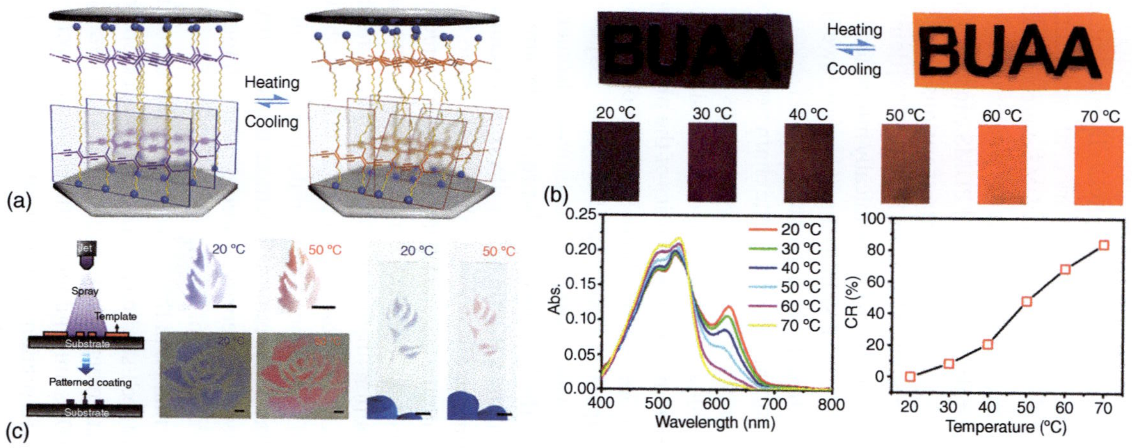

Figure 9.9 Thermochromic artificial nacre based on montmorillonite. (a) Illustration of thermochromism of the artificial nacre. (b) Reversible thermochromism, UV–vis spectra, and CR values between 20 and 70 °C. (c) Illustration of the spray coating via a spray gun, and different artificial nacre patterns inkjet-printed on paper, glass slide, and steel plate. All these artificial nacres show reversible thermochromism between 20 and 50 °C. Source: Peng et al. [56]/with permission from American Chemical Society.

The backbone of PDA is affected by the intermolecular forces caused by mechanical deformation in the polymeric matrix, which results in a mechanochromism and a shortening of the conjugation length. This characteristic can be used to identify certain analytes. For instance, because of the nonpolarity of the solvent molecules and the dearth of functional groups that may interact with the PDA probes, saturated aliphatic hydrocarbons (SAHCs) are challenging to directly create a solvatochromism. The color change of PDA that was physically integrated into PDMS films and microbeads was effectively produced by Park et al. by taking advantage of the mechanical stress caused by the expansion of a PDMS matrix. Additionally, they found that the PDA molecules were triggered to exhibit a color change from blue to red more quickly because of the quicker swelling of PDMS caused by the hydrocarbons with a shorter alkyl chain. It is useful to contrast various alkyl chain lengths using this stimuli-chromic phenomenon. Moreover, Seo et al. [58] demonstrated that PDA nanofibers with pendant groups that were chemically integrated into the hydrogel ions may exhibit a considerable color shift as a result of the swelling caused by the interaction between a dry hydrogel matrix and water.

By grafting photo-responsive substances on the side chain headgroups and doping photo-sensitive moieties that can interact with the headgroups into the matrix, PDA may also be photo-triggered. A blue-to-red colorimetric transition is caused by the alteration in the physical and chemical characteristics of the photo-responsive moieties caused by light irradiation. Reversible photochromism of a PDA crystal with azobenzene-modified headgroups was reported by Baek et al. [59]. A color change and crystal tearing phenomena were caused by the trans-cis isomerization of azo groups exposed to UV light and the hydrogen bonds provided by amines.

Poor electrical conductivity makes it difficult for electrons to cause a backbone deformation, making it challenging for an external electric field to induce an electrochromism of PDA. By causing the polymerization of diacetylenic precursors on carbon nanotube (CNT) surfaces, a CNT/PDA composite with a greater conductivity may be created. The interactions between PDA side chains are altered by the hopping of electrons between CNTs when current passes through the CNTs, which results in a distortion of the PDA backbone and a reduction in conjugation length [53]. PDA-PMMA/graphene composite materials demonstrated an irreversible electrochromism that was generated by a current, according to Zhang et al. [60]. To modify the stimuli-chromic properties of the composites, they found a linear connection between the graphene mass and the critical responsive current.

An affinochromism, another well-studied and commonly used property of PDAs, may be produced by the interaction between the headgroups and analytes, which can disrupt the PDA backbones. PDAs are equipped with excellent selectivity for detecting the targets thanks to the decoration of headgroups aimed at certain chemical and biological analytes. By using PDA nanoparticles that have been functionalized with peptides, Song et al. [61] were able to quickly observe the detection of the H1N1 influenza virus. The PDA backbones were altered by the unique interaction between peptide groups and the HA1 protein on the virus. The nanoparticles' distortion also became more noticeable as the virus concentration rose, and as a result, their hue changed from blue to purple to red.

9.3.3 Self-Reporting Materials Based on Encapsulated Systems

When employing encapsulated systems for self-reporting materials, the encapsulating containers separate the indicator solutions from the polymeric matrix, minimizing any negative interactions between the microencapsulated indicators and the media. The encapsulating containers are ruptured by the microcracks in bulk polymers and the bruising on their surfaces, releasing the indicator solutions into the area of damage or overflowing over the damaged surfaces. External stimuli then cause the disengaged indicators to be activated, which results in a change in color, fluorescence, or luminescence [62]. This kind of material is frequently used in aerospace, automotive, civil engineering, and wind turbine applications to observe mechanical damage in polymers, bruising on surfaces, and metal corrosion for microcrack identification, structural health monitoring, and catastrophic failure prevention.

The encapsulated systems are split into three groups based on the kind of encapsulating containers: microcapsules, hollow fibers, and vascular systems. When it comes to giving materials mechanochromic and self-reporting capabilities, microcapsules made of hollow spherical polymer shells have received a lot of attention, research, and usage [20]. The essential steps in the production of microcapsules are the polymerization of the shell's outer shell and the emulsion of the cores [63]. The hydrophobic core ingredients are mixed quickly with the aqueous during this procedure to create an o/w emulsion. Then, to create a shell wall, polymerization is initiated at the water-oil phase contact. The polymeric matrix may be used to create a self-reporting material with the ability to recognize mechanical damage by inserting indicator-filled microcapsules inside of it. Another common container used to hold and release indications is hollow glass fiber, which is often included in non-reinforced polymers [64]. The fiber can improve the mechanical characteristics of the matrices in addition to having outstanding compatibility with a variety of polymers. Hollow fibers can report and repair damage more effectively than microcapsules because they can store more indicators and healing chemicals in them. More complex 2D or 3D structures in microchannel and vascular networks enable many autonomous repairs of damage on a larger scale [65]. All indications and healing agents are linked, similar to circulatory systems, allowing for free movement and effective replenishment.

White et al. [66] created microcapsule/polymer composites to achieve the autonomous healing of polymers. Before the agent's polymerization was initiated by a catalyst built into the polymers beforehand, the encapsulated healing agent was liberated from the ruptured microcapsules. A red dye was simultaneously enclosed in the containers to demonstrate the self-healing mechanism. Despite this, because both microcracks and microcapsules had a red tint, it was challenging to tell them apart with the naked eye. A method of encapsulating switchable indications was suggested to improve the color contrast between damaged and undamaged parts. According to this plan, the inactivated indicator that was contained in the containers was meant to be colorless when it was discharged into the damaged areas and activated by environmental factors including pH, UV radiation, and catalyst.

To detect microcracks in fiber-reinforced composite and acrylic resin, Vidinejevs et al. [67, 68] presented a pH-induced self-reporting technique. By using silica gel or methyl 4-hydroxybenzoate to create an acidic environment, the microencapsulated crystal violet lactone leuco dye in this method isomerizes to a blue ring-opened form. A sort of material with the capacity to autonomously report damage was created by Li et al. [69] using 2′,7′-dichlorofluorescein (DCF) as a pH-switchable indicator. At the damage site, a color change from pale yellow to brilliant red resulted from the basic environment created by residual amines forcing the acidic DCF released into the resin matrix.

A UV-induced self-reporting substance was created by Credico et al. [70] by inserting microcapsules with a UV-responsive core and a UV-absorbing shell into the polymer matrix. While the encapsulated cores were shielded from activation by the shells, the unrestrained indicators discharged by containers were directly exposed to UV light. As a result, the degraded location underwent photochromism that was easily visible to the naked eye. A sort of self-sensing material based on microcapsules made of UV-protecting shells and fluorescent liquids was produced by Postiglione et al. [71]. Similar to the photochromic cores described above, the photofluorescent process occurred without the need for shells only if the core components were liberated from the ruptured capsules. Consequently, a UV-induced blue fluorescence emission could be seen coming from the microcrack spots.

By utilizing a catalyst to start the polymerization process and produce colored conjugated polymers, Odom et al. [72] presented a unique damage-detecting technique. To be exact, the microcapsules containing the Grubbs–Love (GL) catalyst and 1,3,5,7-cyclooctatetraene (COT) monomer were disseminated throughout the polymeric matrix. Induced by the GL catalyst, the virtually colorless COT monomers underwent ring-opening metathesis polymerization to create the red polyacetylene, which provided a clear visual distinction between damaged and healthy polymers. Robb et al. [73] developed a physically activated self-reporting material for mechanical damage indication using the aggregation-induced emission (AIE) phenomenon. When dissolved in a suitable solvent, the microencapsulated AIE luminogens showed no fluorescence, but since their intramolecular movements were constrained when they precipitated from the solvent, an emission began to occur. As a result, after solvent evaporation, the excurrent core materials exposed to UV light released a blue fluorescence. A charge-transfer complex-based microencapsulating self-reporting approach for viewing mechanical deformation and damage was suggested by Lavrenova et al. [74] Each donor and acceptor were placed inside a microcapsule before being evenly distributed throughout PDMS. The color of the polymeric matrix changed from yellow to red as a result of the two core ingredients being included after deformation or damage.

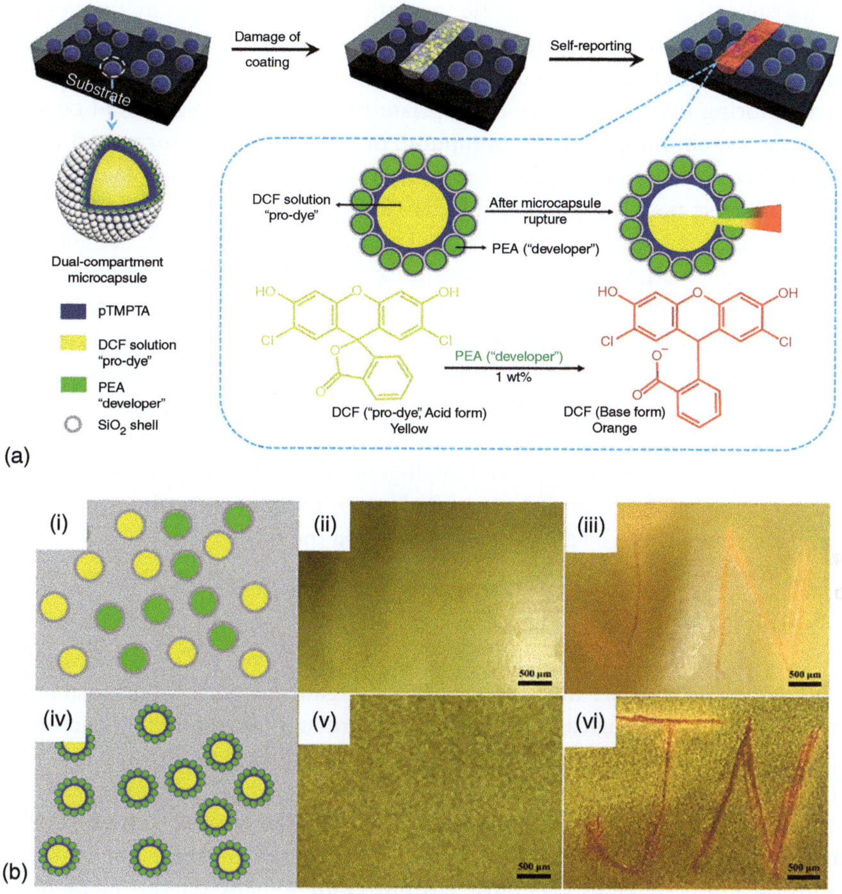

Figure 9.10 Robust damage-reporting strategy enabled by dual-compartment microcapsules. (a) Schematic diagram of the self-reporting material with dual-compartment microcapsules and its damage visualization mechanism. (b) Damage visualization effects of the self-reporting material based on double microcapsules (iv) and dual-compartment microcapsules (i). Source: Chen et al. [75]/with permission from American Chemical Society.

Recently, Chen et al. [75] created a self-reporting material with dual-compartment microcapsules for the viewing of mechanical stress and microcracks. They used identical microcapsules to encapsulate two different payloads in various compartments. In contrast to the conventional double-microcapsule self-reporting strategies, the two-component reaction's exceptionally small diffusion lengths caused the color to shift at the damaged region more quickly and noticeably (Figure 9.10).

9.3.4 Outlook for SRDs

The best SRDs should be capable of quickly reporting, significantly reacting, and accurately diagnosing. SRDs are expected to exhibit changes in color, fluorescence, or luminescence as soon as unacceptable insulation degradation or environmental deterioration happens to alert users to potential catastrophic failure risks. The variation in optical characteristics should be simple to detect with the unaided eye or optical monitoring instruments. An appropriate reaction threshold should be set, and the activated chromophore or fluorophore of SRDs should be turned off after material performances or ambient circumstances have returned to normal levels to eliminate both false-positive and false-negative findings.

The following three stages of SRDs, from concepts to applications, are shown: SRDs for diverse circumstances, such as exterior insulation of insulators, internal insulation of cables, and substantial insulation of electric machinery, are anticipated to be created in highly regulated and optimized laboratory settings. Maintaining the self-reporting capabilities of SRDs in challenging external conditions, as well as high enough stability and fatigue resistance of the polymeric matrices throughout their lifespan, takes a lot of work in order to access the commercial market. SRDs are also anticipated to be used with artificial intelligence tools like computer vision and machine perception. The performance of dielectrics in various conditions can be monitored and forewarned by an improved system in the future.

9.4 Self-Healing Dielectrics

Self-healing mechanisms are ubiquitous in living organisms, making it possible to maintain a stable physiological environment and rebuild normal vital functions. Since damage resistance and long lifetime are also desirable properties of nonliving materials, many efforts have been made to introduce self-healing properties into synthetic materials. To date, a variety of self-healing methods have been proposed, including microencapsulation, reversible bonding and interactions, and physical interdiffusion [76, 77]. However, most current studies focus on the healing of mechanical injuries, where fracture loading and other mechanical properties are used to assess healing efficiency. With the emerging trend of designing smart high-temperature dielectric polymers, self-healing strategies for dielectrics have received increasing attention. Recent advances in restoring dielectric properties and dielectric strength show great potential for improving dielectric reliability and durability and reveal possible industrial applications in the near future.

In this section, we first compare mechanical and electrical damage. The special nature of electrical damage is highlighted, which should be taken into account when designing self-healing dielectrics, especially when attempting to exploit traditional self-healing strategies. Next, we take a closer look at some recent developments, focusing on improvements that make them applicable to dielectric polymers. We conclude the discussion by summarizing possible paths and next challenges for designing self-healing dielectrics.

9.4.1 Expectations and Challenges in Developing Self-Healing Dielectrics

In addition to mechanical damage, including cracking and fracture, dielectric polymers may also suffer electrical damage when operating under high electric fields. The typical form of electrical damage is electrical treeing, which seriously affects the reliability and life of the application [78]. Direct breakdown occurs almost simultaneously with overvoltage, and the treeing process in solid dielectrics usually occurs at normal operating voltages and lasts longer. Electrical treeing usually starts at locations with high electrical stress, especially where interstitial moisture is present [9]. When the local electric field at these locations exceeds the electrical strength of the polymer matrix [79], some electrons are accelerated and collide with the polymer chains, resulting in bond scission and free radicals. As the degradation process continues, microcavities appear where partial discharges can occur, leading to the rapid growth of tree channels [78]. Ultimately, the electric tree will be a fractal system with dendritic hollow channels. During treeing, the dielectric and insulating properties of the polymer are adversely affected, while the risk of catastrophic electrical breakdown is increased [9].

Apart from electrical trees, water trees and corona discharges are also common types of electrical damage. Water treeing is a degradation phenomenon that creates dendritic structures in polymer dielectrics under the influence of both moisture and electrical field, featuring a permanent hydrophilic characteristic and a reduced breakdown strength [80]. Corona discharging, which involves interactions between polymer chains and high-energy particles or radiations, can also do harm to the surface layer of dielectrics, leading to a lower tensile strength and an elevated dielectric loss [81, 82].

When developing self-healing insulating materials, an ideal self-healing system should be resistant to mechanical and electrical damage. Electrical trees are usually measured in microns, while mechanical damage self-healing is always tested in millimeters or larger. Therefore, when adopting current dielectric polymer self-healing methods, considering whether the technique is capable of healing electrical damage should be one of the most important aspects.

In addition to the aforementioned factors, additional challenges in designing dielectric polymers with self-healing properties include the elimination of chemical degradation products and the ability to heal after a long waiting time (the time period from injury to the application of the conditions required for healing) [83]. It is also desirable that self-healing dielectrics be easy to fabricate and recycle for full life-cycle control. However, these issues have not received much attention so far, indicating that further research is still needed in this emerging field.

9.4.2 Melting Interdiffusion by Magnetic/Microwave Heating of Nanoparticles

When subjected to an oscillating magnetic field (OMF), magnetic nanoparticles undergo energy dissipation processes and generate heat due to Brownian and Néel relaxations [84, 85]. Mechanical damage self-healing of various thermoplastic polymers has been demonstrated by introducing magnetic nanoparticles into dielectric polymer matrices [86–89], a schematic diagram of which is shown in Figure 9.11. At high temperatures under OMF treatment, the polymer melts and flows to close the cracks, allowing chain diffusion and randomization to restore mechanical properties [89, 90]. At least 3.5 wt% [87] or 1 vol% [88] of magnetic nanoparticles

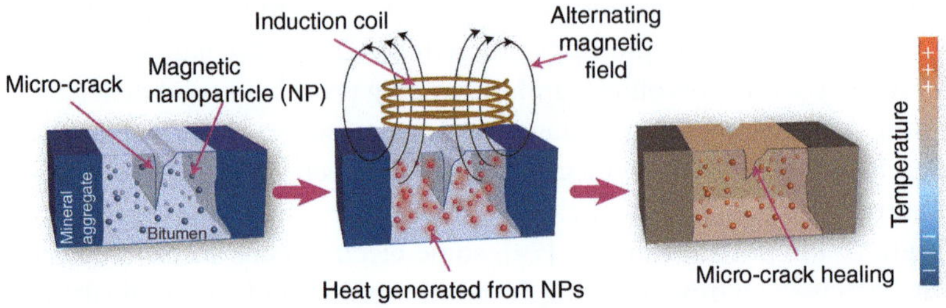

Figure 9.11 A schematic of healing mechanical damage by incorporating magnetic nanoparticles in bitumen matrix. Source: Jeoffroy et al. [88]. Reproduced with permission of Elsevier.

can obtain satisfactory self-healing properties, which may be detrimental to the insulating properties of polymers.

On the other hand, a self-healing method that directly uses nanoparticles as patches has been proposed, which is based on entropic depletion. In polymer nanocomposite systems, it is observed that nanoparticles may be affected by an entropic depletion attraction and thus aggregate into microcracks [91–93]. The conformational entropy of polymer chains that are close to nanoparticles is reduced due to the confinements by nanoparticles [93]. Thus, it will be entropy-favorable to "squeeze out" the particles, resulting in the segregation of nanoparticles into cracks. However, both enthalpic and entropic interactions between polymer chains and nanoparticles would be accountable for the behavior of particles. When miscibility between nanoparticles and polymer matrix is poor, enthalpic interactions will dominate and suppress the formation of a uniform nanocomposite [92]. On the contrary, good compatibility will make it possible for entropy interactions to exert influence. The size of nanoparticles also plays a significant role, since the entropy penalty must be strong enough for the migration of particles to proceed. The segregation of nanoparticles into cracks could only be observed for particles of at least a size comparable to the gyration radius of polymer chains [93].

The formation of nanoparticle aggregation at the surface of cracks is analogous to the hemostasis mechanism of vascular injury in our bodies. Inspired by the hemostasis process, the self-healing method utilizing nanoparticle migration has been theoretically verified by computer simulations. As shown by Monte Carlo simulations [94] and molecular dynamics simulations [95], nanoparticles added to polymers could migrate and become localized at cracks, forming patches to repair the damaged regions. It is also presented that the healing efficiency of mechanical properties could reach 75–100% [94]. However, there are still several limitations to this method. The simulation temperature was assumed to be higher than the polymer's glass transition temperature so that polymer chains would possess higher mobility for nanoparticles to migrate. With only nanoscale cracks or notches simulated, whether this self-healing method is suitable for larger damage is not certain. Moreover, as nanoparticles are consumed during the healing process, the fraction of particles will decrease with more defects to be healed, leading to a possible decline in healing efficiency and a maximum healing capacity.

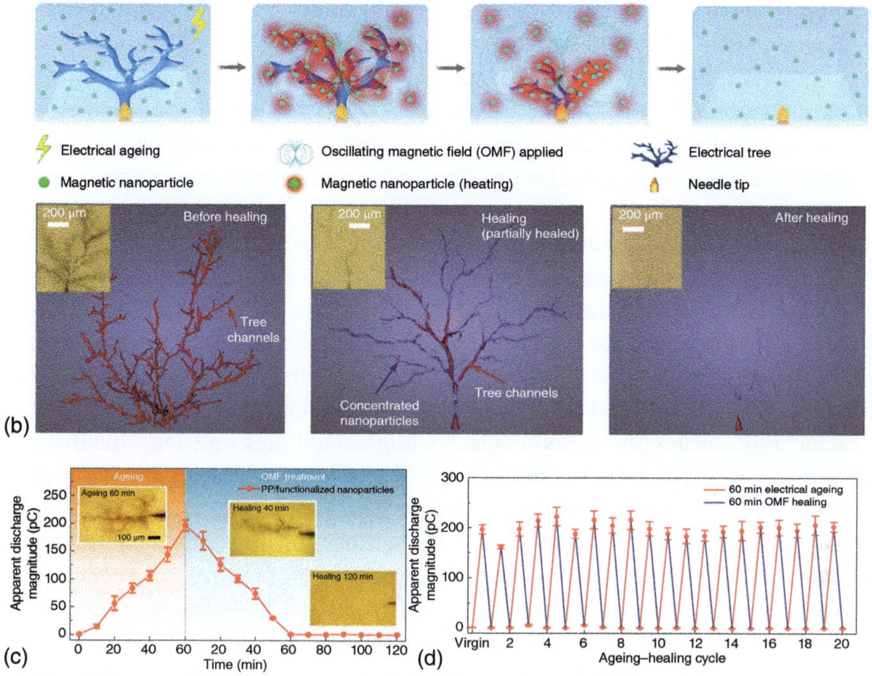

Figure 9.12 Defect-targeted self-healing dielectric polymers with superparamagnetic nanoparticles. (a) A schematic of healing electrical trees by incorporating superparamagnetic nanoparticles. When treated by an oscillating magnetic field, the nanoparticles will migrate toward tree channels and generate heat, leading to melting interdiffusion at the damaged site. (b) The reconstruction of electrical trees in polypropylene (PP) during the self-healing process by computed micro-X-ray tomography. (c) Apparent partial discharge magnitude of PP specimens with magnetic nanoparticles during aging and healing. (d) Apparent partial discharge magnitude during multiple aging-healing cycles. Source: Yang et al. [96]/Reproduced with permission form Springer Nature.

By combining the two abovementioned self-healing techniques, Yang et al. developed a defect-targeted self-healing method for thermoplastic polymers (Figure 9.12a) that is competent for both electrical and mechanical damage [96]. Superparamagnetic nanoparticles, with the ability to migrate to defect sites, were incorporated into commercial polypropylene (PP). After electrical treeing, the specimen was treated with OMF for self-healing. The nanoparticles were observed to concentrate around the tree channels, generating heat and resulting in melting interdiffusion (Figure 9.12b). After healing of the damage, the nanoparticles could redisperse in the matrix, enabling the next healing phase. Electrical properties including leakage current, partial discharge magnitude, and tree inception voltage effectively recovered to the same level before electrical aging (Figure 9.12c,d). Further investigations confirmed the validity of this defect-targeted method in healing mechanical damage and in different thermoplastic polymers, such as polymethyl methacrylate (PMMA) and perfluorosulfonic acid (PFSA) [97].

It is worth mentioning that surface functionalization was quite essential in this emerging approach [96]. The type of surface modifier was chosen to enhance the compatibility between nanoparticles and polymers, which is a prerequisite for entropic depletion. The thickness of functioning layers was optimized to match the gyration radius of the polymer matrix while maintaining relatively high mobility. The surface-functioning layers also provided steric repulsion to prevent permanent aggregation.

By combining the advantages of magnetic interdiffusion and nanoparticle migration, the content of nanoparticles as low as 0.09 vol% can meet the self-healing requirement without affecting the dielectric and insulating properties of the polymer [96]. This advantage comes from the tendency of the particles to migrate, as the nanoparticles and thermal energy can be concentrated at the damaged site. The nanoparticles act only as heaters, so they are not consumed or depleted, giving the nanocomposite highly reproducible healing capabilities and long incubation periods. Even after 100 cracking-healing cycles or a 100-day waiting time, it is still possible to achieve a healing efficiency of more than 90% [97].

The main limitations of this defect-targeted self-healing approach lie in the polymer matrix and healing stimulus. Melting interdiffusion requires a thermoplastic matrix so that molten polymer can flow into microcracks. The healing temperature must exceed the glass transition temperature of the polymer to ensure the fluidity of the nanoparticles. OMF with high intensity and frequency may also pose an obstacle to the practical application of this method. Nonetheless, this emerging approach remains promising in power transmission systems or power electronics where OMFs exist or are readily applied [96]. Increasing the heating power can also make this self-healing mechanism more widely applicable by fine-tuning the composition and size distribution of nanoparticles [85, 98].

For thermoset polymer dielectrics, Sun et al. have proposed a no-shell microcapsule self-healing system by constructing a thermoplastic-thermoset composite polymer [99]. As shown in Figure 9.13, Fe_3O_4@SiO_2 nanoparticles are coated onto paraffin wax microspheres, giving magnetic field control and microwave heating functionalities to the healing spheres. A magnetic field can be applied during the curing process of epoxy resins, leading to a greater concentration of microspheres in vulnerable areas that are supposed to bear higher electrical stress. After electrical treeing, epoxy resin samples are irradiated by microwaves. The paraffin wax microspheres are melted, refilling the hollow tree channels. Both mechanical damage (Figure A1b) and electrical trees (Figure A1c) can be healed by this method, opening up new routes toward self-healing of dielectric composite polymers (Figure 9.13).

By constructing a thermoplastic-thermoset composite structure, this approach has made it possible to utilize melting interdiffusion in the self-healing of thermoset dielectrics with its long-term and repeatable healing capability. However, microwave radiation may still be a hindrance to practical use. As the microspheres can be regarded as a no-shell type of microcapsule, some inherent disadvantages of capsule-based self-healing methods are also applicable to this microsphere approach, which will be discussed in Section 9.4.3.

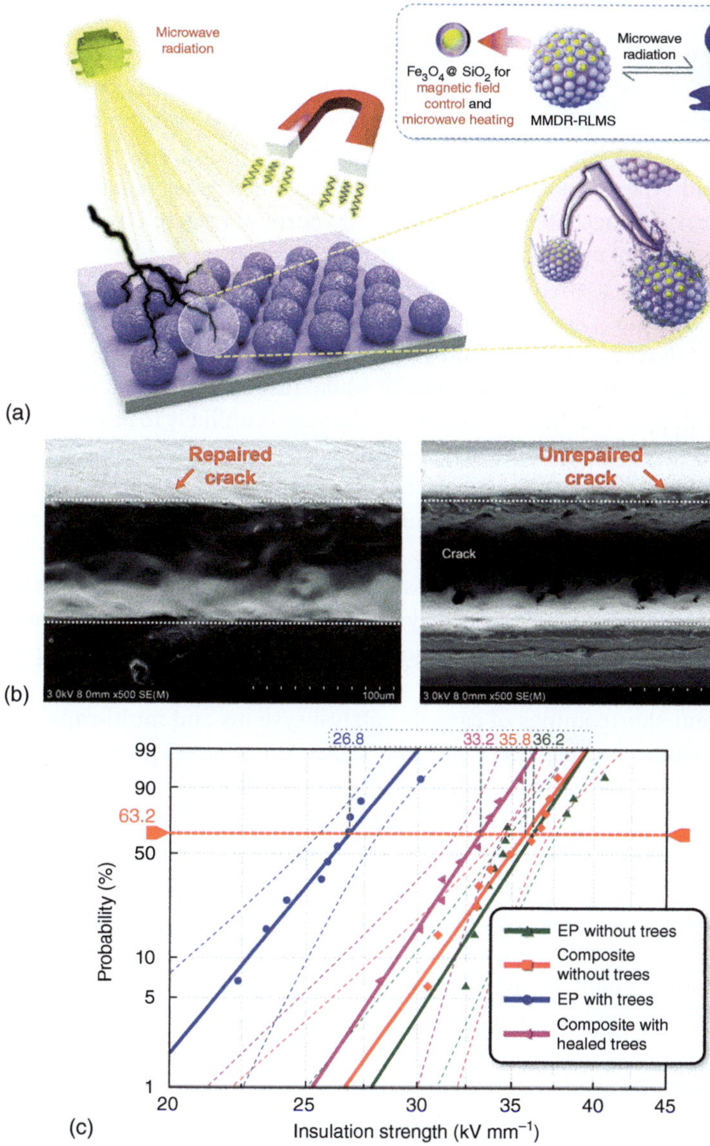

Figure 9.13 Thermoplastic-thermoset composite self-healing dielectrics by incorporating microwave heating microspheres. (a) A schematic of the self-healing strategy. (b) SEM images of healed and unhealed mechanical damage. (c) Weibull distributions of insulation strength during aging and healing of the self-healing epoxy resin. Source: Sun et al. [99]/Reproduced with permission from Royal Society of Chemistry.

9.4.3 Microcapsule-Based Self-Healing Dielectrics

Beginning with the design of self-healing polymer composites using dicyclopentadiene (DCPD) and Grubbs' catalysts [66], external self-healing methods based on microcapsules have been extensively and intensively studied. When the capsule ruptures due to injury, the encapsulated healing agent is released and begins to aggregate. Almost all types of polymerization chemistry show great potential for repairing mechanical damage [100, 101]. However, current research is limited regarding the self-healing of electrical injuries.

Microcapsule self-healing methods can be further divided into different categories based on the form of healing agents (monomers) and catalysts or initiators [100]. Although capsule catalyst systems and multi-capsule systems perform well in repairing mechanical damage, they are not as beneficial when dealing with much smaller electrical trees, especially since the site of electrical damage is unlikely to be exposed to all catalyst or microcapsule mechanical types of damage [83]. Moreover, the problem cannot be solved by increasing the amount of microcapsules or catalysts. When microcapsule concentrations exceed certain limits (usually about 5 wt%), a sharp decline in mechanical and electrical properties, including tensile strength, dielectric loss, and breakdown strength [102–104], is observed. On the other hand, the addition of catalysts and other small molecules may distort local electric fields and act as defects and contaminants, resulting in increased susceptibility to electrical damage and poorer dielectric properties [105–107].

Due to the inherent shortcomings of capsule-catalyst systems and multi-capsule systems, current methods of electrical damage self-healing based primarily on microcapsules include potential functional strategies for aggregation and environmental stimulus triggering. Instead of catalysts, some polymeric chemistry requires external stimuli to proceed with or without corresponding initiators. These reactions can be used to self-heal and eliminate the adverse effects of catalyst incorporation. The latter method uses residual reactive functional groups in the polymer matrix as polymerization agents [100]. As we will see later, these residual functional groups can actually act as hardeners for the polymer matrix, ensuring that the self-healing material has satisfactory dielectric properties.

9.4.3.1 Polymerization Triggered by Environmental Stimuli

Various external stimuli, such as water, oxygen, and sunlight, have been used to trigger the polymerization of monomers to design self-healing or other smart polymers [108–112]. In considering the appropriate stimulation of electrical damage healing, researchers are very concerned with the phenomenon and byproducts of electrical aging. Photoemission and radical generation are common effects associated with electric trees. Radicals are produced as a result of the collision of hot electrons with polymer molecules, and light emission comes from several different sources. When excited molecules return to the ground state and when opposite charge carriers recombine, various kinds of light can be produced [78]. Partial discharge also produces light emission that peaks in the ultraviolet range [113].

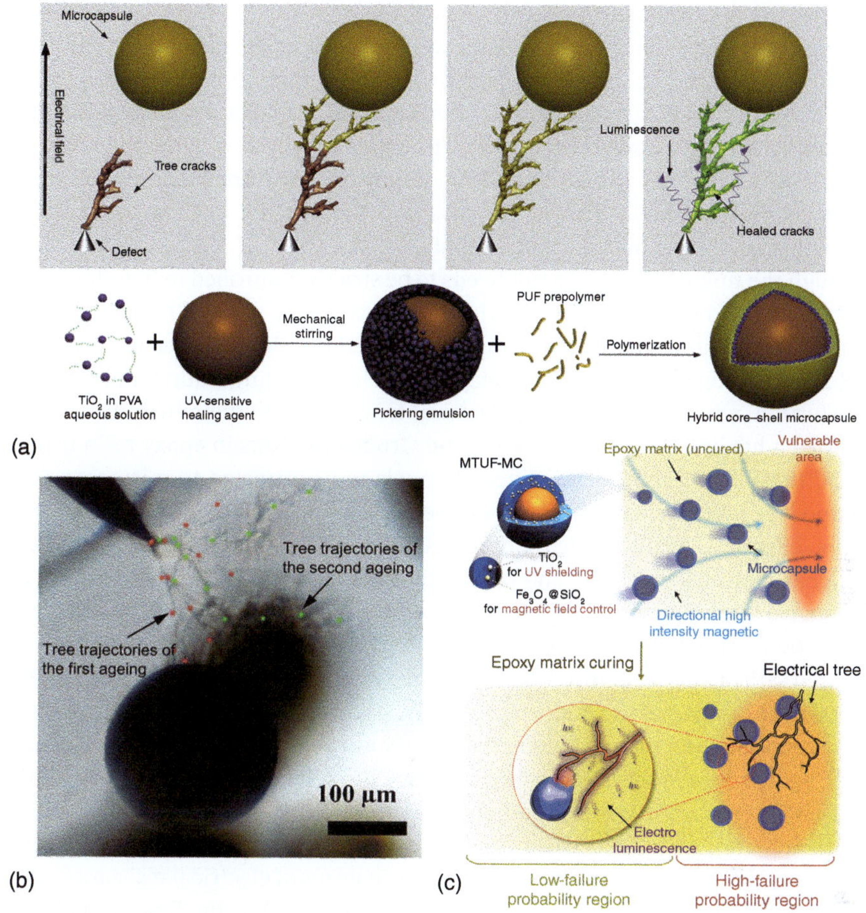

Figure 9.14 Microcapsule-based dielectric polymers with the self-healing capability of electrical trees triggered by environmental stimuli. (a) The self-healing mechanism using polymerization is triggered by UV light during the electrical treeing process and the preparation of microcapsules. TiO_2 nanoparticles in capsule shells serve as light-absorbing materials. Source: Gao et al. [114]/Reproduced with permission from Elsevier. (b) Electrical trees developed in new paths in the re-treeing process rather than along the trajectories of the first aging phase. Source: Gao et al. [114]. Reproduced with permission of Elsevier. (c) Schematic of UV-triggered self-healing method with magnetic targeting. The microcapsules can be guided toward regions with higher electrical stress and failure possibility during manufacturing. Source: Sima et al. [115]. Reproduced with permission of the American Chemical Society.

Gao et al. proposed a self-healing method for electrical defects based on ultraviolet light derived from electric trees (Figure 9.14a) [114]. Epoxy monomers (bisphenol A epoxy acrylate) and reactive monomers (trimethylolhexane triacylate), as well as photoinitiators (1-hydroxycyclohexyl phenyl ketone) and SiO_2 nanoparticles, were encapsulated by one-step Pickering emulsion polymerization. Then different proportions of microcapsules are mixed into epoxy resins to form self-healing, insulating composites. SiO_2 nanoparticles were introduced to adjust the dielectric and insulating properties of microcapsules, thus reducing the negative effects on the properties of the epoxy resin matrix. Microcapsules lead to acceptable decreases in breakdown strength and resistivity of self-healing composites when the volume fraction is less than 5%. In addition, TiO_2 nanoparticles were incorporated into the capsule shell to shield it from UV rays and prevent the healing agent from polymerizing before the microcapsules burst.

Although the microcapsule content needs to be strictly controlled to maintain satisfactory properties of the polymer matrix, the microcapsule method has the unique advantage of attracting the electrical tree route, making the capsules more prone to rupture. The doped microcapsules may cause charge accumulation and electric field distortion, which may affect the electrical tree routing in their vicinity. Lesaint et al. tested a healing system with DCPD and Grubbs catalysts in epoxy resin using electrical trees, and a tendency for tree channels to be attracted to microcapsules was found [116]. Wang et al. studied the same system in polyethylene [102], and the attraction effect was again verified. Computer simulation also showed that capsules with higher dielectric constant and conductivity than the substrate would have a long-distance attraction effect on tree tracks, while capsules with lower breakdown strength could attract electric trees within a short distance [114]. This characteristic ensured satisfactory healing efficiency despite a rather low doping fraction. After electrical trees punctured the microcapsules, the healing agents would flow into hollow tree channels, and radical polymerization would be initiated by photoinitiators and UV light from the electrical aging process. The whole process of self-healing needed no external intervention and could achieve more than 95% volume ratio of healed tree in terms of electrical resistance and breakdown voltage were channeled ultimately [114]. The insulating properties in terms of electrical resistance and breakdown voltage were recovered upon self-healing, which is the first example of high-efficiency restoration of electrical performance by using microcapsule-based self-healing approaches.

Microcapsule-based self-healing methods are regarded as being able to heal damage at each site once because the repair agent can only be exposed and reacted once [101]. This problem is partly solved by a new aspect of the self-healing approach triggered by ultraviolet light. By adjusting the healing agent, the local insulation of the healed area is even better than that of the undamaged area. Therefore, the new electrical tree will tend to develop along the new path rather than along the original path (Figure 9.14b) [114]. Thus, the risk of running out of healing agents in one area and losing self-healing ability can be effectively reduced.

Another attempt of developing self-healing dielectrics was made by Sima et al. (Figure 9.14c) [115], using the same stimulus of UV light originating from electrical aging but with completely different healing reactions. Monomers (3,4-epoxycyclohexyl-methyl 3,4-epoxycyclohexanecarboxylate) and photoinitiators (triarylsulfonium hexafluorophosphate salt) were encapsulated and would undergo a cationic polymerization process in the presence of UV light. In addition to TiO_2 nanoparticles, Fe_3O_4 nanoparticles with SiO_2 coatings were also incorporated into

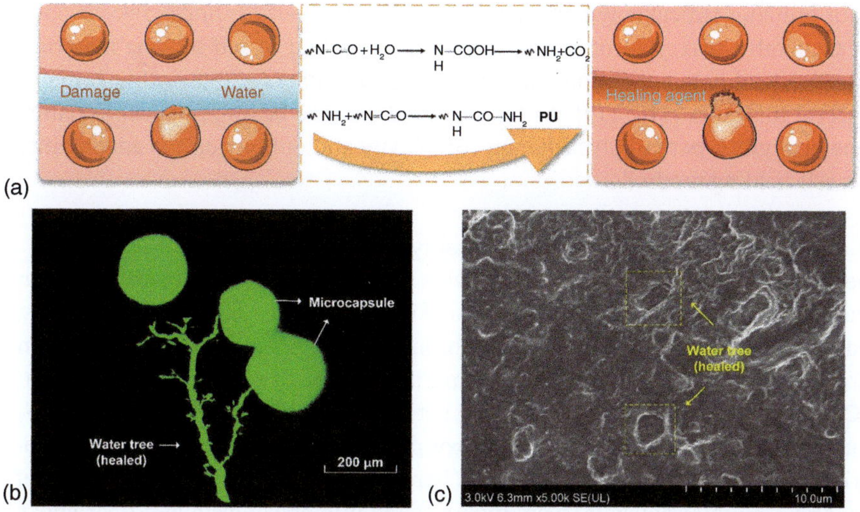

Figure 9.15 Microcapsule-based dielectric polymers with self-healing capability of water trees. (a) A schematic of the self-healing strategy. (b) Fluorescence microscopy image of healed water tree in dielectric composites. Fluorescent dye was added to the healing agent to label the microcapsules. (c) Cross-section SEM image of healed water tree channels. Source: Sima et al. [117]/Reproduced with permission from American Chemical Society.

the capsule shells, endowing the microcapsules with targeted mobility under a magnetic field. During the hardening process of the epoxy matrix, the microcapsules could be attracted toward more vulnerable regions by applying a directional magnetic field. Thus, a controllable distribution of microcapsules could be achieved with higher concentration and better self-healing ability in regions with higher electrical stress or stricter working conditions.

In consideration of the presence of residual water in water tree channels, Sima et al. have also proposed a water-triggered self-healing method for water tree damage in epoxy resins [117]. For water treeing degradation, residual water in tree channels will stop normal healing agents from filling the hollow channels completely. To obtain a self-healing dielectric polymer for water trees, isophorone diisocyanate (IPDI) was encapsulated using microfluidic techniques and introduced into polymers. As shown in Figure 9.15a, IPDI would flow into tree channels and react with water, achieving solidification of healing agents and self-healing of water tree channels. Fluorescence microscopy revealed that IPDI would refill water tree channels completely (Figure 9.15b). A Cross-sectional SEM image (Figure 9.15c) has also proven a decent self-healing efficiency of water tree damage. Although the self-healing reaction can be triggered without any external stimulation, residual water in tree channels or extremely high humidity are necessary, which may limit the versatility of utilizing this method among various damage types or environmental conditions.

The self-healing methods introduced above have exploited UV light or water as stimuli for polymerization, which opens up the prospect of further investigation. For instance, radicals are also generated during electrical treeing, showing a possible path for utilizing radical polymerization to develop new self-healing dielectric polymers. Generally, two types of microcapsules are required in self-healing materials based on radical polymerization: one containing monomers and the other containing radical initiators [118, 119]. As for electrical damage healing, radical initiators may be negligible, potentially reducing complexity and enhancing the adaptability of this self-healing method.

9.4.3.2 Polymerization by Latent Functionality

In the latent functionality route, polymerizers are derived from residual reactivity in dielectric polymers [100], such as residues of epoxy resins or additional hardeners. Based on the potential hardener approach, imidazoles and Lewis acids have been used in self-healing epoxies that are primarily concerned with mechanical damage. Cationic ring-opening polymerization between epoxy resins and Lewis acid BF_3 was used to design self-healing fiber-epoxy polymer composites [120]. Microcapsules containing epoxy resin and potential hardener $CuBr_2(2\text{-MeIm})_4$ (a complex of $CuBr_2$ and 2-methylimidazole) were incorporated into epoxy resin and fiber-epoxy resin composites to develop new self-healing materials [121, 122]. Since imidazoles are widely used in industry and have good commercial availability and chemical stability, imidazole-based self-repair schemes show great potential for self-healing dielectrics [123].

Based on the reaction principle of 2-ethyl-4-methylimidazole (EMI) and epoxy resin, Xie et al. designed a self-healing method for electrical damage (Figure 9.16a,b) [124]. The first hardening process was performed at relatively low temperatures by mixing the epoxy monomer (diglycidyl ether of bisphenol A, DGEBA), primary curing agent (phenol-aldehyde amine, PAA), and latent hardener (EMI) with microcapsules containing healing agents (another epoxy monomer, diglycidyl ether of bisphenol F, DGEBF). Amine hardeners play a major role in this stage, while EMI can still maintain its reactivity at high temperatures until the self-healing stage. Less than 3 wt% EMI is sufficient to cure while retaining the excellent mechanical and dielectric properties of the epoxy matrix (Figure 9.16c).

Similar to UV-induced self-healing methods, microcapsules were also observed to attract tree tracks in latent functionality strategies, and new trees tended to grow in new areas rather than healed areas [124]. In addition, the long-term effectiveness was tested by accelerated thermal aging experiments. Although severe oxidation can be inferred from visual color changes and FTIR results, EMI-modified epoxy resins can still induce self-healing reactions [124], showing good long-term effectiveness.

Since epoxy-based dielectric polymers usually operate at very high temperatures, the EMI-epoxy curing process can be carried out autonomously under appropriate working conditions. In a broad sense, maximum reactivity can be obtained at different temperatures by using different derivatives of imidazole. It is, therefore, possible and has great potential to design a variety of self-healing systems that can maximize the reactivity of potential functionality at the estimated operating temperature of the self-healing dielectrics.

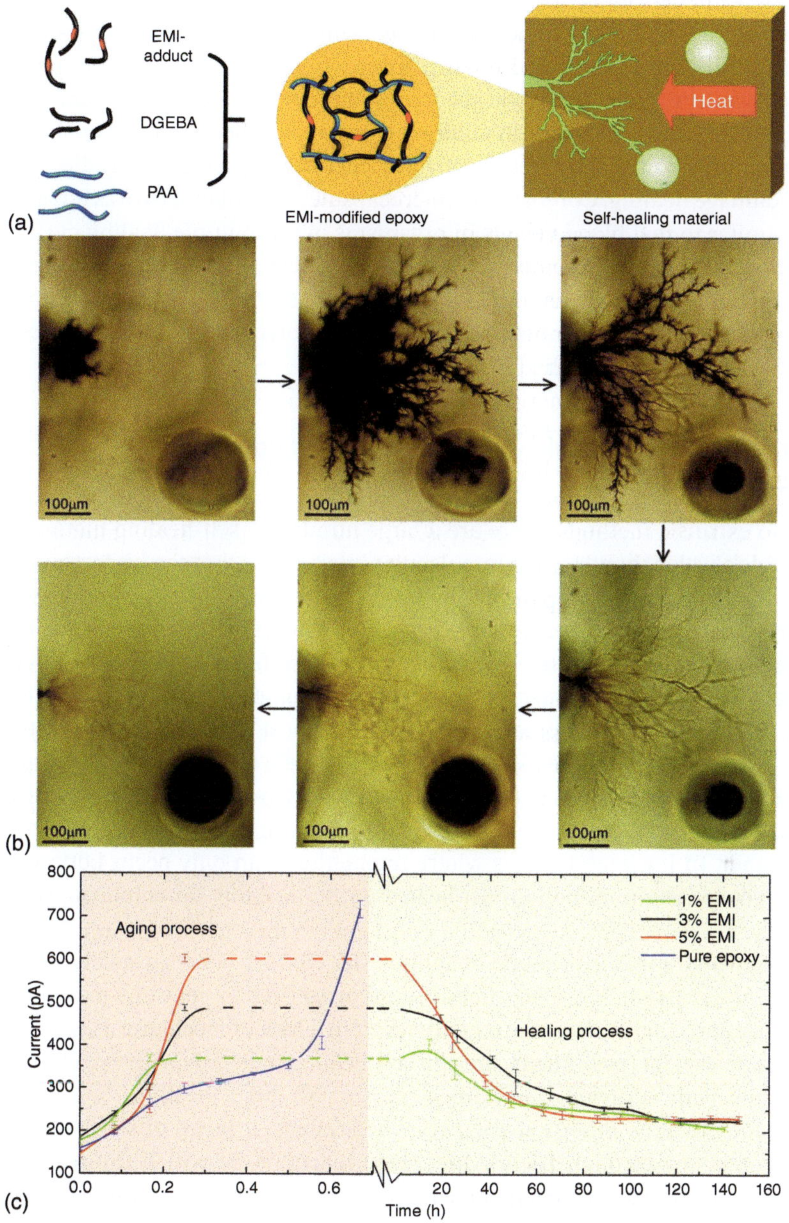

Figure 9.16 Microcapsule-based self-healing dielectric polymers triggered by latent functionality. (a) Schematic of fabricating self-healing epoxy by latent functionality. Epoxy monomer DGEBA, primary curing agent PAA, and latent hardener EMI were mixed to produce EMI-modified epoxy. Self-healing was obtained by cross-linking EMI and epoxy monomers from microcapsules. (b) Optical microscope images of the aging and healing process of self-healing epoxy samples. (c) Leakage current of epoxy specimens with different concentrations of EMI during electrical aging and healing. Source: Xie et al. [124]/Reproduced with permission from American Chemical Society.

In conclusion, the advantage of microencapsulation is that the healing process can be performed automatically under normal working conditions without manual intervention. While the very limited repair capacity of a particular location is an inherent disadvantage, better dielectric properties of the repair area may reduce the risk of treeing in the same location. In addition to microcapsule-based self-healing, external approaches to vascular networks have also been extensively studied in mechanical damage healing. Compared to microcapsules, vascular networks appear to be more similar to real blood vessels in organisms, with multiple healing cycles in one place and the ability to replenish healing agents externally [100]. However, vascular self-healing networks are more expensive to design and maintain, posing additional challenges in the manufacture of dielectric polymers, so more research is needed before they can be applied in practice.

9.4.4 Intrinsic Self-Healing Dielectrics by Reversible Bonds or Interactions

In addition to extrinsic methods, there are a large number of self-healing methods using reversible covalent bonds or supramolecular interactions that can be destroyed and rebuilt in multiple healing cycles [125, 126]. When using intrinsic methods to repair electrical damage, the first challenge will be to enable chemical interactions to bond the fractured surfaces together. In general, the crack healing process has five stages: surface rearrangement, surface proximity, wetting, diffusion, and randomization [90, 127]. The surface approach is negligible in most self-healing work with mechanical damage, where broken surfaces are usually held together manually, and pressure is applied in some cases to ensure tight contact. However, in the context of an electrical tree, it is not an easy problem for separated surfaces to approach and touch each other. In intrinsic systems where interactions can only occur between adjacent molecules, chain mobility must bear the responsibility for achieving surface proximity.

A major challenge for intrinsic self-healing is the conflict between mechanical stiffness and chain dynamics. The effectiveness of self-repair requires excellent chain flexibility and fluidity, but they usually result in a lack of mechanical stability [101, 128]. Therefore, it is beneficial to design self-healing materials with two states: a working state with sufficient mechanical properties and a healing state with increased chain mobility. Based on the two-state strategy, a series of self-healing systems were realized through Diels–Alder (DA) reactions, in which the reverse DA reaction depolymerizes the cross-linked polymer at high temperatures and the DA reaction reconstructs covalent bonds to repair cracks at lower temperatures [126, 129–132]. However, these self-healing polymers have the same disadvantage of being completely uncrosslinked in the repair state, which can lead to shape changes and performance degradation, especially in applications such as insulating dielectrics.

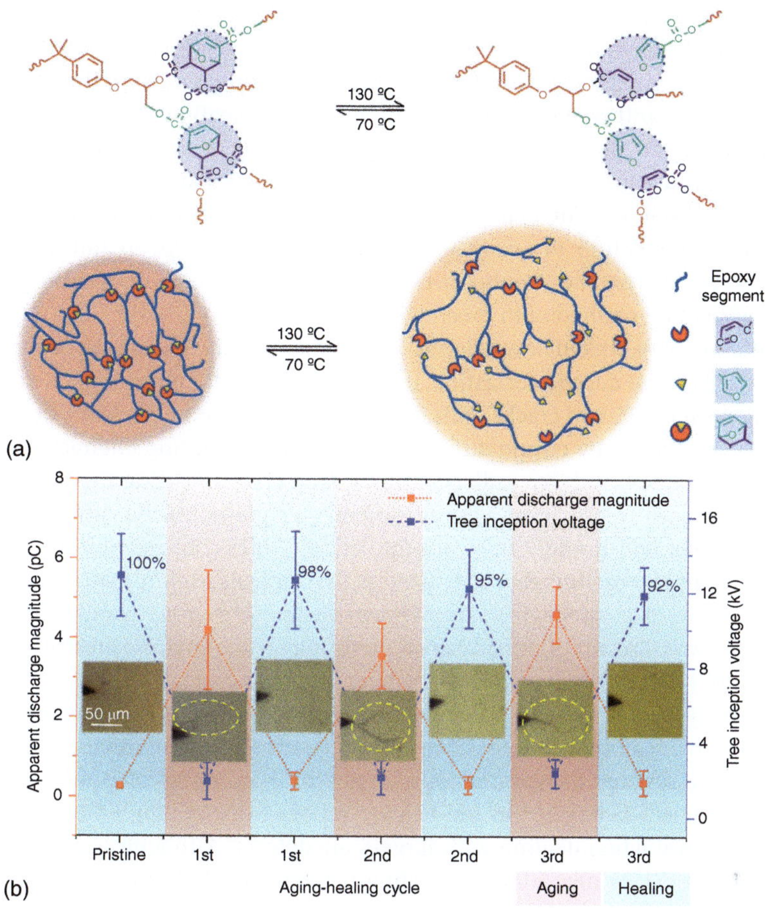

Figure 9.17 Inspiration and mechanisms of self-healing by reversible crosslinking networks. (a) Schematic of partially reversible cross-linking network the polymer network may undergo the conversion between a fully cross-linked state and a partially cross-linked state under different temperatures. (b) Dielectric and insulating properties (apparent partial discharge magnitude and tree inception voltage) and optical images of partially reversible cross-linking epoxy in aging-healing cycles. Source: Xie et al. [133]/Reproduced with permission from Royal Society of Chemistry.

To avoid a completely uncrosslinked state of healing, Xie et al. proposed a partially reversible cross-linked epoxy network (Figure 9.17a) [133]. The first step is to synthesize a dynamic cross-linking agent by DA reaction of 3-furfuric acid and maleic anhydride. Self-healing epoxy networks are manufactured by curing commercial epoxy monomers with dynamic cross-linkers. Each cross-linker molecule can connect three epoxy monomers to form a fully cross-linked network. After internal damage, the cross-linker undergoes retro-DA reaction by heating to

130 °C and decomposes into furoic acid and maleic anhydride, which have only one and two cross-linking points with the epoxy chain, respectively, forming a partially cross-linked polymer. By reducing the degree of cross-linking, higher chain fluidity and more free volume will provide good conditions for self-healing. In addition, the overall structure and properties of the epoxy matrix can be well preserved because the matrix is still cross-linked rather than thermoplastic. After the healing phase, fully cross-linked epoxies can be obtained by lowering the temperature and allowing the DA reaction to occur again. This transition between full and partial cross-linked states can be simply controlled by temperature, demonstrating a convenient self-healing method with excellent reversibility.

Partially reversible cross-linked networks increase chain mobility during the healing process through partial depolymerization, making it possible for epoxy chains to diffuse into hollow channels and cracks. This method also shows unique advantages in curing electrical damage by eliminating tree byproducts, including chemically degraded gaseous products and free radicals from broken chains. During the healing phase, the increased free volume within the polymer promotes the diffusion of gaseous byproducts and hinders their accumulation, while conjugated double bonds from the dynamic cross-linkers can react with free radicals to eliminate them. Self-healing tests showed a high degree of recovery in mechanical strength and insulation after more than 10 repair cycles and good performance in repairing electrical trees (Figure 9.17b) [133].

Self-healing epoxy networks with partially reversible cross-linking sites show great potential in practical applications, as only modification of the hardener may be required while commercial epoxy monomers can still be used as-is. However, since chain mobility is a key factor in the healing process, the healing efficiency may not be satisfactory when dealing with large-scale defects. Although applying more heat treatment cycles may improve healing performance [133], it may also reduce the convenience of this self-healing method.

A recent advance in designing self-healing high-temperature polymer dielectrics was made by Wan et al., who used disulfide bond exchange as the reversible interaction [134]. As shown in Figure 9.18a, the damaged surfaces in copolymerized polyimide could be re-bound by exchanging disulfide bonds under hot pressing. Both mechanical damage and electrical damage (specifically corona damage) could be healed by this approach. Copolyimide films showed a higher or equivalent tensile strength after mechanical damage and healing (Figure 9.18b). The surface of copolyimide films was cracked after corona damage (Figure 9.18d), whereas the damaged surface was healed after healing by hot pressing (Figure 9.18e). As shown in Figure 9.18c, the breakdown strength of healed copolyimide films increased after healing, showing that the exchange of disulfide bonds has healed the surface defects.

With a good Young's modulus, high thermal stability, excellent insulation, and dielectric properties, this copolymerized polyimide has shown great potential in high-temperature industrial applications. Further investigations may take place to improve the convenience of healing conditions or enhance the versatility of various damage forms.

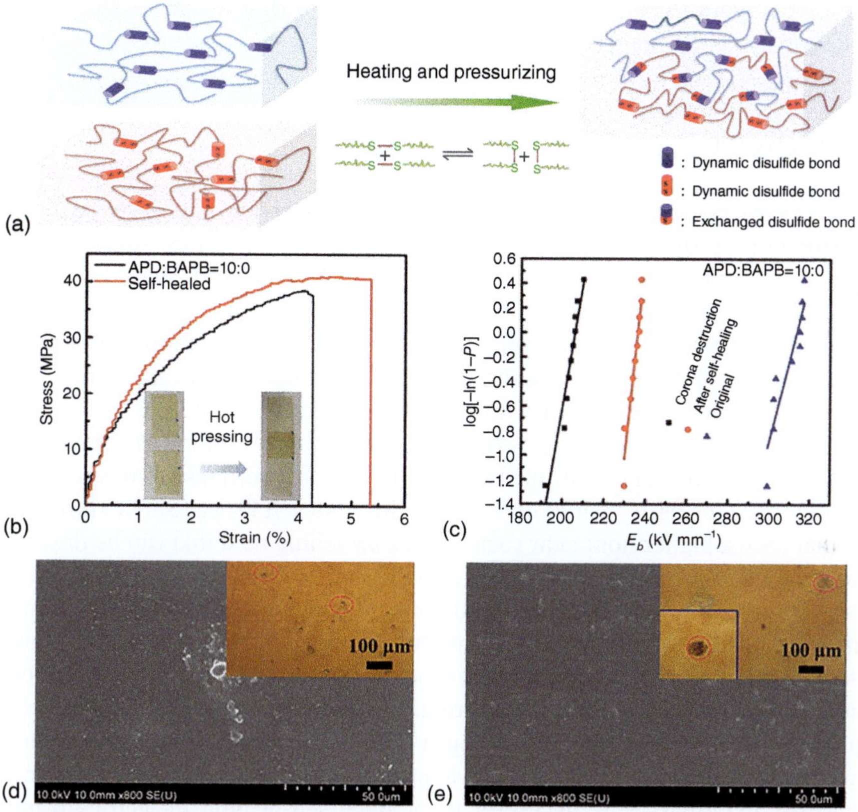

Figure 9.18 Self-healing high-temperature dielectric polymer by disulfide bond exchange. (a) A schematic of disulfide bond exchange and self-healing process. (b) Stress-strain curves of copolyimide films before and after self-healing. (c) Weibull distributions of breakdown strength of copolyimide films. (d–e) SEM and optical images of the copolyimide films after corona damage (d) and after self-healing (e). Source: Xie et al. [133]/Reproduced with permission from Royal Society of Chemistry.

9.4.5 Summary of Self-Healing Dielectrics

In view of the increasing demand for reliability of high-temperature dielectric polymers, several methods for designing self-healing dielectrics have been proposed. Strict requirements for dielectric and insulating properties, as well as special damage forms, such as electric trees, water trees, and corona damage, are the main differences between this field and traditional self-healing materials. Although the above self-healing methods can cure electrical damage, there are still some problems such as inconvenient self-healing conditions, complex manufacturing processes, or inconsistent self-healing efficiency of different damage degrees. In addition, while all of these self-healing methods are currently being tested under laboratory conditions, their repair performance and long-term effectiveness remain to be evaluated in practical applications. In all, these limitations suggest that there is still a lot of work to be done for dielectric polymers with intelligent self-healing functions to become a reality.

9.5 Outlook

Internal electric field grading, insulation deterioration detection, and electrical damage restoration present several difficulties for traditional dielectrics. An inventive strategy for overcoming these obstacles is to create smart dielectric materials, such as self-adaptive, self-reporting, and self-healing dielectrics, with bio-inspired and autonomous functionalities. However, both in the lab and in the commercial world, there are several challenges to creating, producing, and using smart materials. The thoughts should be brought to life with a lot of work.

The following describes three possible evolutionary stages for smart dielectrics. In the beginning, carefully controlled and ideal laboratory settings are meant to be used to enhance the performance of the smart dielectrics. It is necessary to produce SADs that have a higher nonlinear coefficient, a switching field that can be flexibly set over a wider range, and more stable features against a variety of environmental factors including filler concentration and temperature. Both producing functional fillers with increased performance and discovering fresh field grading concepts are viable approaches for reaching this objective. It is anticipated that SRDs will be created that are capable of quickly reporting, significantly reacting, and accurately diagnosing. SRDs may be created using a variety of substances, systems, and techniques, including stimuli-responsive polymers, optical chemosensors, and bio-inspired materials. It is necessary to produce self-healing dielectrics using simpler, more trustworthy, and reproducible processes. While the qualities of the treated region should return to their former level with a highly repeatable healing process, the materials are recommended to autonomously repair damage as soon as feasible without additional treatment.

There are various challenges that must be addressed before smart dielectrics can be commercialized. Functional additives' stability is essential. Functional additives are compounds that are physically or chemically inserted into the polymeric matrices to provide the materials with autonomous functionalities, such as ZnO microvaristors, SP chromophores, and microencapsulated healing agents. The majority of the time, the additives are susceptible to adverse conditions like overheating, UV radiation, mechanical stress, acid rain, ozone, free radicals, etc. In order to retain the autonomous functionalities of smart dielectrics, their ability to withstand aging and fatigue under challenging circumstances is crucial. Additionally, using smart materials in super- or ultra-high voltage engineering is hampered by the performance loss of the matrix caused by unsuitable or excessive additions. On one hand, the creation of sophisticated additives with improved performance and compatibility is necessary to address this issue. The additives must diffuse more readily and uniformly into the polymeric matrix, include less reactive agents that might lead to material degradation, and exhibit better properties including increased resistivity, toughness, and elasticity. The smart dielectrics, on the other hand, should be more sensibly created in order to lower the number of additives and get rid of flaws that are incorporated into the polymeric matrix during processing. The cost of smart dielectrics must also be competitive for practical use in the business sector. Therefore, it is necessary to reduce the price of raw materials and simplify the synthesis process. Toxicants and pollutants are meant to be avoided due to their effects on living things and the environment.

Universality and adaptability are the ultimate goals of smart dielectrics. A wide range of voltage levels, circumstances, and damage scales are covered by the autonomous functions. Regardless of whether a polymeric matrix is glassy or rubbery, thermoplastic or thermoset, the functional additives are compatible with it. To execute autonomous lifetime control, identical dielectrics have been organically combined with self-adaptive, self-reporting, and self-healing capabilities.

References

1 Tanaka, T. (2005). Dielectric nanocomposites with insulating properties. *IEEE Transactions on Dielectrics and Electrical Insulation* 12 (5): 914–928. https://doi.org/10.1109/TDEI.2005.1522186.

2 Lewis, T.J. (1994). Nanometric dielectrics. *IEEE Transactions on Dielectrics and Electrical Insulation* 1 (5): 812–825. https://doi.org/10.1109/94.326653.

3 Stuart-Fox, D., Moussalli, A., and Whiting, M.J. (2008). Predator-specific camouflage in chameleons. *Biology Letters* 4 (4): 326–329. https://doi.org/10.1098/rsbl.2008.0173.

4 Tracey, W.D. (2017). Nociception. *Current Biology* 27 (4): R129–R133. https://doi.org/10.1016/j.cub.2017.01.037.

5 Bloch, R. (1941). Wound healing in higher plants. *The Botanical Review* 7 (2): 110. https://doi.org/10.1007/BF02872446.

6 Liu, Z., Zhang, F., Yu, J. et al. (2018). Research on key technologies in ±1100 kV ultra-high voltage DC transmission. *High Voltage* 3 (4): 279–288. https://doi.org/10.1049/hve.2018.5023.

7 Zhao, X., Hu, J., Yuan, Z., and He, J. (2021). Design of adaptive bushing based on field grading materials. *High Voltage* 6 (4): 625–636. https://doi.org/10.1049/hve2.12090.

8 Saha, T.K. (2003). Review of modern diagnostic techniques for assessing insulation condition in aged transformers. *IEEE Transactions on Dielectrics and Electrical Insulation* 10 (5): 903–917. https://doi.org/10.1109/TDEI.2003.1237337.

9 Eichhorn, R. (1977). Treeing in solid extruded electrical insulation. *IEEE Transactions on Electrical Insulation* EI-12 (1): 2–18. https://doi.org/10.1109/tei.1977.298001.

10 Huang, X., Han, L., Yang, X. et al. (2022). Smart dielectric materials for next-generation electrical insulation. *iEnergy* 1 (1): 19–49. 10.23919/IEN.2022.0007.

11 Glatz-Reichenbach, J., Meyer, B., Strümpler, R. et al. (1996). New low-voltage varistor composites. *Journal of Materials Science* 31 (22): 5941–5944.

12 Robertson, J. and Varlow, B.R. (2003). The AC non linear permittivity characteristics of barium titanate filled acrylic resin. In: *International Conference on Properties & Applications of Dielectric Materials,* IEEE, Nagoya, Japan, pp. 761–764.

13 Wang, X., Nelson, J., Schadler, L., and Hillborg, H. (2010). Mechanisms leading to nonlinear electrical response of a nano p-SiC/silicone rubber composite. *IEEE Transactions on Dielectrics & Electrical Insulation* 17 (6): 1687–1696.

14 Blatter, G. and Greuter, F. (1986). Electrical breakdown at semiconductor grain boundaries. *Physical Review B: Condensed Matter* 34 (12): 8555–8572.

15 Blatter, G. and Greuter, F. (1986). Carrier transport through grain boundaries in semiconductors. *Physical Review B Condensed Matter* 33 (6): 3952.

16 Yang, X., Zhao, X., Li, Q. et al. (2018). Nonlinear effective permittivity of field grading composite dielectrics. *Journal of Physics D Applied Physics* 51 (7): 075304.

17 Robertson, J. and Varlow, B.R. (2005). Non-linear ferroelectric composite dielectric materials. *IEEE Transactions on Dielectrics & Electrical Insulation* 12 (4): 779–790.

18 Christen, T., Donzel, L., and Greuter, F. (2010). Nonlinear resistive electric field grading part 1: Theory and simulation. *IEEE Electrical Insulation Magazine* 26 (6): 47–59.

19 Donzel, L. and Schuderer, J. (2012). Nonlinear resistive electric field control for power electronic modules. *IEEE Transactions on Dielectrics & Electrical Insulation* 19 (3): 955–959.

20 Rifaie-Graham, O., Apebende, E.A., Bast, L.K., and Bruns, N. (2018). Self-reporting fiber-reinforced composites that mimic the ability of biological materials to sense and report damage. *Advanced Materials,* https://doi.org/10.1002/adma.201705483 30 (19): 1705483.

21 Abot, J.L., Song, Y., Vatsavaya, M.S. et al. (2010). Delamination detection with carbon nanotube thread in self-sensing composite materials. *Composites Science and Technology* 70 (7): 1113–1119. https://doi.org/10.1016/j.compscitech.2010.02.022.

22 Donzel, L., Greuter, F., and Christen, T. (2011). Nonlinear resistive electric field grading Part 2: Materials and applications. *IEEE Electrical Insulation Magazine* 27 (2): 18–29.

23 Yang, X., Zhao, X., Hu, J., and He, J. (2018). Grading electric field in high voltage insulation using composite materials. *IEEE Electrical Insulation Magazine* 34 (1): 15–25. https://doi.org/10.1109/MEI.2018.8246118.

24 He, J.L. and Hu, J. (2007). Discussions on nonuniformity of energy absorption capabilities of ZnO varistors. *IEEE Transactions on Power Delivery* 22 (3): 1523–1532.

25 Huang, Z., Yuan, Z., Sun, G. et al. (2022). Thickness effect on threshold electric field of ZnO microvaristors/silicone rubber composite with nonlinear conductivity. *Composites Part A: Applied Science and Manufacturing* 158: 106969. https://doi.org/10.1016/j.compositesa.2022.106969.

26 Saltzer, M., Jonsson, E., and Gäfvert, U. (2009). Response of non-linear materials to applied voltages of arbitrary wave form. In: *Proceedings of the 21st Nordic Insulation Symposium (NORD-IS 09)*, Gothenburg, Sweden, pp. 1–4, 15-17 June 2009.

27 Mårtensson, E. and Gäfvert, U. (2003). A three-dimensional network model describing a non-linear composite material. *Journal of Physics D: Applied Physics* 37 (1): 112. https://doi.org/10.1088/0022-3727/37/1/019.

28 Huang, Z., Yang, X., Yuan, Z. et al. (2022). The degradation of silicone rubber composites with ZnO microvaristors under impulse voltage. *Journal of Physics D: Applied Physics* 55 (35): 355501. https://doi.org/10.1088/1361-6463/ac767d.

29 Yang, X., Hu, J., Wang, S. et al. (2022). A dielectric polymer/metal oxide nanowire composite for self-adaptive charge release. *Nano Letters* 22 (13): 5167–5174. https://doi.org/10.1021/acs.nanolett.2c00887.

30 Huang, Z. et al. (2023). Time-domain response model of nonlinear dielectric properties for electrical field grading composites. *Composites Science and Technology* 232: 109879. https://doi.org/10.1016/j.compscitech.2022.109879.

31 Yang, X., Wang, S., Zhao, X. et al. (2022). Stability of self-adaptive dielectrics under various electrical-thermal stress. *IEEE Transactions on Dielectrics and Electrical Insulation* 30 (2): 1, 625–1, 633. https://doi.org/10.1109/TDEI.2022.3228522.

32 Yang, X., Hu, J., Chen, S., and He, J. (2016). Understanding the percolation characteristics of nonlinear composite dielectrics. *Scientific Reports* 6 (1): 30597. https://doi.org/10.1038/srep30597.

33 Eda, K. (1989). Zinc oxide varistors. *IEEE Electrical Insulation Magazine* 5 (6): 28–30. https://doi.org/10.1109/57.44606.

34 Tao, M., Ai, B., Dorlanne, O., and Loubiere, A. (1987). Different "single grain junctions" within a ZnO varistor. *Journal of Applied Physics* 61 (4): 1562–1567. https://doi.org/10.1063/1.338091.

35 Wang, G., Deng, Y., Xiang, Y., and Guo, L. (2008). Fabrication of radial ZnO nanowire clusters and radial ZnO/PVDF composites with enhanced dielectric properties. *Advanced Functional Materials* 18 (17): 2584–2592. https://doi.org/10.1002/adfm.200800109.

36 Patrick, J.F., Robb, M.J., Sottos, N.R. et al. (2016). Polymers with autonomous life-cycle control. *Nature* 540 (7633): 363–370. https://doi.org/10.1038/nature21002.

37 Caruso, M.M., Davis, D.A., Shen, Q. et al. (2009). Mechanically-induced chemical changes in polymeric materials. *Chemical Reviews* 109 (11): 5755–5798. https://doi.org/10.1021/cr9001353.

38 Calvino, C. (2021). Polymer-based mechanochromic composite material using encapsulated systems. *Macromolecular Rapid Communications*, https://doi.org/10.1002/marc.202000549 42 (1): 2000549.

39 Ma, L., Ren, C., Wang, J. et al. (2021). Self-reporting coatings for autonomous detection of coating damage and metal corrosion: a review. *Chemical Engineering Journal* 421: 127854. https://doi.org/10.1016/j.cej.2020.127854.

40 Klajn, R. (2014). Spiropyran-based dynamic materials. *Chemical Society Reviews* 43 (1): 148–184. https://doi.org/10.1039/C3CS60181A.

41 Lokshin, V.A., Samat, A., and Metelitsa, A.V. (2002). Spirooxazines: synthesis, structure, spectral and photochromic properties. *Russian Chemical Reviews* 71 (11): 893–916. https://doi.org/10.1070/RC2002v071n11ABEH000763.

42 Bandara, H.M.D. and Burdette, S.C. (2012). Photoisomerization in different classes of azobenzene. *Chemical Society Reviews* 41 (5): 1809–1825. https://doi.org/10.1039/C1CS15179G.

43 Logtenberg, H., van der Velde, J.H.M., de Mendoza, P. et al. (2012). Electrochemical switching of conductance with diarylethene-based redox-active polymers. *The Journal of Physical Chemistry C* 116 (45): 24136–24142. https://doi.org/10.1021/jp307892s.

44 Schenderlein, H., Voss, A., Stark, R.W., and Biesalski, M. (2013). Preparation and characterization of light-switchable polymer networks attached to solid substrates. *Langmuir* 29 (14): 4525–4534. https://doi.org/10.1021/la305073p.

45 Davis, D.A., Hamilton, A., Yang, J. et al. (2009). Force-induced activation of covalent bonds in mechanoresponsive polymeric materials. *Nature* 459 (7243): 68–72. https://doi.org/10.1038/nature07970.

46 Kosuge, T., Imato, K., Goseki, R., and Otsuka, H. (2016). Polymer–inorganic composites with dynamic covalent mechanochromophore. *Macromolecules* 49 (16): 5903–5911. https://doi.org/10.1021/acs.macromol.6b01333.

47 Kosuge, T., Zhu, X., Lau, V.M. et al. (2019). Multicolor mechanochromism of a polymer/silica composite with dual distinct mechanophores. *Journal of the American Chemical Society* 141 (5): 1898–1902. https://doi.org/10.1021/jacs.8b13310.

48 Zhu, S., Li, M., Tang, S. et al. (2014). Electrochromic switching and microkinetic behaviour of oxazine derivatives and their applications. *European Journal of Organic Chemistry*, https://doi.org/10.1002/ejoc.201301182 2014 (6): 1227–1235.

49 Rosario, R., Gust, D., Hayes, M. et al. (2003). Solvatochromic study of the microenvironment of surface-bound spiropyrans. *Langmuir* 19 (21): 8801–8806. https://doi.org/10.1021/la0344332.

50 Florea, L., McKeon, A., Diamond, D., and Benito-Lopez, F. (2013). Spiropyran polymeric microcapillary coatings for photodetection of solvent polarity. *Langmuir* 29 (8): 2790–2797. https://doi.org/10.1021/la304985p.

51 Abdollahi, A., Mouraki, A., Sharifian, M.H., and Mahdavian, A.R. (2018). Photochromic properties of stimuli-responsive cellulosic papers modified by spiropyran-acrylic copolymer in reusable pH-sensors. *Carbohydrate Polymers* 200: 583–594. https://doi.org/10.1016/j.carbpol.2018.08.042.

52 Fries, K., Samanta, S., Orski, S., and Locklin, J. (2008). Reversible colorimetric ion sensors based on surface initiated polymerization of photochromic polymers. *Chemical Communications* 47: 6288–6290. https://doi.org/10.1039/B818042C.

53 Shiraishi, Y., Adachi, K., Itoh, M., and Hirai, T. (2009). Spiropyran as a selective, sensitive, and reproducible cyanide anion receptor. *Organic Letters* 11 (15): 3482–3485. https://doi.org/10.1021/ol901399a.

54 Charych Deborah, H., Nagy Jon, O., Spevak, W., and Bednarski Mark, D. (1993). Direct colorimetric detection of a receptor-ligand interaction by a polymerized bilayer assembly. *Science* 261 (5121): 585–588. https://doi.org/10.1126/science.8342021.

55 Wang, H., Han, S., Hu, Y. et al. (2017). Polydiacetylene-based periodic mesoporous organosilicas with colorimetric reversibility under multiple stimuli. *Colloids and Surfaces A: Physicochemical and Engineering Aspects* 517: 84–95. https://doi.org/10.1016/j.colsurfa.2017.01.006.

56 Peng, J., Cheng, Y., Tomsia, A.P. et al. (2017). Thermochromic artificial nacre based on montmorillonite. *ACS Applied Materials & Interfaces* 9 (29): 24993–24998. https://doi.org/10.1021/acsami.7b07953.

57 Park, D.-H., Heo, J.-M., Jeong, W. et al. (2018). Smartphone-based VOC sensor using colorimetric polydiacetylenes. *ACS Applied Materials & Interfaces* 10 (5): 5014–5021. https://doi.org/10.1021/acsami.7b18121.

58 Seo, S., Lee, J., Kwon, M.S. et al. (2015). Stimuli-responsive matrix-assisted colorimetric water indicator of polydiacetylene nanofibers. *ACS Applied Materials & Interfaces* 7 (36): 20342–20348. https://doi.org/10.1021/acsami.5b06058.

59 Baek, W., Heo, J.-M., Oh, S. et al. (2016). Photoinduced reversible phase transition of azobenzene-containing polydiacetylene crystals. *Chemical Communications* 52 (97): 14059–14062. https://doi.org/10.1039/C6CC08606C.

60 Zhang, W., Xu, H., Chen, Y. et al. (2013). Polydiacetylene-polymethylmethacrylate/graphene composites as one-shot, visually observable, and semiquantative electrical current sensing materials. *ACS Applied Materials & Interfaces* 5 (11): 4603–4606. https://doi.org/10.1021/am401099s.

61 Song, S., Ha, K., Guk, K. et al. (2016). Colorimetric detection of influenza A (H1N1) virus by a peptide-functionalized polydiacetylene (PEP-PDA) nanosensor. *RSC Advances* 6 (54): 48566–48570. https://doi.org/10.1039/C6RA06689E.

62 Calvino, C. and Weder, C. (2018). Microcapsule-containing self-reporting polymers. *Small* 14 (46): 1802489. https://doi.org/10.1002/smll.201802489.

63 Wang, J.-p., Zhao, X.-p., Guo, H.-l., and Zheng, Q. (2004). Preparation of microcapsules containing two-phase core materials. *Langmuir* 20 (25): 10845–10850. https://doi.org/10.1021/la0490902.

64 Pang, J.W.C. and Bond, I.P. (2005). 'Bleeding composites'—damage detection and self-repair using a biomimetic approach. *Composites Part A: Applied Science*

and Manufacturing 36 (2): 183–188. https://doi.org/10.1016/j.compositesa.2004.06.016.

65 Patrick, J.F., Hart, K.R., Krull, B.P. et al. (2014). Continuous self-healing life cycle in vascularized structural composites. *Advanced Materials* 26 (25): 4302–4308. https://doi.org/10.1002/adma.201400248.

66 White, S.R., Sottos, N.R., Geubelle, P.H. et al. (2001). Autonomic healing of polymer composites. *Nature* 409 (6822): 794–797. https://doi.org/10.1038/35057232.

67 Vidinejevs, S., Aniskevich, A.N., Gregor, A. et al. (2012). Smart polymeric coatings for damage visualization in substrate materials. *Journal of Intelligent Material Systems and Structures* 23 (12): 1371–1377. https://doi.org/10.1177/1045389X12447289.

68 Vidinejevs, S., Strekalova, O., Aniskevich, A., and Gaidukov, S. (2013). Development of a composite with an inherent function of visualization of a mechanical action. *Mechanics of Composite Materials* 49 (1): 77–84. https://doi.org/10.1007/s11029-013-9323-9.

69 Li, W., Matthews, C.C., Yang, K. et al. (2016). Autonomous indication of mechanical damage in polymeric coatings. *Advanced Materials* 28 (11): 2189–2194. https://doi.org/10.1002/adma.201505214.

70 Di Credico, B., Griffini, G., Levi, M., and Turri, S. (2013). Microencapsulation of a UV-responsive photochromic dye by means of novel UV-screening polyurea-based shells for smart coating applications. *ACS Applied Materials & Interfaces* 5 (14): 6628–6634. https://doi.org/10.1021/am401328f.

71 Postiglione, G., Colombo, A., Dragonetti, C. et al. (2017). Fluorescent probes based on chemically-stable core/shell microcapsules for visual microcrack detection. *Sensors and Actuators B: Chemical* 248: 35–42. https://doi.org/10.1016/j.snb.2017.03.136.

72 Odom, S.A., Jackson, A.C., Prokup, A.M. et al. (2011). Visual indication of mechanical damage using core–shell microcapsules. *ACS Applied Materials & Interfaces* 3 (12, 12): 4547–4551. https://doi.org/10.1021/am201048a.

73 Robb, M.J., Li, W., Gergely, R.C.R. et al. (2016). A robust damage-reporting strategy for polymeric materials enabled by aggregation-induced emission. *ACS Central Science* 2 (9): 598–603. https://doi.org/10.1021/acscentsci.6b00198.

74 Lavrenova, A., Farkas, J., Weder, C., and Simon, Y.C. (2015). Visualization of polymer deformation using microcapsules filled with charge-transfer complex precursors. *ACS Applied Materials & Interfaces* 7 (39): 21828–21834. https://doi.org/10.1021/acsami.5b05797.

75 Chen, Y., Li, W., Luo, J. et al. (2021). Robust damage-reporting strategy enabled by dual-compartment microcapsules. *ACS Applied Materials & Interfaces* 13 (12): 14518–14529. https://doi.org/10.1021/acsami.0c20276.

76 Wu, D.Y., Meure, S., and Solomon, D. (2008). Self-healing polymeric materials: a review of recent developments. *Progress in Polymer Science* 33 (5): 479–522. https://doi.org/10.1016/j.progpolymsci.2008.02.001.

77 Wang, S. and Urban, M.W. (2020). Self-healing polymers. *Nature Reviews Materials* 5 (8): 562–583. https://doi.org/10.1038/s41578-020-0202-4.

78 Shimizu, N. and Laurent, C. (1998). Electrical tree initiation. *IEEE Transactions on Dielectrics and Electrical Insulation* 5 (5): 651–659. https://doi.org/10.1109/94.729688.

79 Ieda, M. (1980). Dielectric breakdown process of polymers. *IEEE Transactions on Electrical Insulation* EI-15, no. 3: 206–224. https://doi.org/10.1109/tei.1980.298314.

80 Ross, R. (1998). Inception and propagation mechanisms of water treeing. *IEEE Transactions on Dielectrics and Electrical Insulation* 5 (5): 660–680. https://doi.org/10.1109/94.729689.

81 Jianguo, W., Yang, L., Chunhua, F. et al. (2009). Performance influence of high temperature vulcanization silicone rubber by alternating current corona discharge. *Acta Polymerica Sinica* 4: 331–337.

82 Gao, Y., Wang, J., Liang, X. et al. (2016). Influence of AC and DC corona on high temperature vulcanized silicone rubber. *Proc. Chin. Soc. Electr. Eng* 36: 274–284.

83 Yang, Y., Dang, Z.M., Li, Q., and He, J. (2020). Self-healing of electrical damage in polymers. *Advanced Science* 7 (21): 2002131. https://doi.org/10.1002/advs.202002131.

84 Rosensweig, R.E. (2002). Heating magnetic fluid with alternating magnetic field. *Journal of Magnetism and Magnetic Materials* 252: 370–374. https://doi.org/10.1016/s0304-8853(02)00706-0.

85 Fortin, J.P., Wilhelm, C., Servais, J. et al. (2007). Size-sorted anionic iron oxide nanomagnets as colloidal mediators for magnetic hyperthermia. *Journal of American Chemical Society* 129 (9): 2628–2635. https://doi.org/10.1021/ja067457e.

86 Ahmed, A.S. and Ramanujan, R.V. (2015). Curie temperature controlled self-healing magnet–polymer composites. *Journal of Materials Research* 30 (7): 946–958. https://doi.org/10.1557/jmr.2015.59.

87 Yoonessi, M., Lerch, B.A., Peck, J.A. et al. (2015). Self-healing of core–shell magnetic polystyrene nanocomposites. *ACS Applied Materials & Interfaces* 7 (31): 16932–16937. https://doi.org/10.1021/acsami.5b04314.

88 Jeoffroy, E., Koulialias, D., Yoon, S. et al. (2016). Iron oxide nanoparticles for magnetically-triggered healing of bituminous materials. *Construction and Building Materials* 112: 497–505. https://doi.org/10.1016/j.conbuildmat.2016.02.159.

89 Corten, C.C. and Urban, M.W. (2009). Repairing polymers using oscillating magnetic field. *Advanced Materials*, https://doi.org/10.1002/adma.200901940 21 (48): 5011–5015.

90 Wool, R.P. (2008). Self-healing materials: a review. *Soft Matter* 4 (3): 400–418. https://doi.org/10.1039/B711716G.

91 Lee, J.-Y., Zhang, Q., Emrick, T., and Crosby, A.J. (2006). Nanoparticle alignment and repulsion during failure of glassy polymer nanocomposites. *Macromolecules* 39 (21): 7392–7396. https://doi.org/10.1021/ma061210k.

92 Gupta, S., Zhang, Q., Emrick, T. et al. (2006). Entropy-driven segregation of nanoparticles to cracks in multilayered composite polymer structures. *Nature Materials* 5 (3): 229–233. https://doi.org/10.1038/nmat1582.

93 Balazs, A.C., Emrick, T., and Russell, T.P. (2006). Nanoparticle polymer composites: where two small worlds meet. *Science* 314 (5802): 1107–1110. https://doi.org/10.1126/science.1130557.

94 Lee, J.Y., Buxton, G.A., and Balazs, A.C. (2004). Using nanoparticles to create self-healing composites. *The Journal of Chemical Physics* 121 (11): 5531–5540. https://doi.org/10.1063/1.1784432.

95 Tyagi, S., Lee, J.Y., Buxton, G.A., and Balazs, A.C. (2004). Using nanocomposite coatings to heal surface defects. *Macromolecules* 37 (24): 9160–9168. https://doi.org/10.1021/ma048773l.

96 Yang, Y., He, J., Li, Q. et al. (2019). Self-healing of electrical damage in polymers using superparamagnetic nanoparticles. *Nature Nanotechnology* 14 (2): 151–155. https://doi.org/10.1038/s41565-018-0327-4.

97 Yang, Y., Gao, L., Xie, J. et al. (2020). Defect-targeted self-healing of multiscale damage in polymers. *Nanoscale* 12 (6): 3605–3613. https://doi.org/10.1039/c9nr09438e.

98 Lee, J.H., Jang, J.-t., Choi, J.-s. et al. (2011). Exchange-coupled magnetic nanoparticles for efficient heat induction. *Nat Nanotechnol* 6 (7): 418–422. https://doi.org/10.1038/nnano.2011.95.

99 Sun, P., Zhao, M., Sima, W. et al. (2022). Microwave-magnetic field dual-response raspberry-like microspheres for targeted and repeated self-healing from electrical damage of insulating composites. *Journal of Materials Chemistry C* 10 (28): 10262–10270. https://doi.org/10.1039/D2TC01693A.

100 Blaiszik, B.J., Kramer, S.L.B., Olugebefola, S.C. et al. (2010). Self-healing polymers and composites. *Annual Review of Materials Research* 40 (1): 179–211. https://doi.org/10.1146/annurev-matsci-070909-104532.

101 Diesendruck, C.E., Sottos, N.R., Moore, J.S., and White, S.R. (2015). Biomimetic self-healing. *Angewandte Chemie International Edition* 54 (36): 10428–10447. https://doi.org/10.1002/anie.201500484.

102 Wang, Y., Li, Y., Zhang, Z., and Zhang, Y. (2019). Effect of doping microcapsules on typical electrical performances of self-healing polyethylene insulating composite. *Applied Sciences* 9 (15): https://doi.org/10.3390/app9153039.

103 Li, Y., Wang, Y., Zhang, Y. et al. (2021). Effect of microcapsules on dielectric properties of self-healing low density polyethylene insulation composites. *IEEE Transactions on Dielectrics and Electrical Insulation* 28 (3): 924–931. https://doi.org/10.1109/tdei.2021.009393.

104 Zhang, Y., Wang, Y., Li, Y. et al. (2021). Self-healing of mechanical damage of polyethylene/microcapsules electrical insulation composite material. *Journal of Materials Science: Materials in Electronics* 32 (22): 26329–26340. https://doi.org/10.1007/s10854-021-06953-9.

105 Li, J.Y., Zhang, L., and Ducharme, S. (2007). Electric energy density of dielectric nanocomposites. *Applied Physics Letters* 90 (13): https://doi.org/10.1063/1.2716847.

106 Calame, J.P. (2003). Evolution of Davidson–Cole relaxation behavior in random conductor–insulator composites. *Journal of Applied Physics* 94 (9): 5945–5957. https://doi.org/10.1063/1.1615302.

107 Li, J.Y., Huang, C., and Zhang, Q. (2004). Enhanced electromechanical properties in all-polymer percolative composites. *Applied Physics Letters* 84 (16): 3124–3126. https://doi.org/10.1063/1.1702127.

108 Di Credico, B., Levi, M., and Turri, S. (2013). An efficient method for the output of new self-repairing materials through a reactive isocyanate encapsulation. *European Polymer Journal* 49 (9): 2467–2476. https://doi.org/10.1016/j.eurpolymj.2013.02.006.

109 Suryanarayana, C., Rao, K.C., and Kumar, D. (2008). Preparation and characterization of microcapsules containing linseed oil and its use in self-healing coatings. *Progress in Organic Coatings* 63 (1): 72–78. https://doi.org/10.1016/j.porgcoat.2008.04.008.

110 Song, Y.K., Jo, Y.-H., Lim, Y.-J. et al. (2013). Sunlight-induced self-healing of a microcapsule-type protective coating. *ACS Appl Mater Interfaces* 5 (4): 1378–1384. https://doi.org/10.1021/am302728m.

111 Gao, L., He, J., Hu, J., and Wang, C. (2015). Photoresponsive self-healing polymer composite with photoabsorbing hybrid microcapsules. *ACS Appl Mater Interfaces* 7 (45): 25546–25552. https://doi.org/10.1021/acsami.5b09121.

112 Guo, W., Jia, Y., Tian, K. et al. (2016). UV-triggered self-healing of a single robust SiO_2 microcapsule based on cationic polymerization for potential application in aerospace coatings. *ACS Applied Materials and Interfaces* 8 (32): 21046–21054. https://doi.org/10.1021/acsami.6b06091.

113 Laurent, C., Massines, F., and Mayoux, C. (1997). Optical emission due to space charge effects in electrically stressed polymers. *IEEE Transactions on Dielectrics and Electrical Insulation* 4 (5): 585–603. https://doi.org/10.1109/94.625646.

114 Gao, L., Yang, Y., Xie, J. et al. (2020). Autonomous self-healing of electrical degradation in dielectric polymers using in situ electroluminescence. *Matter* 2 (2): 451–463. https://doi.org/10.1016/j.matt.2019.11.012.

115 Sima, W., Liang, C., Sun, P. et al. (2021). Novel smart insulating materials achieving targeting self-healing of electrical trees: high performance, low cost, and eco-friendliness. *ACS Applied Materials and Interfaces* 13 (28): 33485–33495. https://doi.org/10.1021/acsami.1c07469.

116 Lesaint, C.; Vetle Risinggård; Jorunn Hølto *et al.* (2014). Self-healing high voltage electrical insulation materials. In: *Presented at the 2014 IEEE Electrical Insulation Conference (EIC),* IEEE, Philadelphia, USA, pp. 241–244.

117 Sima, W., Fan, K., Sun, P. et al. (2022). Magnetically targeted, water-triggered, self-healing microcapsules based on microfluidic techniques enabling targeted healing of water tree damage in epoxy resins. *ACS Applied Materials & Interfaces* 14 (43): 49128–49139. https://doi.org/10.1021/acsami.2c13588.

118 Zhang, C.Y., Jiang, X.B., Rong, M.Z., and Zhang, M.Q. (2012). Free radical polymerization aided self-healing. *Journal of Intelligent Material Systems and Structures* 25 (1): 31–39. https://doi.org/10.1177/1045389x12453961.

119 Dailey, M.M., Silvia, A.W., McIntire, P.J. et al. (2014). A self-healing biomaterial based on free-radical polymerization. *Journal of Biomedial Materials Research Part A* 102 (9): 3024–3032. https://doi.org/10.1002/jbm.a.34975.

120 Xiao, D.S., Yuan, Y.C., Rong, M.Z., and Zhang, M.Q. (2009). A facile strategy for preparing self-healing polymer composites by incorporation of cationic catalyst-loaded vegetable fibers. *Advanced Functional Materials* 19 (14): 2289–2296. https://doi.org/10.1002/adfm.200801827.

121 Yin, T., Rong, M., Zhang, M., and Yang, G. (2007). Self-healing epoxy composites – preparation and effect of the healant consisting of microencapsulated epoxy and latent curing agent. *Composites Science and Technology* 67 (2): 201–212. https://doi.org/10.1016/j.compscitech.2006.07.028.

122 Yin, T., Zhou, L., Rong, M.Z., and Zhang, M.Q. (2008). Self-healing woven glass fabric/epoxy composites with the healant consisting of micro-encapsulated epoxy and latent curing agent. *Smart Materials and Structures* 17 (1): https://doi.org/10.1088/0964-1726/17/01/015019.

123 Hart, K.R., Sottos, N.R., and White, S.R. (2015). Repeatable self-healing of an epoxy matrix using imidazole initiated polymerization. *Polymer* 67: 174–184. https://doi.org/10.1016/j.polymer.2015.04.068.

124 Xie, J., Gao, L., Hu, J. et al. (2020). Self-healing of electrical damage in thermoset polymers via anionic polymerization. *Journal of Materials Chemistry C* 8 (18): 6025–6033. https://doi.org/10.1039/c9tc06989e.

125 Thakur, V.K. and Kessler, M.R. (2015). Self-healing polymer nanocomposite materials: a review. *Polymer* 69: 369–383. https://doi.org/10.1016/j.polymer.2015.04.086.

126 Yang, Y. and Urban, M.W. (2013). Self-healing polymeric materials. *Chem Soc Rev* 42 (17): 7446–7467. https://doi.org/10.1039/c3cs60109a.

127 Wool, R.P. and O, Connor, K.M. (1981). A theory crack healing in polymers. *Journal of Applied Physics* 52 (10): 5953–5963. https://doi.org/10.1063/1.328526.

128 Oberhausen, B. and Kickelbick, G. (2021). Induction heating induced self-healing of nanocomposites based on surface-functionalized cationic iron oxide particles and polyelectrolytes. *Nanoscale Advances* 3 (19): 5589–5604. https://doi.org/10.1039/d1na00417d.

129 Chen, X., Dam, M.A., Ono, K. et al. (2002). A thermally re-mendable cross-linked polymeric material. *Science* 295 (5560): 1698–1702. https://doi.org/10.1126/science.1065879.

130 Plaisted, T.A. and Nemat-Nasser, S. (2007). Quantitative evaluation of fracture, healing and re-healing of a reversibly cross-linked polymer. *Acta Materialia* 55 (17): 5684–5696. https://doi.org/10.1016/j.actamat.2007.06.019.

131 Postiglione, G., Turri, S., and Levi, M. (2015). Effect of the plasticizer on the self-healing properties of a polymer coating based on the thermoreversible Diels–Alder reaction. *Progress in Organic Coatings* 78: 526–531. https://doi.org/10.1016/j.porgcoat.2014.05.022.

132 Turkenburg, D.H. and Fischer, H.R. (2015). Diels-Alder based, thermo-reversible cross-linked epoxies for use in self-healing composites. *Polymer* 79: 187–194. https://doi.org/10.1016/j.polymer.2015.10.031.

133 Xie, J., Yang, M., Liang, J. et al. (2021). Self-healing of internal damage in mechanically robust polymers utilizing a reversibly convertible molecular network. *Journal of Materials Chemistry A* 9 (29): 15975–15984. https://doi.org/10.1039/D1TA03512F.

134 Wan, B., Dong, X., Yang, X. et al. (2022). High strength, stable and self-healing copolyimide for defects induced by mechanical and electrical damages. *Journal of Materials Chemistry C* 10 (31): 11307–11315. https://doi.org/10.1039/D2TC01605B.

155 Xie, K., Yang, M., Fang, M. et al. (2021). Self-healing of [illegible] mechanically robust [illegible] [illegible] [illegible]

156 Wang, F., Zhang, [illegible] (2022). [illegible]

10

The Future Development of High-temperature Polymer Dielectrics

Qi-Kun Feng, Yong-Xin Zhang, Xin-Jie Wang, and Zhi-Min Dang

Tsinghua University, State Key Laboratory of Power System and Department of Electrical Engineering, Shuangqing Road, Beijing 100084, China

10.1 Introduction

High-temperature polymers are widely used in energy storage capacitors, aerospace electrical equipment, electronic appliances, transportation, and other fields due to their excellent performance [1–3]. The development of high-temperature polymer materials has become a hot research topic for academic and industry scientists and has a close relationship with people's daily life. Although a variety of high-temperature polymer materials with different chemical structures and excellent properties have emerged in recent decades, only a few of them have been put into industrial production so far [4, 5]. This phenomenon is mainly affected by the price factor, but also depends on the production process, material performance, and research depth of high-temperature polymer dielectrics [6]. Only materials with excellent comprehensive performance, wide application, and easy preparation processes could be industrialized and developed into large-scale varieties. Therefore, the present development and challenges for high-temperature polymer dielectrics, such as low energy density and complex preparation processes, will be discussed in detail in Section 10.2. The future perspectives and trends for high-temperature polymer dielectrics are proposed in Section 10.3.

10.2 Present Development and Challenges

10.2.1 Temperature Stability

Temperature stability forms the basis of high-temperature polymers and corresponding composites and is a prerequisite for their application at high-temperature conditions. Herein, temperature stability is mainly divided into two aspects – one is the temperature stability of the material, which is reflected by the thermal

High Temperature Polymer Dielectrics: Fundamentals and Applications in Power Equipment, First Edition.
Edited by Jun-Wei Zha and Zhi-Min Dang.

decomposition temperature; the other is the temperature stability of the properties, which is mainly reflected in the degradation of specific properties of the material (dielectric properties, breakdown strength, mechanical properties, energy storage properties, etc.) at high-temperature conditions.

The relatively successful high-temperature resistant polymers and their composite materials that have been developed so far mainly include polyimide (PI), polytetrafluoroethylene (PTFE), polyphenylene sulfide (PPS), and polyethylene terephthalate (PET) [7–11]. There are many varieties of PIs with stable aromatic heterocyclic structure, which leads to them having excellent high-temperature stability and can be used for a long time in the air of 260 °C. PTFE has excellent temperature stability within the temperature range of −200 to 260 °C and short-term use temperature reaching 300 °C. Besides, PTFE also possesses excellent corrosion resistance, thermal stability, and aging characteristics. PPS has good heat resistance, chemical resistance, and hydrolysis resistance, with long-term use temperature of −200 to 260 °C and short-term use temperature of 300 °C. In addition, polyaryletherketone polymers (polyether ketone ketone [PEKK], polyether ether ketone [PEEK], etc.) are widely used in engineering plastics and other fields due to their excellent heat resistance, water resistance, chemical resistance qualities, and good processability. Among them, the composite material based on PEEK can be operated for a long time at 250 °C conditions.

Although the above-mentioned high-temperature polymers have excellent high-temperature stability, their long-term use temperature is usually lower than 300 °C. With the continuous expansion of application scenarios for high-temperature polymers, the temperature stability of high-temperature polymers is becoming more and more stringent [12, 13]. For example, the long-term working time for high-temperature polymers in the aerospace field is usually higher than 400 °C.

In addition to the thermal stability of high-temperature polymer materials, the temperature stability of the corresponding properties is also crucial in the practical applications of high-temperature materials. For high-temperature dielectric film capacitors, for example, PIs usually meet the temperature requirements (150–200 °C). However, serious leakage current and conduction loss will occur within PI films under high-temperature conditions, which causes the energy density of PI films at high temperatures to be extremely low and not meet the energy storage needs of static capacitors. Such a serious conduction loss is largely dependent on the barrier height at the electrode-dielectric interface, which is determined by the band structure at the interfaces instead of the thermal stability of high-temperature polymers [14–16]. Therefore, to improve the temperature stability of high-temperature polymers, it is necessary to not only improve the temperature resistance of the material itself, but also improve the temperature stability of specific properties. The temperature stability of properties of high-temperature polymers could be improved by modifying their molecular structures. In addition, the combination of polymer and functional nanomaterials to form the nanocomposites may overcome some common limitations of high-temperature polymers and obtain favorable features for high-temperature applications.

10.2.2 Polymer-based High-temperature Composites

Polymer-based composite material is a cross-category, multi-disciplinary, and inter-disciplinary research field with broad research prospects. Composite materials can overcome the shortcomings of single materials, so as to give full play to the advantages of each component material, expand the application scope of materials, and improve the economic benefits of materials [17–19]. Nevertheless, high-temperature polymer-based composite materials, especially high-temperature nanocomposites, are still in the stage of exploration and accumulation of experience. Many research results are still limited to the laboratory, and there is still a certain distance from actual industrialization. At present, the challenges faced in the field of high-temperature polymer composites mainly include the following:

(1) Functional fillers, including organic and inorganic fillers, play a decisive role in the overall properties of polymer-based high-temperature composites. However, there is a lack of engineering research on the synthesis process of functional nanoparticles. Besides, the equipment capable of industrial production of nanoparticles needs to be further improved, so as to enhance the yield of functional particles and improve the qualified rate of obtained products. In addition, the collection and storage of microparticles, especially nanoparticles, is also an urgent problem to be solved.
(2) At present, the existing preparation technology for polymer-based high-temperature composites is still immature, and the research on specific process conditions in the preparation technology is not enough. Moreover, the research results obtained about polymer-based high-temperature composites only stay in the laboratory and small-scale production stages and there are a few pieces of research on the problems involved in the expansion of the production scale.
(3) The research on the practical technology of polymer-based high-temperature nanomaterials has not been systematically investigated. In addition, for the performance testing and characterization methods of nanocomposites, especially the characterization of the filler-matrix interface in nanocomposites, more microscopic testing methods are urgently needed to reveal the conformational relationship between the properties and structures of high-temperature polymer composites.

10.2.3 Cost and Scale-up Production

Although high-temperature polymers as well as corresponding polymer-based composites have achieved some large-scale applications, it is still difficult to achieve further scale-up production and development. One of the main challenges is that the understanding and research on the key, basic scientific issues of high-temperature polymers are still insufficient. As a result, it is impossible to accurately design and manufacture high-temperature materials, so as to predict and control their performance and service behavior. At present, some high-temperature nanocomposites have been applied in small-scale industries, but they have not yet fully developed their huge potential. With overall progress in the field of materials, new high-performance and multi-functional high-temperature composites also show a huge development trend. In recent years, scientific and technological workers have gradually realized that to further develop the technologies and scale-up production of high-temperature polymers and composites, they must have a deeper understanding of the basic scientific problems in chemistry, physics, and processing in this complex system. In addition, the scale-up production of high-temperature polymers and composites is not only chemical synthesis but also coordination in physical properties, characterization technology, processing, molding, and other aspects. Only by organizing multidisciplinary joint research can it be made possible to achieve a breakthrough in large-scale applications of high-temperature polymers and composites.

Another challenge is the cost of industrialization of high-temperature polymers and composites. Although high-temperature polymers with new structures synthesized in the laboratory displayed outstanding performances, the novel chemical synthesis method for new high-temperature polymers usually requires strict processing technology and expensive production equipment, which will hinder their application in large-scale industrialization. In addition, for the polymer-based high-temperature composites, high yield and large-scale production of high-quality fillers, especially inorganic fillers, is still in demand and is critical for the development of polymer composites with high qualities. For example, for the high thermal conductivity polymer composites, high-intrinsic thermally conductive fillers (Ag NWs, SiC NW, BNNS, BNNT, CNT, and graphene) are far more expensive, especially in large-scale industrial applications [20]. Besides, almost all the methods developed for large-scale fabrication of high intrinsic thermally conductive fillers involve severe conditions, yielding defective samples. Hence, more attention should be paid to the engineering problems and related basic research of the high-temperature polymer composites in industrial applications, promoting the upgrading of the industrial product.

10.3 Future Perspectives and Trends

10.3.1 Intrinsic High-temperature Polymers

The exploration of intrinsic polymers with novel structures and excellent properties is always the goal pursued in the field of high-temperature polymer materials, and also the main driving force for the continuous development of high temperature materials. More efforts should be focused on the research and development of novel intrinsic high-temperature polymer materials with higher temperature resistance and better comprehensive performance, such as introducing functional molecular structures into the molecular chains of high-temperature materials to improve performance. What is more, in order to improve the feasibility of new materials in industrial preparation, it is of great importance to avoid potential problems as much as possible during the process of preparing new materials, such as poor processing properties, great brittleness, and obvious internal stress.

In addition, the prosperity of computer-aided simulation is reverberating within the research of high-temperature polymers. For example, machine learning has shown great potential in the accelerated discovery of advanced materials in the field of computational molecular design [21–23]. Moreover, computer-aided computing, especially data-driven machine learning and materials genome, has broad development space in rational design and performance prediction of high-temperature polymer materials [24, 25]. However, synthesizing the calculated ideal polymer is full of challenges, which hinders the computational and experimental development of high-temperature dielectric polymers. In order to better utilize computer-aided computation to promote the development of intrinsic high-temperature polymers, future research should pay attention to the following aspects – , more data on various molecular structures need to be calculated and summarized to improve the accuracy of material design and performance prediction. In addition, the real description of polymeric materials is much more complicated than the ideal model used to describe them. More effort is needed from the design of materials to actual industrial applications. More importantly, optimizing the design of feasible processing routes to simplify industrial manufacturing will greatly shorten the required time for rationally designed materials from research to applications.

10.3.2 Polymer-based High-temperature Composites

Composite materials are also the future of high temperature polymers toward which their development is directed, and their properties are closely related to the polymer matrix and filler materials. The following aspects should be stressed in the future direction of polymer-based high-temperature composites.

(1) Researchers should further use the composite process to introduce novel functional fillers, including inorganic and organic fillers, into high-temperature polymers, so as to further improve the thermal properties, electrical properties, mechanical properties, and stability properties of high-temperature polymers.
(2) The micro-interface characteristics and adhesion properties between the fillers and matrix can be further studied with high-precision instruments, to continuously improve the testing, characterization technology, and evaluation standards of various performance indicators of high-temperature polymer materials.
(3) For the industrial production of functional nanofillers, more efforts should be made to further improve the production equipment of fillers, thereby increasing the production efficiency and production yield and reducing the costs of nanofillers.

In a word, compounding is the trend of modern high-temperature material development. Through the compounding of various functional materials, the complementation and optimization of high-temperature polymer properties are realized, so as to prepare high-temperature materials with excellent performances.

10.3.3 Large-scale Industrial Production

With the development of society, the total demand for high-temperature polymers will further expand in automobiles, electronics, and other new high-tech industries, which has brought new opportunities and challenges to global high-temperature polymer producers. New technologies and new processes can often bring revolutionary changes, and there is no exception in the field of high-temperature polymers.

Looking for better preparation technology, reducing the preparation cycle and cost, and developing new mold processing technology are needed to further improve the performance of high-temperature materials. Using the chemical modification of high-temperature polymers and the adjustment of the recipe system of the composite material to prepare high-temperature polymer materials with better performance is an ideal strategy to meet the requirements of high-tech fields. In addition, it is desirable to synthesize high-temperature polymer materials with high purity and high performance under normal temperature and pressure conditions.

In order to achieve large-scale industrial production, it should carry out industrial experiment research, optimize craftsmanship, and realize industrialization as soon as possible. At the same time, we can find new ways to reduce costs from monomer synthesis and aggregation methods, which is helpful to open up the civil and industrial markets and expand the uses of high-temperature polymers. At present, the key factors for the development of high-temperature polymers as well as composite materials have changed from functional performance to industrial prices, namely economic tolerance. Because price is the key factor in developing high-temperature polymers and composite material markets, the production of industrial-grade high-temperature polymers is the direction of development of high-temperature polymers and composite materials in the future.

10.4 Summary

In summary, our research and understanding of advanced high-temperature polymers and corresponding composites continue to evolve. In view of the major technical requirements of high performance, high functionalization, and low-cost preparation of intrinsic high-temperature polymers as well as composites, the development of high-temperature polymers will be guided based on the research of basic scientific issues such as structure-property relationships. New methods, new theories, and new technologies for the industrial production of high-temperature polymers are being developed to solve key technical problems in power systems, aerospace, transportation, and other fields. The design and development of new materials should be combined with the diversification of market demand to improve the comprehensive performance and cost-effectiveness of polymers and composite materials. The successful development of high thermal stability, high thermal conductivity, low dielectric loss, and high insulating properties in high-temperature polymer material requires theoretical and experimental research in interdisciplinary and multidisciplinary cooperation among physicists, chemists, materials scientists, electrical engineers, and computer engineers. It is necessary to mobilize the strength of the scientific community, the technical community, and the industrial community to form the integration of production, study, and research and jointly develop advanced high-temperature polymer materials.

Acknowledgments

This work was supported by the National Natural Science Foundation of China (Nos. 51937007 and 51921005), the National Key Research and Development Program of China (No. 2021YFB2401502), and the Zhongguancun National Independent Innovation Demonstration Zone open laboratory concept verification project.

References

1 Feng, Q.K., Zhong, S.L., Pei, J.Y. et al. (2022). Recent progress and future prospects on all-organic polymer dielectrics for energy storage capacitors. *Chemical Reviews* 122 (3): 3820–3878.

2 Li, H., Zhou, Y., Liu, Y. et al. (2021). Dielectric polymers for high-temperature capacitive energy storage. *Chemical Society Reviews* 50 (11): 6369–6400.

3 Zhang, Y.X., Feng, Q.K., Zhong, S.L. et al. (2021). Digital twin accelerating development of metallized film capacitor: key issues, framework design and prospects. *Energy Reports* 7: 7704–7715.

4 Wan, B.Q., Dong, X.D., Yang, X. et al. (2022). High strength, stable and self-healing copolyimide for defects induced by mechanical and electrical damages. *Journal of Materials Chemistry C* 10 (31): 11307–11315.

5 Guo, Y.Q., Ruan, K.P., Shi, X.T. et al. (2020). Factors affecting thermal conductivities of the polymers and polymer composites: a review. *Composites Science and Technology* 193 (15): 108134.

6 Singh, S., Ramakrishna, S., and Singh, R. (2017). Material issues in additive manufacturing: a review. *Journal of Manufacturing Processes* 25: 185–200.

7 Liu, X.J., Zheng, M.S., Wang, G. et al. (2022). High energy density of polyimide films employing imidization reaction kinetics strategy at elevated temperature. *Journal of Materials Chemistry A* 10: 10950–10959.

8 Wu, J.R., He, J., Yin, K. et al. (2021). Robust hierarchical porous PTFE film fabricated via femtosecond laser for self-cleaning passive cooling. *Nano Letters* 21 (10): 4209–4216.

9 Feng, Q.K., Zhang, Y.X., Liu, D.F. et al. (2022). Dielectric and energy storage properties of all-organic sandwich-structured films used for high-temperature film capacitors. *Materials Today Energy* 29: 101132.

10 Han, X.T., Yang, D., Yang, C.C. et al. (2019). Carbon fiber reinforced PEEK composites based on 3D-printing technology for orthopedic and dental applications. *Journal of Clinical Medicine* 8 (2): 240.

11 Zuo, P.Y., Tcharkhtchi, A., Shirinbayan, M. et al. (2019). Overall investigation of poly(phenylene sulfide) from synthesis and process to applications—a review. *Macromolecular Materials and Engineering* 304 (5): 1800686.

12 Liaw, D.J., Wang, K.L., Huang, Y.C. et al. (2012). Advanced polyimide materials: syntheses, physical properties and applications. *Progress in Polymer Science* 37 (7): 907–974.

13 Das, A., Chatham, C.A., Fallon, J.J. et al. (2020). Current understanding and challenges in high temperature additive manufacturing of engineering thermoplastic polymers. *Additive Manufacturing* 34: 101218.

14 Li, Q., Chen, L., Gadinski, M.R. et al. (2015). Flexible high-temperature dielectric materials from polymer nanocomposites. *Nature* 523: 576–579.

15 Chiu, F.C. (2014). A review on conduction mechanisms in dielectric films. *Advances in Materials Science and Engineering* 2014: 578168.

16 Zhou, Y., Li, Q., Dang, B. et al. (2018). A scalable, high-throughput, and environmentally benign approach to polymer dielectrics exhibiting significantly improved capacitive performance at high temperatures. *Advanced Materials* 30 (49): 1805672.

17 Salunke, D.R. and Gopalan, V. (2021). Thermal and electrical behaviors of boron nitride/epoxy reinforced polymer matrix composite—a review. *Polymer Composites* 42 (4): 1659–1669.

18 Yang, X.T., Guo, Y.Q., Luo, X. et al. (2018). Self-healing, recoverable epoxy elastomers and their composites with desirable thermal conductivities by incorporating BN fillers via in-situ polymerization. *Composites Science and Technology* 164 (18): 59–64.

19 Fan, L.Z., He, H.C., and Nan, C.W. (2021). Tailoring inorganic-polymer composites for the mass production of solid-state batteries. *Nature Reviews Materials* 6: 1003–1019.

20 Huang, X.Y., Zhi, C.Y., Lin, Y. et al. (2020). Thermal conductivity of graphene-based polymer nanocomposites. *Materials Science and Engineering: R: Reports* 142: 100577.

21 Liu, D.F., Feng, Q.K., Zhang, Y.X. et al. (2022). Prediction of high-temperature polymer dielectrics using a Bayesian molecular design model. *Journal of Applied Physics* 132 (1): 014901.

22 Mannodi-Kanakkithodi, A., Pilania, G., Huan, T.D. et al. (2016). Machine learning strategy for accelerated design of polymer dielectrics. *Scientific Reports* 6: 20952.

23 Ma, R., Sharma, V., Baldwin, A.F. et al. (2015). Rational design and synthesis of polythioureas as capacitor dielectrics. *Journal of Materials Chemistry A* 3 (28): 14845–14852.

24 Zhu, J.L., Chu, M., Chen, Z.W. et al. (2020). Rational design of heat-resistant polymers with low curing energies by a materials genome approach. *Chemistry of Materials* 32 (11): 4527–4535.

25 Kim, C., Batra, R., Chen, L.H. et al. (2021). Polymer design using genetic algorithm and machine learning. *Computational Materials Science* 186: 110067.

Index

High Temperature Polymer Dielectrics: Fundamentals and Applications in Power Equipment, First Edition.
Edited by Jun-Wei Zha and Zhi-Min Dang.

i

p